Lecture Notes in Artificial Intelligence 10935

Subseries of Lecture Notes in Computer Science

More information about this series at http://www.springer.com/series/1244

Petra Perner (Ed.)

Machine Learning and Data Mining in Pattern Recognition

14th International Conference, MLDM 2018 New York, NY, USA, July 15–19, 2018 Proceedings, Part II

Editor
Petra Perner
Institute of Computer Vision and Applied Computer Sciences
Leipzig
Germany

ISSN 0302-9743 ISSN 1611-3349 (electronic)
Lecture Notes in Artificial Intelligence
ISBN 978-3-319-96132-3 ISBN 978-3-319-96133-0 (eBook)
https://doi.org/10.1007/978-3-319-96133-0

Library of Congress Control Number: 2018948144

LNCS Sublibrary: SL7 – Artificial Intelligence

Printed on acid-free paper

This Springer imprint is published by the registered company Springer International Publishing AG part of Springer Nature
The registered company address is: Gewerbestrasse 11, 6330 Cham, Switzerland

Preface

The 14th event of the International Conference on Machine Learning and Data Mining MLDM 2018 was held in New York (www.mldm.de) running under the umbrella of the World Congress Frontiers in Intelligent Data and Signal Analysis, DSA2017 (www.worldcongressdsa.com).

After the peer-review process, we accepted 86 high-quality papers for oral presentation. The topics range from theoretical topics for classification, clustering, association rule and pattern mining to specific data-mining methods for the different multimedia data types such as image mining, text mining, video mining, and Web mining. Extended versions of selected papers will appear in the international journal *Transactions on Machine Learning and Data Mining* (www.ibai-publishing.org/journal/mldm).

The tutorial days rounded up the high quality of the conference. Researchers and practitioners got excellent insight into the research and technology of the respective fields, the new trends, and the open research problems that we would like to study further.

A tutorial on "Data Mining," a tutorial on "Case-Based Reasoning," a tutorial on "Intelligent Image Interpretation and Computer Vision in Medicine, Biotechnology, Chemistry, and the Food Industry," and a tutorial on "Standardization in Immunofluorescence" were held before the conference.

We would like to thank all reviewers for their highly professional work and their effort in reviewing the papers. We would also like to thank the members of Institute of Applied Computer Sciences, Leipzig, Germany (www.ibai-institut.de), who handled the conference as secretariat. We appreciate the help and cooperation of the editorial staff at Springer, and in particular Alfred Hofmann, who supported the publication of these proceedings in the LNAI series.

Last, but not least, we wish to thank all the speakers and participants who contributed to the success of the conference. We hope to see you in 2019 in New York at the next World Congress (www.worldcongressdsa.com) on Frontiers in Intelligent Data and Signal Analysis, DSA2018, which combines the following three events: International Conferences on Machine Learning and Data Mining, MLDM (www.mldm.de), the Industrial Conference on Data Mining, ICDM (www.data-mining-forum.de), and the International Conference on Mass Data Analysis of Signals and Images in Medicine, Biometry, Drug Discovery Biotechnology, Chemistry, and Food Industry, MDA.

July 2018

Petra Perner

Organization

Program Chair

Petra Perner	IBaI Leipzig, Germany

Program Committee

Sergey Ablameyko	Belarus State University, Belarus
Reneta Barneva	The State University of New York at Fredonia, USA
Michelangelo Ceci	University of Bari, Italy
Patrick Bouthemy	Inria VISTA, France
Xiaoqing Ding	Tsinghua University, P.R. China
Christoph F. Eick	University of Houston, USA
Ana Fred	Technical University of Lisbon, Portugal
Giorgio Giacinto	University of Cagliari, Italy
Makato Haraguchi	Hokkaido University of Sapporo, Japan
Dimitris Karras	Chalkis Institute of Technology, Greece
Adam Krzyzak	Concordia University, Canada
Thang V. Pham	University of Amsterdam, The Netherlands
Linda Shapiro	University of Washington, USA
Tamas Sziranyi	MTA-SZTAKI, Hungary
Francis E. H. Tay	National University of Singapore, Singapore
Alexander Ulanov	HP Labs, Russia
Zeev Volkovich	ORT Braude College of Engineering, Israel
Patrick Wang	Northeastern University, USA

Additional Reviewers

Mohammad Daneshzand	University of Bridgeport, UK
Yong Jin	Wuhan FiberHome Potevio IT Ltd., P.R. China
Walid Atwa	Walid Atwa, Egypt
Soheila Abrishamin	Florida State University, USA
Piyush Kumar	Florida State University, USA
Aminata Kane	Concordia University, Canada
Yunlong Wang	University of Minnesota, USA
Huan Huo	University of Shanghai for Science and Technology, P.R. China
Terrence Fries	Indiana University of Pennsylvania, USA
Carlos Escobar	General Motors/Tecnológico de Monterrey, USA
Olga Krasotkina	Tula State University, Russia
Juliane Perner	Cancer Research Cambridge, UK
Jason Wang	New Jersey Institute of Technology, USA

Turki Turki	King Abdulaziz University, Saudi Arabia
Mohamed Bouguessa	University du Quebec a Montreal, Canada
Anit Bhandari	University Hospital Schleswig-Holstein, Germany
Carolina Carvalho	Fluminense Federal University, Brazil
Rekar Mohammed	Salahaddin University-Erbil, Iraq
Francky Fouedjio	Stanford University, USA
Vahid Behzadan	Kansas State University, USA
Evgeny Burnaev	Skolkovo Institute of Science and Technology, Russia
Alexander Bernstein	Skolkovo Institute of Science and Technology, Russia
Dymitr Ruta	Bournemouth University, UK
Dan Simovici	University of Massachussetts Boston, USA
Zeev Volkovich	ORT Braude College of Engineering, Israel
Markus Vattulainen	University of Tampere, Finland
Xiaobing Liu	Oklahoma State University, USA
Lumin Zhang	Max Plank Institute for Intelligent Systems, Germany

Contents – Part II

Contents – Part I

Fusing Dimension Reduction and Classification for Mining Interesting Frequent Patterns in Patients Data

Catherine Inibhunu[1(✉)] and Carolyn McGregor[1,2]

[1] University of Ontario Institute of Technology, Oshawa, ON, Canada
catherine.inibhunu@uoit.ca
[2] University of Technology Sydney, Ultimo, NSW, Australia

Abstract. Vast amounts of data are collected about elderly patients diagnosed with chronic conditions and receiving care in telehealth services. The potential to discover hidden patterns in the collected data can be crucial in making effective decisions on dissemination of services and lead to improved quality of care for patients. In this research, we investigate a knowledge discovery method that applies a fusion of dimension reduction and classification algorithms to discover interesting patterns in patient data. The research premise is that discovery of such patterns could help explain unique features about patients who are likely or unlikely to have an adverse event. This is a unique and innovative technique that utilizes the best of probability, rules, random trees and association algorithms for; (a) feature selection, (b) predictive modelling and (c) frequent pattern mining. The proposed method has been applied in a case study context to discover interesting patterns and features in patients participating in telehealth services. The results of the models developed shows that identification of best feature set can lead to accurate predictors of adverse events as well as effective in generation of frequent patterns and discovery of interesting features in varying patient cohort.

Keywords: Dimension reduction · Predictive modelling · Pattern mining

1 Introduction

Discovery of meaningful patterns in data is a problem attempted by researchers in varying applications and in many domains [1]. In customer analytics, the discovery of patterns have been demonstrated to be successful using market basket analysis where customer buying habits are quantified by evaluation of products they purchase. These patterns are then used by companies in targeted advertisements or marketing promotions [2]. In health care, in particular in remote patient monitoring services, the identification of unique patterns prior to patients lengthy hospitalization or multiple emergency room visits (ER Visits) could be crucial in effective management of healthcare resources and potential improvement in quality of life for those patients. As noted in our earlier work in [3], the cost of caring for patients in hospital settings can place a huge burden on any healthcare system, and accurately identifying patterns in patients before an adverse event occurs could be crucial in effective dissemination of

P. Perner (Ed.): MLDM 2018, LNAI 10935, pp. 1–15, 2018.
https://doi.org/10.1007/978-3-319-96133-0_1

healthcare services. We propose that facilitation of this can be achieved by a combination of feature selection, classification and frequent pattern methodologies through; (a) selecting key features about patients, (b) classifying patients based on likelihood of having an adverse event such as hospitalization or emergency room visit and (c) quantifying frequent patterns based on the occurrence of an adverse event.

Most of the techniques that have attempted to perform frequent pattern discovery are rule based which originate from the Apriori described in [2]. However there are many shortcomings in algorithms that use the Apriori principles as the search space is too large resulting in many patterns that are challenging to choose which may be important. One approach for reducing the number of patterns generated is to first evaluate the input fed into the mining algorithms and then apply techniques for identifying interesting patterns.

In this research we propose such a method for the discovery of patterns in patient data. This is performed by combining dimension reduction algorithms to address the input fed to data mining algorithms through generation of key feature sets, quantifying these features through evaluation of their predictive power and then generating frequent patterns and frequent features from patient data. The research premise is that discovery of search patterns could be useful in effective decision making for the provision of health care service resulting in improved quality of life of patients and reduction of cost of care especially on an aging population.

The rest of the paper is as follows: Sect. 2 discusses related work, Sect. 3 details the proposed method, Sect. 4 presents the case study and provides the experimental setup and evaluation, Sect. 5 the discussion summary and concludes in Sect. 6.

2 Related Work

Identifying patterns from data has been attempted in many domains such at the work described in [4] where association rules are utilized to identify high frequency itemsets and its application in retail to identify frequent items bought together. This work does not consider the infrequent itemsets that might also be useful in classifying rare patterns that would convey hidden knowledge, [4].

A similar approach is adopted in [5] where Apriori principles are used to first extract patterns which are then fed to naive bayes for probabilistic classification. A similar approach is described in [6] where association rules and weighted naive bayes for feature identification for clinical decision making. Although both approaches demonstrate potential for combination of different techniques for pattern discovery, the known problems of exponential patterns generated by association algorithms are not addressed.

Bayesian models are adopted in [7, 8] where a combination of bayes classification and association rules are used to infer interestingness in data through generation of frequent itemsets. Their approach gives minimal details on minimizing the search space as well as quantification of interestingness in multiple rulesets generated.

Lattice high utility itemsets are demonstrated in [9] where rules above a given threshold are used to create a graphical representation of rulesets. This approach can generate multiple lattices and therefore a new step would need to be introduced to prune out redundant lattices.

A different approach is demonstrated in [10] by adding a process for evaluation of rules generated using a ranking technique to prune out low ranked rules. The researches claim the method applied had better accuracy on predictions compared to other algorithms i.e. such as C4.5 decision trees but showed poor performance as the extra evaluation step introduced more iterations for processing.

All these algorithms are promising when applied on small datasets, however, they do not address the shortcomings found in association rule mining techniques when handling multiple dimensions of data such as; large search space, too many rules generated and most of rules are hard to interpret. With respect to knowledge discovery using patient data, addressing these challenges would greatly enhance the ability to discover and quantify interesting patterns and features in patients. This information would be valuable to healthcare providers in effective provision of healthcare services.

3 Methods

This research adopts a case study context to apply the proposed method for the discovery of interesting patterns on data collected from patients who participated in a remote patient monitoring program (RPM) [3]. RPM is a telehealth service where a vast amount of data was collected about patients such as; demographic (age, gender), environmental assessment, clinical assessment, adverse events records (hospital admissions, ER Visits) and medical history. Other datasets contained infrequently collected vital status about a patient; pulse rate, blood pressure, SPo2 and weight.

Using the data collected, this research investigates the question on whether the identification and quantification of adverse events predictors can lead to efficient discovery of frequent and interesting patterns in patient data.

To help answer this question, there are two factors to be addressed, first given multiple dimensions of patients' data how can one select a subset of these dimensions in order to quantify predictors and second, how can one find interesting patterns in these dimensions.

To address both factors we follow an innovative method which augments several principles as follows:

(a) Dimension reduction techniques to identifying key features in multiple dimensions,
(b) Building classification models to quantify predictors,
(c) Using high ranked predictors as variables to mine frequent patterns and
(d) Quantifying the mined patterns on two different cohorts.

3.1 Case Study Context

The clinical motivation for this research lies in the need for cost effective methods for dissemination of healthcare services to an aging population. In our earlier work described in [11], we noted that as the average age of Canadians increases so is the exponential increase in healthcare costs associated with aging as well as increase in comorbidities. The cost of caring for patients with chronic conditions can place a substantial burden to healthcare systems. As such, effective dissemination of telehealth services can greatly reduce the cost of similar services provided in hospital settings as well as providing patients care close to home.

The research premise is that, discovery of frequent patterns on patients before an adverse event happens is knowledge that would be crucial for healthcare providers as they make decisions on best care for patients leading to better outcomes and cost effect provision of services.

A detailed account of the principles applied in the proposed method are discussed next.

3.2 Dimension Reduction

There are several techniques utilized in machine learning to identify the most significant contributors of a predictive problem. In our earlier work we applied a subset of patient data to Bayesian models in order to identify and quantify key factors contributing to hospitalization or multiple ER Visits [11]. Several factors were identified but there was no evaluation of how effective those factors were in accurately predicting an adverse event.

In this paper, we recognize that an evaluation of other available patient datasets is necessary and have taken a different approach by linking more patient datasets. This results in multiple dimensions conveying a patient story and therefore the total number of variables collected is quite large. Selecting which dimensions are best predictors is not a trivial problem.

To address this problem, dimension reduction (DR) techniques are applied to capture most relevant features for further processing using the following techniques:

Best First: This technique uses a graphical path finding process for exploring the best attributes to a specific target. This method performs a sequential search starting with no features and adding features one by one and keeping the version of the features with the best performance.

Greedy Stepwise (forwards): This approach uses a greedy hill climbing technique of finding the optimal path. The algorithm starts with a solution and incrementally changes to a new solution until no more improvements can be made to the found solution. This algorithm may start with an empty set of attributes searching forward or backwards for all possible additions of deletions of an attribute depending on their contribution to an optimal feature set thus avoiding the problem of getting trapped in a local optima.

Ranker Information Gain: This technique uses the entropy and calculates the information gain for each attribute with respect to the class variable. The output generates a ranking of each variables with a range of 0 to 1. Attributes that contribute more have a higher entropy and are then selected and those with lower scores are excluded.

Given an original dataset D containing variables $\left(V_{1,}, \ldots, V_{n}.\right)$, then each dimension reduction algorithm DR_i is a function that produces a smaller set of variables (features) $(V_{i1},\ldots\ldots,V_{ik})$ where k is the length of each DR_i and $k<n$. This results to feature sets $FS_z=(DR_1,DR_2,\ldots,DR_z)$, where z is the number of DR algorithms. In this case study, three DR algorithms generates $FS_3=(DR_1,DR_2,DR_3)$ such that, DR_1 is the set of variables generated by best first, DR_2 by greedy stepwise and DR_3 by ranker algorithm. There is potential for the sets to contain similar variables or have equal lengths.

3.3 Classification Algorithms

The next step feeds each of the three feature sets (DR_1,DR_2,DR_3) into a different classification system. In this research we have utilized the following systems which are highly applicable to binary and categorical data:

Bayes Networks: This technique uses probability theory to represent and calculate relationships in data and then making predictions on likely hood of a reality.

For example: A join probability function of the form is adopted;

$$\Pr(G,V,E)=\Pr(G|V,E)\Pr(V|E)\Pr(E)$$

where G = Patient Gender (M/F), V = Vital Status (Pulse = (abnormal_low/normal/abnormal_high)) and E = Adverse Event(Hospital Admission, (Yes/No)). The Bayes Network would then calculates the probability of hospitalization given gender is male, and pulse rate is abnormal.

Rules: Partial C4.5: This algorithm uses a partial decision tree strategy in classification. The technique applied uses an aggressive feature space reduction with a minimal classification accuracy loss. A ruleset is induced on a training dataset using two steps. In particular the C4.5 approach creates an unpruned tree and then rules are adjusted with optimization. We used the implementation of the learning algorithm as provided in WEKA toolkit [12].

Trees: Random Forest: This approach uses ensemble learning method for classification where multiple decision trees are created during training and then outputs predictions of each tree. This approach reduces the problems exhibited in decisions trees with respect to overfitting through the bootstrapping aggregating to tree learners.

Given training set $I=i_1,\ldots.,i_n$ in with responses $J=j_1,\ldots,j_n$ with bagging repeatedly till K times such that a random sample of training set fits trees to the samples. For $k=1,\ldots,K$; first a replacement sampling on n training sets from sets I and J generates sets I_kJ_k which are then used to train the classification tree f_k on each I_kJ_k.

The results of the predictions for unseen sample I' are made by averaging predictions from each of the individual trees on I' such that;

$$f'=\frac{1}{K}\sum_{k=1}^{K}f_k(i')$$

3.4 Frequent Pattern Mining

Discovery of frequent patterns from patient data using a select set of features tackles the problems of association rule mining by:

(a) Reducing the number of iterations needed in a huge dataset compared to a select set of data
(b) A reduction of candidate sets as a small subset of features is processed resulting to a manageable rule set.

To facilitate this process, an adoption of the principles described in [13] for mining closed patterns is investigated allowing for limited features and data fed to mining process. The predictors in the classification model with the highest accuracy and lower error rate are selected and that feature set supplied as input to the association rule mining algorithm. This process is performed on two separate sets of data, one data has cohort of all patients and the other has cohort of patients with an actual adverse event. The goal is to identify a set of frequent features on the general cohort population and those with adverse events.

The proposed approach seeks to generate rule sets (patterns) of the form A => B, where A is the antecedent and B is the consequence. When using patient data, a rule set is of the form:

$$\mathrm{P(A : Patient, W)^{\wedge}\, Q(A, Y) {\rightarrow} Event(A, B)},$$

Where A is a factor of a patient relation, P and Q are predictor variables and W, Y, B are variables that can take values of their predictors. The quantification on the frequency to which the values in W, Y and B occurs in resulting rules sets, forms the interesting features of a given dataset.

The trustworthiness of the rule set is identified by a confidence rate calculated as;

$$Confidence(A \Rightarrow B) = \frac{(\mathrm{No\ \ of\ \ features\ \ containing\ \ both\ \ A\, and\, B})}{\mathrm{No\ \ of\ features\ \ containing\, A}}$$

The other factor is the interestingness of a rule defined as support, this quantifies how true the pattern is and calculated as:

$$Support(A \Rightarrow B) = \frac{(\mathrm{No\ \ of\ features\ containing\ \ both\ A\ and\ B})}{Total\ No\ of\ Features}$$

For example; given a patient data set with demographics, vital status and adverse events. A rule set of the form, Gender (A, “Male”) ^ Pulse (A, “High”) → Event (A, “ER Visit”), is generated with some confidence, i.e. 90%, and the values (gender = male, pulse = high) are the features quantified as frequent on patients where ER Visit is found with some support i.e. 70%.

3.5 The Applied Method Process Flow

To support the proposed method, principles explored are shown in Fig. 1. These are depicted as 5 components starting from data preparation to association rule mining. Details of each component are discussed next.

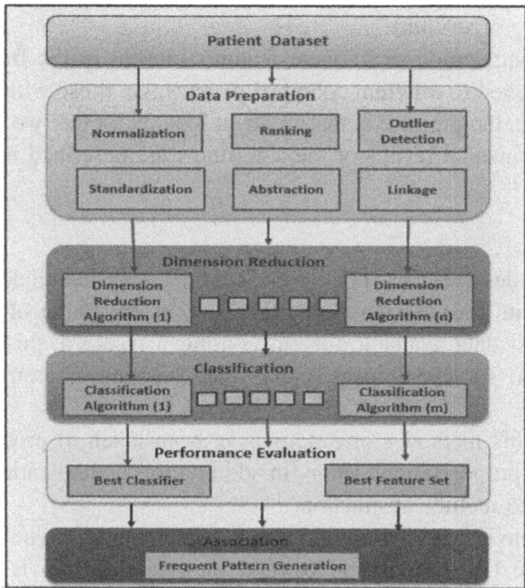

Fig. 1. The proposed method

Component A: Data Preparation - A substantial amount of work is dedicated to processing the data to a point where it's in a state for feeding to the dimension reduction algorithms. The details of the techniques involved in data preparation are discussed further in Sect. 3.6.

Component B: Dimension Reduction - In this component, linked and cleaned data is fed as input to the dimension reduction algorithms i.e. best first, greedy stepwise and ranker information gain algorithms. These algorithms evaluates the input variables against a defined class. In this research, an adverse event class variable has been derived from hospitalization and emergency room visits. This is then used for quantifying the impact of other patients attributes have on a potential adverse event. As three algorithms are processed, each generate a feature set of potential predictors.

Component C: Classification Algorithms - After the dimension reduction algorithms perform feature selection and generate best feature sets, these are then sent to the classification algorithms where models predict adverse events.

Component D: Performance Evaluation - In this component an evaluation of all the 9 models generated is completed by comparison of the models performance on prediction accuracy and error rates. This component results to identification of the best predictive algorithm and best predictive feature set.

Component E: Generation of Interesting Patterns - This is the final component that utilizes the best feature set identified in component D as input to association rule mining algorithm where frequent patterns are generated. This approach adopts the closed feature set mining approach which reduces the number of iterations and search space for the mining algorithms.

In this part, identification of frequent features present in the frequent patterns is identified, when applied to different cohort of patients, i.e. those with an adverse event and those with none, the premise is that frequent features in the two cohorts would be different. The experimental results of the algorithms are described in next Sect. 4.

3.6 Data

In our earlier work described in [11], we built models that identified features that are highly correlated with hospitalization or ER Visits. In this phase of the analytics, we include more patient data available for our continued research, this includes; patient health profile, client survey, clinical frailty scale, nursing assessment and environmental assessments.

To implement this method a few steps have been taken to process the data thus making it fit for the proposed application. In addition to standardization and linkage, an extra step is taken to include abstraction.

Abstraction: Conversion of the vital status numerical data which includes; pulse rate, blood pressure (BP) and blood oxygen saturation (SPo2) is performed using abstraction. This process transforms the numerical data to a higher granularity and quantifies what a value specify in clinical setting i.e. pulse rate: normal/abnormal. Abstraction applied utilizes the work described in [14, 15].

Other techniques involved in handling categorical variables were elimination of elements with low variance, outliers and those with 40 to 50% missing values.

Variable Ranking: To apply the proposed methods, a variable transformation is performed on numerical values to categorical formats. This involves ranking variables based on context representation of underlying data such as frequency distribution. Some data is filtered out if missing or no meaningful variation on the content across patients.

Deriving an Adverse Event: A merge of the hospital admissions and emergency room visits data is completed to establish a unifying target variable to represent an adverse event. In this case we are able to proceed in modeling an adverse event as either a hospital admission (HA) or an Emergency room visit (ER Visit).

After all this processing, the resulting dimensions of data are linked together resulting to a unified dataset ready for the process described in Fig. 1 and experimented as shown in Fig. 2.

4 Experimental Setup and Evaluation

4.1 Experimental

To support the proposed method, an experimental process is adopted as shown in Fig. 2. This process is facilitated by two crucial tools, WEKA a data mining tool detailed in [12] and SAS a statistical and predictive model tool popular in Analytics [16].

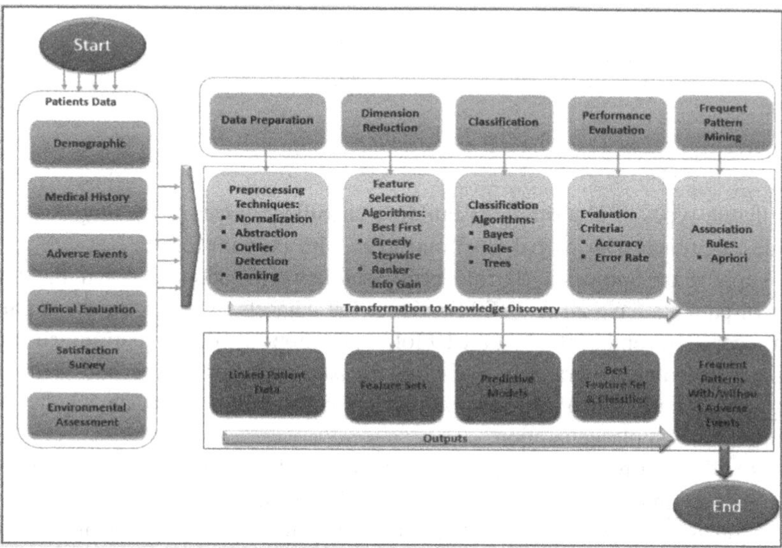

Fig. 2. Experimental process flow (Orange indicates the 5 processes, green the transformation algorithms, and purple the results from each process. Blue is the multiple patient data dimensions sources). (Color figure online)

4.2 Feature Selection with Dimension Reduction

The WEKA tool includes an attribute evaluation technique that checks the impact to a class variable, in this research, best first, greedy stepwise and ranker information gain are utilized for dimension reduction. The results of the three techniques generated are presented in Fig. 3. The total number of variables in best first are in Fig. 3(a), the variables from greedy stepwise search are in Fig. 3(b) and the variables selected using the information gain ranker algorithm are presented in Fig. 3(c).

4.3 Classification

The results of the three feature selection algorithms are then used in building classifiers. This allows testing of how accurate the variables selected in each algorithm accurately predicts the likely hood of a patient having an adverse event. This research utilized

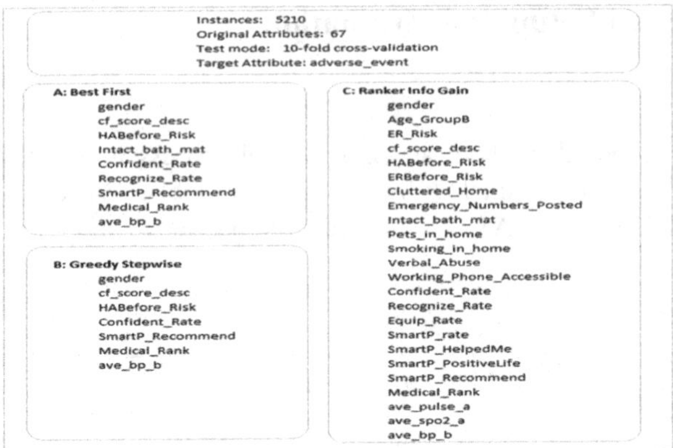

Fig. 3. Feature selection

three different types of algorithms in building classifiers; bayes probability with bayes nets, rules approach with partial C4.5 and trees with random forest.

To support this process, 10 fold cross validation is adopted given the unique number of patients have an n < 100, however the datasets representing those patients contain records beyond 5000. With a 10 fold validation, the data is partitioned into 10 equal parts and during iterations, each fold is used once for testing and 9 times for training. By default Weka uses a stratified cross-validation approach which allows randomness on sample generation.

Cross validation is adopted given its ability to test all possible ways to divide an original dataset into a training set and a validation set. This approach allows the classification to overcome the problem of overfitting and therefore generates more general predictions. In the same process a machine learning algorithm is applied on the training set data and prediction of the algorithm is tested on the test data. We use a 66% split resulting to 66% of the data for training and 34% for testing.

4.4 Classification Results

The results of the 9 models are shown in Fig. 4. The feature set 1 represents the features generated using best first, feature set 2 generated by greedy stepwise and feature set 3 generated by ranker information gain. The rules and trees algorithms shows a higher classification accuracy compared to the models generated using bayes algorithm. Detailed evaluation on all model performance are discussed in Sect. 4.5.

4.5 Performance Evaluation

There are two sets of evaluation applied on the classification algorithms. The first evaluation looks at the predictive strength of models based on 4 measures: accuracy, specificity, sensitivity and precision (Fig. 5). The second evaluation looks at the error rate using 5 measures; kappa statistic, mean absolute error, root mean squared error, relative absolute error and root relative squared error (Fig. 6). The partial C4.5 and

	Bayes: Bayes Nets	Rules : Partial C4.5	Trees: Random Forest
Feature Set 1	Accuracy: 96.05% a b <-- classified as 4651 28 | a = No 178 353 | b = Yes	Accuracy: 97.77% a b <-- classified as 4674 5 | a = No 111 420 | b = Yes	Accuracy: 97.87% a b <-- classified as 4679 0 | a = No 111 420 | b = Yes
Feature Set 2	Accuracy: 95.82% a b <-- classified as 4632 47 | a = No 175 356 | b = Yes	Accuracy: 97.87% a b <-- classified as 4679 0 | a = No 111 420 | b = Yes	Accuracy: 97.87% a b <-- classified as 4679 0 | a = No 111 420 | b = Yes
Feature Set 3	Accuracy: 94.51% a b <-- classified as 4463 216 | a = No 70 461 | b = Yes	Accuracy: 98.79% a b <-- classified as 4679 0 | a = No 68 463 | b = Yes	Accuracy: 98.73% a b <-- classified as 4679 0 | a = No 66 465 | b = Yes

Fig. 4. Classification results from Bayes, rules and trees algorithms (a = no adverse event, b = yes adverse event)

random forest algorithms showed better predictive accuracy at above 98% compared with the bayes models. The two models also have 100% specificity and precision. Next the classification error rate is evaluated.

Algorithms	Models	Accuracy	Sensitivity	Specificity	Precision
	Feature Set 1	96.05%	66.48%	99.40%	92.65%
	Feature Set 2	95.82%	72.48%	99.00%	90.75%
Bayes: Bayes Nets	Feature Set 3	94.51%	86.82%	95.38%	68.09%
	Feature Set 1	97.77%	79.10%	99.89%	98.82%
	Feature Set 2	97.87%	79.10%	100.00%	100.00%
Rules: Partial C4.5	Feature Set 3	98.79%	88.02%	100.00%	100.00%
	Feature Set 1	97.87%	79.10%	100.00%	100.00%
	Feature Set 2	97.87%	79.10%	100.00%	100.00%
Trees: Random Forest	Feature Set 3	98.73%	87.57%	100.00%	100.00%

Best Performing Models

Fig. 5. Predictive strength

Error Rates: As shown in Fig. 6, the random forest using the feature set 3 is the best performing models as it displays the lowest errors rates. In particular the kappa statistics indicates this model is close to the real world as the rate is closest to 1. This model also has the lowest mean, absolute and square classification error rates. As a result the best feature set 3 is chosen as the best predictive feature set and the random forest model is chosen as the best predictor of an adverse event for the examined dataset.

Algorithms	Models	Kappa statistic	Mean absolute error	Root mean squared error	Relative absolute error	Root relative squared error
	Feature Set 1	0.7531	0.0609	0.1875	33.22%	61.98%
	Feature Set 2	0.7394	0.0608	0.1843	33.20%	60.93%
Bayes: Bayes Nets	Feature Set 3	0.7327	0.0608	0.2130	33.16%	70.41%
	Feature Set 1	0.8666	0.0402	0.1388	21.97%	45.88%
	Feature Set 2	0.8717	0.0428	0.2334	15.22%	47.68%
Rules: Partial C4.5	Feature Set 3	0.9244	0.0226	0.0971	12.32%	32.11%
	Feature Set 1	0.8717	0.0342	0.1281	18.69%	42.32%
	Feature Set 2	0.8717	0.0344	0.1286	18.80%	42.50%
Trees: Random Forest	Feature Set 3	0.9268	0.0164	0.0792	8.97%	26.17%

Best Model

Fig. 6. Predictive error rates

The next process takes in the best feature set as input to frequent pattern mining by applying association rules algorithm described in [13].

Frequent Pattern Mining - The results of the application of the association algorithm when presented with data from patients with an adverse event is presented in Fig. 7. We have evaluated the top 100 generated rules. In the 10 top patterns, there are four frequent features noted {average pulse = normal, average BP = bp ideal, recognize rate = yes, gender = male}. On the 40 to 50 patterns, two more frequent features are discovered, {cf_score = mildly frail, ER before risk = high}.

Applying association on the cohort of all patients or those with no adverse events generates a different set of frequent features in the top 10 rules; {intact bath mat = yes, accessible phone = yes, cluttered home = no, verbal abuse = no}, another feature is noted in the bottom part of top 50, {HA before risk = no}.

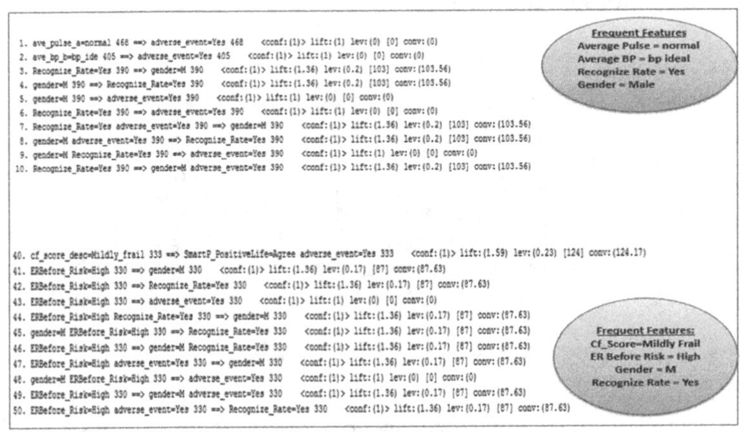

Fig. 7. Top 10 and 50 frequent patterns: with adverse events.

Another evaluation looks at the bottom 100 rules, rule 90 to 100 and presented in Fig. 8. One of the frequent features noted on the cohort of patients with an adverse event was quite unique compared to patients with no adverse event i.e. {Emergency Numbers Posted = N/A}. This is identified as an infrequent but interesting feature in the top 100 rules.

5 Discussion

The proposed method has facilitated an investigative case study approach to discovery of interesting patterns from patients data.

Starting from the rigorous data preparation process described in Sect. 3, a linked dataset was supplied to dimension reduction algorithms that provided a quantifiable approach to reducing the number of variables (features) needed for prediction of an adverse event. This process generated three sets of features from best first, greedy

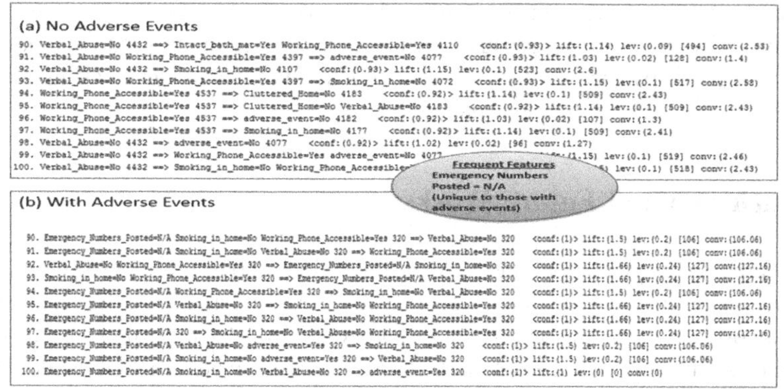

Fig. 8. Top 100 patterns on the general cohort and those with adverse events.

stepwise and ranker information gain DR techniques. These feature sets were then supplied to classification algorithms resulting in 9 different models with varying accuracy and error rates.

The random forest algorithm using the information gain feature set produced the best classification system as it shows the lowest error rates, kappa statistic is the closest to 1 at 0.93 and the highest classification accuracy rate at 98.73%. The partial C4.5 algorithm also performed well on accuracy but the classification error rate was much higher than the random forest. Bayes nets was the lowest performing algorithm, in particular, where the feature set with more variables resulted to an increase in predictive accuracy in the rules and trees algorithms, this was the opposite for bayes nets.

The application of the best feature set to association algorithm results to top 100 rules with over 90% confidence and support rates. Note that it's possible to limit the number of frequent rules generated within the WEKA tool by supplying several parameters [12]. For investigation purposes, this paper selected top 100 rules.

The results on the rules generated from cohort of patients with no adverse events are quite different from those with adverse events. As a result the frequent features found in each cohort is different and could indicate a clear attribute distinction on patients who had an adverse event compared with those who had none. Exploration of this assumption is completed by reviewing the frequent feature identified in the top 100 frequent patterns on the cohort of patients with an adverse event, i.e. ("Emergency Number Posted = N/A"), see (Fig. 8). To support this process, further investigation of the original patient data reveals a unique factor on the patient cohort. Out of 6 clients with an adverse event, male clients who provided no feedback on whether they had emergency number posted were 66.7% higher than female clients. In addition, 83.3% of all clients with an adverse event did not provide an answer on whether they had emergency number posted. These are unique findings that will be explored further with subject matter experts in telehealth services for clinical validation.

The proposed method demonstrates the potential for knowledge discovery in many data domains. This is enabled by an effective and quantifiable approach for mining frequent patterns and interesting features in vast amount of data. Selection of measureable predictors of a given event is shown as effective by fusing principles in dimension reduction and classification techniques. This results to reduced inputs provided to frequent pattern mining algorithms thus significantly decreasing the data processed in discovery of frequent patterns.

6 Conclusion and Future Works

This research has provided an innovative method for discovery of interesting patterns in patients using a fusion of dimension reduction, classification and frequent pattern mining algorithms. The application of the proposed method on a patient case study shows this is a promising approach to predicting adverse events as well as identifying interesting patterns and frequent features in patient data.

We note that there is a potential for the best predictor feature set to contain large variable set, in the future we will investigate if further reduction of this set have any impact on predictive accuracy. Additionally, we will investigate application of the proposed method to real-time data streamed from neonatal health monitoring systems for discovery of unique patterns in neonates.

Acknowledgments. We would like to thank Alaya Care, We Care and Southlake Regional Hospital for their collaboration on the research grant from Ontario Centres of Excellence Advancing Health (MIS#23823) that generated data that was used as experimental data for this research.

References

1. Xiao, S., Hu, Y., Han, J., Zhou, R., Wen, J.: Bayesian networks-based association rules and knowledge reuse in maintenance decision-making of industrial product-service systems. Procedia CIRP **47**, 198–203 (2016)
2. Liu, B., Hsu, W., Ma, Y.: Mining, integrating classification and association rule. In: KDD (1988)
3. Inibhunu, C., Schauer, A., Redwood, O., Clifford, P., McGregor, C.: The impact of gender, medical history and vital status on emergency visits and hospital admissions: a remote patient monitoring case study. In: IEEE LSC, Sidney, Australia (2017)
4. Weng, C.-H.: Discovering highly expected utility itemsets for revenue prediction. Knowl.-Based Syst. **104**, 39–51 (2016)
5. D'Angelo, G., Rampone, S., Palmieri, F.: Developing a trust model for pervasive computing based on Apriori association rules learning and Bayesian classification. Soft Comput. **21**, 6297–6315 (2017)
6. Gao, Y., Xua, A., Hua, P.J.-H., Cheng, T.-H.: Incorporating association rule networks in feature category-weighted naive Bayes model to support weaning decision making. Decis. Support Syst. **96**, 27–38 (2017)
7. Fowkes, J., Sutton, C.: A Bayesian network model for interesting itemsets. In: Frasconi, P., Landwehr, N., Manco, G., Vreeken, J. (eds.) ECML PKDD 2016. LNCS (LNAI), vol. 9852, pp. 410–425. Springer, Cham (2016). https://doi.org/10.1007/978-3-319-46227-1_26
8. Yang, T., et al.: Improve the prediction accuracy of Naïve Bayes classifier with association rule mining. In: IEEE 2nd International Conference on Big Data Security on Cloud (2016)

9. Mai, T., Vo, B., Nguyen, L.T.: A lattice-based approach for mining high utility association rules. Inf. Sci. **399**, 81–97 (2017)
10. Song, K., Lee, K.: Predictability-based collective class association rule mining. Expert Syst. Appl. **79**, 1–7 (2017)
11. Inibhunu, C., Schauer, A., Redwood, O., Clifford, P., McGregor, C.: Predicting hospital admissions and emergency room visits using remote home monitoring data. In: IEEE LSC, Sidney, Australia (2017)
12. Frank, E., Hall, M.A., Witten, I.H.: Data Mining: Practical Machine Learning Tools and Techniques, 4edn. Morgan Kaufmann, Burlington (2016)
13. Vreeken, J., Tatti, N.: Interesting patterns. In: Aggarwal, C.C., Han, J. (eds.) Frequent Pattern Mining, pp. 105–134. Springer, Cham (2014). https://doi.org/10.1007/978-3-319-07821-2_5
14. BPUK. http://www.bloodpressureuk.org/BloodPressureandyou/Thebasics/Bloodpressure chart
15. HeartORG. http://www.heart.org/HEARTORG/Conditions/HighBloodPressure/KnowYour Numbers/Understanding-Blood-Pressure-Readings_UCM_301764_Article.jsp#.WlOkP7en GM8
16. SAS. https://www.sas.com/en_ca/home.html

Memory Efficient Frequent Itemset Mining

Nima Shahbazi(✉), Rohollah Soltani(✉), and Jarek Gryz(✉)

Department of Computer Science and Engineering, York University, Toronto, Canada
{nima,rsoltani,jarek}@cse.yorku.ca

Abstract. Frequent itemset mining has been one of the most popular data mining techniques. Despite a large number of algorithms developed to implement this functionality, there is still room for improvement of their efficiency. In this paper, we focus on memory use in frequent itemset mining. We propose a new approach in which transactions are represented in a compact graph with the number of nodes equal to the number of distinct items in a database. Our experimental results confirm the efficiency of memory use without significantly sacrificing the execution time of the mining algorithm.

Keywords: Data mining · Frequent pattern · Memory efficient Association rule · Incremental mining

1 Introduction

Frequent itemset mining has been deployed for a wide range of applications since its introduction by [1] in 1993, including bioinformatics [15], web mining [9], software bug mining [11], system caching [16], and numerous other fields. Given a data set of transactions (each containing a set of items), frequent itemset mining finds all the sets of items that satisfies the *minimum support*, a parameter provided by a user or an application. The value of that parameter determines the number of itemsets discovered by the mining algorithm. If a small minimum support threshold is chosen for a database, a large number of frequent itemsets may be found. On the other hand, a large minimum support threshold may result in missing interesting itemsets. Thus, the mining process usually needs to be run multiple times before a satisfactory result can be achieved. In either case, the size of the data structure used by the mining algorithms to store temporary results can vary widely and is difficult to predict in advance. When dealing with large databases, the representation and storage of the data becomes a critical factor in the processing time of the mining algorithms. The existing techniques deploy either list-based [5,6,13] or tree based [8,10,12,14] structures to store data. The problem with both structures, however, is that with a large number of itemsets processed by the algorithm, the application may run out of memory.

The most popular structure for frequent pattern mining is FP-tree [8], which is a prefix tree representing the transactions with only the frequent items in a

P. Perner (Ed.): MLDM 2018, LNAI 10935, pp. 16–27, 2018.
https://doi.org/10.1007/978-3-319-96133-0_2

compact way. In an FP-tree, the number of nodes required to create the tree can be substantially greater than the total number of distinct items in the dataset. This happens because, in general, items can be repeated multiple times in the tree. In the worst case (admittedly unlikely), the prefix tree can grow to a maximum of 2^n nodes, where n is the number of distinct items in a database. If a dataset holds 100 distinct items, to mine such a dataset a computer would require storage capable of holding 2^{100} or approximately 10^{30} nodes. However, with the deployment of the method we propose in this paper, we would only require a storage capacity capable of storing a total of 100 nodes, which is much more feasible.

The main objective of this work is to propose a memory efficient method to store the data processed by the frequent itemset mining algorithm. Instead of a prefix tree, we store the data in a compact directed graph whose nodes represent items. In this way, the size of the graph is bounded by the number of distinct items present in the database.

Our algorithm was tested on six different data sets from the UCI Machine Learning Repository [2] and the results were analyzed and compared against results obtained from two state-of-the-art methods: FP-Growth [8] and CanTree [10]. The results not only showed a dramatic reduction in the amount of memory required to store the data structure required for mining, but also showed that running time of our algorithm was comparable to that of tree based frequent item set mining methods. In fact, in four of the tested datasets (Connect, Accident, T10I4D100K and Pumsb) our algorithm is superior with respect to runtime performance to the CanTree method.

The rest of this paper is organized as follows. In Sect. 2, we present a brief background on the related work and algorithms developed for frequent itemset mining. Section 3 offers the details of the data structure and the mining algorithm. Experimental results are presented in Sect. 4, while the conclusions in Sect. 5.

2 Related Work

Data structures used in frequent itemset mining algorithms can mainly be divided into array-based and tree-based. PrePost, PrePost+, and the FIN algorithm are all examples of successfully deployed array-based algorithms for frequent pattern mining. The PrePost algorithm [6] deploys a novel vertical data structure N-list for mining frequent itemsets. However, this algorithm consumes more memory than most algorithms for sparse datasets. PrePost+ [5] algorithm was later proposed. It deploys a pruning system to reduce search space. Even though this method requires less memory than PrePost it still uses more memory than FP-Growth (described below). FIN was then proposed to improve the results of PrePost+ by proposing a new data structure called Nodesets which only require the pre-order of each node, thus again reducing memory use [4].

One of the most popular tree-based methods is the FP-Growth algorithm. The FP-Growth algorithm is a divide-and-conquer algorithm designed to improve results obtained by Apriori-based methods. It divides the task of obtaining the frequent itemsets into tree-building and mining phases. The tree-building phase retrieves the data from the database and stores it in an efficient prefix tree structure, after which the mining phase is performed on the prefix tree. Within the tree-building phase a prefix tree called the FP-tree is created from transactions. In order to construct the FP-tree only two passes over the database are required. The first pass accounts for obtaining the support of each distinct item within the database. From all the data in the database, only the frequent items and their support counts are preserved for later processing. A second database scan is then completed in order to generate the FP-Tree.

When the database is large it may be impossible to process and construct the FP-tree [8]. One of the methods proposed to overcome this issue is to partition the database into smaller databases and to mine each of those databases separately. Variations of this algorithm were designed to be deployed in parallel on separate nodes [3,7].

The CFP-Tree (Compressed Frequent Pattern) structure was proposed [12] to address the issue of explosively growing frequent itemset. CFP-Tree reduces memory requirements of the prefix tree by deploying a sharing system to reduce the required number of nodes. CFP-Tree method facilitates frequent itemset retrieval by not requiring decompressing the data. CFP-Trees are compact and are comparable to other concise representations used for frequent itemsets [12].

CanTree proposed in [10] was designed for incremental mining purposes. A user defined canonical order is used to arrange items and create the prefix tree which encapsulates all database content. This method requires only one database scan as opposed to the two required by the FP-growth algorithm. In the CanTree, items are arranged according to a certain canonical order, which is unaffected by frequency changes. This method does not require merging, splitting, or swapping of nodes. CanTree deploys a similar mining method to FP-Growth by using a divide and conquer strategy. Conditional trees are created for frequent items by traversing the tree path upwards only. Only frequent items are included in the traversal step and they can be easily looked up through a simple header table.

A variation of CanTree, called CanTries, reduces further memory requirements of CanTree by forming nodes on the same pathway with similar frequencies into mega nodes. This can be done without increasing the runtime of the mining algorithm. Still, all of the tree-based algorithms require data structures much larger than the number of items in a database and cannot be guaranteed to be memory resident.

3 Methodology

As indicated above, the goal of our method is to decrease memory requirements for the data structure on which the mining for frequent patterns is performed. Thus, we replace the tree structure utilized in previous methods [8,10] with a labeled directed graph.

3.1 Graph Construction

Let $A = \{a_1, \cdots, a_n\}$ be a set of items, and $D = \{T_1, \cdots, T_m\}$ is a database containing transactions. Each transaction contains a set of items (an itemset) so the transaction is just a subset of A. We require, however, that all items in each transaction are sorted with respect to their indices, that is, $T_k = <a_{k_1}, \cdots, a_{k_l}>$, where $a_{k_i} \in A, 1 \leq i \leq l, l > 1$. Transactions with $l = 1$, only increase the total counts of an item and do not add an edge to the graph. We define an edge-labeled directed graph[1] as $G = (N, E)$ where:

- $N = A$
- $E = \bigcup_{k=1}^{m}(\{<a_{k_i}, a_{k_{i+1}}> | T_k = <a_{k_1}, \cdots, a_{k_l}> \in D, 1 \leq i \leq l-1\}$ $\cup \{<a_{k_l}, a_{k_1}> | T_k = <a_{k_1}, \cdots, a_{k_l}> \in D\})$

We call the edges $<a_{k_i}, a_{k_{i+1}}>$, $1 \leq i \leq l-1$ *forward edge* s, and the edge $<a_{k_l}, a_{k_1}>$ a *backward edge*. Labels in our graph are binary strings, and we call them edge codes. Label on a backward edge is called *backward edge code* or *BEC*, and label on a forward edge is called *forward edge code* or *FEC*. For a transaction with l items, $l-1$ forward edge between two contiguous items and one backward edge between the last and the first items in the transaction will be created.

Consider the following example.

Example 1: Let $T_1 : \{a_3, a_2, a_4, a_1\}$, $T_2 : \{a_4, a_3, a_1\}$, and $T_3 : \{a_3, a_2\}$ be three transactions in a database, with four distinct items $\{a_1, a_2, a_3, a_4\}$. In order to add a transaction to the graph, the items within each transaction are first sorted based on their indices. Therefore, items in T_1 are sorted so that $T_1 = <a_1, a_2, a_3, a_4>$. Then the following set of edges $E = \{<a_1, a_2>, <a_2, a_3>, <a_3, a_4>, <a_4, a_1>\}$ is created for transaction T_1. In this example three forward edges and one backward edge are created for the transaction (as shown in Fig. 1a with red edges). Then T_2 is added to the graph with two forward edges from a_1 to a_3 and from a_3 to a_4 and with one backward edge from a_4 to a_1 (as shown in Fig. 1b with green edges). Finally, T_3 is added to the graph. Since it shares its forward edge with T_1 and has a distinct backward edge, no new label is created for it (the labeling method is discussed below). The graph with all three transactions is shown in Fig. 1c.

Each transaction requires a unique backward edge which completes a loop for that transaction in the graph. When T_1 was added to the graph a label r was assigned to it. Since T_2 shares a common backward edge with T_1 we need to distinguish it from T_1 s backward edge, hence we labeled it g . For the overlapping edges such as these two we merge the labels ($r + g$ in this example). When T_3 is added to the graph, we create a new backward edge $<a_3, a_2>$ in the graph and since that edge is not yet in the graph an overlap of codes will not occur on that edge and the existing code (r) is used for it. We will show later that each transaction can be extracted from the graph with its unique backward edge code.

[1] Edge labeling will be defined later.

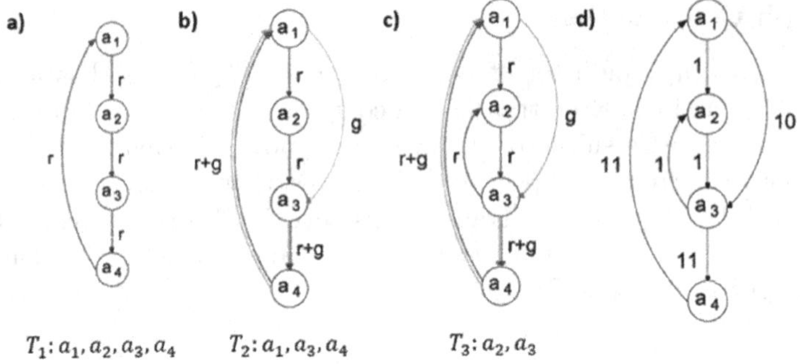

Fig. 1. Resulting graph after adding each of the transactions: T_1 (a) T_2 (b) T_3 (c), and (d) the completed graph after adding T_1 T_2 T_3 with the new coding structure.

3.2 Edge Labeling

For efficiency of implementation, we designed a coding structure based on binary strings. Each transactions code will be binary representations of powers of 2 values $\left(2^0,\ 2^1, 2^2, \cdots\right)$. In our example, r is represented by 2^0 $(r \rightarrow 1)$ and g is represented by 2^1 $(g \rightarrow 10)$. We call the code assigned to each transaction a *unique backward code* (UBC) as this code is unique for a backward edge of every transaction. In our example, T_1 and T_2 share a backward edge so they need two distinct UBCs; 1 is used for T_1 and 10 is used for T_2. (Notice that T_3 does not share a backward edge $<a_3, a_2>$ with any other transaction hence it does not require a distinct UBC.) When two transactions with different UBCs share an edge, we need to label that edge with their two distinct UBCs. Again, for efficiency of implementation, we can do better: we use the logical OR of their UBCs. T_1 and T_2 have two common edges in the graph, $<a_3, a_4>$ and $<a_4, a_1>$. The UBCs of these transaction: 1 and 10 respectively, are ORed together to create FEC and BEC with the value 11. On the other hand, knowing the value of the edge code to be 11 we can deduce that this it is a combination of the UBC 1 and the UBC 10. The complete labeled graph for the three transactions is shown in Fig. 1d.

We now show formally how transactions are assigned their UBCs and how these UBCs are represented as graph labels. For the smallest transactions containing just two items, the corresponding UBC is calculated as follows:

$$UBC\left(<a_i, a_j>\right) = \begin{cases} 1 & j - i = 1 \\ 10^{\frac{(j-2)(j-3)}{2}}.10^i & otherwise \end{cases} \tag{1}$$

Thus, in Example 1, the UBC of T_3 is equal to 1.

For transactions with more than two items, the maximum UBC of consecutive pairs within that transaction is calculated as follows:

$$\begin{aligned} UBC\left(<a_1, a_2, \cdots, a_{k-1}, a_k>\right) = \\ max\left(UBC\left(<a_1, a_2>\right), \cdots, UBC\left(<a_{k-1} \cdots a_k>\right)\right) \end{aligned} \tag{2}$$

In Example 1, the UBCs for T_1 and T_2 respectively, are 1 and 10. Note that the UBCs calculated via Eq. 2 are not necessarily unique for transactions. But any clashes of UBCs of two different transactions will be identified *before* these transactions are inserted into the graph (this is taken care in Line 3 of Procedure 1 below).

After assigning each transaction a UBC, we need to create labels (edge codes) of the graph edges, that is, the BECs and the FECs. As shown in Example 1, if two or more transactions share a backward edge, the labels of the edges shared by these transactions will be created via a logical OR of their UBCs. We say that the labels generated in this way *contain* the UBCs of their transactions. In fact, there is a straightforward way of telling when an edge code contains a given UBC.

Observation 1: BEC L contains UBC X, if and only if, the logical AND between L and X is not zero.

We now ready to present an algorithm for inserting a transaction into a graph.

Procedure 1: Insert Transaction T in to graph G

1. Calculate the UBC of T using Eqs. 1 and 2; call it X.
2. Create forward and backward edges (unless they are already in G) for T and assign 0 for all its BECs and FECs.
3. If BEC of T already contains X, generate a new UBC for T: X = binary representation of $\left(2^{number\_of\_bits(BECofT)}\right)$.
4. Perform a logical OR on BECs and FECs of T with X.

Two elements of the procedure require an explanation. First, assigning 0 in Line 2 initializes the values of the edge codes for edges in the graph that did not exist before so that the operations in Line 4 could be performed. Second, Line 3 verifies whether the UBC of the new transaction is unique on the backward edge of that transaction: if the BEC of that edge contains that UBC, it means that some other transaction with the same backward edge has the same UBC. To make the UBC of the new transaction unique, we assign it a new UBC as specified in Line 3.

We now present an example where we show how transactions are identified in the graph generated via Procedure 1.

Example 2: Consider a database with four items and all possible transactions with length greater than one. Figure 2a shows the transactions and their corresponding UBCs. The graph representing the transactions with appropriate edge codes is shown in Fig. 2b. Consider as an example the edge $<a_1, a_2>$. This edge is shared by transactions T_1, T_7, T_8, and T_{11} with their respective UBCs 1, 1, 1000, 1. A logical OR between these codes results in 1001. Note that even in this relatively complex example, the condition in Line 3 of Procedure 1 is never satisfied and the transactions keep their original UBCs.

3.3 Identifying Transactions in the Graph

The ultimate goal for creating the graph was to enable mining for frequent itemsets. For this, we need to identify the transactions in the graph and create conditional trees and then the same FP-Growth mining algorithm will be used. As we have observed before, for every transaction there exists a loop in the resulting graph. In order to extract all transactions from the graph, a traversal of all backward edges is required. When a given backward edge is traversed, all UBCs of its BEC are extracted. These are the UBCs of all transactions sharing that backward edge. Procedure 2 describes how the nodes of a transaction with a given UBC is identified. The procedure starts with node v which is the in-vertex of the given backward edge.

Procedure 2: Extract nodes of transaction with UBC X originating from node v

1. Perform the logical AND between X and all FECs of edges outgoing from v.
2. Follow the edge with non-zero result to reach the next node.
3. Repeat steps 1 and 2 until a loop is completed.

Consider again the graph shown in Fig. 2b and assume we are considering the backward edge which connects a_4 to a_1. BEC of that edge is 1111 which contains four UBCs: 1, 10, 100, and 1000. Suppose that the transaction with UBC 10 is chosen for traversal. We initialize Procedure 2 with UBC 10 and node a_1. Of the three outgoing edges, only the one from a_1 to a_3 with the FEC = 10 succeeds ($10\ AND\ 10 \neq 0$). For the remaining two outgoing edges the AND result of their FECs and UBC 10 is zero ($(1001\ AND\ 10) = 0$ $(100\ AND\ 10) = 0$). After reaching a_3 we can only go to a_4 ($10\ AND\ 11\ \neq 0$). The loop is then complete and transaction $T_9 : <a_1,\ a_3,\ a_4>$ is retrieved.

In order to extract all transactions within a graph in Fig. 2b, a traversal of all backward edges with their containing UBC is required. For example, a_4 has three backward edges with edges codes 1, 1001 and 1111. Code 1 contains UBC 1. Code 1001 contains UBC 1000 and 1. Code 1111 contains UBC 1, 10, 100 and 1000. By traversing all the corresponding UBCs on a_4 backward edges, all seven transactions containing item a_4 are extracted ($T_3, T_5, T_6, T_8, T_9, T_{10}\ and\ T_{11}$). These extracted transactions (all transactions containing item a_4) are used to create conditional trees and then the same FP-Growth mining algorithm find the frequent itemsets [8]. UBCs are used to obtain transactions, but do not reflect the number of times that transaction has been repeated. If a transaction $T_9 : <a_1,\ a_3,\ a_4>$ appears more than once (ex. three times) then an integer three is assigned for its corresponding UBC, as shown in Fig. 3. Therefore, when a new transaction is parsed, if it already exists in the graph its integer count is incremented by one. In other words, the corresponding count of a transaction indicates its frequency within the database.

To speed up conditional tree [8] creation another step which can be taken is when we have a transaction such as $T_{11} : <a_1, a_2, a_3, a_4>$ not only do we add

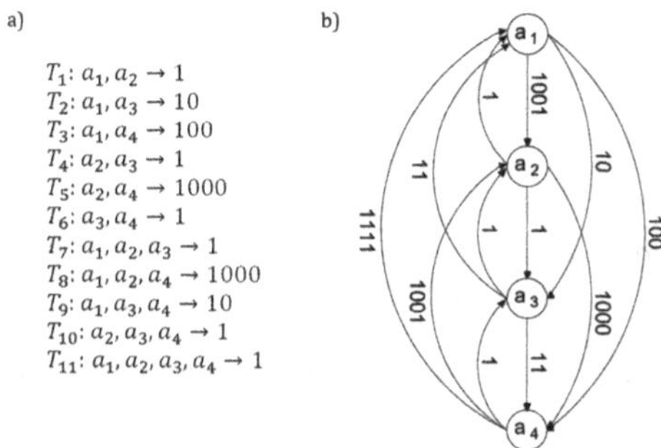

a)
b)
T_1: $a_1, a_2 \rightarrow 1$
T_2: $a_1, a_3 \rightarrow 10$
T_3: $a_1, a_4 \rightarrow 100$
T_4: $a_2, a_3 \rightarrow 1$
T_5: $a_2, a_4 \rightarrow 1000$
T_6: $a_3, a_4 \rightarrow 1$
T_7: $a_1, a_2, a_3 \rightarrow 1$
T_8: $a_1, a_2, a_4 \rightarrow 1000$
T_9: $a_1, a_3, a_4 \rightarrow 10$
T_{10}: $a_2, a_3, a_4 \rightarrow 1$
T_{11}: $a_1, a_2, a_3, a_4 \rightarrow 1$
a1
a2
a3
a4
1001
1
11
10
1111
100
1000

Fig. 2. (a) A transactional dataset with four distinct items and all transactions with a length of greater than 2 (b) the proposed methods resulting graph

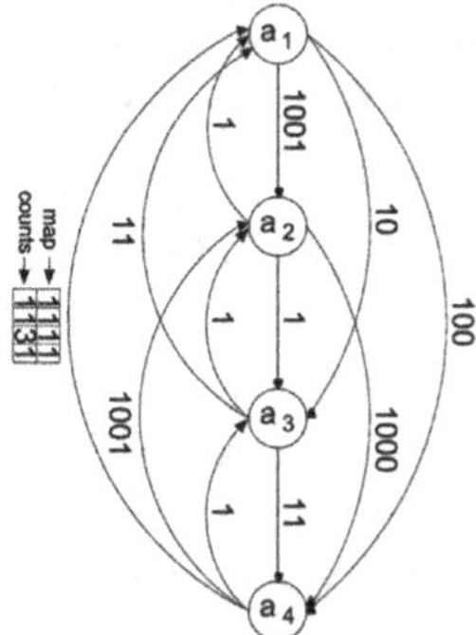

Fig. 3. For transactions which happen more than once we can set an integer to represent its count

a directional link from a_4 to a_1 with a code of 1, but we do this for all other nodes connected to a_1 (as shown in Fig. 4). This will help in speeding up the conditional tree creation. This is because the conditional tree which is based on T_{11} : $<a_1, a_2, a_3, a_4>$ also includes transactions $<a_1, a_2, a_3>$, $<a_1, a_2>$. This extra step does not stall transaction processing, and is not a requirement.

4 Performance Results

In this section we present performance evaluation of our method (we call it Compact Graph) in comparison to FP-Growth and CanTree is given. FP-Growth and CanTree demonstrate good performance in incremental frequent item set mining and were chosen as a baseline to assess the Compact Graphs results. All computational aspects were conducted in JAVA on a Linux Centos 6 with Intel Core 2 Duo 3.00 GHz CPU and 4 GB memory. From the UCI Machine Learning Repository the Chess, Connect, Mushroom, Accidents, Pumsb and T10I4D100K datasets were used. Seven runs were conducted and reported runtime figures represent the average of their execution times.

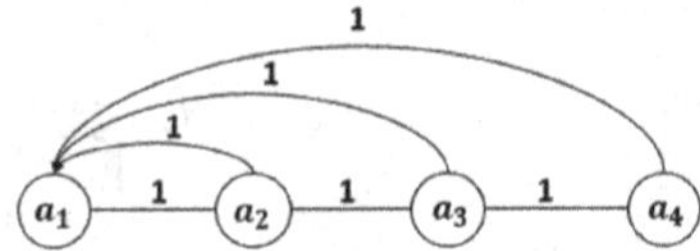

Fig. 4. In order to speed up conditional tree creation, we can add a directional link for all codes connecting to the first node as well.

Table 1. Number of nodes used for data representation by FP-Growth, CanTree, and the compact graph.

Dataset	FP-Growth	Can-Tree	Compact graph
Chess	38610	39551	75
Mushroom	27122	34004	119
Accident	4243242	5262556	468
Connect	359292	1092209	129
Pumsb	1126155	2773441	2088
T10I4D100k	714731	748409	870

Table 1 shows the number of nodes required to generate the prefix tree for CanTree, FP-Growth and the Compact Graph. Tests were also conducted on the total memory consumption of CanTree and FP-growth in comparison to the Compact Graph with the results compiled for each of the tested datasets as shown in Table 2. Compact Graph performs better in all cases, and other than the mushroom dataset, outperforms the other algorithms by a large margin. Due to its alphabetically structured tree and high node count, CanTree consumes the most amount of memory. FP-Growth has a more efficient tree structure which leads to smaller memory utilization and fewer nodes. However, Compact Graph utilizes the least number of nodes possible with a compact structure, and therefore is the most memory efficient.

Figure 5 shows runtime results for the Compact Graph, FP-growth, and CanTree utilizing different support thresholds on the tested datasets. The purpose of this experiment was to measure effects of the minimum support threshold on the runtime of algorithms and to confirm that the Compact Graph is not sacrificing the execution time of the mining algorithm. Total execution time consist of the amount of time required for prefix tree or graph construction in addition to the frequent item set mining step (Eq. 3):

$$TotalTime = ConstructionTime + MiningTime. \quad (3)$$

The construction phase is shortest for CanTree in comparison to FP-Growth and the compact graph method; this is due to CanTree not requiring tree reconstruction and deploying an alphabetical order to add transactions. Despite

Table 2. Memory Comparison.

(a) Chess

Minimum support	0.0	0.4	0.5	0.6	0.7
FPgrowth/compact	12	20	21	17	11
Can_Tree/compact	14	26	35	42	86

(b) Mushroom

Minimum support	0.0	0.1	0.2	0.3	0.4
FPgrowth/compact	4	8	7	5	3
Can_Tree/compact	5	17	29	67	98

(c) Connect

Minimum support	0.0	0.7	0.75	0.8	0.85
FPgrowth/compact	28	7	6	3	2
Can_Tree/compact	73	987	1067	1146	1463

(d) Accident

Minimum support	0.0	0.1	0.12	0.14	0.16
FPgrowth/compact	48	120	116	118	122
Can_Tree/compact	69	195	215	223	234

(e) Pumsb

Minimum support	0.0	0.55	0.6	0.65	0.7
FPgrowth/compact	9	12	26	23	18
Can_Tree/compact	17	323	376	485	523

(f) T10I4D100k

Minimum support	0.0	0.01	0.02	0.03	0.04
FPgrowth/compact	12	27	52	66	34
Can_Tree/compact	14	43	86	165	436

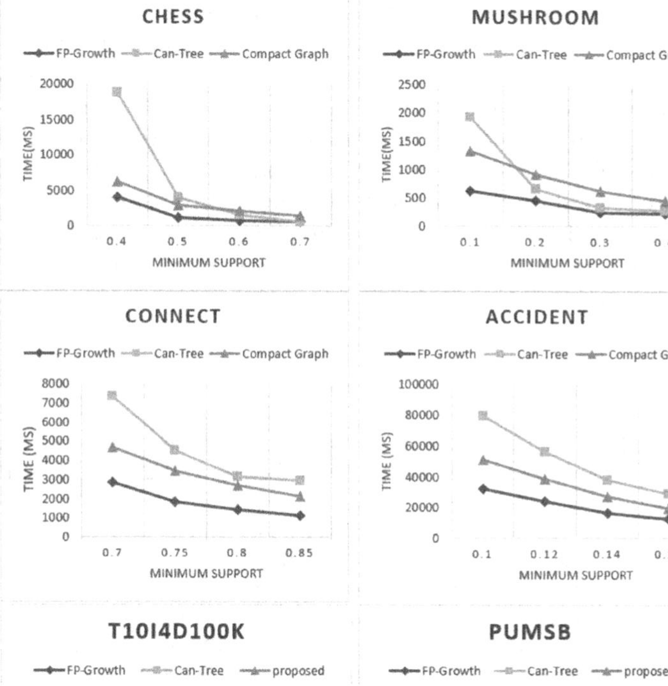

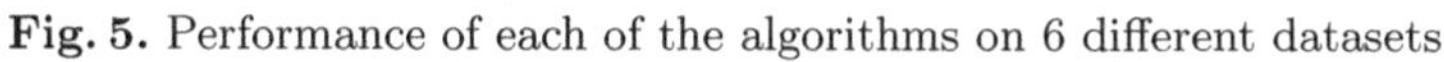

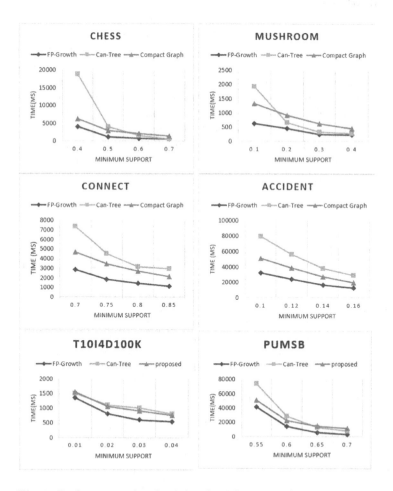

Fig. 5. Performance of each of the algorithms on 6 different datasets

having the fastest prefix tree construction phase, CanTree generates a larger tree than the ones generated by FP-Growth or the Compact Graph. Mining time is highly correlated to the number of nodes in the pruned prefix tree (pruned based on the minimum support threshold), therefore larger prefix trees result in larger mining times. This causes mining time to become a dominant factor, whereas when we have a high minimum support, the construction time will be dominant.

The Compact Graph has a higher construction time than both CanTree and FP-Growth, which is due to the need for encoding data in the graph. In the mining phase the Compact Graph is slower than FP-Growth, but based on the support threshold used it may prove a better option to CanTree. This is due to the complex structure of CanTree; with a lower support threshold, the Compact Graph mines the data more efficiently.

5 Conclusion

The size of memory becomes a limiting factor in mining large databases. The solutions offered so far, such as those proposed in incremental mining methods, called for dividing the database into subsets and mining each of them separately. In this paper, we took a different approach and faced the problem head-on. We proposed a new approach for storing large collections of frequent itemsets in a compact graph. This method utilizes only one node per each distinct item in the database, therefore reducing dramatically the amount of required memory. Our experimental results have shown that our method performs much better than both FP-Growth and Can-Tree in terms of memory usage and is comparable to them in terms of runtime.

References

1. Agrawal, R., Imieliński, T., Swami, A.: Mining association rules between sets of items in large databases. In: ACM SIGMOD Record, vol. 22, pp. 207–216. ACM (1993)
2. Blake, C.L., Merz, C.J.: UCI repository of machine learning databases. Department of Information and Computer Science, University of California, Irvine, p. 55 (1998). http://www.ics.uci.edu/~mlearn/mlrepository.html
3. Buehrer, G., Parthasarathy, S., Tatikonda, S., Kurc, T., Saltz, J.: Toward terabyte pattern mining: an architecture-conscious solution. In: Proceedings of the 12th ACM SIGPLAN Symposium on Principles and Practice of Parallel Programming, pp. 2–12. ACM (2007)
4. Deng, Z.-H., Lv, S.-L.: Fast mining frequent itemsets using nodesets. Expert Syst. Appl. **41**(10), 4505–4512 (2014)
5. Deng, Z.-H., Lv, S.-L.: Prepost+: an efficient n-lists-based algorithm for mining frequent itemsets via children-parent equivalence pruning. Expert Syst. Appl. **42**(13), 5424–5432 (2015)
6. Deng, Z.H., Wang, Z.H., Jiang, J.J.: A new algorithm for fast mining frequent itemsets using N-lists. Sci. China Inf. Sci. **55**(9), 2008–2030 (2012)

7. El-Hajj, M., Zaiane, O.R.: Parallel leap: large-scale maximal pattern mining in a distributed environment. In: 12th International Conference on Parallel and Distributed Systems, ICPADS 2006, vol. 1, pp. 8–pp. IEEE (2006)
8. Han, J., Pei, J., Yin, Y.: Mining frequent patterns without candidate generation. In ACM SIGMOD Record, vol. 29, pp. 1–12. ACM (2000)
9. Kosala, R., Blockeel, H.: Web mining research: a survey. ACM SIGKDD Explor. Newsl. **2**(1), 1–15 (2000)
10. Leung, C.K.-S., Khan, Q.I., Li, Z., Hoque, T.: CanTree: a canonical-order tree for incremental frequent-pattern mining. Knowl. Inf. Syst. **11**(3), 287–311 (2007)
11. Li, Z., Zhou, Y.: PR-miner: automatically extracting implicit programming rules and detecting violations in large software code. In: ACM SIGSOFT Software Engineering Notes, vol. 30, pp. 306–315. ACM (2005)
12. Liu, G., Hongjun, L., Yu, J.X.: CFP-tree: a compact disk-based structure for storing and querying frequent itemsets. Inf. Syst. **32**(2), 295–319 (2007)
13. Pei, J., Han, J., Lu, H., Nishio, S., Tang, S., Yang, D.: H-mine: hyper-structure mining of frequent patterns in large databases. In: Proceedings IEEE International Conference on Data Mining, ICDM 2001, pp. 441–448. IEEE (2001)
14. Shahbazi, N., Soltani, R., Gryz, J., An, A.: Building FP-tree on the fly: single-pass frequent itemset mining. Machine Learning and Data Mining in Pattern Recognition. LNCS (LNAI), vol. 9729, pp. 387–400. Springer, Cham (2016). https://doi.org/10.1007/978-3-319-41920-6_30
15. Wang, J.T.L., Zaki, M.J., Toivonen, H.T.T., Shasha, D.: Introduction to data mining in bioinformatics. In: Wu, X., et al. (eds.) Data Mining in Bioinformatics, pp. 3–8. Springer, London (2005). https://doi.org/10.1007/1-84628-059-1_1
16. Yan, X., Han, J., Afshar, R.: CloSpan: mining: closed sequential patterns in large datasets. In: Proceedings of the 2003 SIAM International Conference on Data Mining, pp. 166–177. SIAM (2003)

Fuzzy Networks Model, a Reliable Adoption in Corporations

John Velandia(✉), Gustavo Pérez, and Holman Bolivar

Engineering Faculty, Universidad Católica de Colombia, Bogotá, Colombia
{javelandia, gperezh, hdbolivar}@ucatolica.edu.co

Abstract. Computing huge amounts of information and performing complex operations in a unique fuzzy logic system is a challenge in the field of fuzzy logic. This paper presents a Knowledge engineering application whereby a Fuzzy Network (FN) is used to build a complex computing model to reproduce corporate dynamics and to implement a Model Reference Adaptive Control (MARC) strategy for Corporate Control [2]. This model is used as a What If? Environment to explore future consequences of actions planned within a strategic scenario context in terms of KPIs displayed in a Balanced ScoreCard (BSC) control board. Corporation's strategy map is required to plan the Knowledge Identification and Capture Activity (KICA) required to obtain the knowledge to be represented in the FN's nodes rule bases. KICA produces linguistic variables as well as the qualitative relationships amongst them. A FN appears as a natural solution to model the knowledge distributed within the members participating in all analysis and decision making tasks along the organization. Additionally, as proof of concept a prototype which capable of designing and simulating networks of fuzzy systems is presented based on the standard IEC 61131-7.

Keywords: Fuzzy logic · Fuzzy networks · Corporate model · Fuzzy system

1 Introduction

Fuzzy Logic (FL) is used when it is difficult or even impossible to construct precise mathematical models [1]. FL is a logical system which is an extension of multivalued logic to serve as a logic model of human mind [2]. Complex traffic and transportation problems are example of having subjective knowledge, commonly known as linguistic information, which is not solve using classical mathematical techniques, since they are hard to quantify [3].

Information systems with complexity in multivariable control are oriented to diagnose the knowledge management in companies, among many others. It is computationally impracticable to concentrate all heuristics in one fuzzy system, due to the number of variables and rules that are considered in a whole system [4]. Currently, exists a trend to build Networks of Fuzzy Systems (NFS), as it has been noted in the last World Congresses presided by the International Fuzzy Systems Association, IFSA and the North American Fuzzy Information Processing Society, NAFIPS [5]. Despite,

P. Perner (Ed.): MLDM 2018, LNAI 10935, pp. 28–41, 2018.
https://doi.org/10.1007/978-3-319-96133-0_3

commercial tools are available to build NFS, they are limited by parameters setting [6]. Moreover, commercial tools are expensive, and there is no official free tool available to build NFS.

Since BSC officially appeared [7] it has become an important corporate control tool; however the feedback it provides through the KPIs takes a good time after an action has been taken. This occurs because the time constants involved in corporate dynamics are rather long ranging from weeks to months depending of the strategic deep of decisions and the corresponding actions.

This paper presents an enhancement of the BSC control strategy by means of a knowledge based computing model, representing the corporate dynamics and implementing a Model Reference Adaptive Control (MRAC) strategy [8]. Since the strategy map [9] displays the inner causality in an organization's dynamic, it is used to conduct the KICA required to identify the involved linguistic variables and the qualitative relationships amongst them. The model is constructed using a FN which stores the knowledge gathered through a KICA. This model is used to provide the **What If?** Environment to test different strategic scenarios and to identify the best actions to be taken on order to obtain desired KPI's values in a given time horizon. FN have already been used and reported [4, 10] as a tool to build expert systems using distributed knowledge.

Additionally, as proof of concept of the proposed model a design of a prototype that builds NFS is presented. The prototype is named FuzzyNet, and its objective is to design and simulate networks of fuzzy systems. It is based on the standard IEC 61131-7, ensuring the basic components of the programming methodology, environment and functional characteristics of Fuzzy Logic Control (FLC) [11, 12]. This research is part of the project "RESIDIF", its aim is to deploy a free tool for simulating scenarios that require NFS for corporations.

The novelty of this research comprises:

- A model that encompasses a formal path to create fuzzy networks. This implies a business process definition, including its activities.
- A software prototype named FuzzyNet that allows materialize the proposed model. This prototype is presented by mean of software components and graphical interfaces. Additionally, JFuzzyLogic library is presented as an important software integration for achieving real simulations.
- A corporate model application is defined based on the fuzzy network model, in this way the software prototype and real business processes are integrated to demonstrate a real application of this model in the industry.

This paper is organized as follows: Sects. 1 and 2 present background of FLCs and corporate concepts. In Sect. 3 up to date related work is presented, in Sect. 4 the proposed model is defined, including a comparison of non-commercial fuzzy software is presented, in order to define a candidate framework that would be integrate to FuzzyNet. Section 5 presents the application of the proposed model for corporations. Finally, Sects. 6 and 7 presents conclusions and future work.

2 Fundamentals

2.1 Fuzzy Logic

The concept of fuzzy logic is deeply related to how people perceive the environment. People constantly make ambiguous statements that depend on how observer perceives the physical or chemical effect, e.g., if someone says "the bus station is going to be congested", it would be interested to assess if this statement is true, and when it would happen. The reasoning based on information that is not accurate is also a common procedure, and it is precisely what fuzzy logic does [13]. Observation of the environment, the formulation of logical rules and mechanisms of decision-making [14].

In order to automate and standardize processes arising from the theory of fuzzy logic, a basic model was defined for application development [11]. A typical FLC architecture is composed of four principal components: a fuzzifier, a fuzzy rule base, an inference engine and a defuzzifier [15, 16].

The first component of the model is known as fuzzifier, which is the gateway of the fuzzy model. This component performs a mathematical procedure that consists of converting an element of the universe of discourse in fuzzy values. The defuzzifier is based on a set of rules antecedents and consequences, which are known as linguistic expressions. "yes", is the antecedent and "then", the consequent.

2.2 Corporate Concepts

2.2.1 BSC Control Structure

The current BSC control loop is displayed in Fig. 1, no matter how good the measurement system is, but the values entered for the KPIs in the BSC Board, since they will reflect the consequences of the taken actions only after a good amount of time since the time constants involved in the corporate dynamics are sometimes weeks or even months long. The *What If?* environment introduced here allows to test groups of actions, in fact, whole strategies, to examine the future KPI's values and so to determine the best strategy in terms of future KPI's values.

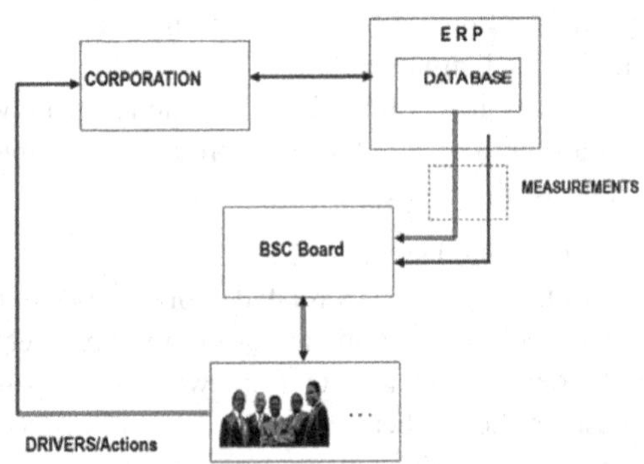

Fig. 1. Current BSC control loop structure.

2.2.2 Model Reference Adaptive Control Structure

The *What if?* environment is obtained using the mentioned MRAC control strategy and this is showed in Fig. 2. The Corporate Model allows to test actions showing future KPI's values in the BSC Model Board.

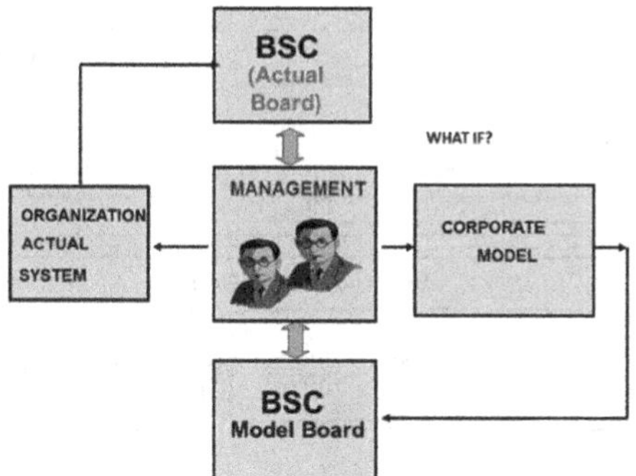

Fig. 2. MRAC structure with What If? Environment

Once the structure depicted in Fig. 2 is constructed, a concrete simulation is willing to start testing strategies and actions that would be the input of the proposed model. Afterwards, an observation is taking place over the BSC Model Board, in order to identify and analyze the resultant KPIs in the specified time horizon.

3 Related Work

Fuzzy networks have not been an ongoing discussion in fuzzy logic, despite this is a need for processing complex and parallel fuzzy systems. None of the indexed scientific databases provide research works, for instance Science Direct, Elsevier, ACM and Google Scholar. The closer research work to this topic are studies in fuzzy networks with aggregation of rule bases for decision-making problem solving, which is a model proposal [17].

There some additional studies are performed regarding cognitive and neural networks [18, 19], however they do not focus on fuzzy systems, which means that these sort of studies are not related to fuzzy networks, for example, these studies lack of information regarding inference engine, rules and some of the fundamental components of fuzzy systems. In conclusion, this research topic is novel and therefore further work is pending to develop in several scenarios.

4 Model for Building Networks of Fuzzy Systems

This section describes the architecture of FuzzyNet. The proposed architecture comprises: (A) Definition of the business process model. (B) The main components of the prototype to be developed. (C) Integration to fuzzy logic implementation.

4.1 Process Workflow to Design and Simulate Fuzzy Networks

Considering that any information system is created to support business processes, Fig. 3 describes the process to design and simulate NFS. Business Process Model Notation (BPMN) is used as convention. This process is accomplished in two phases:

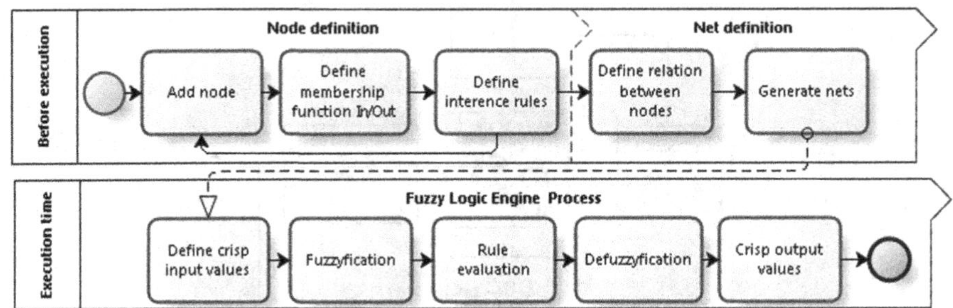

Fig. 3. Simulation process to create fuzzy networks.

(1) Before execution time: This phase consists of defining number of nodes needed for building a fuzzy network, and the network itself. The workflow consists of adding a node that contains a node's name, and linguistic variables to define fuzzy sets. Then, it is defined membership function to use with crisp input and output values. Followed by definition of inference rules that are provided by an expert. Once set of nodes (fuzzy systems) is defined, a relation among nodes is fixed, in order to build dependencies and prioritize nodes execution. The last activity in the workflow comprises generation of a fuzzy network. During this phase users only perform a network setting.

(2) During execution time: This phase encompasses the classic workflow of fuzzy logic systems, however in this process model acts as engine, since it is the core of the process. Once the network is designed, users provide crisp input values, which would be unique input for entire network, but for every node. Afterwards, the engine executes the matching of crisps values with the linguistic terms, this is named fuzzyfication. Then, the inference activity is performed, which comprises the assessment of rules, and a fuzzy set as result. Finally, the result of the inference process has to be converted into crisp numerical values, activity known as defuzzification, then information is presented throughout graphical interface.

4.2 Software Components Model

According to Fig. 4 *Net management component* is responsible for the definition of the fuzzy network. Network's name is set and relationships among nodes are stablished to accomplish the fuzzy network simulation. The setting is saved in a XML file to guarantee future interchange of information by mean of generic language.

Node management component is in charge of creating, updating and removing nodes from the fuzzy network; this implements a logic such that guarantee minimum

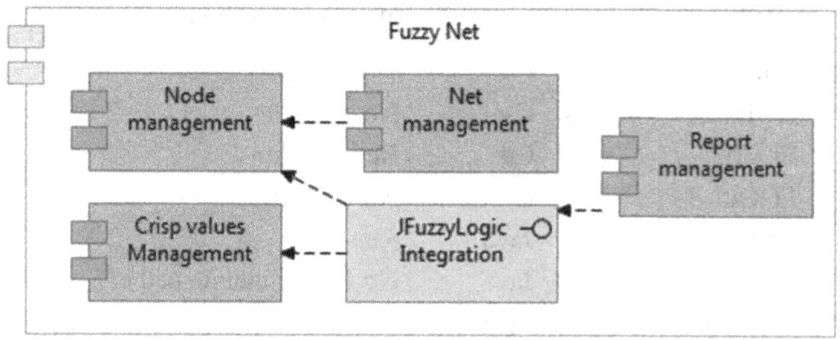

Fig. 4. Set of components for simulating fuzzy networks.

two nodes with their respective linguistic variables, linguistic qualifiers and membership function, i.e., triangular, Gauss, Cosine, Sigmoidal, Difference of sigmoidal, Trapezoidal and Generalized bell.

Crisp values management incorporates the variables and values that willing to use in the fuzzification process. These values are associated to the membership function automatically; these crisp values are transformed to fuzzy values after the simulation of the fuzzy network is performed.

JFuzzyLogic Integration is an external software package that allows using the membership functions automatically. This package acts as plugin, which means that the prototype may call any functionality natively. The following subsection specifies the reason of using this particular tool in this research.

Report management is an additional component that is not part of the model, however it is considered for obtaining general information regarding the simulations. For instance, this component provides an outline of the number of nodes that conform the network and their relationships, crisp values, linguistic qualifiers setting, inference engine method and rules that depend on the context, in this case corporate control model. Meanwhile Figure 6 presents a sampling interfaces of this model by mean of the prototype FuzzyNet.

4.3 Integration to JFuzzyLogic

IEC 61131 is a standard developed to integrate fuzzy control applications This standard is implemented among different providers to allow systems to connect with each other in a common way. The principal objective of this section is to find a tool that supports the IEC standard for being integrated into the software prototype that is used as proof of concept.

The main functionality of the tool that would be selected is the Fuzzy Controller (FC), because it allows to simulate and to run diffuse networks by mean of implemented libraries. FCL is a key functionality since it defines the parameters that are set in IEC for using adequately fuzzy controllers. The following criterion are considered to select the tool: FCL is provided, Programming Language - PL, functionality, and number of memberships functions - MF. Java language is the preference because the prototype is developed in Java, due to the portability and independency of operating system. Hence, Table 1 presents a comparison of noncommercial fuzzy logic tools:

Table 1. Comparison of open fuzzy logic software tools.

Name	FCL	PL	Functionality	MF
AwiFuzz	Yes	C++	No compiles	2
FLUtE	No	C#	Beta version	1
FOOL	No	C	No compiles	5
Funzy	No	Java	Run	2
FuzzyJToolkit	No	Java	No longer maintained	15
Jfuzzynator	No	Java	Run	2
JFCM	No	Java	Run	–
jFuzzyLogic	Yes	Java	Run	25
jFuzzyQt	Yes	C++		8
Libai	No	Java	Run	3
Nefclass	No	C++ Java	Run	1
Nxtfuzzylogic	No	Java	Run	1
UNFuzzy 2.0	Yes	C++	No compiles	8

According to the Table 1 a detailed analysis is performed for each criterion to select an adequate software tool:

FCL. The outcome after these tools were assessed is that only four of them are suitable for the required integration, specifically these tools are AwiFuzz, jFuzzyLogic, jFuzzyQt and Unfuzzy. It signifies that only these few tools remain in this analysis, because the surplus does not follow the FCL standard.

PL (Programming Language). As it was previously mentioned, the ideal language would be Java, given that the software prototype is developed in this language. Thereby, jFuzztLogic is currently the unique tool that meets this specific criterion.

Functionality. Before the aforementioned criteria were evaluated, a validation regarding whether each tool run on Windows properly was made, and the result shows that most of the tools work adequately in the operating system Windows. Another result to highlight is that four of them run properly on Windows, due to the lack of support and maintenance.

MF (Number of Memberships Functions). The membership functions are directly related with the inputs and outputs, having more memberships there more flexibility in its definition. The numbers with asterisks are membership functions which are built from the combination of membership functions. jFuzzyLogic is the most outstanding tool thanks to the 25 memberships functions implemented, which contributes to the enhancement of features in the proposed prototype.

Among the metrics established to choose a software tool, it is decided to opt for a tool that provide functionality for the development of fuzzy systems, also that implements Fuzzy Control Language specifications and that encompasses the highest number of membership functions, thus one could conclude that jFuzzyLogic is the most suitable tool for this research, specifically for the proposed prototype.

4.4 Principles of NFS

In the development phase they were established the principles for building networks fuzzy controllers, which describe the features that have the bindings between fuzzy controllers.

Binding. The binding between fuzzy nodes, it is established by means of linguistic variables, that play role of inputs or outputs in the nodes. Thus when a binding is established between nodes, the linguistic variable through which the operation is binding to the two nodes, in the origin node like an output and in the destination node like an input to the fuzzy controller.

Simulation. The simulation process of fuzzy networks consists in the individual simulations of nodes that compose the network, in a specific order defined by how they are configured the bindings, obeying the premise that to simulate a node should be considered if you have inputs function as outputs from another node, and if so the latter must first be simulated.

4.5 The Prototype

This phase is developed based on the architecture designed by diagrams obtained in the development phase, this architecture was validated through building prototyping, which involves the creation of a prototype software that will serve as the desired application archetype. This process is necessary to prove that the architecture designed to comply with the main objective, create networks of fuzzy logic.

Figure 5 shows the final FN designed architecture. Five layers were required in order to implement the network using FuzzyNet, a software tool to implement and simulate FN, developed at Universidad Católica de Colombia.

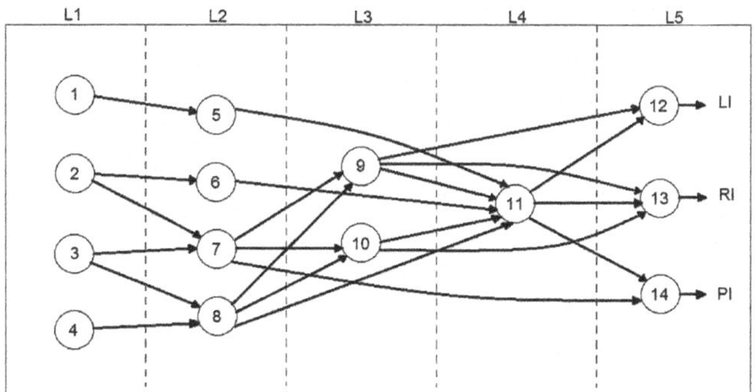

Fig. 5. FN architecture, a sample network.

In the previous figure, the simulation encompasses the following characteristics: all nodes employ Mandani (Minimum) for the fuzzy inference; membership functions implemented are the standard L, Lambda (Triangle) and Gamma Functions, available in majority fuzzy systems software tools; defuzzification is performed with center of gravity defuzzifier.

As result of testing the sample network, one could conclude that software components work correctly when the simulation is performed using the proposed model. In details, the process consists of definition of linguistic variable by typing the values by the user, including its possible results the risk that it is supposed to be assumed by the company.

Figure 6 presents a sampling interfaces of the prototype.

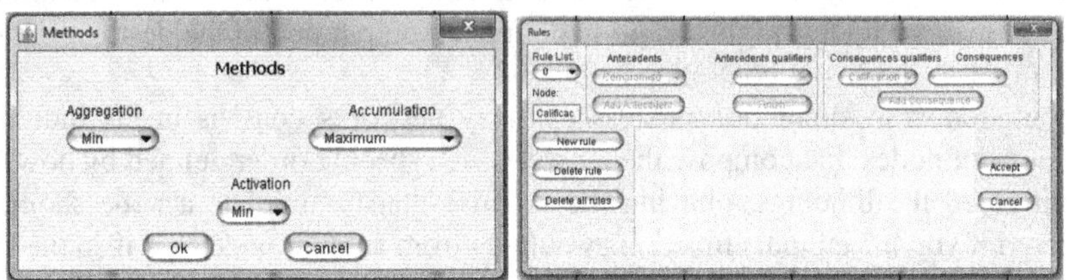

Fig. 6. Prototype interfaces sampling

5 Corporate Model Application

5.1 Corporate Model

Figure 7 shows a typical FN structure for the corporate model.

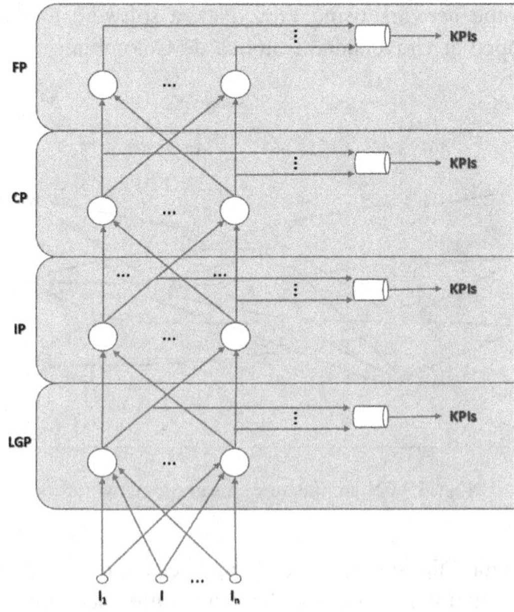

Fig. 7. Corporate model application

The FN structure is distributed along the four BSC Perspectives: Financial Perspective (FP), Customer Perspective (CP), Internal Perspective (IP) and Learning and Grow Perspective (LGP). Every Node in the FN is separately created and tuned using the MRAC interface included in the MANAGEMENT Block (Fig. 2). Design parameters for each node are determined through the tuning procedure using criteria obtained in the KICA. Figure 7 shows the structure of a node with the design parameters for every module. Depending on the particular dynamic associated to a company the FN could be either Feedforward or Recurrent. The Rule Base for every Node will contain the fuzzy rules obtained after analyzing the results obtained with the KICA. The nodes producing the KIP's values are explicitly shown. All the fuzzy rules for every node contain the time as a linguistic variable in the Antecedent. Rule k is then explicitly written as (1). Although the Inputs go into nodes in the Learning and Growing Perspective in Fig. 7, inputs can actually go into any node in any perspective.

Rk: IF X1 is LX_1 *AND…AND Xm is LXm AND*
Time is LT THEN Y1 is LY1 AND … AND Yn is LYn
Where:
X1…Xm are the inputs to the node and Y1…Yn are the outputs,
LXi Ɛ {L, M, H}, LYi Ɛ {L, M, H}
LT Ɛ {VS, S, M, L, VL}
With: L = Low, H = High, VS = Very Short, S = Short, M = Medium, L = Long, VL = Very Long.

Although only one layer is shown for the FN in each perspective, it is the particular dynamics for each corporation's value-creating processes which determines the FN's nodes structure for every perspective. The particular FN structure constructed for any particular organization will reflect the particular dynamics and the inner causality within the value-creating processes in that corporation.

5.2 The What If? Environment

The User Interface also contains the simulation environment so that managers and/or planners can perform the simulation tasks required to test any strategy. This environment keeps the record of all simulations to facilitate the supervision tasks required to identify the best strategies (Fig. 8).

When a corporation has a comprehensive and well maintained Data Base with a few years of data, rules can be identified using a mining procedure with the help of Adaptive (Trainable) Fuzzy Systems [8]. This capability is built in the UNFUZZY tool [6] used for the developments reported in [9, 10].

Implementing this MRAC strategy with *What If?* Environment requires the committed participation of all members in the organization. KEA in particular requires the open and patient collaboration since many times it is necessary to go over some

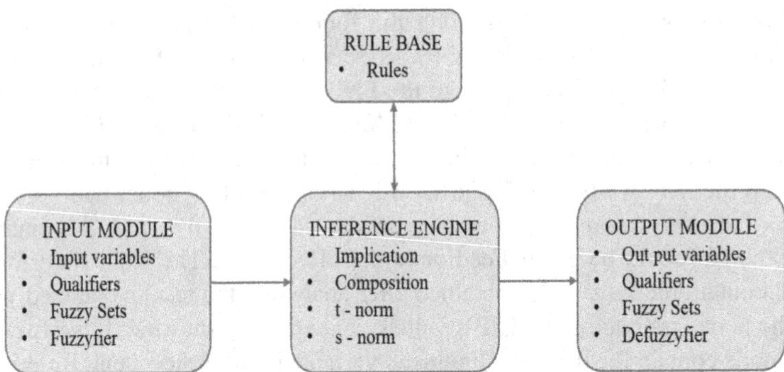

Fig. 8. Node's structure with design parameters.

particular subjects in order to clearly identify the linguistic variables as well as the nature of relationships.

MRAC with *What If?* Environment was implemented in a local utility. It is currently used in the planning tasks and it has been taken as a pilot experience to scout new possibilities. The task force organized for this accomplishment is now working on model refinement to include risks, using the experience gathered in a recent work [4]. This means identifying the associated risks to every strategic goal and to actually acquire the heuristics associated to its assessment. This approach intends to reach the point where we can modify the fuzzy rules for the KPI nodes in the corporate model to include risk values. The BSC Boards, Corporate Board as well as Model Board, will also be modified to include the column corresponding to risk values.

5.3 Utility Company Application

Figure 9 shows the strategy map for a utility company, where the inputs are also indicated. As a result of KICA the input and output linguistic variables for every process in the map were identified as well as the proper fuzzy relationship relating each input-output pair in order to elaborate the fuzzy rules for every node. Identifying the proper fuzzy rules includes detecting the adequate set of Qualifying Linguistic Terms (QLT) as well as the Trend relationships, i.e. whether a variable appearing in the consequent of a particular fuzzy rule is growing or decaying when a variable in the antecedent is growing.

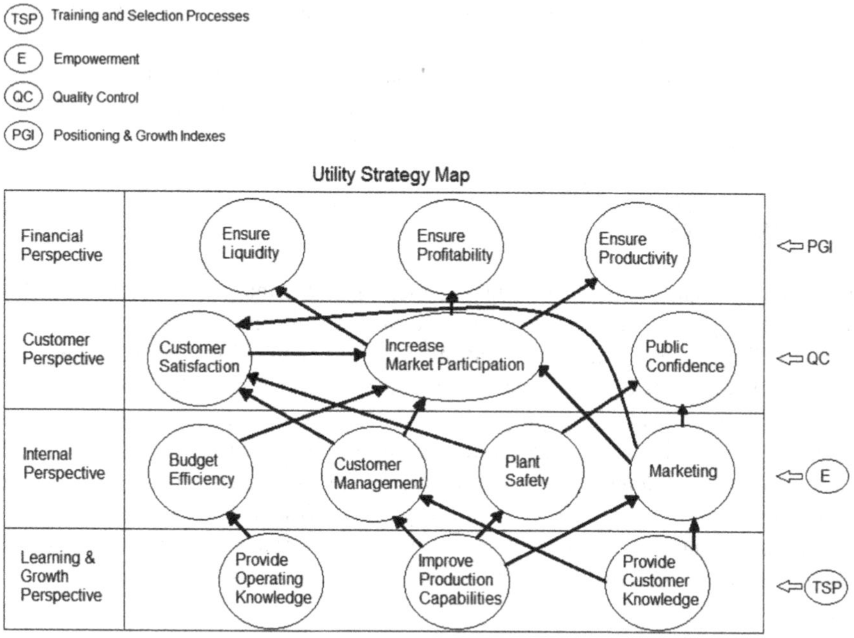

Fig. 9. Strategy map.

6 Conclusion

A deduction based on the proposed model is that network fuzzy systems attempt to bind individual fuzzy systems which are known in this research as nodes. Therefore, a fuzzy system is not seen a *silo* anymore, but a set of systems that form a fuzzy network. Thus, in a fuzzy network every node is able to send a receive values to simulate scenarios.

A limitation that arose during the corporate model application was the lack of performing stress and load tests. This would be a future work to measure the quality attributes in the hardware and software system, specially, when nodes require execute in parallel with complex operations.

The proposed model implies a scenario with at least two dependent fuzzy systems, hence, one fuzzy system must always execute first, then the second one is executed using the previous result from the prior system. In scenarios where there more than two systems, parallelism is applicable.

The validation of the proposed model concludes that the Corporate Model constructed after the KICA using a Fuzzy Network is an excellent example of a working asset obtained through capitalizing organization knowledge.

A Fuzzy Network is a versatile way to model corporate knowledge since this is distributed all over the people working in analysis and decision making activities within the organization's value-creating processes.

A What If? Environment provides the BSC corporate control strategy a powerful tool to explore different strategic scenarios.

Identifying the risks associated to every strategic goal as well as the associated heuristic to its assessment would allow to add the Risk column to the BSC Board, adding Risk Management to the MRAC STRATEGY.

7 Future Work

This work is the starting point for ongoing fuzzy systems that are required to work in parallel or complex operations, even with different objectives. The proposed fuzzy model is willing to use in academic and industry implementations, because the model definition encompasses a general process and activities that any other fuzzy network would use it. Specifically, the proposed model is due to incorporate it into a real business process, thus in that way the model would be proven.

The prototype implemented for this research is able to future improvement, adding more components or activities regarding fuzzy sets or systems. Also, integrating some other plugins related to inference engines, in case more membership functions are required. hardware performance, software processing.

Disclosure

- **Funding:** This study was funded by Universidad Católica de Colombia (grant number is not public).
- **Conflict of Interest:** The authors declare that they have no conflict of interest.

References

1. Laskri, M.T., Beggas, M., Médini, L., Laforest, F.: Towards an ideal service QoS in fuzzy logic-based adaptation planning middleware. J. Syst. Softw. **92**, 71–81 (2014)
2. Zadeh, L.A.: Fuzzy logic: issues, contentions and perspectives. In: 1994 IEEE International Conference on Acoustics, Speech, and Signal Processing, ICASSP 1994, vol. 6, p. VI/183 (1994)
3. Sarkar, A.: Application of fuzzy logic in transport planning. Int. J. Soft Comput. **3**(2), 1 (2012)
4. Hoyos, G.P.: Pipeline risk assessment using a fuzzy systems network. In: 2013 Joint IFSA World Congress and NAFIPS Annual Meeting (IFSA/NAFIPS), pp. 1495–1498 (2013)
5. Seising, R., Trillas, E., Kacprzyk, J. (eds.): Towards the Future of Fuzzy Logic, vol. 325. Springer, Cham (2015). https://doi.org/10.1007/978-3-319-18750-1
6. Duarte, O.G., Pérez, G.: Unfuzzy: fuzzy logic system analysis, design, simulation and implementation software. In: Proceedings of the EUSFLAT-ESTYLF Joint Conference, Palma de Mallorca, Spain, 22–25 September 1999, pp. 251–254 (1999)
7. Norton, D.P., Kaplan, R.S.: Transforming the Balanced Scorecard from Performance Measurement to Strategic Management, 15th edn. Harvard Business School Publishing Corporation, Boston (2001)

8. Wang, L.-X.: Adaptive Fuzzy Systems and Control. PTR Prentice Hall, Upper Saddle River (1994)
9. Norton, D.P., Kaplan, R.S.: Strategy Maps, Converting Intangible Assets into Tangible Outcomes. Harvard Business School Publishing Corporation (2004)
10. Gustavo, P.: A fuzzy logic based expert system for short term energy negotiations. In: 18th International Conference of the North American Fuzzy Information Processing Society - NAFIPS (Cat. No. 99TH8397), pp. 149–152 (1999)
11. Commission, International Electrotechnical technical committee industrial process measurement and control. Programmable Controllers, Part 7 - Fuzzy Control Programming. Commission, International Electrotechnical technical committee industrial process measurement and control (1997)
12. Cingolani, P., Alcalá-Fdez, J.: jFuzzyLogic: a Java library to design fuzzy logic controllers according to the standard for fuzzy control programming. Int. J. Comput. Intell. Syst. **6** (Suppl. 1), 61–75 (2013)
13. Zadeh, L.A.: Is there a need for fuzzy logic? Inf. Sci. **178**(13), 2751–2779 (2008)
14. Alfaro-Garcia, V.G., Gil-Lafuente, A.M., Klimova, A.: A fuzzy approach to competitive clusters using moore families. In: Rutkowski, L., et al. (eds.) ICAISC 2015. LNCS (LNAI), vol. 9119, pp. 137–148. Springer, Cham (2015). https://doi.org/10.1007/978-3-319-19324-3_13
15. Camastra, F., Ciaramella, A., Giovannelli, V., Lener, M., Rastelli, V., Staiano, A., Staiano, G., Starace, A.: A fuzzy decision system for genetically modified plant environmental risk assessment using Mamdani inference. Expert Syst. Appl. **42**(3), 1710–1716 (2015)
16. Meschino, G.J., Nabte, M., Gesualdo, S., Monjeau, A., Passoni, L.I.: Fuzzy tree studio: a tool for the design of the scorecard for the management of protected areas. In: Espin, R., Pérez, R.B., Cobo, A., Marx, J., Valdés, A.R. (eds.) Soft Computing for Business Intelligence. SCI, vol. 537, pp. 99–112. Springer, Heidelberg (2014). https://doi.org/10.1007/978-3-642-53737-0_6
17. Yaakob, A.M., Gegov, A., Rahman, S.F.A.: Decision making problem solving using fuzzy networks with rule base aggregation. In: 2017 IEEE International Conference on Fuzzy Systems (FUZZ-IEEE), pp. 1–6 (2017)
18. Cruz-Vega, I., Garcia-Limon, M., Escalante, H.J.: Adaptive-surrogate based on a neuro-fuzzy network and granular computing. In: Proceedings of the 2014 Annual Conference on Genetic and Evolutionary Computation, pp. 761–768 (2014)
19. Nápoles, G., Mosquera, C., Falcon, R., Grau, I., Bello, R., Vanhoof, K.: Fuzzy-rough cognitive networks. Neural Netw. **97**, 19–27 (2018)

Detection of Computer-Generated Papers Using One-Class SVM and Cluster Approaches

Renata Avros and Zeev Volkovich(✉)

Department of Software Engineering, ORT Braude College, 21982 Karmiel, Israel
{ravros, vlvolkov}@braude.ac.il

Abstract. The paper presents a novel methodology intended to distinguish between real and artificially generated manuscripts. The approach employs inherent differences between the human and artificially generated wring styles. Taking into account the nature of the generation process, we suggest that the human style is essentially more "diverse" and "rich" in comparison with an artificial one. In order to assess dissimilarities between fake and real papers, a distance between writing styles is evaluated via the dynamic dissimilarity methodology. From this standpoint, the generated papers are much similar in their own style and significantly differ from the human written documents. A set of fake documents is captured as the training data so that a real document is expected to appear as an outlier in relation to this collection. Thus, we analyze the proposed task in the context of the one-class classification using a one-class SVM approach compared with a clustering base procedure. The provided numerical experiments demonstrate very high ability of the proposed methodology to recognize artificially generated papers.

Keywords: Artificial papers detection · Text mining · SVM

1 Introduction

Latterly, various fake senseless scientific papers written by computer programs were over and over again accepted for publication in various scientific conferences and journals. At first glance, such documents being completely meaningless look like human composed manuscripts, since they effectively imitate standard scientific papers by their structure, grammar, and other attributes.

Producing of fake documents has, evidently, begun after the establishment of an open artificial papers generator "SCIGen" innovated in 2005 by three graduate students of the Massachusetts Institute of Technology - Jeremy Stribling, Maxwell Krohn, and Dan Aguayo. The "SCIGen" software is intended to generate synthetic manuscripts in the computer science discipline. Later, other scientific document generators such as "SCIgen-Physic" focusing on physics, "Mathgen" concentrating on math and the "Automatic SBIR" (Small Business Innovation Research) Proposal Generator dealing with grant proposal fabrication were suggested.

It is indeed impressive that computer programs are now good enough to create a "tolerable gibberish", but the computerized recognition of these works by some

P. Perner (Ed.): MLDM 2018, LNAI 10935, pp. 42–55, 2018.
https://doi.org/10.1007/978-3-319-96133-0_4

authoritative journals has taken on a wide scale and has become a serious problem. In order to detect fake scientific papers, many approaches have been proposed. The fact that counterfeit articles created by "SCIgen" or by other mentioned generators are always structured in a similar way suggests that a detection would be possible.

For example, in order to detect fake generated papers, the authors in [1] suggested measuring the keywords occurrences in the title, abstract and paper body, which is expected to be "uniform" in real papers. Moreover, it is natural to assume that the mentioned keywords are expected to appear frequently in real documents; yet rarely in fake articles.

An inter-textual similarity was used in [2] by counting the differences in word frequencies of two texts aiming to differ real and fake documents. This idea was extended in [3] by counting occurrences of word combinations in phrases. The list of references and keywords of the cited articles can serve as additional criteria to recognize artificial papers. As shown in [4], the various reference in a forged paper cannot be found on the Internet.

Another approach [5] is based on analyzing the texts' compression profiles suggests that there is a meaningful disagreement in the compression ratio between human-written and computer-generated texts.

As was found in [6], there is a significant dissimilarity in the topological properties of the natural and the generated texts. Various methods to identify synthetic scientific papers were discussed in [7, 8].

In this paper, we present a novel methodology intended to distinguish between real and artificially generated manuscripts. This approach relies on inherent differences between the human and artificially generated wring styles. Namely, taking into account the nature of the generation process, we can suggest that the human style is essentially more "diverse" and "rich" in comparison to an artificial one. In fact, a counterfeit paper is formed by randomizing existing prechosen components. In consequence, the separate parts of a fake document are almost unrelated one to another, although they are obviously associated within a real text. Thus, the generated papers are much similar in their own style and significantly differ from the human written documents, which can be varied among themselves in their own style, but more significantly differ from fake texts.

In order to assess dissimilarities between fake and real papers, a distance between writing styles is evaluated in this paper using the methodology discussed in [9–13]. A set of documents generated by SCIGen is captured as the training data so that a real document is expected to appear as an outlier in relation to this collection. Therefore, we analyze the proposed task in the context of the one-class classification.

One class classification can be considered as a special case of two-class classification problem in the situations where only data of one class is accessible and well described, while the data from other class is hard or impossible to obtain. The usage of this methodology is reasonable in our approach since we suppose that the artificially created papers are grouped into a target class sharing the common characteristics, while the real manuscripts "fall out" of it. A real document, seemingly not belonging to the target class, is expected to be labeled as "outlier" representing the second class.

In our methodology, all texts under consideration are divided into chunks and the distance values between them are calculated according to the replica introduced in

[9–13] intending to quantify the difference in the documents' writing style. Afterwards, a One-class Support Vector Machine aiming to allocate where the chunks of the tested manuscript are placed, is constructed. If the majority of them are located outside of the training class then it is concluded that the document is real. Otherwise, the text is recognized as a fake one.

An additional approach presented in this paper is based on as a partition of all attained chunks into clusters provided by means of the mentioned earlier distance using the Partition Around Medoids (PAM) algorithm. A manuscript is recognized as a real or a fake one according to the majority voting of its own chunks. As we will see, applying this distance makes it possible to detect almost surely the fake documents by means of the two proposed methods.

This paper is organized as follows. In Sect. 1, we present the introduction with related approaches aiming at the identification of fake scientific manuscripts. The background is presented in Sect. 2. The methods employed for the characterization and classification of texts are presented in Sect. 3. The numerical experiments and obtained results are presented in Sect. 4, and the conclusion is presented in Sect. 5.

2 Preliminary

2.1 Dynamic Dissimilarity Between Texts' Styles

To pattern the human writing process the following model was proposed in [9].

Let us take a group of texts Δ, where each document $D \in \Delta$ is considered as a sequence of its own chunks

$$D = \{D_1, \ldots, D_m\}$$

of the same size L. In order to evaluate the development of a text within the writing process, the Mean Dependency characterizing the mean relationship between a chunk D_i, $i = T+1, \ldots, m$ and the set of its T "precursors" is outlined:

$$ZV_T\left(D_i, \Delta_{i,T}\right) = \frac{1}{T} \sum_{D_j \in \Delta_{i,T}} s\left(D_i, D_j\right),\ i = T+1, \ldots, m, \tag{1}$$

where $\Delta_i = \left\{D_{i-j}, j = 1, \ldots, T\right\}$ is the set of T "precursors" of D_i, and s is a similarity measure. In this model, s is constructed using the common Vector Space Model. Every chunk is expressed as a terms' frequency vector of terms taken from a given dictionary $\mathcal{D}$, and s is calculated as the Spearman's ρ (see, e.g. [14]):

$$\rho = 1 - \frac{6 \sum_{i=1}^{n} \boldsymbol{d_i}}{\boldsymbol{n}(\boldsymbol{n}^2 - 1)},$$

where d_i, $i = 1,..n$ are the differences between the corresponding ranks in the scales, and n is the dictionary size. ρ handles the frequency tables as a kind of ordinal data related to the frequencies rank placing. Rank ties take a rank equal to the mean of their position in the ascending order of the values. The Spearman's ρ is actually the famous Pearson's correlation amid the ranks.

Resting upon the ZV_T measure, a distance (dynamic dissimilarity) between documents' styles is produced as:

$$DZV_T(D_i, D_j) = abs\left(ZV_T\left(D_i, \Delta_{i,T}\right) + ZV\left(D_j, \Delta_{j,T}\right) - ZV\left(D_i, \Delta_{j,T}\right) - ZV\left(D_j, \Delta_{i,T}\right)\right).$$

Subsequently, the association of a text with the "precursors" of another text is derived from the association with its own neighbors. It is easy to see that it is merely a semi-metric. Even though, if the documents are almost consistently associated with their own "precursors" and the "precursors" of another document, then they are practically indistinguishable from the writing style standpoint.

An example presented in Fig. 1 is a 3D scatter plot of three main components of DZV_T computed for a collection of two books "2010: Odyssey Two" by A.C. Clarke and "Harry Potter and the Philosopher's Stone" by J.K. Rowling.

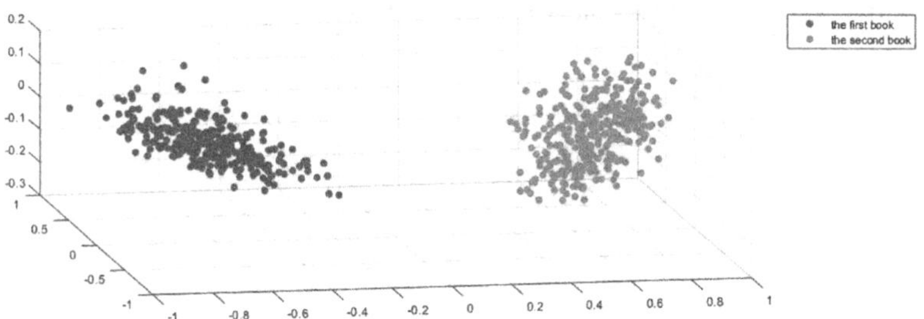

Fig. 1. Scatterplot plot of three main components of DZV_T of a two books' collection.

Two well separated bunches composed of the corresponding chunks unambiguously reflect a significant difference in the styles. This illustration demonstrates the ability of the proposed distance to distinguish the styles of distinct authors.

2.2 One-Class SVM

The One Class SVM (One Class Support Vector Machine) methodology was actually proposed in [15] as adjustment of the classical SVM (Support Vector Machine) methodology to the one-class classification problem. Here, after converting the data using a kernel transform just the origin is treated as the member of the second class, and the usual two-class SVM methodology is applied. Formally, the problem is described as follows.

Let us consider a training sample $\{x_1, x_2, \ldots, x_k\}$ from a group X, where X is a compact subset of a Euclidean space R^n, together with a kernel map $\Phi : R^n \rightarrow H$. Feature space H is implicit (and usually unknown) in all kernel methods. Aiming to isolate the data from the origin, the subsequent quadratic programming problem is solved:

$$\min_{w\in H,\xi_i\in R^n,\gamma\in R} \frac{1}{2}\|w\|^2 + \frac{1}{k\eta}\sum_{i=1}^{k}\xi_i - \gamma$$

subject to

$$(w \cdot \Phi(x_i)) \geq \gamma - \xi_i,\ i = 1, 2, \ldots, k,\ \xi i \geq 0.$$

Here

- ξ_i - slack variables (one for each point in the sample);
- γ - the distance to the origin in feature space;
- w - the parametrization of the hyperplane separating the origin from the data in H;
- η - the expected fraction of data points outside the estimated one class support.

If w and ρ are a solution, then the decision rule

$$f(x) = sign((w \cdot \Phi(x)) - \gamma)$$

takes positive values for most points located in the training set. The dual form of the problem is

$$\min_{\alpha_i\in R} \frac{1}{2}\sum_{i,j=1}^{k} \alpha_i\alpha_j K(x_i, x_j)$$

subject to

- $0 \leq \alpha_i \leq \frac{1}{k\eta}$;
- $\sum_{i=1}^{k} \alpha_i = 1.$

where $K(x_i, x_j)$ is a kernel function defined as $\Phi(x_i)\bullet \Phi(x_j)$. One obtains the decision function

$$f(x) = \sum_{i=1}^{k} \alpha_i K(x_i, x) - \gamma.$$

Analogously to the discussed earlier primal form, a negative value of this function reveals outliers. The data points satisfying

$$0 < \alpha_i < \frac{1}{k\eta}$$

are the support vectors located exactly on the separating hyperplane.

3 Approach

We base our consideration on a common view on the human writing process (see, e.g. [16]) proposing that this development consists of four main elements: planning, composing, editing and composing the final draft. Such an assumption leads to the evident conjecture about the natural connection of the sequentially written parts of a human created manuscript. On the other hand, SCIgen based on context-free grammar methodology randomly produces nonsense texts in the form of computer science research papers including graphs, diagrams, and citations. Thus, these artificial texts do not pass the mentioned steps of the human writing process. It is very natural to expect that this contrived style is essentially different from one inherent to people. However, all generated papers are supposed to be very close in their writing style. The proposed model is created resting upon these matters in the framework of the one-class classification methodology.

3.1 One-Class SVM Classifier

Applying a One-Class SVM approach, the target class is constructed using the artificially generated papers, whereas outliers are associated with the humanly written papers. In this fashion, a collection D_0 of artificially papers is generated, and a dictionary $\mathcal{D}$, the delay parameter T, and size of the chunks L are selected planning to apply the Dynamic Similarity model presented in Sect. 2.1.

Afterwards, a tested document is divided into chunks, and each chunk is transformed using the Vector Space Model in an occurrences vector. These vectors together with the target class vectors form a collection under consideration. A quadratic matrix of the DZV_T pairwise distances between all chunks in this expanded group having at least T "precursors" is calculated:

$$Dis = \left\{ DZV_T(D_{i_1}^{(m_1)}, D_{i_2}^{(m_2)}) \right\},$$

where m_1, m_2= $1,\ldots, M + 1$. i_1 and i_2 are the sequential indexes of the documents' chunks taking values from 1 to $n_1 - T$ and $n_2 - T$, correspondently. n_1 and n_2 are numbers of chunks in the documents. We create the target class from the rows of this matrix using apparently an embedding of the chunks set into a Euclidian space with the dimensionality:

$$Dim = \sum_{i=1}^{M+1} n_i - (M+1)T.$$

This embedding procedure increases the resolution of the method so that each piece is associated with a vector possessing coordinates corresponding to its distances to all other chunks.

A one-class SVM classifier is constructed on the matrix Dis. An example of such a matrix is given in Fig. 2.

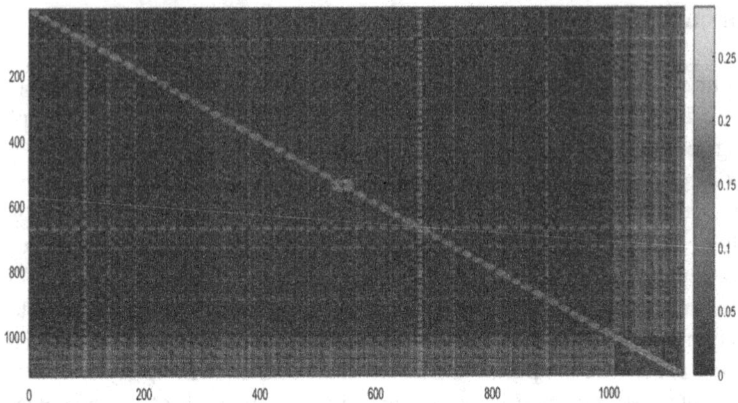

Fig. 2. An example *Dis* matrix.

The brighter areas indicate the distances between the tested texts and the elements of the target class. They are obviously larger than the corresponding ones inside distances. Therefore, it is natural to expect that the document in question is not fabricated.

A conclusion according to the total allocation of the tested document is made in our methodology using the following rules:

- **Majority voting of the chunks (SVM Rule 1).** The document is assigned to the target class (a fake paper case) or recognized as an outlier (a real paper case) if more than half of the chunks belong to the consistent class.
- ***P*-value rule (SVM Rule 2).** The average score of the decision function is calculated for each paper from the target class and for the tested paper. After, the sample *p*-value of the mean corresponding to the text in question is compared with the fraction of its chunks in the total their number. The article is accepted as a real one if the found *p*-value is greater of the fraction.

3.2 Clustering Based Classifier

Clustering is as a dependable unsupervised process for discovering prototypes in unlabeled data. The key benefit of a general clustering procedure is the proficiency to uncover intrusions in the checked data without any prior information. In our approach, a type of clustering is employed to reveal differences between an artificial and the human writing styles.

Our perspective suggests that the chunks of a tested real paper could produce a group located sufficiently far from the target class, i.e. to be a separate cluster. In this manner, a meaningful partition of the analyzed collection into two clusters indicates a difference in the styles and approve the authenticity of the verified paper.

The procedure is implemented in the following way. The matrix ***Dis*** of the pairwise *DZV* distances is constructed by the described in the previous subsection technique for chunks of the target class papers and chunks of a paper being examined.

In the next step, we partition the rows of this matrix into two clusters aiming to turn out a cluster corresponding to the artificial articles and a cluster corresponding to the tested manuscript. Each document is apportioned to a cluster (i.e. style) with the highest winning rate of the matching chunks, and the conclusion is made consistent with this cluster assignment. Therefore, if the majority of the target class texts are located in a group different from one containing the majority of the tested text's chunks then one concludes that this article is not false. Otherwise, the document is recognized as an artificial one.

4 Experiments

In order to evaluate the ability of our proposed methodology, we provide several numerical experiments performed in the MATLAB environment through the chunk sizes $L = 200$ and 400 with the delay parameter T value equal to 10.

4.1 Experiments Setup

4.1.1 Material

One hundred artificial papers are constructed by means of the SCIGen procedure and fifty genuine manuscripts are drawn from the "arXiv" repository [17]. Fifty artificial papers form the target class, and the residual ones are tested together with the drawn real manuscripts. The results are represented via the True Positive Rate (TPR, otherwise known as Sensitivity or Recall), which evaluates the part of papers (false or real) that are rightly recognized as such.

4.1.2 Dictionary Construction

Two different types of the dictionary of terms $\mathcal{D}$ are used.

- **The N-grams model.** Here, within a preprocessing procedure, all uppercase letters are transformed to the corresponding lowercase letters, and all other characters are omitted. The vocabulary contains all N-grams appearing in the trading class documents. Recall that an N-gram is an adjoining N-character piece formed by the characters occurring in a sliding window of length N. We use in our experiments no more 10000 of the most common N-grams with lengths of three, four and five.
- **The content-free word model.** Content-free words do not express any semantic meaning on their own. This kind of terms can be associated with a stylistic "glue" of the language appearing to set up the connection between all terms that. Joint occurrences of the content-free words offer a stylistic indication for authorship verification [18, 19]. This approach was essentially applied in a study of quantitative patterns of stylistic influence [20] and modeling of the writing style evolution [13].

4.1.3 SVM Parameters

Discovering an appropriate data representation from their primary attributes is an essential part of any classification problem. As a rule, not each characteristic is constructive for classification and can even harm the result. With the intention, just the necessary information has to be retained for further study. The following figure represents an example of a scatterplot of the two-dimensional principal subspace found in the matrix ***DIS*** (Fig. 3).

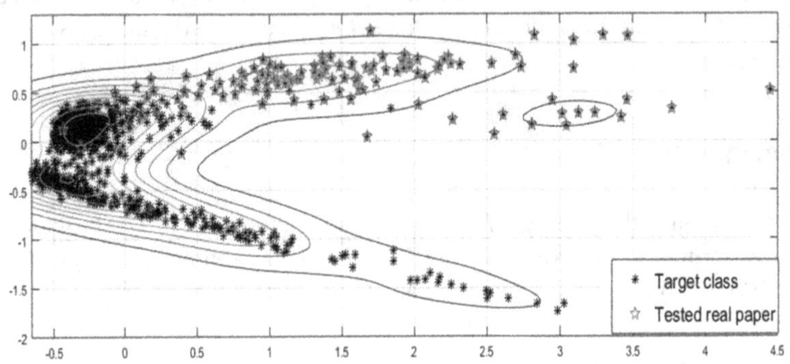

Fig. 3. Scatter plot of the two leading components of ***DIS***.

The two leading components explain 64.5% of the total variation in the data. The plot shows that a linear separation between the target class and its outliers is hardly expected. Thus, in our experiments the Gaussian Radial Basis Function kernel (the RBF kernel)

$$K(x_i, x_j) = exp\left(-\frac{(x_i - x_j)^2}{2\sigma^2}\right)$$

is employed. The standard MATLAB 2017b SVM toolbox is applied. All columns of the data are standardized. The width parameter σ is estimated using the inbuilt heuristic procedure.

The threshold η corresponds to the expected fraction outliers among the artificially generated papers. Picking a suitable level for η is very important for constructing an effective one-class SVM model. Small values can enlarge the false positive rate (the ratio between the number of artificial papers mistakenly documented as real ones and the total number of actual fabricated texts). On the other hand, a large η can obviously increase the false negative rate, calculated as the fraction of real papers recognized as fabricated.

To estimate its value we chose 20 synthetic paper and 20 real papers and run the parameter η starting at 0.01 and gradually increasing the value with an increment of 0.01 until 0.3. Each selection of η provides a pair of the false positive rate and the false negative rate. An appropriate value of η is chosen a balance point between these characteristics likewise to the ROC methodology. In our case, it is 0.15.

4.2 Results

4.2.1 N-grams Model

Dealing with an *N*-gram based Vector Space Model we can naturally to expect that for sufficiently small lengths the chunks the obtained vector representation tends to be more and more independent of the text, de facto random. To illustrate it, let us consider histograms of the ZV_T values constructed for $N = 3$, 4 and 5 and $L = 200$ within the training class. The mean values are 0.15, 0.06 and 0.04 correspondently (Fig. 4).

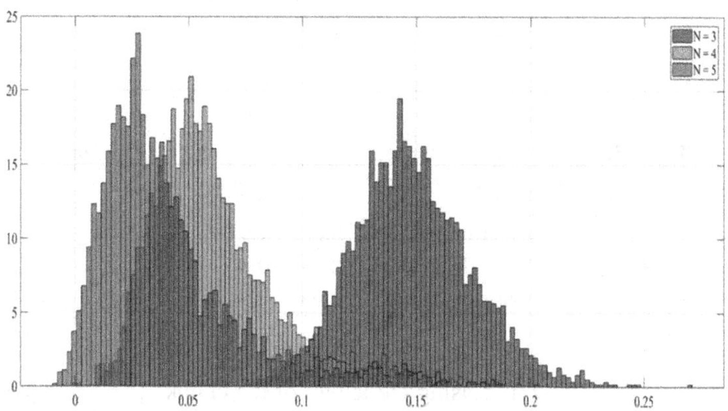

Fig. 4. Histograms of the ZV_T values within the training class calculated for $N = 3$, 4 and 5 and $L = 200$.

The histograms lean to concentrate around the origin. Therefore, the association between the sequential parts text vanishes. A similar situation appears once real papers are considered (Fig. 5).

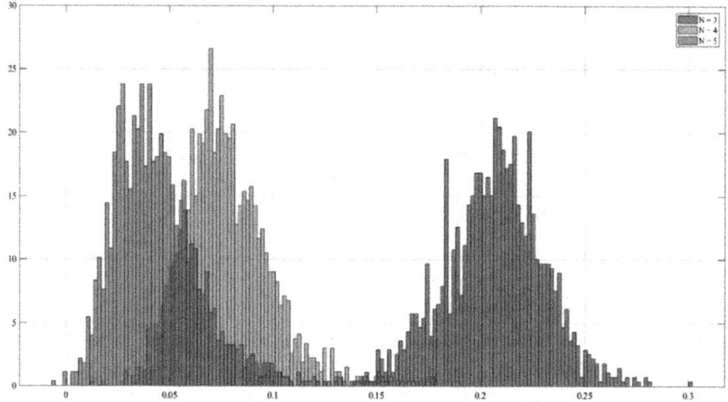

Fig. 5. Histograms of the ZV_T values within 50 real papers calculated for $N = 3$, 4 and 5 and $L = 200$.

However, the histograms of real papers seek much slower. These observations make possible to suggest that the methods are expected sufficiently well to recognize the fake papers even for small values of L, but can fail with the real articles. The following tables approve this suggestion (Tables 1 and 2).

Table 1. TRP calculated for fake papers for the chunk size $L = 200$.

	$N = 3$	$N = 4$	$N = 5$
Clustering	1	1	1
SVM rule 1	1	1	1
SVM rule 2	1	1	1

Table 2. TRP calculated for real papers the chunk size $L = 200$.

	$N = 3$	$N = 4$	$N = 5$
Clustering	0.86	0.82	0.64
SVM rule 1	0.96	0.88	0.80
SVM rule 2	0.92	0.90	0.90

This can be corrected by increasing the length of the chunk. The subsequent tables provide outcomes obtained for the chunk size $L = 400$ (Tables 3 and 4).

Table 3. TRP calculated for fake papers the chunk size $L = 400$.

	$N = 3$	$N = 4$	$N = 5$
Clustering	1	1	1
SVM rule 1	1	1	1
SVM rule 2	1	1	1

Table 4. TRP calculated for real papers the chunk size $L = 400$.

	$N = 3$	$N = 4$	$N = 5$
Clustering	1	1	1
SVM rule 1	1	0.96	0.98
SVM rule 2	0.92	0.92	0.92

4.2.2 Content-Free Word Model

As was mentioned earlier, the content-free word model is being a kind of statistical glue. While it does not have any sense, it helps to join the words into meaningful texts (Fig. 6).

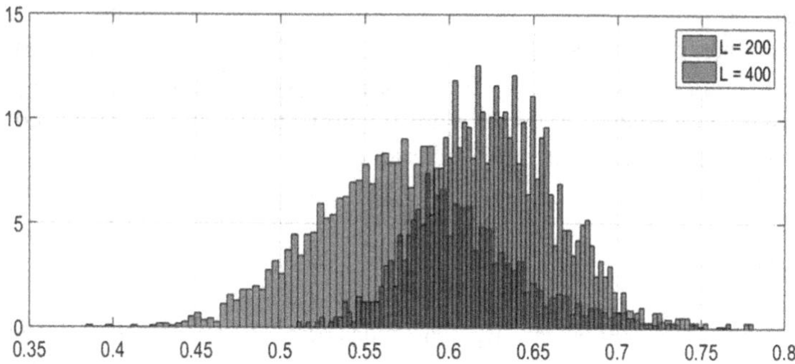

Fig. 6. Histograms of the ZV_T values within the training class calculated for *L* = *200* and *L* = *400* using the content-free word.

This figure represents histograms of the Mean Dependency found for two different values of the chunks size *L*. The corresponding means are 0.57 and 0.63. The fake papers are surely recognized again (Tables 5 and 6):

Table 5. TRP calculated for fake papers for the chunk size *L* = *200* and *L* = *400.*

	L = *200*	*L* = *400*
Clustering	1	1
SVM rule 1	1	1
SVM rule 2	1	1

However, detection of real papers is successful just for sufficiently large values of *L*.

Table 6. TRP calculated for real papers the chunk size *L* = *200, L* = *400,* and *L* = *500.*

	L = *200*	*L* = *400*	*L* = *500*
Clustering	0.62	0.96	0.96
SVM rule 1	0.58	1	0.98
SVM rule 2	0.78	0.78	0.78

5 Conclusion

This paper proposes a novel method constructed to make a distinction between computers generated scientific and human written papers. The problem, treated in the framework of the one class classification methodology, is considered using the one-class SVM methodology compared with a clustering approach. N-grams based and the content-free word based dictionaries are applied for the building of vector representations of texts. The key issue providing the highly reliable results is the dynamic distance measuring dissimilarity between particular implementations of the writing process. The fake papers are surely identified for all configurations of the system parameters. The approach also classifies sufficiently well the human written papers for suitably chosen parameters values. We are planning to extend the proposed research aiming to test other outliers detection techniques.

References

1. Lavoie, A., Krishnamoorthy, M.: Algorithmic detection of computer generated text. arXiv: 1008.0706, August 2010
2. Labbe, C., Labbe, D.: Duplicate and fake publications in the scientific literature: how many SCIgen papers in computer science? Scientometrics **94**(1), 379–396 (2013)
3. Fahrenberg, U., et al.: Measuring global similarity between texts. In: Besacier, L., Dediu, A.-H., Martín-Vide, C. (eds.) SLSP 2014. LNCS (LNAI), vol. 8791, pp. 220–232. Springer, Cham (2014). https://doi.org/10.1007/978-3-319-11397-5_17
4. Xiong, J., Huang, T.: An effective method to identify machine automatically generated paper. In: Pacific-Asia Conference on Knowledge Engineering and Software Engineering, KESE 2009, pp. 101–102. IEEE (2009)
5. Dalkilic, M.M., Clark, W.T., Costello, J.C., Radivojac, P.: Using compression to identify classes of inauthentic texts. In: Proceedings of the 2006 SIAM Conference on Data Mining (2006)
6. Amancio, D.R.: Comparing the topological properties of real and artificially generated scientific manuscripts. Scientometrics **105**(3), 1763–1779 (2015)
7. Williams, K., Giles, C.L.: On the use of similarity search to detect fake scientific papers. In: Amato, G., Connor, R., Falchi, F., Gennaro, C. (eds.) SISAP 2015. LNCS, vol. 9371, pp. 332–338. Springer, Cham (2015). https://doi.org/10.1007/978-3-319-25087-8_32
8. Nguyen, M.T., Labbe, C.: Engineering a tool to detect automatically generated papers. In: Mayr, P., Frommholz, I., Cabanac, G. (eds.) BIR@ECIR, ser. CEUR Workshop Proceedings, vol. 1567, pp. 54–62. CEURWS.org (2016)
9. Volkovich, Z., Granichin, O., Redkin, O., Bernikova, O.: Modeling and visualization of media in Arabic. J. Informetr. **10**(2), 439–453 (2016)
10. Volkovich, Z.: A time series model of the writing process. In: Perner, P. (ed.) Machine Learning and Data Mining in Pattern Recognition. LNCS (LNAI), vol. 9729, pp. 128–142. Springer, Cham (2016). https://doi.org/10.1007/978-3-319-41920-6_10
11. Volkovich, Z., Avros, R.: Text classification using a novel time series based methodology. In: 20th International Conference on Knowledge Based and Intelligent Information and Engineering Systems, KES 2016, York, United Kingdom, 5–7 September 2016 (2016). Procedia Comput. Sci. **96**, 53–62 (2016)
12. Korenblat, K., Volkovich, Z.: Approach for identification of artificially generated texts. In: HUSO 2017: In the Third International Conference on Human and Social Analytics (2017)
13. Amelin, K., Granichin, O., Kizhaeva, N., Volkovich, Z.: Patterning of writing style evolution by means of dynamic similarity. Pattern Recogn. **77**, 45–64 (2018)
14. Kendall, M.G., Gibbons, J.D.: Rank Correlation Methods. Edward Arnold, London (1990)
15. Schölkopf, B., Williamson, R., Smola, A., Shawe-Taylor, J., Platt, J.: Support vector method for novelty detection. In: Solla, S.A., Leen, T.K., Müller, K. (eds.) Proceedings of the 12th International Conference on Neural Information Processing Systems (NIPS 1999), pp. 582–588. MIT Press, Cambridge (1999)
16. Harmer, J.: How to Teach Writing. Pearson Education, Delhi (2006)

17. www.arXiv.org/archive/cs. Accessed 2 July 2017
18. Juola, P.: Authorship attribution. Foundations and Trends in Information Retrieval, vol. 1, no. 3, pp. 33–334 (2006)
19. Binongo, J.: Who wrote the 15th book of Oz? An application of multivariate analysis to authorship attribution. Chance **6**(2), 9–17 (2003)
20. Hughes, J.M., Foti, N.J., Krakauer, D.C., Rockmore, D.N.: Quantitative patterns of stylistic influence in the evolution of literature. Proc. Natl. Acad. Sci. **109**, 7682–7686 (2012)

Understanding Customers and Their Grouping via WiFi Sensing for Business Revenue Forecasting

Vahid Golderzahi(✉) and Hsing-Kuo Pao

Department of Computer Science and Information Engineering,
National Taiwan University of Science and Technology, Taipei 10607, Taiwan
golderzahi@gmail.com, pao@mail.ntust.edu.tw

Abstract. Emerging technologies provide a variety of sensors in smartphones for state monitoring. Among all the sensors, the ubiquitous WiFi sensing is one of the most important components for the use of Internet access and other applications. In this work, we propose a WiFi-based sensing for store revenue forecasting by analyzing the customers' behavior, especially the grouped customers' behavior. Understanding customers' behavior through WiFi-based sensing should be beneficial for selling increment and revenue improvement. In particular, we are interested in analyzing the customers' behavior for customers who may visit stores together with their partners or they visit stores with similarly patterns, called group behavior or group information for store revenue forecasting. The proposed method is realized through a WiFi signal collecting AP which is deployed in a coffee shop continuously for a period of time. Following a procedure of data collection, preprocessing, and feature engineering, we adopt Support Vector Regression to predict the coffee shop's revenue, as well as other useful information such as the number of WiFi-using devices, the number of sold products. Overall, we achieve as good as 7.63%, 11.32% and 14.43% in the prediction on the number of WiFi-using devices, the number of sold products and the total revenue respectively if measured in Mean Absolute Percentage Error (MAPE) from the proposed method in its peak performance. Moreover, we have observed an improvement in MAPE when either the group information or weather information is included.

Keywords: Customer behavior · Group behavior
Received Signal Strength Indicator (RSSI) · Revenue forecasting
WiFi sensing

1 Introduction

Nowadays most stores provide WiFi services for customers who are equipped with WiFi functioning smartphones and interested in accessing Internet. It is well known that WiFi signals, along with other video or non-video-based technology

P. Perner (Ed.): MLDM 2018, LNAI 10935, pp. 56–71, 2018.
https://doi.org/10.1007/978-3-319-96133-0_5

may be helpful in understanding people's behavior for people located in a smart space, or in particular the customers' behavior in stores [3,12]. In this work, we propose a method based on WiFi sensing given customers' behavioral inputs for store revenue forecasting. In particular, we are interested in the customers' group information where we can observe friends who find each other to go for a drink together or different individuals may share similar visiting behavior even they do not know each other.

Compared to online shopping where all the surfing and purchase behaviors from customers are automatically logged, the marketing in brick-and-mortar business usually face the challenges as they need to deploy the customer analytics framework to the physical realm. In one way or the other, the traditional stores must find solutions to keep track of customers such as when customers may visit the stores, what they prefer to own and what they really purchase in the end based on their judgment between the product quality and price. To understand targeted customers as much as they can to boost the stores' revenue, two major technologies offer the answers: the video and non-video-based approaches. To avoid the privacy leaking issues, the non-video-based approaches are generally favored from the customers' side because they keep the customers' information to its minimum for business analytics. Among the various non-video-based approaches, WiFi-sensing is a major choice due to its popularity. Existing WiFi sensing, which is a cost-effective as well as privacy-preserving technology, can be appropriate for customer behavior analysis [1,3].

Signal-based indoor sensing for human tracking and business analytics are generally categorized into several categories [14]. A rich set of IoT (Internet of Things) technology with sensors such as passive infrared sensor (PIR), ultrasound, temperature sensors, as well as various vision-based devices can be deployed in the indoor environment for human counting, tracking and activity recognition to name a few. In general, we need to spend efforts on the device deployment physically and the device calibration and threshold setting may not be straightforward for this kind of technology. On the other hand, there are also some devices that we need the humans located in the indoor environment to carry to make the sensing possible. Some wearable devices and smartphones fall into this category. Apparently, we prefer a scenario that is: (1) easy to deploy in the indoor environment, (2) providing high sensing accuracy, and (3) with enough covering rate among people. In another word, we look for a sensing technology where we can: (1) easily implement both in its hardware and software, (2) find convincing tools for analytics and (3) detect as high percentage of people as possible in a given environment where each of the targeted people carries a device that is necessary for sensing. We propose a WiFi-based sensing method [13] where we only assume smartphone carrying from the customers for the indoor customer detection and tracking for business revenue forecasting. By having the technology, we keep the deployment efforts to the minimum and at the same time, we enjoy a decent sensing performance.

The proposed method is realized in a coffee shop where we track and analyze customers' behavior and the associated group information with a WiFi AP. For customers who visit the coffee shop with functioning WiFi, we can collect the

WiFi related information and use it to summarize customers' behavior. One of the reasons why we choose the coffee shop for our study is because drinking coffee and visiting the coffee shop is considered not a mandatory but an optional activity for people where we may choose to have with our friends and when we have certain mood for relaxation or doing business in the environment. This coffee shop is located in Da'an district in Taipei City and close to a university area. Most of its customers are students who may spend their time to have fun with their friends, or work on their homework/projects individually or with a group. From time to time, the coffee shop owner may provide some special discounts to students to encourage them coming to the shop, which could lead to revenue increment.

We use the WiFi related information to track the coffee shop's customers and analyze their behavior using RSSI signals captured via the WiFi AP. We monitor the coming and leaving time for each customer as well as their duration of stay given the RSSI signals. Occasionally, the AP may grab some data from people who pass by the coffee shop or stay in a store nearby. We address these noise data by applying some filters on RSSI signals and the duration of customers' stay. Furthermore, we will extract frequent customers and analyze their behavior to detect the groups of frequent customers. Customers may form a group if they come to the shop together. On the other hand, we also consider a group if customers from the group often come to the coffee shop at some similar time or stay for similar duration. For instance, some people may come the shop before going to work or stay in the shop for almost the whole day long. We believe that they could have similar working patterns or share similar income levels and should behave similarly in their visiting and purchase behavior. In the end, we discuss both of the cases where we may not include or may include the group information as described above in the feature set for prediction. We take turn to predict the total number of customers' devices, the total number of sold products and the total revenue. The prediction model is Support Vector Regression (SVR).

We should emphasize the main contributions and what differentiate the proposed method from the previous solutions for indoor human sensing and business revenue estimation as follows:

- The proposed method is based on an easy-to-deployed scenario where we only assume smartphone carrying from the customers. Moreover, the proposed approach is a *passive* approach where we need customers to open no special software to activate the sensing. On the indoor environment, we need only a tuned AP for WiFi signal collection. By having this property, we can easily convince business stores for its realization.
- We focus on using the customers' group information for revenue forecasting. The group information separates customers from different groups, such as loyal customers, customers with different vocations, customers with different product preferences and customers with different daily or weekly schedule. Knowing the above information may improve the business revenue as the business should have more understanding about its customers.

- The proposed method respects the privacy issue. Unlike many indoor tracking strategies, we collect the information only the part for *signal broadcasting* from customers. Usually, we can assume customers have no objection on releasing the information. It could be hard to hide the broadcasting information in general when a handshaking communication is needed.

The remaining of the paper is organized as follows. An overview of WiFi-based and non-WiFi-based sensing approaches is provided in Sect. 2. Afterwards, we discuss the proposed method along with all the necessary procedures in Sect. 3, which is followed by the experiment results and evaluation in Sect. 4. Finally, we conclude our work in Sect. 5.

2 The Past Work

The goal is to adopt indoor sensing on customers for business revenue forecasting. There are a variety of technology that has been developed for this purpose. As we briefly described, the major strategies can be separated into several groups based on whether we need to deploy certain devices or system on the indoor side and whether we assume any devices from the customers to carry to make the sensing possible. In this section, other than the research that we have discussed in Sect. 1, we mainly discuss the approaches that are directly related to this work. We emphasize that what we plan to detect and track is more than a handful customers where we may not assume any limit for the number of customers. Moreover, identifying the tracked customers is valuable to have in this application. Therefore, the IoT solutions such as PIR, ultrasound and temperature sensing are not precise enough to solve the problem. On the other hand, the vision-based methods may not be the best choice due to the privacy concerns from the general public. We turn our attention to the approaches where we assume customers carrying devices and the devices provide enough information for detection, tracking and analytics.

The user-carrying device approach can be divided into smartphone and non-smartphone categories. The former represents a scenario in which users carry their own smartphones, thereby they are trackable and their identifiable information would be extractable through the smartphones [1]. The latter relies on additional wearable devices that should be carried by users such as bracelets, smart glasses, RFID, etc. For instance, Han et al. [3] implemented a Customer Behavior IDentification (CBID) system based on passive RFID tags. Their system includes three main parts; discovering popular items, revealing explicit correlations, and disclosing implicit correlations to understand customers' purchase behavior. The technology is mainly focused on a small set of people and may have difficulty when we have a large number of unknown people to track and therefore hard to implement in the crowded situation [4,7].

On the category of smartphones, we have all-in-one devices which have the identifiable information as well as a various set of equipments, sensors and apps for information collection and environmental monitoring. People may prefer to carry smartphones simply because the smartphones play such a role of combining many functionalities in a single device [6,8]. That implies using smartphones as the assumed carrying device for customer sensing should provide enough covering rate when we use smartphone-related signals to estimate the existence of customers. Among all possibilities, WiFi-equipped smartphones can be considered one of the best solutions to be carried by unknown people or a large number of customers who intend to communicate with public devices due to the built-in identifiable characteristics in the smartphones. By having that, we aim to detect, track and analyze people with their existence and group behavior [6,11].

Zeng et al. [12] proposed WiWho, which is a method to identify a person using walking gait analysis through the WiFi signals. WiWho consists of two endpoints, a WiFi AP and any WiFi-equipped device for communicating and collecting Channel State Information (CSI). It has some limitations such as assuming the straight walking paths from customers and should have the performance limit while the tracked person turns. Vanderhulst et al. [6,11] discussed a framework to detect human spontaneous encounters in which spontaneous and short-lived social interactions between a small set of individuals have been detected. It leverages existing WiFi infrastructure and the WiFi signals, so-called "probe" can periodically be radiated by a device to search for available networks. The probes are used to capture radio signals transmitted from users' devices to detect human copresence. The limitations of the proposed method include device variety, a limited number of participants to be allowed for high accuracy detection, and the required application to be installed on users' smartphones.

An extended Gradient RSSI predictor and filter was proposed by Subhan et al. [10]. It is a predictive approach to estimate RSSI values in presence of frequent disconnections. The approach predicts users' positions and movements in terms of their current situations and movements. The distance changes between users' devices and the AP lead to the increase and decrease of the RSSI values and therefore the targeted users as well as their movements can be detected.

As other similar research, Maduskar and Tapaswi [6] proposed an approach to trace people's positions and movements using an RSSI measurement of WiFi signals from several APs in predetermined locations. The RSSI-based approaches have the minimum complexity compared to other signal-based indoor localization techniques. In their approach, the larger size of the APs results in more accurate location estimation. The weakness of the approach is that a careful initialization is necessary given a new environment, e.g., customer sensing given an indoor store. Du et al. [2] proposed algorithms for fine-grained mobility classification and structure recognition of social groups using smartphones through their embedded sensors. They have utilized embedded accelerometer to detect group mobility behavior. Afterward, a supervised learning algorithm is applied to recognize different levels of group mobility, such as stationary, walking, strolling, and running. The method can also be used to recognize the relations and structures of a group by monitoring the leader-follower, the left-right relations and distances using smartphones' sensor data. To compare the above two research work, the localization technique is basically not a must to have in our scenario because the main purpose of the proposed method is on understanding when customers visit a store instead of what customers prefer to own. Therefore, it is the visiting behavior not the purchase behavior that interest us. In the next section, we discuss the proposed method in details.

3 Proposed Method

The goal is to predict the number of customers and revenue on each day given the past selling and customers' behavior history. What is different from previous approaches is that we rely on the WiFi signal collection to help us know further about the visiting customers where the WiFi information may tell us the information from the macro scope such as the total number of customers to the micro scope such as the customers' identifiable information. We demonstrate the whole prediction scenario starting from data collection, data preprocessing to prediction model itself in the next few subsections.

3.1 Data Collection

The data for the proposed method includes two parts: the WiFi-based data and others. The WiFi-based data has been collected using a WiFi collecting device, TP-Link TL-WR703N WiFi router in the coffee shop. It is an access point (AP) which operates in IEEE 802.11n mode to collect the data from customers' devices such as smartphone, laptop, tablet, etc. The extracted data from the received WiFi signals per customer includes:

- the physical address (MAC address),
- service set identifier (SSID), and
- received signal strength indicator (RSSI).

Given the MAC address information, we can calculate the number of devices for each day. Usually the number of devices may be close to the number of customers per day if each customer carries only one smart device (further discussed below in the assumption part). The WiFi data has been extracted using the Wireshark packet analyzer[1]. Based on the collected WiFi information, we also derive some information which may be important for the prediction:

- the come-in time of a customer,
- the leaving time of a customer, and
- the way a customer was served, such as "staying in" or "prepared to go".

The come-in time and leaving time are recored based on the first and last signals that we can collect for each specific MAC address (identity). How a customer was served is estimated based on the duration of the WiFi signals that we received per MAC address, such as above or below a predefined threshold (further discussed below). Furthermore, there are customer considered as frequent

[1] https://www.wireshark.org.

customers. We add a set of group information features, which are extracted via *frequent customers' behavior analysis.* Note that the above three are individual based, collected for each MAC address. On the other hand, the SSID and RSSI are collected following a predefined sampling rate. In addition to the WiFi related information, we also collect some other information which may have influence on the coffee shop revenue. The information includes:

- temperature, and
- rain probability.

On the side, the analyzed dataset also consists of the number of various sold products and the total revenue for each day. The owner of the analyzed coffee shop is kind to provide the valuable information for us to confirm the performance of the proposed method. Some correlation between the number of sold products, the revenue and the number of devices (MAC address) is further discussed below. The dataset was collected from 2016/09/02 to 2016/12/04 in which the training data is set from 2016/09/02 to 2016/11/20 including 78 days, and the test data is from 2016/11/21 to 2016/12/04 including 14 days. Due to some technical difficulty, we have a few days of missing data. The longest of data missing is a gap of eight days from the 6th to the 7th weeks, shown in Fig. 1. We have consulted the average ratio between the number of devices and the number of sold products to fill the missing values for the study.

Privacy Issues. We need to emphasize that we try our best to respect the customers' privacy. The data collection procedure is focused only on the part that the customers broadcast to the environment. We do not attempt to construct a data collection procedure where the customers' browsing history, browsing URLs, etc. may be collected through our AP. That is, we do not trick customers by creating an AP where we may have the above information or even the username or password information from customers.

3.2 Data Preprocessing

The first issue of customer behavior analysis is to identify real customers. In this study, we filter the devices (identified via the MAC address) detected by the WiFi AP by setting thresholds of duration from five minutes to three hours. That is, we assume that each customer stays in the coffee shop no shorter than five minutes and no longer than three hours (5 min $\leq$ staying time $\leq$ 3 h). The detected devices with duration shorter than five minutes are assumed to be passing by devices and the devices with duration longer than three hours are likely to be the staff of the coffee shop or neighboring shops. The thresholds are decided based on our visual estimation when we visited the coffee shop. Also, we set a threshold applied to RSSI where we include only the RSSI greater than -70 dBm in the data collection (-70 dBm $<$ RSSI).

As customers reach the entrance door of the coffee shop, they are in the range of our data collection AP and the RSSI keeps increasing as customers moving into the coffee shop. We record the devices and identify the come-in time when the devices show the above pattern. Recording the customers' leaving time is the opposite. The stay duration for customers can be used for customers' behavior analysis such as the reasons they visit the coffee shop (for study, meeting friends or web surfing, etc.) and whom they go with.

3.3 The Proposed Prediction Method

The complete step-by-step procedure of the proposed method includes: (1) data collection, (2) feature extraction, (3) clustering for group information extraction, (4) model building and (5) prediction. We start by collecting the WiFi data through our modified AP. The WiFi data are compiled into several features and they are combined with the non-WiFi features such as weather information to form a complete feature set for model training. On the side, we have additional information provided by the coffee shop owner such as revenue related information to confirm our evaluation.

Analyzing customers' behavior or more specifically finding customers' group information is a key contribution of the proposed method. We assume customers who are classmates, partners, or colleagues may go to the coffee shop together frequently as a group. On the other hand, some customers, even they may not know each other can behave similarly such as they may visit the shop at similar time or on similar days (all coming in the morning, after lunch, after work or coming during weekdays or weekend), or with similar frequency (once per day or once per week). Given the above group behavioral inputs, we would like to extract a set of features called *group information* to describe different customers. By including those features, we may have a better chance to understand different customers and thus a better chance to predict the revenue of a business.

Given a set of customers' features, we adopt a hierarchical clustering method called Unweighted Pair Group Method with Arithmetic mean (UPGMA) to find customers' group information. Specifically, we have a set of features to describe customer i as:

$$\mathbf{z}_i = (z_{i1}, z_{i2}, \ldots, z_{iK}) \tag{1}$$

where we have K days to consider in our customer analysis and we should use K binary attributes to indicate the presence of customer i in the coffee shop on different days. That is,

$$z_{ik} = \begin{cases} 1 & \text{if customer } i \text{ visits the coffee shop on day } k, \\ 0 & \text{otherwise.} \end{cases} \tag{2}$$

Given the above inputs, UPGMA builds a rooted tree (dendrogram) that reflects the structure of pairwise similarities between different customers [5]. To describe the similarity between two clusters C_i and C_j, we utilize a proportional averaging formulation written as:

$$\sigma_{ij} = \frac{1}{|C_i| \cdot |C_j|} \sum_{p \in C_i} \sum_{q \in C_j} \sigma_{pq}, \tag{3}$$

where $|C_i|$ and $|C_j|$ represent the cardinality of the set (i.e., the size) for C_i and C_j respectively; also, σ_{pq} measures the similarity between two entities p and q from C_i and C_j respectively. We measure the similarity between two customers p and q as:

$$\sigma_{pq} = \sum_k \delta(z_{pk}, z_{qk}) \cdot \delta(z_{pk}, 1), \tag{4}$$

where the function $\delta(x, y)$ outputs 1 if $x = y$ and outputs 0 if $x \neq y$. That is, we count 1 when two show up in the coffee shop on the same day and count 0 otherwise. All pairs of customers are compared through the pairwise computation to form a similarity matrix in the end. Then, a pair of elements with the maximum similarity are recognized and clustered together as a single grouped pair first. Afterwards, the similarity between this pair and all other elements are recalculated to form a new matrix. We go on to find the pair with the maximum similarity for grouping step by step until all are combined into one in the end [4,5]. The output of UPGMA is a dendrogram and we can find the final grouping result by setting an appropriate number of clusters. In the end, the group information shall be used in building the Support Vector Regression (SVR) model [9] for the prediction on the number of customers' devices, the number of sold products and the total revenue.

3.4 Assumptions and Limitations

The goal is to analyze customers' behaviors that are related to coffee consumption. Due to the WiFi-based data collection nature, we first assume that all customers carry WiFi-based devices and their WiFi signals can be detected easily by the deployed AP. That is, the WiFi function must be on at all times when the customers visit the coffee shop, starting from entering to leaving the coffee shop, for all customers. Based on the assumption, we could capture customers' existence, in particular, we know when customers come to the coffee shop and leave the coffee shop. That is, as soon as we detect RSSI signals for each customer's device, it will be assumed that this is the exact entrance time for the customer. The leaving time of a customer is also assumed to be the time of losing or dropping off of the RSSI signal received from the customer's device. There are also some limitations in this research, such as using just one WiFi AP leads to weak distinguishment of the exact coming and leaving time for each customer. Moreover, we cannot detect the exact location and position of each customer. In the data cleaning phase, removing noisy or irrelevant data is hard especially in a crowded area[2].

4 Experiment Results

We would like to predict the number of customers' devices, the number of sold products and the total revenue given a set of WiFi-based and non-WiFi-based features. There are two scenarios that we discuss:

[2] There is a convenient store right next to this coffee shop.

Table 1. The statistics of frequent and non-frequent customers.

	The frequent customer (%)	Non-frequent customer (%)
The number of customers	11%	89%
The number of visits	27%	73%

1. In the first scenario, we take turn to work on three prediction tasks given a sliding window of size L as well as other features such as the day in a week, the weather information, which consists of the temperature and rain forecasting to build the learning model.
2. In the second scenario, we consider additional features, the group information with the same sliding window as described in the first scenario to build the learning model.

We attempt to analyze how the past presence or purchase records can be used to predict the future presence or purchase. In particular, between the first and the second scenarios, we discuss how the group information can help us for better prediction. We utilized Support Vector Regression (SVR) [9] as the predictive model. The size of sliding window L is set as $L = 14$ for this work. The detail result shall be shown below.

4.1 Statistics of Frequent and Non-frequent Customers

Before going on to demonstrate the effectiveness of the proposed method, we first study some basic statistics of the data set. In many retail stores, the transactions from the frequent customers may usually dominate the store revenue. In this case, we also would like to understand the contribution from the frequent and non-frequent customers separately. In Table 1, we show the numbers of frequent and non-frequent customers, which are 11% and 89% out of the whole group of customers who visited the shop during the data collection period. Interestingly, we also observe that this 11% frequent customers contribute 27% of the visiting times in the coffee shop, compared to 73% of the visits from non-frequent customers. It implies a relatively large consumption from the frequent customers compared to the non-frequent ones. When we aim to find a predictive model with good performance, we better to focus more on the prediction of the frequent customers rather than the non-frequent ones. Fortunately, the frequent customers are likely to come to the coffee shop in a regular manner and could be predicted easily if compared to the non-frequent group. Moreover, the prediction on the frequent rather than the non-frequent customers may be easy simply because we usually have a relatively large frequent customers' data in the training set. We discuss more along these two aspects below.

We also analyze the daily visits (in %) from the frequent and non-frequent customers, shown in Fig. 1. From the beginning to the end of data collection period, we observe that the percentage of the frequent customers increases slightly as time moving forward. It may due to that the major group of customers includes a significant percentage of students from a nearby university. The students may know each other better and better starting from September (the beginning of the semester) through November and they may have more chances to go for a coffee when they know each other better. Note that we have a few days of missing values due to data collection difficulty in the period.

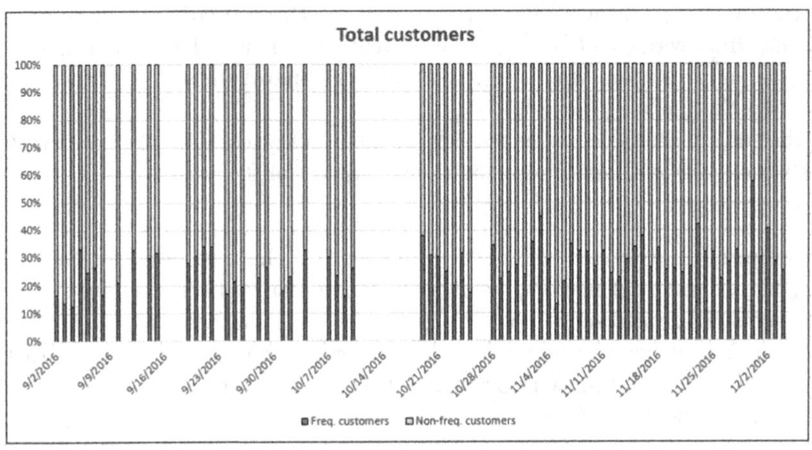

Fig. 1. The percentage of frequent and non-frequent customers per day.

4.2 Features and Results

The Feature Set. We first discuss the features that are used in this study. Following the description in the beginning of this section, let us use τ_t, ρ_t and day_t to describe the information such as the temperature forecasting, the probability of raining and the day in a week on the t-th day respectively. The sliding window of size L for the above information (except the day in a week) can be written as:

$$\begin{aligned} \textbf{temp}_{t,L} &= (\tau_{t-L}, \dots, \tau_{t-1}), \\ \textbf{rain}_{t,L} &= (\rho_{t-L}, \dots, \rho_{t-1}). \end{aligned} \tag{5}$$

To speak of the group information, we set the number of groups for group information extraction as $K = 4$. Given the assignment, we have the group features written as:

$$\mathbf{g}_t = (g_{t,1}, g_{t,2}, \dots, g_{t,k}, \dots, g_{t,K}), \tag{6}$$

where $g_{t,k}$ records the number of customers from group k who visit the coffee shop on the t-th day. We can describe the group information with the sliding window of size L as:

$$\mathbf{g}_{t,L,k} = (g_{t-L,k}, \dots, g_{t-1,k}), \quad \forall k \in \{1, \dots, K\}. \tag{7}$$

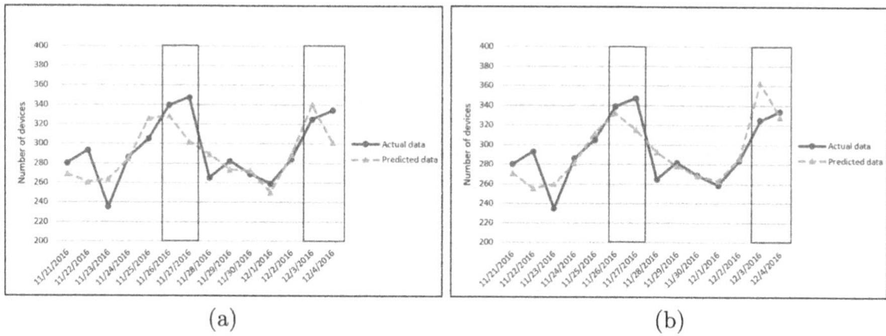

Fig. 2. The prediction on the total number of customers' devices: (a) without and (b) with the group information (weather information not included). The X-axis is the UTC (Epoch time) format and the Y-axis represents the number of customers' devices.

After all, we also write down the sliding window of size L for the target value that we want to predict:

$$\mathbf{y}_{t,L} = (y_{t-L}, \ldots, y_{t-2}, y_{t-1}). \tag{8}$$

In the end, the overall feature set in this study can be written as:

$$\begin{aligned} \mathbf{D}^{g}_{t,L} &= (\text{day}_t, \mathbf{temp}_{t,L}, \mathbf{rain}_{t,L}, \mathbf{y}_{t,L}; y_t), \\ \mathbf{D}^{\text{gp}}_{t,L} &= (\text{day}_t, \mathbf{temp}_{t,L}, \mathbf{rain}_{t,L}, \mathbf{y}_{t,L}, \mathbf{g}_{t,L,1}, \ldots, \mathbf{g}_{t,L,K}; y_t), \end{aligned} \tag{9}$$

for the case without or with the group information included, respectively. The target value y_t that we want to predict could be the number of customers' devices, the number of sold products or revenue as described before. From time to time, we may have the feature set described in Eq. 9 too large to create the risk of overfitting. To avoid the situation, we reduce the dimensionality by shrinking the feature size of sliding window as follows. For each sliding window, e.g., the sliding window for temperature, we may choose a pre-defined function such as the mean function to compress a long sliding window to a scalar such as:

$$\text{temp}_t = (\tau_{t-L} + \cdots + \tau_{t-1})/L. \tag{10}$$

Some other possible functions for shrinkage include minimization, maximization. Now the complete feature set is shrunk to:

$$\begin{aligned} \mathbf{d}^{g}_{t,L} &= (\text{day}_t, \text{temp}_{t,L}, \text{rain}_{t,L}, \mathbf{y}_{t,L}; y_t)\,, \\ \mathbf{d}^{\text{gp}}_{t,L} &= (\text{day}_t, \text{temp}_{t,L}, \text{rain}_{t,L}, \mathbf{y}_{t,L}, \text{g}_{t,L,1}, \ldots, \text{g}_{t,L,K}; y_t), \end{aligned} \tag{11}$$

for the cases of not including the group information or including the group information respectively. Note that we choose the mean function for the shrinkage on all the sliding windows except for the sliding window for the target value.

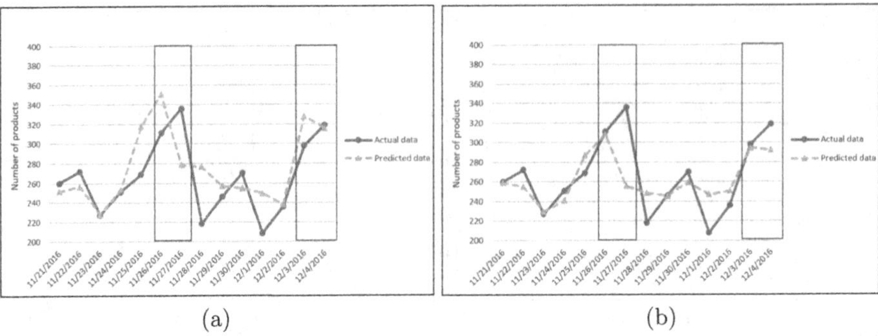

Fig. 3. The prediction on the total number of sold products: (a) without and (b) with the group information (no weather information included). The X-axis is the time and the Y-axis represents the number of products.

Result. The first experiment is to predict the number of customers' devices given the features described in Eq. 11. In Fig. 2, we demonstrate the effectiveness of the proposed method by showing the difference between the actual and the predicted result on the total number of customers' devices spanning two weeks[3]. First, we notice the ups and downs on the selling between different days where we usually have high selling and revenue during the weekends (Nov. 26, 27, and Dec. 3, 4). In fact, those are the days that we have significant gaps between the actual and predicted result. In (a), we have the prediction given the features without the group information and we include the group information for prediction in (b). Overall, we obtain an improvement from 9.11% to 7.63% in MAPE (Mean Absolute Percentage Error) from (a) to (b) if the group information is included and the weather information is not included (also in Table 2).

In the second experiment, we aim to predict the number of sold products. Other than the number of devices which may not be 100% identical to the number of customers, the amount of sold products could be a better quantity to reflect the business profit. In Fig. 3, we can compare between the actual number of sold products and the prediction. Again, we observe more selling during the weekends rather than during the weekdays. The weekend period is also the time that we have worse prediction if compared to the prediction on the weekdays.

To compare between the scenario when we include no group information and the scenario when we do include group information, we found out that including the group information can improve the prediction result from 15.06% to 11.32% in MAPE (without the weather information). It implies that including group information can help us understand more about customers' behavior on visiting the coffee shop. Intuitively speaking, people often visit coffee shops with their partners. The decision about whether people visit a coffee shop or not may be highly influenced by their partners. On the other hand, the group information

[3] We add random numbers in the Y-axis for Figs. 2, 3 and 4 due to a concern from the coffee shop.

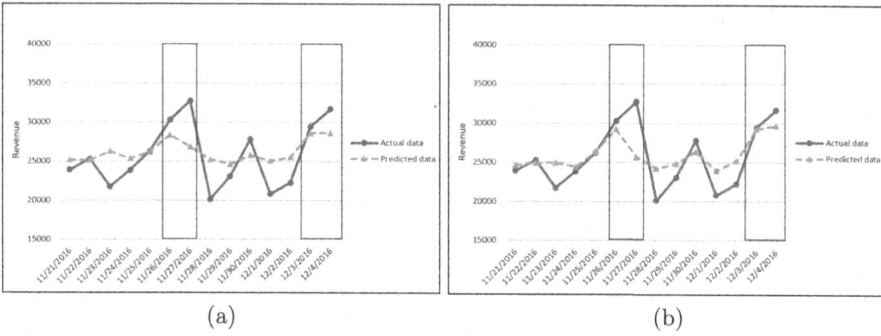

Fig. 4. The prediction on the total revenue: (a) w/o and (b) w. the group information (no weather information). The X-axis is the time and the Y-axis shows the revenue.

Table 2. The summary of all predictions in MAPE. The boldface numbers show the best performance in each group and the underline, boldface numbers show the best performance across all settings.

	w/o weather		w. weather	
	w/o group info.	w. group info.	w/o group info.	w. group info.
	MAPE (%)	MAPE (%)	MAPE (%)	MAPE (%)
# of devices	9.11	**7.63**	**7.95**	8.43
# of sold products	15.06	**11.32**	**11.60**	12.25
Revenue	18.10	**14.43**	14.58	**14.51**

may also imply a similar behavior on visiting the coffee shop such as the people in the same group may choose to visit the coffee shop on similar days or at similar moments. This kind of group information could reflect the vocations that the customers have or the living style they share. Knowing such information may give us more hints on predicting whether or not certain people visit the coffee shop on a particular day or at a particular moment.

In the end, we discuss the revenue prediction as described in Fig. 4. Again, we have similar result like the prediction on the number of devices and the prediction on the number of sold products. We have the prediction errors improved from 18.10% to 14.43% in MAPE when the weather is not included and from 14.58% to 14.51% in MAPE when the weather is included. Overall, we have the improvement when the group information is included in four out of six different settings, as shown in Table 2 given the settings such as without or with the weather information and for different prediction tasks. In the table, we also noticed the improvement from including the weather information in many of the occasions. Note that including both the weather information and group information may not produce the best result. We believe that too many features may harm the performance due to overfitting and the problem could be eased when more data are collected in the near future.

5 Conclusion

We proposed an easy-to-deployed, low cost and privacy-preserving method for business revenue forecasting based on WiFi sensing. A WiFi collection AP was installed in an indoor environment to collect related WiFi signals for us to understand more about customers who visit the business. The case study was done in a coffee shop where we analyzed the WiFi-based and non-WiFi-based data for 12 weeks for the evaluation. We worked on three prediction tasks such as the prediction on the number of devices, the number of sold products and the total revenue. In the experiment study, we found out the improvement when the weather information is included; more importantly, when the group information is included in most of the prediction tasks even with a limited data collection period. The prediction on the number of devices, the number of sold products and the revenue can reach 7.63%, 11.32%, and 14.43% in MAPE in their peak performance. A large scale data collection and study is on the way for more extensive study in the near future.

References

1. Draghici, A., Steen, M.V.: A survey of techniques for automatically sensing the behavior of a crowd. ACM Comput. Surv. **51**(1), 21:1–21:40 (2018)
2. Du, H., Yu, Z., Yi, F., Wang, Z., Han, Q., Guo, B.: Recognition of group mobility level and group structure with mobile devices. IEEE Trans. Mob. Comput. **17**(4), 884–897 (2018)
3. Han, J., Ding, H., Qian, C., Xi, W., Wang, Z., Jiang, Z., Shangguan, L., Zhao, J.: CBID: a customer behavior identification system using passive tags. IEEE/ACM Trans. Netw. **24**(5), 2885–2898 (2016)
4. Lau, E.E.L., Lee, B.G., Lee, S.C., Chung, W.Y.: Enhanced rssi-based high accuracy real-time user location tracking system for indoor and outdoor environments. Int. J. Smart Sens. Intell. Syst. **1**(2), 534–548 (2008)
5. Loewenstein, Y., Portugaly, E., Fromer, M., Linial, M.: Efficient algorithms for accurate hierarchical clustering of huge datasets: tackling the entire protein space. Bioinformatics **24**(13), i41–i49 (2008)
6. Maduskar, D., Tapaswi, S.: RSSI based adaptive indoor location tracker. Sci. Phone Apps Mob. Devices **3**(1), 3 (2017)
7. Nguyen, K.A.: A performance guaranteed indoor positioning system using conformal prediction and the WiFi signal strength. J. Inf. Telecommun. **1**(1), 41–65 (2017)
8. del Rosario, M.B., Redmond, S.J., Lovell, N.H.: Tracking the evolution of smartphone sensing for monitoring human movement. Sensors **15**(8), 18901–18933 (2015)
9. Smola, A.J., Schölkopf, B.: A tutorial on support vector regression. Stat. Comput. **14**(3), 199–222 (2004)
10. Subhan, F., Ahmed, S., Ashraf, K., Imran, M.: Extended gradient RSSI predictor and filter for signal prediction and filtering in communication holes. Wirel. Pers. Commun. **83**(1), 297–314 (2015)
11. Vanderhulst, G., Mashhadi, A.J., Dashti, M., Kawsar, F.: Detecting human encounters from WiFi radio signals. In: Proceedings of the 14th International Conference on Mobile and Ubiquitous Multimedia, Linz, Austria, 30 November–2 December 2015, pp. 97–108 (2015)

12. Zeng, Y., Pathak, P.H., Mohapatra, P.: WiWho: WiFi-based person identification in smart spaces. In: 2016 15th ACM/IEEE International Conference on Information Processing in Sensor Networks (IPSN), pp. 1–12, April 2016
13. Zeng, Y., Pathak, P.H., Mohapatra, P.: Analyzing shopper's behavior through WiFi signals. In: Proceedings of the 2nd Workshop on Workshop on Physical Analytics, WPA 2015, pp. 13–18. ACM, New York (2015)
14. Zhang, D., Xia, F., Yang, Z., Yao, L., Zhao, W.: Localization technologies for indoor human tracking. In: 2010 5th International Conference on Future Information Technology, pp. 1–6, May 2010

Evaluation of Hybrid Classification Approaches: Case Studies on Credit Datasets

Erkan Cetiner[1(✉)], Vehbi Cagri Gungor[2], and Taskin Kocak[1]

[1] Bahcesehir University, Istanbul, Turkey
ecetiner87@hotmail.com, taskin.kocak@eng.bau.edu.tr
[2] Abdullah Gul University, Kayseri, Turkey
cagrigungor@gmail.com

Abstract. Hybrid classification approaches on credit domain are widely used to obtain valuable information about customer behaviours. Single classification algorithms such as neural networks, support vector machines and regression analysis have been used since years on related area. In this paper, we propose hybrid classification approaches, which try to combine several classifiers and ensemble learners to boost accuracy on classification results. We worked with two credit datasets, German dataset which is a public dataset and a Turkish Corporate Bank dataset. The goal of using such diverse datasets is to search for generalization ability of proposed model. Results show that feature selection plays a vital role on classification accuracy, hybrid approaches which shaped with ensemble learners outperform single classification techniques and hybrid approaches which consists SVM has better accuracy performance than other hybrid approaches.

Keywords: Credit-risk · Hybrid-classifier · Feature selection

1 Introduction

Credit is a widespread and ossified concept which is one of the major term for modern finance. Credit suppliers desire all people to use it and unfortunately some customers find themselves in a credit loop trouble. People use credit for buying a house, a car and so on as physical needs and also for education, vacation and so on as career or entertainment purposes. Without a good credit management plan, people, companies or governments can find themselves in crisis as Argentina in 2002 or Asia 1997 [1].

Naturally, some debtors could not fulfill their credit payback requirements, creating a credit loss, sometimes even making the creditor bankrupt. That is the reason why credit risk analysis is an emerging issue. A need of well managed credit risk analysis is crucial to determine probability of credit losses [2].

Credit Risk Analysis (CRA) manages supplier risk assessment according to its product portfolio served to customers [3]. Finance institutions aim at using credit risk management is minimize potential future losses and maximize bank`s rate of return. For a successful business life of a finance organization, credit risk management is vital task and needs to be processed carefully.

P. Perner (Ed.): MLDM 2018, LNAI 10935, pp. 72–86, 2018.
https://doi.org/10.1007/978-3-319-96133-0_6

In general, there are two main models for assessing credit risk, traditional statistical models and artificial intelligence models. Traditional models have power of explanation clearly but requires too strict preconditions for assessment. Machine learning techniques continuously develop in time, and used now for credit risk assessment. For example, Gallo showed that artificial neural networks represent an easily customizable tool for studying many problems which are very difficult to analyze with standard economic models in 2006 [4]. Shachmurove showed that artificial neural networks have the ability to analyze complex patterns quickly with high accuracy [5]. Satchidananda compared the efficiency of the decision trees with logistic regression [6]. Yu et al. developed a work on a comparative study for data mining algorithms, such as logistic regression, support vector machines and decision trees, on credit risk evaluation and they stated that logistic regression outperforms other classification techniques [7]. Wang et al. worked on a new fuzzy support vector machine to evaluate credit risk [8]. Hao et al. worked also on support vector machine on fuzzy hyper-plane and searched for its application on credit risk [9]. Also, Kaya et al. obtained good results for PD estimation by logistic regression on German Credit Dataset [10].

Zhang et al. focused on credit scoring on Brazilian credit card dataset [11]. Huang et al. compared the performance of the support vector machines and neural networks on Taiwan's and USA's banks datasets [12]. Doumpos and Zopounidis combined different classification techniques in a stacked generalization approach [13]. Gaganis, Pasiouras, Spathis and Zopounidis compared the classification performance of kNN with discriminant analysis and LR where kNN outperformed both them on the FAME dataset obtained from Bureau Van Dijk's Company [14]. Campos et al. worked on innovative soft computing techniques for financial credit risk measurement [15]. Kotsiantis, proposed a hybrid model to evaluate credit risk which is also a basis for this project work [16].

Recent studies cover ensemble learning models. Hybrid learning approaches generated to obtain better performance results in terms of accuracy, precision and recall metrics [17]. Also it has shown that feature selection algorithms play vital role on overall success rate of proposed models. All these studies have great influence on credit-scoring era. In a hybrid approach, one can combine feature selection and base classifiers, or base classifiers can be used along with ensemble learners [18].

The remaining sections are organized as follows. Section 2 describes the methods which are used in this study. In Sect. 3, the dataset description and data pre-processing, and development of the hybrid credit scoring model have been introduced. Section 4 evaluates experiments' results. Finally, Sect. 5 concludes the paper.

2 Background

In this section, the major components for our work are described briefly. In Sect. 2.1, feature selection algorithms are discussed. In Sect. 2.2, classifiers which are used during experiments are mentioned. Finally Sect. 2.3 describes clustering mechanisms.

2.1 Feature Selection Algorithms

Feature Selection is a process which automatically searches for the best subset of attributes in the dataset. The notion of "best" is application-dependent, but typically means the highest accuracy. In the literature, it is shown that feature selection algorithms can reduce overfitting, improve accuracy, and reduce training time [19]. InfoGain, GainRatio and ChiSquare algorithms are used in this work.

2.2 Classifiers

Base Classifiers
Single classification algorithms used for experiments are described in this section.

ANN
Neural networks use 3-layered model, including input layer, hidden layer and output layer. Input layer is responsible for information gathering and passing it through hidden layer. Information flow is controlled by weighted links. In addition, neural networks have different approach to the problems. With neural networks; large amount of elements work as parallels. They are learned with samples and are not programmed for specific tasks. They can process unexpected patterns and can be able to generalize them [21].

SVM
SVM works according to hyper plane separation concept. Unique classes separated from each other by an optimal distance. SVM is widely used for regression and classification problems. Using kernel function is key component to finding optimal distance. It enables optimal lines to be flexible and this brings advantage of SVM.

Naïve Bayes
Naïve Bayes is a probabilistic approach and based on Bayesian Theorem. According to this algorithm; there is no relation between feature presences of classes. Algorithm does not require re-deployment for new training sets and results obtained fast. However; its weakness is the axiom of independency of dataset attributes.

Hidden Markov Model
It is an extension of Markov models. It considers probabilistic outcomes of states. Scalability, initial parameter setup and training data size matters for success rate of HMM.

Self-organizing Map
SOM, a branch of ANN, works as an unsupervised learning mechanism and creates two-dimensional results. Instead of error-correction learning; SOMs use competitive learning which differs them from other ANN techniques. This gathers opportunity of visualizing low dimension views of high dimension data.

Decision Trees

ADTree

ADTree works on the concept of rotation of decision nodes; according to verification of condition and related nodes. Instances are classified among traversing true nodes and rules defined then this traversal [22].

C4.5 -J48-

It works in same manner as ID3; takes information entropy as decision metric. Training data is already classified. Each sample includes a row consisted by attributes and values and also class attribute where sample falls. In each node iteration; subsets are determined according to attributes value where splits entire samples most effectively.

RandomForest

RandomForest constructs number of decision trees at training time and pointing the class which is mode of classes according to individual trees. It is a simple combination of tree predictors and those trees depends on random sampling of same distribution for all trees in forest.

Ensemble Classifiers

Boosting

Boosting is an algorithm used to reduce bias and variance in supervised learning. A weak learner is a classifier which is slightly correlated with the true classification. On the other hand, a strong learner is a classifier that is arbitrarily well-correlated with the true classification.

Bagging

Bagging commonly used for statistical classification and regression problems to improve stability and accuracy of machine learning algorithms. It also helps to avoid overfitting. Although it is usually applied to decision tree algorithms, it can be used with any other types of algorithms.

ClassificationViaClustering

Uses clusters for classification. Cluster numbers correspond to different class labels of the dataset. If there is no cluster found for a particular instance; missing value is returned during prediction.

2.3 Clustering Algorithms

Clustering refers to gathering similar type of data points which have same attitude in terms of classification purposes. Following sections describe three different clustering algorithms; K-Means, EM and DBSCAN.

K-Means
K-means clustering is based on quantization of vectors. Main purpose is distribution of observations to clusters where each observation of a cluster is a prototype of that cluster.

EM
EM is popular iterative improvement algorithm. It can be seen as extension of k-means algorithm. Each instance corresponds to a different cluster with a weight according to probabilistic distribution.

DBSCAN
(DBSCAN) is a data clustering algorithm. It calculates distances of data instances and their neighbors. Areas with more instances than predetermined threshold are grouped into clusters.

3 Experimental Design

In this section, datasets used in paper work, performance evaluation metrics and proposed methodology described in sequence.

3.1 Datasets

There are two datasets used in our work, the first one is obtained from a Turkish Bank and second one is public German Dataset.

Turkish Bank Dataset
Dataset is obtained from a corporate finance organization [23]. Dataset has continuous (numerical) and discrete (categorical or symbolic) attributes. Dataset has 15470 data instances regarding credit applications. There are 96 attributes regarding 2 different layouts. Layout-1 contains personal information such as occupation; age; income; property; marriage status and so on. Layout-2 contains all credit history of related customer.

To use the dataset for the proposed model, data pre-processing needs to be done. Without any sequence in the operations, there following data pre-processing methods have been applied for data preparation and cleaning:

- Removing some of the records because of having missing values;
- Integrating data;
- Attributes with continuous values have been categorized for better classification and regression purposes.

After data pre-processing, 38 features have been chosen for the final experiment (Table 1).

Table 1. Turkish credit dataset features

Feature	
AccountID	DownPayment
FinanceType	CreditProduct
SocialSecurityInfo	Maturity
HaveKids	AccountStatus
Salary Income	PaybackType
Property Income	PerformanceStatus
Stock/Sharehold Income	TotalAccountNumber
Freelance Income	TotalDebtLevel
Total Montly Family Income	MonthlyDebtLevel
Gender	WorstPaymentStatus12month
ApplicantAgePeriod	WorstPaymentStatus6month
HasCar	CreditInfo
CreditType	TotalBalanceWithMortgage
EducationLevel	TotalBalanceWithoutMortgage
Relationship Status	CreditScoringPoint
HasHouse	AccountLegalProceedingJobTyp
HasPhone	JobType
BankInfo	DecisionFlag
Total WorkingTime	
Total ResidenceTime	

German Dataset

German Credit Dataset is a public access credit database that this study utilizes [24]. German Credit Dataset has continuous (numerical) and discrete (categorical or symbolic) attributes. Dataset has 1000 data instances regarding retail credit applications. 300 of the instances are of the bad borrowers and 700 of them are of the good borrowers. Dataset contains a total of 20 attributes 7 of which are numerical and 13 of which are categorical (Table 2).

Table 2. German credit dataset features

Feature
Status of existing checking account
Duration in month
Credit history
Purpose
Credit amount
Savings account/bonds
Present employment since
Installment rate in percentage of disposable income

(continued)

Table 2. (*continued*)

Feature
Personal status and sex
Other debtors/guarantors
Present residence since
Property
Age in years
Other installment plans
Housing
Number of existing credits at this bank
Job
Number of people being liable to provide maintenance for
Telephone
Foreign worker
Decision

3.2 Performance Evaluation

Our aim in this study is comparing performances of different data mining classifiers. By using corporate dataset, experiments have been applied and outcomes were calculated. To make comprehensible comments on those outcomes, it is a crucial step to define performance metrics used in this work. Since comparison of performances based on these metrics, without understanding them properly, it is impossible to see improvements for future work analysis [16]. Following metrics are used in this study. An example confusion matrix is shown on Fig. 1. Elements of confusion matrix:

- True positive (TP): Customer is good and prediction is good.
- False negative (FN): Customer is good but predicted as bad.
- False positive (FP): Customer is bad but predicted as good.
- True negative (TN): Customer is bad and predicted as bad.

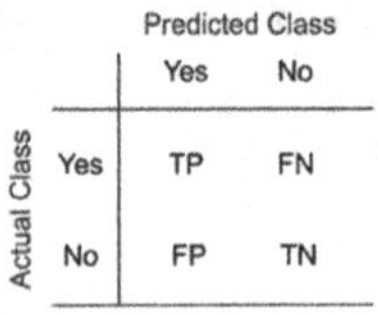

		Predicted Class	
		Yes	No
Actual Class	Yes	TP	FN
	No	FP	TN

Fig. 1. Confusion matrix

Classifiers are compared according to 2 performance metrics which are accuracy and type–1 accuracy. Accuracy of an algorithm represents the effectiveness level for a classifier and shows how it asses data qualitatively.

- Accuracy = (TP + TN)/(total instance number) – How accurately an algorithm will classify future data.
- TYPE-1 Accuracy = FP/(total instance number) – Percentage represents classing a customer as "good" when they are actually "bad".

3.3 Proposed Method

Proposed model has 3 main steps. At first; data pre-processing techniques applied to raw dataset to have more comprehensive and clean results at the end. At second stage; the best feature selection algorithm is looked for by comparing accuracy based on ANN classifier. At third stage; bagging, boosting, classificationviaclustering and base classifiers are compared along the chosen feature selection algorithm selected at step-2 (Fig. 2).

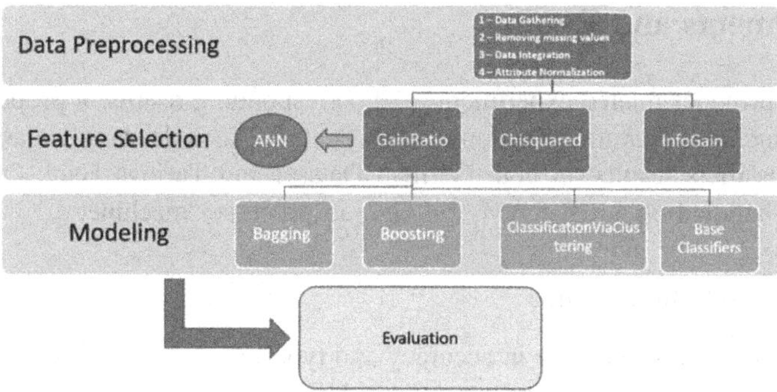

Fig. 2. Proposed model

In this paper, we propose hybrid classification approaches, which try to combine several classifiers and ensemble learners to boost accuracy on classification results. To construct our model, the following mechanisms have been employed:

1. Dataset is preprocessed with normalization, discretization and outlier analysis operations.
2. Three feature selection algorithms are compared along with ANN; which are InfoGain, GainRatio and ChiSquared.
3. From the second step, the best feature selection method among others is chosen for the next experiments.
4. Ensemble learners are tested along base classifiers and performance evaluations are obtained in terms of accuracy and Type-1 accuracy. In this step, several base classifier techniques are employed with ensemble learners.
5. Performance results from step-4 are obtained and the best techniques (top 4) in terms of accuracy have been applied to German Dataset; to verify generalizability of models.

Overall, in this study, we have compared the efficiency of different feature selection algorithms for credit scoring problem using German and Turkish credit datasets. We have also compared the accuracy of general base classifiers and proposed hybrid models by combining base classifiers along ensemble learners. In addition, we have determined the appropriate feature selection algorithm as pre-process step for parameter tuning of datasets. Furthermore, detailed performance evaluations of base classifiers and hybrid models on Turkish bank dataset and German Credit Dataset have been conducted to have an idea about the generalizability of their applicability on credit scoring domain. Finally we included computational complexity in terms of duration for each repeated experiment. To the best of our knowledge, this is the first study, which compares the performance of hybrid classification approaches based on their classification accuracy on both German credit dataset and Turkish bank dataset as a case study.

4 Experiments and Results

This section provides related experiments and corresponding results of proposed work. Section 4.1 includes Feature Selection results, Sect. 4.2 includes single classifiers and ensemble classifiers results on both German Dataset and Turkish Bank Dataset. All experiments handled on 8 GB-RAM, 2.4 GHz i5 processor machine.

4.1 Feature Selection Results

Feature selection plays vital role in accuracy and type-I accuracy results. According to experiments based on MultiLayerPerceptron -ANN- on WEKA, **GainRatio** with 40 selected attribute have better results among **InfoGain** and **ChiSquared** selection algorithms. Thus, it is selected for the rest of the model.

InfoGain Results

InfoGain experiments are tested with WEKA up to 50 number of attributes along the ANN -MultilayerPerceptron-. Following Table 3 shows accuracy level for each attribute number up to 50 attribute.

Table 3. Experiment on Turkish bank dataset with ANN based on InfoGain

Model	# of Attributes	Accuracy (%)
1	10	74
2	15	74.2
3	20	74.6
4	25	74.6
5	30	74.6
6	40	74.9
7	50	74.9

GainRatio Results

GainRatio experiments are tested with WEKA along the ANN –MultilayerPerceptron-. Following Table 4 shows accuracy level for each attribute number up to 50 attribute.

Table 4. Experiment on Turkish bank dataset with ANN based on GainRatio

Model	# of attributes	Accuracy (%)
1	10	81.1
2	15	81.2
3	20	81.2
4	25	81.2
5	30	81.9
6	40	81.9
7	50	81.9

ChiSquared Results

ChiSquared experiments are tested with WEKA along the ANN -MultilayerPerceptron-. Following Table 5 shows accuracy level for each attribute number up to 50 attribute.

Table 5. Experiment on Turkish bank dataset with ANN based on ChiSquared

Model	# of attributes	Accuracy (%)
1	10	78.9
2	15	78.9
3	20	77.9
4	25	77.9
5	30	77.8
6	40	77.8
7	50	77.5

4.2 Ensemble Classifiers

Boosting

In Table 6; it is shown that Boosting-SVM-RBF discards other among accuracy and type-I accuracy. Furthermore, ANN and NaïveBayes have better results when we compare them with Decision Tree Classifiers. Thus, we can say that base classifiers with boosting have better classification ability than decision tree classifiers when applied with boosting.

Table 6. Experiments on Turkish bank dataset with boosting

Model	Ensemble classifier	Performance measures		Computational complexity (min)
		Accuracy	Type I accuracy	
1	ANN	81.9	0.079	58
2	SVM-RBF	83.7	0.082	56.3
3	NaiveBayes	80.1	0.074	12.4
4	RandomForest	79.7	0.072	16.7
5	ADTree	79.7	0.072	16.7
6	J48	80	0.074	15.3

Bagging

As it is compared with boosting results; we can say that bagging has minor improvements among boosting ensemble learner according to accuracy results. Again SVM-RBF outperformed other hybrid classifiers when we look at Table 7. Relation with decision tree classifiers and base classifiers are same as in boosting case. Base classifiers adopted with bagging have better classification results among decision tree classifiers.

Table 7. Experiments on Turkish bank dataset with bagging

Model	Ensemble classifier	Performance measures		Computational complexity (min)
		Accuracy	Type I accuracy	
1	ANN	82	0.079	58
2	SVM-RBF	83.9	0.082	56.3
3	NaiveBayes	80.1	0.074	12.4
4	RandomForest	79.9	0.072	16.7
5	ADTree	79.9	0.072	16.7
6	J48	80.1	0.074	15.3

ClassificationViaClustering

In Table 8, it can be commented that ClassificationviaClustering techniques have similar results to bagging and boosting algorithms but overperformed by all SVM-RBF algorithms. K-Means gives best result among DBSCAN and EM.

Table 8. Experiments on Turkish bank dataset with ClassificationViaClustering

Model	Ensemble classifier	Performance measures		Computational complexity (min)
		Accuracy	Type I accuracy	
1	EM	81	0.078	23.4
2	K-Means	81.4	0.078	23.4
3	DBSCAN	78.5	0.076	25.6

Base Classifiers

In Table 9, only base classifiers are tested. It is seen that SVM-RBF algorithm has better results. But, it is also seen that; without ensemble learner techniques; base classifiers performance on accuracy has been dropped. Also, the HMM method has slightly outperformed ANN, SOM and NaiveBayes classification methods.

Table 9. Experiments on Turkish bank dataset with base classifiers

Model	Classifier	Performance measures		Computational complexity (min)
		Accuracy	Type I accuracy	
1	SVM-RBF	83.4	0.081	48.5
2	ANN	81.9	0.079	54.4
3	NaiveBayes	79.9	0.078	10.4
4	HMM	82.8	0.08	12.5
5	SOM	80	0.079	35.3

Best 4 Algorithms on German Dataset

In Table 10; best 4 algorithms chosen according to previous experiments and they were applied on German Dataset. Results show similar attitude as in Bank Dataset.

Table 10. Experiments on German dataset

Dataset	Model	Base classifier	Performance measures		Computational complexity (min)
			Accuracy	Type I accuracy	
German	Bagging	ANN	73.1	0.072	4 min
German	Bagging	SVM-RBF	74.1	0.074	3 min
German	Boosting	SVM-RBF	72.6	0.071	3 min
German	Base Classifier	SVM-RBF	73.5	0.073	3 min

ROC Results on Turkish Bank Dataset for Top 4 Algorithms

A receiver operating characteristic (ROC), or ROC curve, illustrates the performance of a binary classifier system when its discrimination threshold changes [25]. The curve shows the true positive rate (TPR) against the false positive rate (FPR). In general, the quality of a ROC curve is summarized using the area under the curve (AUC). The

Table 11. Experiments on Turkish bank dataset with top 4 algorithms based on accuracy performances

Dataset	Model	Base Classifier	Performance measures
			AUC
Bank dataset	Bagging	ANN	0.7725
Bank dataset	Bagging	SVM-RBF	0.6443
Bank dataset	Boosting	SVM-RBF	0.7614
Bank dataset	Base classifier	SVM-RBF	0.671

higher AUC scores are the better. As shown in Table 11; the ANN with bagging gives the best AUC value which comes to following ROC curves. Although it has a slightly lower accuracy value than SVM-RBF with bagging; the ROC measure is one of the most appropriate with neural networks in binary classification problem.

5 Conclusion

Credit risk concept has become a vital issue in finance sector in recent years. In this work, we proposed a combined hybrid method for assessment purposes. By combining several classifiers, it is aimed to gather their single strengths together and create unique, more robust one for individual credit risk evaluation. At the beginning, data preprocessing techniques were applied on raw dataset to discard irrelevant attributes and missing value rows. The proposed method has 3 important findings:

- To strengthen the classification power and also to decrease the level of time and calculation complexity; we defined the exact number of attributes which would have the most impact on decision. For this purpose, we compared feature selection algorithms based on ANN classifier. We repeated these experiments until we reached a maximum marginal gain in terms of accuracy. We found that GainRatio has better results among other feature selection algorithms and we went to the next step based on GainRatio feature selection method.
- After comparing boosting and bagging ensemble learners, we applied the base classifiers and decision tree classifiers along with them. We found that the bagging learners have slightly better results that the boosting ensemble learners. In addition, we found out that the base classifiers along with the bagging and the boosting learners outperform the decision tree classifiers. Further, we experimented classificationviaclustering techniques as a hybrid approach and found that they have similar performance results when compared to boosting and bagging learners. However, the K-Means based method outperforms all other classificationviaclustering algorithms. Also, we tested base classifiers without embedding any ensemble classifier and found that using ensemble classifiers impact classification.
- When we compared time complexity of related algorithms; we see that naïve bayes outperforms others very far. Decision tree classifiers also have better performance among SVM and ANN. ANN performs poorly according to its back propagation nature. Also, it seems that boosting and bagging have negligible effects when they compared to each other, but there is a small overhead with them and base classifiers. Also, clustering techniques outperforms SVM, ANN and Ensemble learners in terms of time complexity.

We conclude that using ensemble classifiers on both datasets gave better performance results in terms of accuracy when compared to base classifiers. To the best of our knowledge, this is the first study, which compares the performance of hybrid classification approaches based on their classification accuracy on both German credit dataset and Turkish bank dataset as a case study. The goal of using such diverse dataset is to show generalization capability of our approaches

Acknowledgement. The work of V.C. Gungor was supported by the Turkish National Academy of Sciences Distinguished Young Scientist Award Program (TUBA-GEBIP) under Grand no. V.G./TUBA-GEBIP/2013-14.

References

1. Joseph, C.: Credit Risk Analysis: A Tryst with Strategic Prudence, Chap. 1 (2006). ISBN 0070581363
2. Joseph, C.: Credit Risk Analysis: A Tryst with Strategic Prudence, Chap. 2 (2006). ISBN 0070581363
3. Joseph, C.: Credit Risk Analysis: A Tryst with Strategic Prudence, Chap. 3 (2006). ISBN 0070581363
4. Gallo, C., Letizia, C., Stasio, G.: Artificial neural networks in financial modelling. Research Gate, pp. 1–21 (2006)
5. Shachmurave, Y.: Applying Artificial Neural Networks to Business Economics & Finance (2002)
6. Sogala, S.S.: Comparing the Efficacy of the Decision Trees with Logistic Regression for Credit Risk Analysis. Head Risk Solutions & Research, HP India
7. Yu, H., Huang, X., Hu, X., Cai, H.: A comparative study on data mining algorithms for individual credit risk evaluation (2010)
8. Wang, Y., Wang, S., Lei, K.K.: A new fuzzy SVM to evaluate credit risk. IEEE Trans. Fuzzy Syst. **13**, 820–831 (2005)
9. Hao, P.-Y., Lin, M.-S., Tsai, L.-B.: A new SVM with fuzzy hyper-plane and its application to evaluate credit risk
10. Kaya, M.E., Gürgen, F., Okay, N.: An analysis of support vector machines for credit risk modeling. In: PAKDD 2007 DMBiz Workshop, China (2008)
11. Zhang, Y., Orgun, M.A., Baxter, R., Lin, W.: An application of element oriented analysis based credit scoring. In: Perner, P. (ed.) ICDM 2010. LNCS (LNAI), vol. 6171, pp. 544–557. Springer, Heidelberg (2010). https://doi.org/10.1007/978-3-642-14400-4_42
12. Huang, Z., Chan, H., Hsu, C.-J., Chan, W.-H., Wu, S.: Credit rating analysis with support vector machines and neural networks: a market comparative study. Decis. Support Syst. **37**, 543–558 (2003)
13. Doumpos, M., Zopounidis, C.: Model combination for credit risk assessment: a stacked generalization approach. Ann. Oper. Res. **151**, 289–306 (2006)
14. Gaganiz, C., Pasioures, F., Spathis, C., Zopounidis, C.: A comparison of nearest neighbors discriminant and logit models for auditing decisions. Intell. Syst. Account. Financ. Manag. **15**, 23–40 (2007)
15. Campos, R., Ruiz, F.J., Agell, N., Angulo, C.: Financial credit risk measurement prediction using innovatiove soft-computing techniques
16. Kotsiantis, S.: Credit risk analysis using a hybrid data mining model. Int. J. Intell. Syst. Technol. Appl. **2**, 345–356 (2007)
17. Cetiner, E.: Classifiers performance comparison on credit risk analysis. MS thesis Study, Computer Engineering Department, Bogazici University (2011)
18. Koutanaie, F.: A hybrid data mining model of feature selection algorithms and ensemble learning classifiers for credit scoring. J. Retail. Consum. Serv. **27**, 11–23 (2015)
19. Keller, D., Sehami, M.: Toward optimal feature selection. In: Proceedings of International Conference on Machine Learning (1996)

20. Narendra, P.M., Fukunaga, K.: A branch and bound algorithm for feature selection. IEEE Trans. Comput. **C-26**, 917–922
21. Shunin, Y.N.: Neural networks modelling of business situations and decision-making analysis. Comput. Modell. New Technol. **9**(2), 17–26 (2005)
22. Beryor, H., Merkl, D., Dittenbach, M.: Exploiting partial decision trees for feature subset selection in e-mail categorization. In: Proceedings of the 2006 ACM symposium on Applied Computing (2006)
23. Turkish Bank Dataset
24. "German Dataset", Statlog Project Databases. ftp://ftp.ics.uci.edu/pub/machine-learning-databases/statlog/german
25. Fawcett, T.: An introduction to ROC analysis. Pattern Recogn. Lett. **27**, 861–874 (2005)
26. http://www.crsouza.blogspot.com/2010/03/kernel-functions-for-machine-learning.html
27. West, M.: Bayesian factor regression models in the "Large p, small n" paradigm. Bayesian Stat. **7**, 723–732 (2003)
28. Bank for International Settlements. http://www.bis.org

An Efficient Two-Layer Classification Approach for Hyperspectral Images

Semih Dinc(✉), Babak Rahbarinia, and Luis Cueva-Parra

Auburn University at Montgomery, Montgomery, AL 36117, USA
{sdinc,brahbari,lcuevapa}@aum.edu
http://www.cas.aum.edu/departments/mathematics-and-computer-science

Abstract. Different from regular RGB images that only store red, green, and blue band values for each pixel, hyperspectral images are rich with information from the large portion of the spectrum, storing numerous spectral band values within each pixel. An efficient, two-layer region detection framework for hyperspectral images is introduced in this paper. The proposed framework aims to automatically identify various regions within a hyperspectral image by providing a classification for each pixel of the image, associating them to distinct regions. The first layer of the system includes two new classifiers, and is responsible for generating probability scores as the "new feature set" of the original dataset. The second layer works as an ensemble classifier and combines the newly generated features to estimate the region of the sample. Experimental results show that the proposed system can produce accurate classifications with an average area under the ROC curve of 0.98 over all regions. This result indicates the higher accuracy of the proposed system compared to some other well-known classifiers.

Keywords: Remote sensing · Spectral features
Hyperspectral imaging · Multilayer classification

1 Introduction

Conventional color cameras can capture limited portion of the electromagnetic radiation known as the "visible light". The camera is sensitive to three primary colors red, green, and blue (RGB), and its output is a digital image where each pixel has an RGB value. An RGB image does not always have sufficient information for detecting the objects with similar color hue and light intensity. Furthermore, in dim conditions, color cameras simply fail as they cannot collect enough visible light to form a meaningful image. Exploiting more spectral features of the scene is a powerful alternative for those challenging conditions. Spectral imaging sensors (or Hyperspectral cameras) can capture a much larger portion of the spectrum with higher spectral resolution (or number of spectral bands), when compared to the conventional cameras. A pixel on the image can be represented by hundreds of features, which are collected from different wavelengths of the spectrum. Since materials show different reflectance property to

P. Perner (Ed.): MLDM 2018, LNAI 10935, pp. 87–102, 2018.
https://doi.org/10.1007/978-3-319-96133-0_7

each wavelength, spectral features yield very strong clue of the material type on the pixel. Hundreds of features are collected for a pixel to form a spectral signature of the material. This rich spectral information gives researchers a new insight for a broad range of problems.

However, hyperspectral imagery comes with some challenges to address. First is the high noise rate. Spectral sensors are susceptible to the general Gaussian noise as well as the spectral noise that is caused by water absorption bands. These bands are known and they are usually removed from the dataset before processing. But the general noise should be handled by the classifiers. Second challenge is the high data dimensionality. Having high number of spectral features for each pixel is very beneficial for detecting the region of interest (ROI). But each feature adds another dimension to the data domain, so data become more sparse and traditional classifiers tend to overfit the data [1].

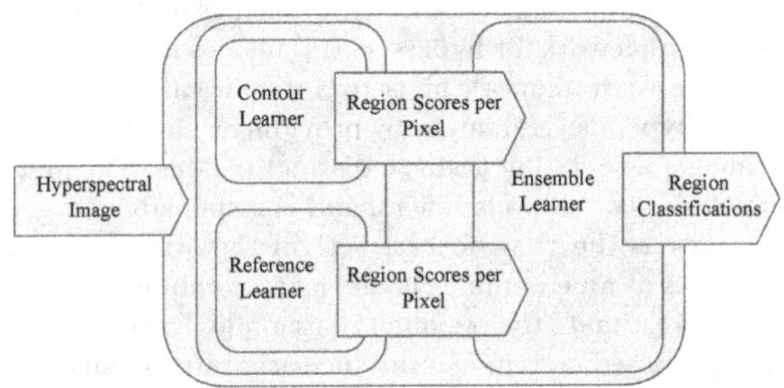

Fig. 1. Overview of the system

The main goal of this research is to design a novel two-layer classification system to *accurately* and *efficiently* identify and categorize various regions in a given hyperspectral image. As it can be seen in Fig. 1, our system is designed in two layers. The first layer's task is to provide classifications for each pixel of the image by employing different techniques which analyze the input from distinct viewpoints. In our design, we will describe two novel components for this layer. Our two individual components in the first layer, namely *Contour Learner* and *Reference Learner*, work independently and in parallel to classify and label each pixel of the hyperspectral image. By generating profiles of each region, these two components produce and attach a list of region scores to each pixel. Each region score in the list identifies the confidence that a given pixel is associated with a specific region. In Sects. 3.1 and 3.2, we outline the details of these two learners.

The second layer is comprised of an *Ensemble Learner* module that receives the separate lists of region scores for each pixel from all the components in the first layer. By combining the obtained information from the previous layer, this module builds a model that aims to improve and enhance the classification results. Essentially, the output of this module would be a list of region probability scores for each pixel of the image. However, the newly generated probability scores from this layer are expected to be more accurate. This is due the properties of ensemble learners where the meta-classifier could potentially learn from the behaviors and more importantly from the mistakes of previous learners to produce more precise estimations. The details of this module are provided in Sect. 3.3.

In summary, this paper makes the following contributions.

- A two-layer region classification framework for region detection in hyperspectral images.
- Two novel pixel classification models in the first layer that efficiently produce a list of region scores for pixels by generating different detection surfaces.
- An ensemble classification model in the second layer that utilizes the region scores produced by the previous layer as meta features to produce accurate region classifications.
- A comprehensive set of evaluations showing the accuracy of the proposed model compared to some well-known classifiers.

2 Literature Survey

Early applications of Hyperspectral imagery (HSI) were tightly connected with remote sensing, where images of a geographic area are captured via satellite or a plane from high altitudes. Usually the goal is to analyze the ground area and detect the target objects or materials [2], agricultural products, concrete surfaces, and water regions. After developments in recent years, HSI became more popular for in-door applications as well. Sample applications can be listed as: detecting freshness of fruits, diagnosing skin illnesses, or detecting tooth decays.

Traditional classifiers such as Support Vector Machines (SVM) and Neural Networks (NN) have been widely adopted and used for HSI classification purposes. Authors of [3], for example, used an SVM classifier and experimented with different kernels to improve their system's accuracy. The cost of training in these methods might be high as these classifiers need to undergo a costly training phase. Furthermore, the generated model of these classifiers cannot be examined manually by field experts due to the internal complexities of the underlying learners. This issue has recently been raised by authors of [4] who explored the importance of interpretability in machine learning systems and suggested that the decisions of such systems should be explainable. In contrast, our system aims to incorporate more "visual" information into the generated models of the modules in the first layer. This additional information might be useful for users and researchers to better understand the output of these modules. Our system benefits from a novel two-layer architecture, where the efficient first-layer modules generate new meta features for the second layer.

Also, deep learning methods have been employed in HSI region classification. For example, the authors of [5] used single-layer autoencoder (AE) and multi-layer stacked AE (SAE) for deep features extractions. Particularly, SAE increases the accuracy of SVM and Logistic Regression (LR) and obtains better accuracy when compared with other extraction methods like PCA, KPCA, and NMF. Additionally in this paper, authors introduces a combined SAE-LR method, which provides an increased statistical accuracy compared with the well-known Radial Basis Function (RBF)-SVM classifier. While the SAE-LR has faster testing times than SVM or KNN, its drawback lies on the longer training times.

Other studies such as [6], which proposes a deep belief network (DBN) combined with logistic regression (LR) framework for spatial and spectral-spatial classification, and [7], which proposes an active learning algorithm based on weighted incremental dictionary learning (WI-DL) with deep network training, found that deep learning techniques for HSI classification still have challenges compared with RBF-SVM and other methods. A marginal gain in accuracy (in the order of 1%–2%) requires longer training and total times (about 5–8 times longer). In general and compared to these works, our system employs a novel two-layer architecture that can produce accurate results efficiently.

Sparsity models have also been utilized to perform HSI classification. The study conducted in [8] is based on sparse linear combination of atoms (i.e. spectral features) to represent pixels. The authors define a sparsity model where a sparse vector is obtained for a test pixel through some optimization approaches. The sparse vectors are generated for each pixel class and then are used to approximate (reconstruct) the test pixels using the training pixels. The final classification output for a test pixel is obtained by a sparse vector that best approximates the test pixel. A drawback associated with these works is the high cost of computations. For example, [8] is expensive since sparse vectors need to be found for every test pixel and for every pixel class. This is a costly process as the sparsity norm is a complicated procedure in optimization computation.

Assigning a probability score to each pixel in the HSI is proposed in some previous studies. For instance, in [9], authors utilize the output of an SVM classifier to generate multiple probability maps for each image region. Then, they apply edge preserving spatial filtering to each map using a guidance image as a reference. And finally they choose the maximum probability class for each pixel. A similar study [10] generates same probability maps using SVM, then applies the random walker optimization technique to each map to enhance probability scores. Our approach differs from these studies in two ways. First, our solution employs multiple techniques to generate region scores rather than relying on only SVM. This is beneficial as later we combine these scores to improve accuracy. So we show that employing multiple classifiers yield more robust decision capability. Second, unlike other studies, we employ an additional ensemble learner layer to merge the region scores into a single probability map.

3 Proposed Method

The proposed two-layer classification system is composed of three modules organized in a two-layer architecture. The first layers contains a *Contour Learner* module and a *Reference Learner* module, while an *Ensemble Learner* module is placed in the second layer.

3.1 Layer 1: Contour Learner Module

The main idea behind Contour Learner is to create individual models for each region of the image. We refer to each region model as a *contour*. Each contour (i.e. model) is then used to estimate the probability that a given test sample belongs to the region represented by a contour.

Learning Contours. Contours will be created for all regions within the training samples. To generate a contour for region r, we extract a set of all training pixels P_r that belong to a region r. Each pixel p in P_r is essentially a vector of spectral bands $B_p = \{b_1, b_2, \ldots, b_d\}$ where d is the total number of spectral bands in the image. Then we create *heatmaps* for all spectral bands k, $k = 1, 2, \ldots, d$, using all the values from the kth band of pixels in P_r. The resulted contour is simply represented by a $s \times d$ matrix C_r, where s is the maximum possible value for a spectral band, d is the total number of spectral bands in an image, and $C_r[i, j]$ shows the percentage of training pixels that share band value i for the jth band. Essentially, the contour C_r highlights the areas of concentration of training pixels' values along the spectral bands for a region. One could imagine that the heatmap creates a contour (or a number contours) over the spectral bands of a region by summarizing the training pixels' values.

Contours Examples and Intuition. Figure 2 shows contours generated for three regions of our dataset (presented in Sect. 4.1). Comparing the contours to each other, one can see slight differences in the concentration of pixel band values among these regions.

Fig. 2. Examples of contours generated from three regions

Computation of Scores. Given a test pixel p, the set of all learned contours from all regions is utilized to produce a list of region scores for p. Each score in the list represents a confidence level that a given pixel belongs to a region represented by a contour. The following formula shows how we compute the score of a pixel p for region r, which is represented by contour C_r.

$$score_{p,r} = \frac{1}{d}\sum_{k=1}^{d} C_r[B_p[k], k]$$

The intuition behind the region score is that the formula estimates how closely a given pixel matches the contours of a group of pixels that belong to a region in the image (i.e. the areas of concentration of region's band values). As a result, the output of this module is a list of region scores for each test pixel. The list of region scores is then attached to the pixel.

3.2 Layer 1: Reference Learner Module

While Contour Learner takes into consideration the similarity of a pixel's B_p, spectral bands' values, to a model to compute the region score, Reference Learner considers the similarity of the *shape* of B_p to approximate region scores. The idea here is to account for variations in a pixel's B_p due to natural and environmental effects. To illustrate, consider a sample pixel that belongs to "grass" region. Ideally, this pixel's B_p should match the contours of region "grass", learned during the training. However, in a real-world scenario, the values of pixel's B_p might shift if the pixel is in a shaded area in the image, for example. As a result, the contour might produce a low probability score indicating that the pixel is not a good match for "grass", even though the shape of the pixel B_p is identical to that of the contour. This problem could potentially occur if the training data is not representative of the entire region and could possibly lead to low classification accuracy.

Basically, Reference Learner aims to produce the region scores for pixels from a different angle. Essentially, Contour Learner and Reference Learner produces similar outputs but from two different viewpoints by harnessing distinct features that are not mutually susceptible to the same shortcomings. In fact, this idea forms the basis of the Ensemble Learner, which we will describe in Sect. 3.3.

Learning References. A reference R_r is a vector that summarizes B_p vectors of all pixels belonging to a region r. During classification, the *shape* of the reference of region r is compared to that of a sample test pixel. The output is a score that indicates how similar these two vectors are in terms of their *trends*. These trends could simply be thought of as the locations and sizes of peeks and valleys of the vectors.

To generate R_r, a reference for region r, we will consider all B_p's of training pixels from region r. One simple approach would be to compute an item-wise average of spectral band values, i.e. $R_r[i] = \frac{1}{n}\sum_{j=1}^{n} B_j[i]$, where $R_r[i]$ indicates the ith band value of the reference for region r and $B_j[i]$ is the ith band value of the jth training pixel, $j = 1, 2, \ldots, n$.

One could note that average is sensitive to outliers. To account for this we compute the item-wise median of band values instead, which is less influenced by outliers. The references will be generated for all regions during training.

Reference Examples and Intuition. Figure 3 shows references generated for three regions of our dataset (Sect. 4.1). While the references look very similar, there are subtle differences in terms of their shapes.

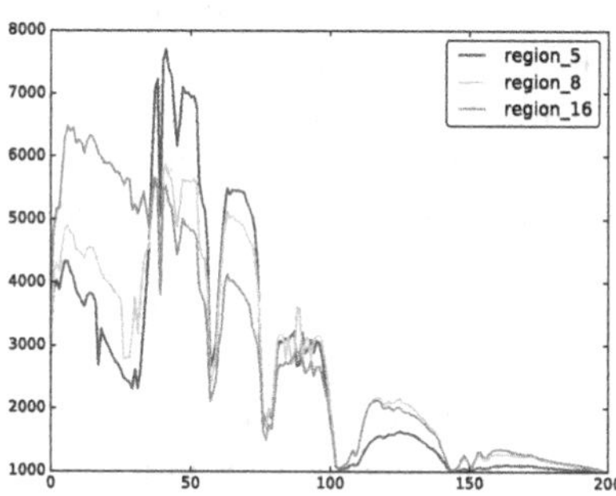

Fig. 3. Examples of three references

Computation of Reference Scores. Having discussed how a reference is computed for a region, we now turn our attention into how we compute region score values for a test pixel in a sample test image.

Given a pixel's B_p and a region r reference R_r, the goal is to compare the shape of the two vectors. It is important to emphasize again that unlike Contour Learner, Reference Learner is not concerned about whether the values of the vectors match or not. Rather it aims to *assess the similarity of the trends of the two vectors*. For example, consider two identical curves in a 2D space, where one of them is shifted up by a constant amount. In this case the Reference Learner should produce a high score for the similarity of the two curves, even though their point-wise distance might be significant.

There are some statistical methods that could achieve this goal. Specifically, we employ Pearson's Correlation Coefficient [11], which is a method that compares two curves (B_p of test pixel and R_r) by measuring their linear correlation and is insensitive to bias and scaling to an extent. The Pearson's Correlation Coefficient has a value between -1 and 1, where 1 indicates complete positive linear correlation, 0 means no linear correlation, and -1 means complete negative correlation.

Similar to Contour Learner, given a test pixel p, the set of all learned references from all regions is utilized to produce a list of scores for p. The output, therefore, would be a list of region scores that is attached to the pixel.

3.3 Layer 2: Ensemble Learner Module

Any classification model should deal with error due to either "bias" or "variance". The error caused by a model that suffers from low bias is because of under-fitting. In this case, the model did not learn the underlying hypothesis in order to produce acceptable predictions. On the other hand, a model suffering from high variance is over-fitted or simply memorized the training data to an extent that it cannot generalize beyond the training samples. The main challenge in training a classification model, therefore, is to find a good trade-off between bias and variance. The concept of ensemble learning is to combine multiple classifiers to combat the low bias and/or high variance of the underlying combined learners, which has been studied extensively in the past [12–16].

In Sects. 3.1 and 3.2, we discussed two different methods to learn distinct models to classify the regions within an image. While these models share the same goal, they aim to recognize the regions employing distinct techniques. As a result, they will learn different models and, consequently, will make different mistakes. We take advantage of this phenomenon and try to combine them in our second layer which contains an *Ensemble Learner* module. The goal of this module is to *learn* from the decisions of the underlying combined learners to enhance the accuracy of the model. As we will demonstrate in evaluation of the two modules in the first layer (Sects. 4.2 and 4.3), the two modules, while having an acceptable accuracy, are not perfect by any means. However, that would be fine because where the Contour Learner is right the Reference Learner might be wrong or vice versa. Therefore, by combining these two models and learning from their behaviors and mistakes, we can improve our overall performance.

Training the Ensemble Learner. In this paper, we employ the Stacking [16] approach as the ensemble learner module. Considering our modules in the previous layer, an ensemble learner is trained based on the two sets of region score lists, from Contour Learner and Reference Learner, as feature vectors. In another words, during training, this learner receives a list of region scores from the modules in previous layer and treats the scores as new features, referred to as *meta features*. Then it fits a model on top of the meta features according to the correct region labels.

Computation of Probability Scores. The output of the Ensemble Learner module would be a list of probability scores for each pixel. Given a test pixel, each region score signifies the confidence of the Ensemble Learner module about that pixel's association to a region in the image. However, as we discussed, the new region scores are expected to be more accurate. Also, it is important to note that in order to obtain probability scores for all regions, a multi-class classifier should be used in this layer. Additionally, note that at this point the region probability scores for a given pixel p would sum to one ($\sum_{r=1}^{n} score_{p,r} = 1$, where n is the total number of regions) due to the usage of a multi-class classifier.

4 Evaluation

In this section, first, we present the dataset that is used to evaluate our system. Next, we provide the individual evaluation results for the two learners in the first layer of our framework, namely Contour Learner and Reference Learner. Then we present the evaluation of the framework as a whole and show the effectiveness of our system. Finally, we compare the classification results of some other well-known methods that are commonly used in previous work to our framework.

4.1 Dataset

The AVIRIS data (or image) comprises of a $145 \times 145 \times 220$ matrix that corresponds to 220 different bands of images having size of 145×145. We transform the matrix data to a vector form as a 21025×220 matrix. This representation indicates that there are 21025 samples with 220 different features. Figures 4 and 5 and Table 1 provide some details on the AVIRIS data.

Fig. 4. RGB view of AVIRIS image

Fig. 5. Ground-truth of AVIRIS image

4.2 Evaluation Results of Contour Learner

To evaluate the Contour Learner we proceed as follows. We consider all the pixels in the dataset and their labels, i.e. their regions, and perform a standard 10-fold cross-validation (CV). To be specific, given a dataset of labeled pixels of size n and in round k of the CV, the set of pixels in the dataset form a training set TR_k, containing 9/10 of the n pixels, and a test set TS_k, containing 1/10 of the n pixels. Then TR_k is used to learn the contours (as it was described in Sect. 3.1), and TS_k is used for evaluation. The same procedure is repeated for all the 10 folds. It is important to note that the test pixels are never used in generating the contours in any of the CV folds.

A test pixel p will be labeled as a specific region r, if $score_{p,r}$ (Sect. 3.1) is highest compared to all the other region scores. In essence, a test pixel's region

Table 1. Class names and number of samples

Class number	Class	Samples
1	Alfalfa	54
2	Corn-notill	1434
3	Corn-mintill	834
4	Corn	234
5	Grass-pasture	497
6	Grass-trees	747
7	Grass-pasture-mowed	26
8	Hay-windrowed	489
9	Oats	20
10	Soybean-notill	968
11	Soybean-mintill	2468
12	Soybean-clean	614
13	Wheat	212
14	Woods	1294
15	Buildings-Grass-Trees-Drives	380
16	Stone-Steel-Towers	95
Total		10366

classification is determined by a region with a contour that more closely matches the test pixel. Table 2 presents the results. In the table and for each region, we show the following standard evaluation metrics in machine learning: (i) true positive rate (TP), also known as sensitivity or recall, and (ii) false positive rate (FP). Please note that other metrics such as precision (true negative rate), accuracy, and etc. are easily computable using TP and FP. So for clarity purposes we only report TP and FP.

As it can be seen in Table 2, while the results are not bad per se, they are not very accurate. For example, Contour Learner works well for some regions, such as regions 1, 13 and 14, but does very poorly on some other regions, such as regions 12 and 15. This was expected as the Contour Learner is a *weak learner* and by itself as a standalone classifier cannot fully separate and distinguish the classification of test pixels. However, in our framework, this learner's results will be combined with Reference Learner and fed to the Ensemble Learner module to produce the final classification. We will show in Sect. 4.4 that our framework produces very accurate results and demonstrate its generalization capabilities. Note that due to low cost of computations, Contour Learner is in fact a very efficient classifier which does not require computationally expensive learning algorithms. Learning the contours, essentially, entails going over the training pixels once.

Table 2. Evaluation results of contour learner

Region	TP [%]	FP [%]
1	91.3	2.5
2	24.2	4.0
3	51.9	7.0
4	12.6	0.4
5	60.6	0.5
6	40.0	1.0
7	85.7	1.9
8	50.2	1.4
9	80.0	3.2
10	47.4	8.5
11	58.0	14.7
12	4.7	0.5
13	92.1	3.7
14	94.5	1.4
15	13.4	0.1
16	65.5	4.6

Table 3. Evaluation result of reference learner

Region	TP [%]	FP [%]
1	89.1	0.9
2	35.6	6.9
3	22.7	1.6
4	46.8	3.1
5	54.6	4.6
6	50.2	0.9
7	92.8	2.1
8	75.3	0.9
9	70.0	2.6
10	56.7	14.4
11	43.0	9.8
12	26.4	5.9
13	93.1	0.8
14	67.7	1.4
15	25.9	2.4
16	89.2	0.0

4.3 Evaluation Results of Reference Learner

Similar to the previous section, here we present the evaluation results of Reference Learner as a standalone classifier. To this end we perform standard 10-fold CV and provide region classifications for test pixels. A test pixel p's region is determined by a reference with a shape more closely resembling that of p's (see Sect. 3.2). Table 3 shows the region classification results. Again, the results indicate that Reference Learner is a weak learner and alone cannot produce very accurate classifications. However, it is worth mentioning that for some regions, such as region 8 or 16 Reference Learner does a better job than Contour Learner, while for some other regions, such as region 14, Contour Learner produces better classifications. The differences between the decision surfaces of these two learners allows our framework to combine these results in the second layer to produce more accurate results as it was discussed in Sect. 3.3.

Similar to Contour Learner, the Reference Learner is also quite efficient as it simply computes an item-wise median of the training samples to generate the references.

4.4 Framework Evaluation

In Sects. 4.2 and 4.3 we showed that the two learners in the first layer of our framework cannot produce accurate results as standalone classifiers. In this section, we demonstrate how our region classification framework produces very accurate results by combining the outputs of Contour Learner and Reference Learner using the Stacking classification algorithm [16]. In essence, the Ensemble Learner module takes advantage of the fact that Contour Learner and Reference Learner produce different decision boundaries and as a result make different errors, and by combining their output, it significantly improves the region classifications (detailed are discussed in Sect. 3.3).

To evaluate our framework, we consider the output of the Ensemble Learner module for all the pixels in the dataset. To do this we follow the standard training and testing evaluation methods for the Stacking classification approach and perform 10-fold CV to report the results. The details of evaluation techniques for Stacking are extensively discussed in the literature (for example, see the algorithm in [17]). Basically, for a test pixel p that was never seen by either of the base learners (Contour and Reference Learners) during training, region scores would be computed according the base learners. The lists of scores generated by each learner are then used as *meta features* for the Ensemble Learner. A full dataset of labeled meta features would be generated when the scores are computed for all test pixels by Contour Learner and Reference Learner. At this point, we proceed to evaluate the Ensemble Learner module using 10-fold CV.

Note also that we can use any classifier, such as Random Forests, plain Decision trees, SVM, and etc., as our Ensemble Learner. We conducted experiments to choose the best classifier based on its performance. Our pilot experiment results indicated that Random Forest performed the best among many other classifiers. Therefore, due to space constraints, we only report the results obtained using the Random Forest classifier as our Ensemble Learner.

The results are presented in Table 4. For each region in the dataset, Table 4 shows various trade-offs between TP and FP. For example, the column "TP [%] $FP = 1\%$" shows the TP results for all regions when the classifier is tuned to produce $FP = 1\%$. We also provide the area under ROC curves in Table 4 (shown in the "AUC" column). Furthermore, it is important for the classifier to produce high TP rates while keeping the FP rates low. As a result, we also report the normalized *partial* area under the ROC curve for $0\% \leqslant FP \leqslant 1\%$ in "PAUC" column.

As it can be observed from Table 4, our framework produces very accurate results for the regions in the dataset while producing very low FPs. Compare, for example, the results that were obtained using Contour Learner or Reference Learner as standalone classifiers to the overall accuracy of our framework. This indicates that simple and computationally efficient weak learners from the first layer could be effectively combined in the second layer to produce outstanding results. For many regions, such as region 6, 8, 9, 13, 14, and 16 our framework achieves TP rates of at least 95% while keeping the FP rate as low as 1%. For some regions that Contour Learner and Reference Learner performed very poorly as standalone classifiers, such as region 15, we can see that by combining their results we can achieve very good results (95% TP and 3% FP). Some regions such as regions 3 and 11 are much more difficult to classify compared to other

Table 4. Evaluation results for our framework (results of ensemble learner)

Region	TP [%] FP=1%	TP [%] FP=2%	TP [%] FP=3%	TP [%] FP=4%	TP [%] FP=5%	AUC	PAUC
1	93.5	93.5	93.5	97.8	97.8	0.97	0.87
2	48.6	58.8	64.7	69.8	74.7	0.94	0.37
3	54.5	64.3	71.1	75.4	79.2	0.96	0.41
4	73.0	83.1	87.8	91.1	92.8	0.98	0.53
5	93.8	96.3	96.5	96.7	97.1	0.98	0.88
6	95.2	97.8	99.0	99.7	99.9	1.00	0.84
7	92.9	92.9	96.4	96.4	96.4	0.96	0.91
8	99.8	100.0	100.0	100.0	100.0	1.00	0.93
9	95.0	95.0	100.0	100.0	100.0	1.00	0.81
10	60.4	73.3	77.9	80.8	84.0	0.97	0.46
11	55.4	63.9	70.7	75.6	79.5	0.96	0.42
12	45.0	58.5	69.3	73.7	76.6	0.95	0.29
13	97.6	99.0	99.0	99.0	99.0	0.99	0.90
14	95.7	97.9	98.7	99.3	99.4	1.00	0.88
15	80.3	92.0	95.3	97.2	97.4	0.99	0.54
16	98.9	100.0	100.0	100.0	100.0	1.00	0.99

regions. However, our framework still produces quite acceptable results for such difficult regions.

Please note that we also experimented with different feature selection algorithms for the two layers in our framework. However, in our test, they did not improve the results.

4.5 Comparison with Traditional Classifiers

As we discussed in Sect. 2, many previous works have used traditional classifiers, such as SVM, for region classification in hypersepctral images. Therefore, in this section we compare our framework with Decision Trees, Naive Bayes, KNN, and SVM classifiers. Table 5 reports the results where for various FP values, we report the average of TP rates, AUCs, and PAUCs for all regions. As it can be observed from the table, our framework's accuracy is higher compared to all the aforementioned classifiers. For example, the area under the ROC curve for $0\% \leqslant FP \leqslant 1\%$ (column "PAUC") is 0.69 in our framework, while SVM's PAUC is equal to only 0.4. This value is much lower for other classifiers. Finally, in Fig. 6, we show the final classification results of our framework and compare it to the other classifiers of Table 4.

Table 5. Comparison of our framework with other classifiers based on average TP, AUC, and PAUC for all regions

Classifier	TP [%] FP=1%	TP [%] FP=2%	TP [%] FP=3%	TP [%] FP=4%	TP [%] FP=5%	AUC	PAUC
Our framework	79.97	85.39	88.74	90.79	92.11	0.98	0.69
Naive Bayes	29.28	39.56	39.93	39.93	47.50	0.56	0.16
Decision trees	49.36	64.47	66.46	70.61	74.80	0.77	0.35
KNN	45.13	54.40	62.11	66.64	70.67	0.74	0.29
SVM	55.65	75.68	79.07	82.11	82.55	0.96	0.40

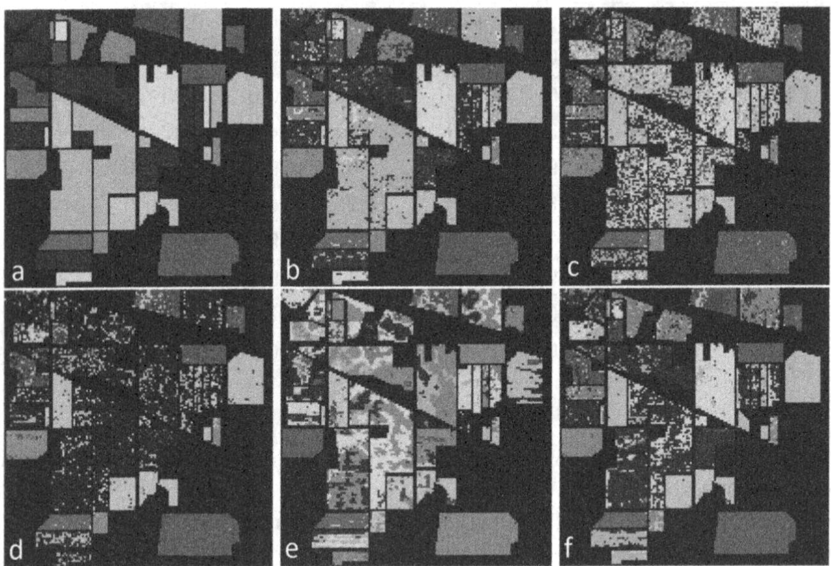

Fig. 6. Final classification results of each classifier. (a) Ground truth image, (b) Proposed Method, (c) Decision Trees, (d) KNN, (e) Naive Bayes, (f) SVM

5 Conclusions and Future Works

This paper presents an efficient two-layer classification method for hyperspectral images. Unlike traditional classifiers that are using original dataset features, our approach classifies samples using newly created probability features. In the first layer, two new classifiers were introduced to create region scores for each sample. Then in the second layer, an ensemble classifier was utilized to classify samples with new features. We performed experiments on well-known Indian Pines dataset and compared our results with four other classifiers. Results show that the proposed method outperforms others significantly particularly on some challenging regions.

In the future, we plan to improve our results. Since the number of regions in the dataset is a factor for our method (they contribute to the number of meta features), our first plan is to experiment on different datasets to see how the performance changes by different number of regions. Our second plan is to improve robustness of the first layer by including more classifiers to the layer. In this way, a more reliable probability feature set may be obtained. Our third plan is to add a third layer to the system to exploit spatial features of the dataset. We believe this will improve the final classification result. Finally, we plan to investigate the contribution of various feature selection techniques. We tested a few techniques for this study, but none of them contributed to the classification performance.

Acknowledgement. This research was supported by the Auburn University at Montgomery's 2018 Faculty Research Grant Support Award.

References

1. Binol, H., Bilgin, G., Dinc, S., Bal, A.: Kernel Fukunaga-Koontz transform subspaces for classification of hyperspectral images with small sample sizes. IEEE Geosci. Remote Sens. Lett. **12**(6), 1287–1291 (2015)
2. Dinc, S., Bal, A.: A statistical approach for multiclass target detection. Procedia Comput. Sci. **6**, 225–230 (2011). Complex Adaptive Sysytems
3. Mercier, G., Lennon, M.: Support vector machines for hyperspectral image classification with spectral-based kernels. In: IEEE International Geoscience and Remote Sensing Symposium, vol. 1, pp. 288–290. IEEE (2003)
4. Doshi-Velez, F., Kim, B.: Towards a rigorous science of interpretable machine learning (2017)
5. Chen, Y., Lin, Z., Zhao, X., Wang, G., Gu, Y.: Deep learning-based classification of hyperspectral data. IEEE J. Sel. Top. Appl. Earth Obs. Remote Sens. **7**(6), 2094–2107 (2014)
6. Chen, Y., Zhao, X., Jia, X.: Spectral-spatial classification of hyperspectral data based on deep belief network. IEEE J. Sel. Top. Appl. Earth Obs. Remote Sens. **8**(6), 2381–2392 (2015)
7. Liu, P., Zhang, H., Eom, K.B.: Active deep learning for classification of hyperspectral images. IEEE J. Sel. Top. Appl. Earth Obs. Remote Sens. **10**(2), 712–724 (2017)
8. Chen, Y., Nasrabadi, N.M., Tran, T.D.: Hyperspectral image classification using dictionary-based sparse representation. IEEE Trans. Geosci. Remote Sens. **49**(10), 3973–3985 (2011)
9. Kang, X., Li, S., Benediktsson, J.A.: Spectral-spatial hyperspectral image classification with edge-preserving filtering. IEEE Trans. Geosci. Remote Sens. **52**(5), 2666–2677 (2014)
10. Kang, X., Li, S., Li, M., Benediktsson, J.A.: Extended random walkers for hyperspectral image classification. In: 2014 IEEE International Geoscience and Remote Sensing Symposium (IGARSS), pp. 1520–1523. IEEE (2014)
11. Pearson, K.: Note on regression and inheritance in the case of two parents. Proc. R. Soc. Lond. **58**, 240–242 (1895)

12. Efron, B.: Bootstrap methods: another look at the jackknife. Ann. Stat. **7**, 1–26 (1979)
13. Polikar, R.: Bootstrap-inspired techniques in computational intelligence. IEEE Signal Process. Mag. **24**, 59–72 (2007). Credit, O. ISSN 1053-5888
14. Jacobs, R.A., Jordan, M.I., Nowlan, S.J., Hinton, G.E.: Adaptive mixtures of local experts. Neural Comput. **3**(1), 79–87 (1991)
15. Kuncheva, L.I.: Combining Pattern Classifiers: Methods and Algorithms. Wiley, Hoboken (2004)
16. Wolpert, D.H.: Stacked generalization. Neural Netw. **5**(2), 241–259 (1992)
17. Tang, J., Alelyani, S., Liu, H.: Data Classification: Algorithms and Applications. Data Mining and Knowledge Discovery Series, pp. 37–64. CRC Press, Boca Raton (2014)

Predicting Social Unrest Using GDELT

Divyanshi Galla[1(✉)] and James Burke[2]

[1] PWC, BG House, Hiranandani Business Park, Powai, Mumbai, India
divyanshi.galla@pwc.com

[2] PWC, 600 13th Street Office, Washington, DC, USA
james.a.burke@pwc.com

Abstract. Social unrest is a negative consequence of certain events and social factors that cause widespread dissatisfaction in society. We wanted to use the power of machine learning (Random Forests, Boosting, and Neural Networks) to try to explain and predict when huge social unrest events (Huge social unrest events are major social unrest events as recognized by Wikipedia page 'List of incidents of civil unrest in the United States') might unfold. We examined and found that the volume of news articles published with a negative sentiment grew after one such event - the death of Sandra Bland - and in other similar incidents where major civil unrest followed. We used news articles captured from Google's GDELT (Global Database of Events, Language, and Tone) table at various timestamps as a medium to study the factors and events in society that lead to large scale unrest at both State and County levels in the United States of America. In being able to identify and predict social unrest at the county level, programs/applications can be deployed to counteract its adverse effects. This paper attempts to address this task of identifying, understanding, and predicting when social unrest might occur.

Keywords: Social unrest · News media · GDELT · Themes · Events
Random forest · Ada boost with random forest · LSTM · County level USA

1 Introduction

Social unrest can be extremely detrimental to society, especially when it escalates into rioting and violent demonstrations. Local communities bear the brunt of the impact, and it can have a lasting effect on their socioeconomic development.

A great deal of research has gone into detecting social unrest using social media data; most of the work centering around Twitter [2], Tumblr [4], Facebook and some research was carried using Google's GDELT event table [5]. Our work focuses on the exploration and analysis of news media data to detect social unrest. News media data contains a wealth of rich insights that can be used to study various social factors and events that take place in our society. By isolating underlying trends of factors that lead to social unrest, we could utilize this power to warn the public of the impending danger of adverse social events. Analysis of this data allows us to identify certain themes and events that are closely associated with a given region. It would be truly remarkable if we could develop a system that will be able to detect when and where a social unrest event is brewing, and alert the appropriate authorities to take necessary action; either in

P. Perner (Ed.): MLDM 2018, LNAI 10935, pp. 103–116, 2018.
https://doi.org/10.1007/978-3-319-96133-0_8

preventing the event all together, or providing adequate safety measures for the people. By developing machine learning models that can adequately predict social unrest using news media data, and by reducing manual effort associated with tracking and monitoring such events, we can move a step closer to achieving this system.

Previous research [5] used GDELT's event table to build a Hidden Markov Model (HMMs) based framework to predict indicators associated with country instability. Identifying these indicators at a more granular geographic level could enable focused and effective action from the government, whether in the form of targeted outreach programs, distribution of additional law enforcement resources, or development of early warning systems. Based on these assumptions, we decided to focus our research on huge social unrest events that occur not only at the state level, but also at the county level (within the contiguous US).

We hypothesized that huge riots that took place in Baltimore in the year 2015 and in Milwaukee and Charlotte in the year 2016 had an ethnic discrimination factor and that the protests associated with the Dakota Access Pipeline had environmental related factors associated with a deteriorating sentiment in news articles indicative of building unrest levels in a region. We also hypothesized that a region that is subject to higher occurrences of events associated with *escalated threatening, coercing, assaults, protests* and other forms of *disapproving* behavior might suffer from a buildup of higher levels of social unrest.

We chose one such event, Sandra Bland's death, to test our hypotheses, using GDELT's Global Knowledge Graph (GKG) and Event tables that monitor print, broadcast, and web news media across world. Sandra's death in prison caused major civil unrest after FBI investigation revealed that the required policies weren't followed. After her death on 14th July, 2015 discussion grew with worsening sentiment till impact subsided. A huge unrest event in the form of a protest occurred on 29th July 2015. This confirmed our hypothesis that news reflects society and can be used to detect building unrest. Figure 1 depicts how sentiment of articles discussing the incident vary with time in the case of Sandra Bland's death. The size of the bubble represents mentions of event in news; the Y axis contains the sentiment value and the X axis is timeline.

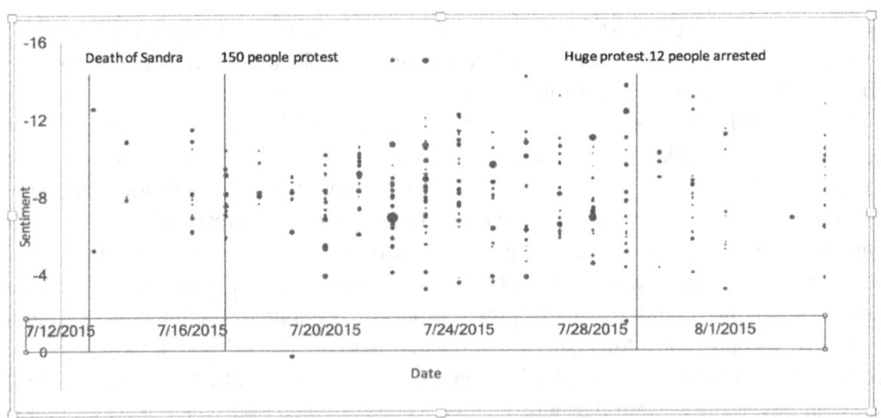

Fig. 1. Figure indicates variation of average tone with respect to time.

2 Related Work

This paper is focused on using news media to predict social unrest at county and state levels for USA. There is a large amount of work done in predicting social unrest using social media. [1] uses social messaging to predict social unrest. [2] works towards automated unrest prediction and filtering the vast volume of tweets to identify tweets relevant to unrest. [3] focuses on detecting emerging civil unrest events by analyzing massive micro-blogging streams of Tumblr. [4] presents an event forecasting model using activity cascades in Twitter to predict the occurrence of protests in three countries of Latin America: Brazil, Mexico, and Venezuela. [5] is using information based on news media from temporal burst patterns in GDELT event streams to uncover the underlying event development mechanics in five countries of South East Asia. Our research aims at measuring social factors prevalent in society using the GKG table of GDELT along with event types gathered from GDELT's event table that might lead to huge social unrest. Our main contribution is to take into account a thoroughgoing picture of society to predict unrest level at the county level of location compared to previous work that looked at larger geographical regions. We are predicting unrest levels in 2918 counties across USA which can be an efficient tool to combat the ill effects of protests.

3 Data

3.1 About GDELT

GDELT is the global database of events, location and tone that is maintained by Google. It contains structured data that is mined from broadcast, print and web news sources in more than 100 languages since 1979. It connects people, organizations, locations, themes, and emotions associated with events happening across the world. It describes societal behavior through eye of the media, making it an ideal data source for measuring social factors and for testing our hypotheses. We have used two GDELT tables, the Global Knowledge Graph (GKG) and the events table.

GDELT's GKG captures what's happening around the world, what its context is, who's involved, and how the world is feeling about it; every single day. Of the 27 fields that it contains, we used the following four: *Location, Date, Themes* with associated *Sentiment*. *Location* field contains all locations found in news article and the average tone of the document is *sentiment* associated with it. The *Date* format was - YYYYMMDD. We divided 472 themes of interest into the following four categories - Crime, Economy, Environment and Health. This was done to observe which category was influencing escalation of unrest. Following are some illustrative examples of themes - examples of Crime themes include *armed conflict, arrest, crimes against humanity*; Economy themes include *democracy, constitution* and *alliance*; Environment themes include *deforestation* and *climate change* and Health themes include *disease*.

The GDELT events table captures information on events in CAMEO format (conflict and mediation event observations) capturing two actors and action performed by Actor1 on Actor2. The events table contains 58 fields. We used *SQLDATE, EventRootCode*, *NumMentions*, *AvgTone* and *Location* fields. SQLDATE is the date in YYYYMMDD format. *EventRootCode* is the root-level category the event falls under. *NumMentions* is total number of mentions of an event across all source documents which helps to measure importance of an event. *AvgTone* is the average tone of all documents containing one or more mentions of an event (the score ranges from −100 (extremely negative) to +100 (extremely positive)). *Location* is the city, state, country description of all locations mentioned in the article. The cameo event codes that were used include *Disapprove, Threaten, Protest, Coerce* and *Assault* as we hypothesized that these event types are more likely to lead to huge social unrest.

3.2 Social Unrest Events Source

Wikipedia was used to identify major social unrest events in the US from 2015–2017 These events are referred to as 'huge social unrest events' throughout the paper. These events were verified using various news sources.

3.3 Creation of Mapping

A mapping here is a time and location representation of USA. The mapping points have two attributes- location and time. Different locations at various points of time have distinct behavior. Hence the sample points were taken every day to train and test the model. A sample window of one month was considered and a performance window of one month was used. Observation windows of values one, three and six months were also used (Fig. 2). 2918 counties on all dates in period 2015–2017 were taken as sample points. The data contained 1,049,840 points. If an unrest event occurred within the performance window of the data point, it has been marked as an event point. Other data points are marked as nonevent points. Unrest events were identified from the list of major unrest events in the USA in the period from 2015 to 2017 extracted with location and time attributes from Wikipedia. Final Data on which model was trained contained 1265 event points. Nonevent points were under sampled to 8735 randomly to create a balanced dataset.

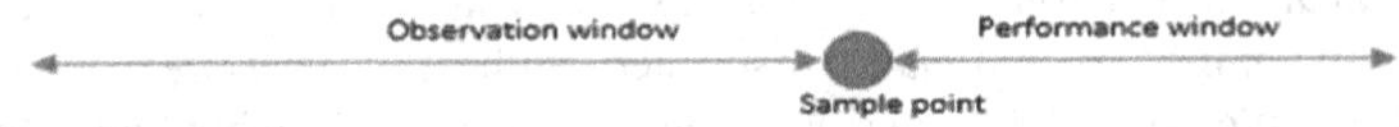

Fig. 2. Positioning of windows with respect to sample point

3.4 Extraction of Data

As GDELT adds news articles every fifteen minutes, each article is concisely represented as a row in the GDELT table. A number of processing steps were carried out to represent the GDELT data with location and time attributes of required format. The GKG table contained 7.4 TB of data and events table had 127 GB of data. We used Google BigQuery to subset this data to obtain data for the USA and carried out cleaning like activities to collect entries of themes and event codes of interest. Later we stored Theme and Event tables in Google cloud storage and performed the cleaning steps outlined below using PySpark. GDELT contained all themes discussed in an article as an entry in a single row. We separated these themes into separate entries. The themes that we targeted were displayed in a number of ways, they were either present isolated from other themes, or in conjunction with other themes. For instance, *agriculture* is one of our themes of interest; there are entries that call out *agriculture* specifically, then there are entries where *agriculture* is combined with another them such as *agriculture and food*. We needed to collect all instances where agriculture is present, so we created a mapping with all possible combinations of each theme of interest to make sure we capture all possibilities and not just the exact matches of theme of interest.

The *location* field within the GDELT tables has a format of city/landmark, state and country level. City to county mapping was used to extract data at the county level. Improving this mapping will also enable us to collect more accurate information at County level. Themes with their respective sentiments were cast at county and state levels of location. For every data point of created mapping, we captured the following features for time periods of one, three, and six months prior to the sample date: Total number of times the themes of interest occurred (GKG table), the overall sentiment associated with themes (GKG table), the total number of times the cameo event codes of interest were mentioned (event table) and the sum of Tone associated with them (Event table). A Total of 1446 features were extracted from the GDELT tables.

4 Methods

4.1 Framework

Huge social unrest events can be preceded with the occurrence of a number of different categories of events and a wide range of the social factors/themes. Using the social unrest events that were captured from Wikipedia (and cross referenced with various news sources), we created a mapping as described previously in Sect. 3.3 for all the unrest event points and non-unrest event points for the 2015–2017 period. The location and time where major social unrest occurred was marked event point, while remaining data points were marked as non-event points. Using this mapping, we collected data from cleaned tables of GDELT.

We trained machine learning models on the data using the 1446 features that were captured for the period 2015–2016 as training data. In order to capture how the variation within the features impacted social unrest over time, we used an LSTM model with the 33 features (out of the 1446) that were identified as most significant factors to explain social unrest (based on the results of the Random Forest analysis). We then tested the models on the 2017 data for out of date validation. Results and metrics have been analyzed and discussed in the sections that follow.

In addition, we also predicted locations that would be subject to higher levels of social unrest for the next month as of 12th November 2017. To achieve this, we collected data from January 2015 to 11th November 2017 and trained machine learning models to predict the levels of social unrest. We then identified counties that had the highest unrest levels and after completion of the performance period, we validated the results of our prediction with the events that actually occurred in the county during that one month performance period by reading news articles. Figure 3 summarizes framework of our research as a diagram.

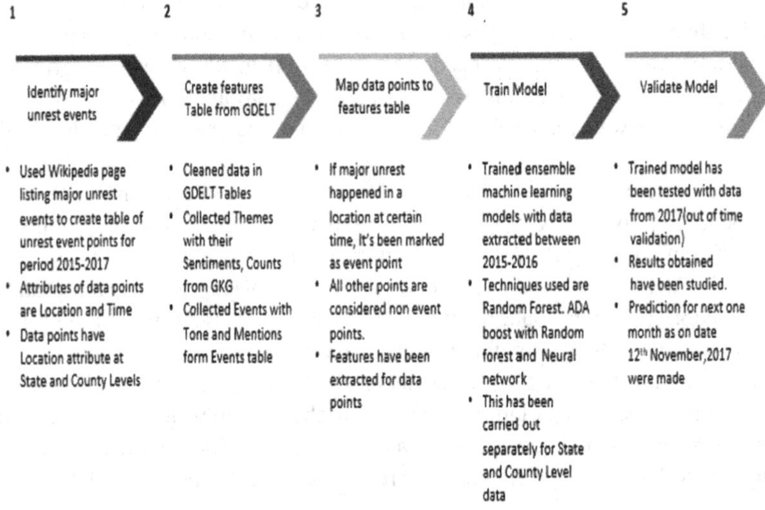

Fig. 3. Diagram explaining framework of analysis

4.2 Models Summary

Random Forest. Random forests are a form of ensemble learning techniques that can be used for classification and regression problems. They operate by constructing a multitude of decision trees at training time and output a predicted class based on a voting mechanism that takes into consideration the most commonly occurring class at each split in the decision tree process. Random forests perform implicit feature selection and provide feature importance which will help us understand the features that have the most significant impact in building of huge unrest event. Having a large data set with many feature variables, we applied this technique to build our model.

ADA Boost with Random Forest. ADA boost is a boosting algorithm developed for classification problems that is less prone to the problem of overfitting. Hence, we explored this technique.

Neural Network. Artificial neural networks are computing systems inspired by the brain's biological neural networks. Neural networks perform particularly well at finding nonlinear relationships in data. We used neural networks to analyze if such patterns exist in our data.

Long Short-Term Memory Networks. Long short-term memory networks are a special kind of Neural Network known as Recurrent Neural Networks (RNNs). They are capable of learning long-term dependencies within a data sequence. In order to capture time variation of the features and their impact on social unrest events, we ran a preliminary LSTM model using 33 most significant features obtained from random forest feature importance. This can be further developed in the future.

5 Experimental Results

5.1 Feature Description and Importance

1446 features were extracted from the GKG and Event tables. Feature importance obtained from the random forest model was used to find top features list. Some of these are *armed conflict, arrest, conflic*t and *violence, corruption* in the Crime category and *alliance, constitution, democracy* in Economy category. To visualize how features vary over time, the Dakota access pipeline unrest event was analyzed at the state level, it is discussed in following paragraphs.

Consider the value of a feature from sample date t to one month prior $t-1$ as F_{t-1}, value from $t-2$ to $t-3$ as F_{t-3} and value from $t-5$ to $t-6$ as F_{t-6}. Let the percentage change in feature from F_{t-3} to F_{t-1} be represented as P2 and percentage change in feature F_{t-6} to F_{t-3} be represented as P1

$$P_2 = \frac{F_{t-1} - F_{t-3}}{F_{t-3}} \times 100 \tag{1}$$

$$P_1 = \frac{F_{t-3} - F_{t-6}}{F_{t-6}} \times 100 \tag{2}$$

Figure 4 is a comparison of how much features vary in case of event points and nonevent points with time, which is a comparison of the value change from P1 to P2. To illustrate this change the Crime armed conflict feature has been used and it can be seen that the feature varies a lot in case of event points compared to nonevent points. This is the case with many top features. Figure 5 depicts how different important features vary with respect to a specific unrest event. The percentage change in the values of features is higher as a major unrest event approaches (P2 > P1). It can also be observed that the feature environment oil count has a high variation from −69.2% to 19.75% indicating news articles have increased coverage of articles talking about oil before a huge unrest event occurred (The Dakota access pipeline unrest incident).

Figure 6 represents the same phenomenon in absolute values where it can be observed that as a social unrest event is building up, the amount of discussion around factors increases (A2 > A1).

$$A_2 = F_{t-1} - F_{t-3} \tag{3}$$

$$A_1 = F_{t-3} - F_{t-6} \tag{4}$$

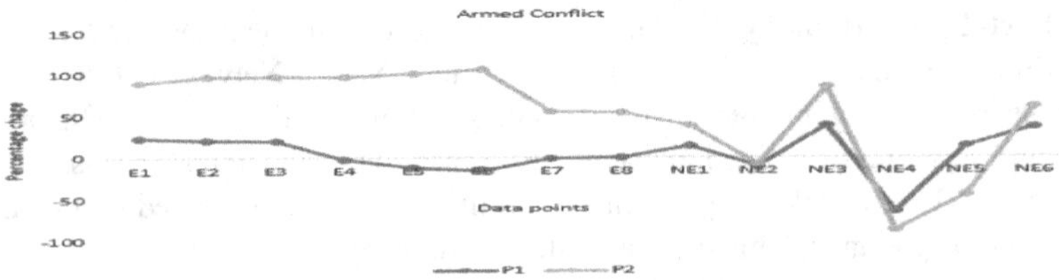

Fig. 4. Comparison of feature variation between event and nonevent points. E is event point and NE is nonevent point

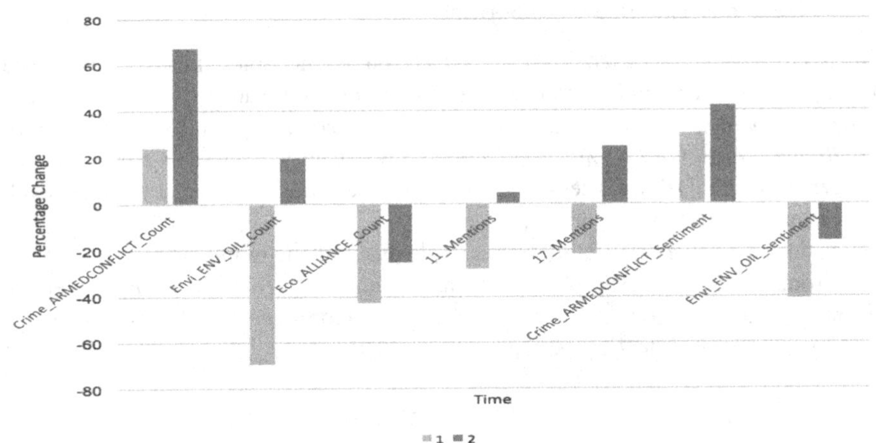

Fig. 5. Comparison amongst features for Dakota access pipeline unrest event.

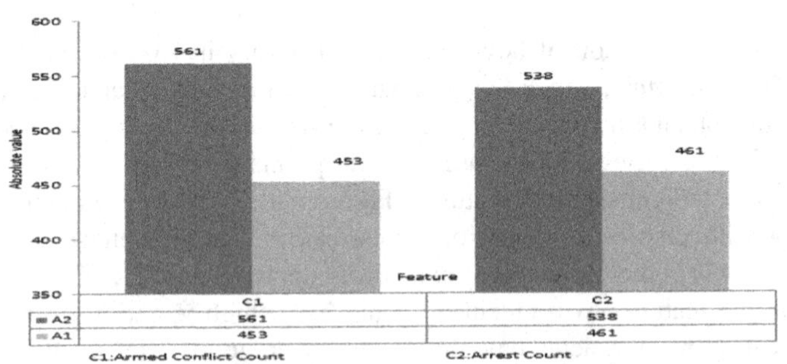

	C1	C2
A2	561	538
A1	453	461

Fig. 6. Comparison of variation of counts of feature with time

To capture this time variation of features, a preliminary model using an LSTM (Long short-term memory network) was run using 33 features that were listed as important by random forest to explain social unrest. Used features are *disapprove* event mentions and tone, *threaten* event mentions and tone, *protest* event mentions and tone, Counts and sentiments associated with themes *protest, armed conflict* and *conflict and violence*. Results obtained are below (Table 1). This performance can be enhanced by increasing features considered.

Table 1. LSTM results

Recall	Precision	F1 score
0.78	0.38	0.51

5.2 Metrics

If a nonevent point is being marked as an event point by our model, it is a false positive. If an actual event point is not being detected by our model it is a false negative. The percentage of misclassification is dependent on the probability threshold that is used during the modeling process. If a smaller cut off is chosen, nonevent points might also cross threshold and lead to false positives and if bigger cutoff is chosen some of event points might fail to meet the mark. An optimum cut off is to be selected. We selected a threshold that maximizes the number of event points that are found, while trying to reduce the number of false alerts. The results obtained at the State (Fig. 7) and County (Fig. 8) levels are discussed below. Metrics discussed below are estimated on only out of date validation set.

State Level Results

Random Forest. 90% of Nonevent points were correctly marked as nonevents. 10% of nonevent points were wrongly marked as event points which fall under false positives. 82% of event points were correctly marked as unrest events. 18% of event points were wrongly marked as nonevent points which fall under false negatives. F1 score is the harmonic mean of precision and recall. For the random forest model - Precision is 0.6, Recall is 0.82 and F1 score is 0.69.

ADA Boost with Random Forest. 90% of Nonevent points were correctly marked as nonevents. 10% of nonevent points were wrongly marked as event points which fall under false positives. 69% of event points were correctly marked as unrest events. 31% of event points were wrongly marked as nonevent points which fall under false negatives. Using ADA boost with Random Forest, Precision is 0.55, Recall is 0.69 and F1 score is 0.61.

Neural Networks. 72% of Nonevent points were correctly marked as nonevents. 28% of nonevent points were wrongly marked as event points which fall under false positives. 72% of event points were correctly marked as unrest events. 28% of event points were wrongly marked as nonevent points which fall under false negatives. Using the neural network model, Precision is 0.31, Recall is 0.72 and F1 score is 0.44.

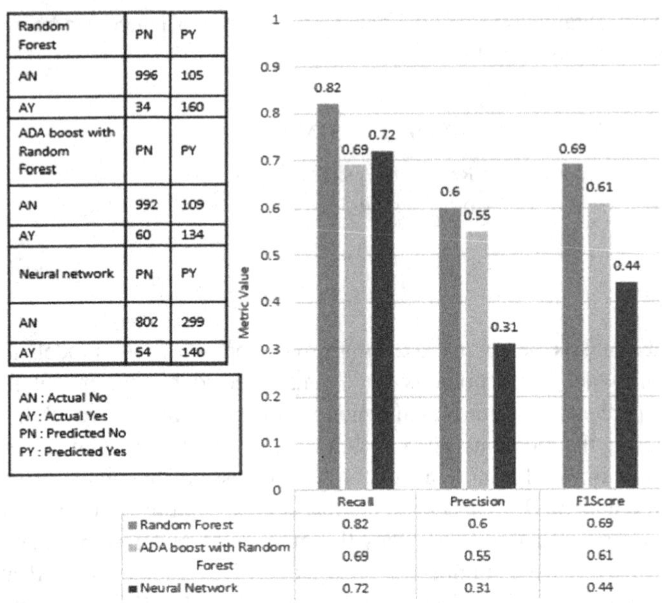

Fig. 7. Model results at state level

County Level Results

Random Forest. 95% of Nonevent points were correctly marked as nonevents. 5% of nonevent points were wrongly marked as event points which fall under false positives. 82% of event points were correctly marked as unrest events. 18% of event points were wrongly marked as nonevent points which fall under false negatives. Using Random Forest, Precision is 0.69, Recall is 0.82 and F1 score is 0.75.

ADA boost with Random Forest. 96% of Nonevent points were correctly marked as nonevents. 4% of nonevent points were wrongly marked as event points which fall under false positives. 79% of event points were correctly marked as unrest events. 21% of event points were wrongly marked as nonevent points which fall under false negatives. Using ADA boost with Random Forest, Precision is 0.76, Recall is 0.79 and F1 score is 0.77.

Neural Networks. 84% of Nonevent points were correctly marked as nonevents. 16% of nonevent points were wrongly marked as event points which fall under false positives. 39% of event points were correctly marked as unrest events. 61% of event points were wrongly marked as nonevent points which fall under false negatives. Using the neural network model, Precision is 0.25, Recall is 0.39 and F1 score is 0.31. It is evident from results that Random forest is performing well with good F1 scores at both State and County levels.

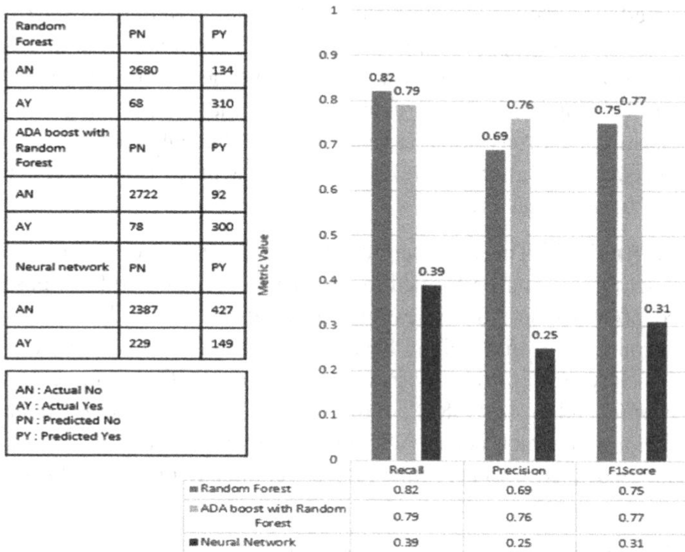

Fig. 8. Model results at county level

5.3 Analysis of Misclassified Cases

To perform out of date validation, Wikipedia data of huge social unrest events in 2017 was considered to test a model that was trained on 2015–2016 data. This contained unrest events related to President Trump, protests against the government on May Day, protest against Milos Yiannopoulos' speech and protests due to improper behavior of police. The majority of these events are due to growing levels of dissatisfaction that the public feel as a result of activities of leaders and government. The values of these themes have been analyzed in the paragraph below for four categories, Event (E), Nonevent (N), False Positive (FP) and False Negative (FN) categories at county level.

A high volume of negative conversations regarding leader and government themes was observed in the news prior to actual unrest event occurrence in associated regions. A high volume of conversation was also observed in locations which are false positives, which could be useful in identifying regions where unrest is boiling up. Not much of a discussion is observed in the case of nonevent points. The average number of times the theme *leader* was discussed is 17174 for event group, 373 for nonevent group, 4554 for false positive group and 123 for false negative group. The average number of times the theme *general government* was discussed was 6614 for event group, 162 for nonevent group, 1684 for false positive group and 51 for false negative group. The average sentiment associated with *local government* is −1.6 for event group, −0.9 for non event group, −1.8 for false positive group and 0.3 for false negative group. false negative cases have lower values like nonevent points and hence are not being detected by model.

Observations made with respect to a selected list of the decision trees is discussed below to understand what the potential root causes of misclassified cases are.

Conditions based on sentiments associated with *armed conflict, protest* and *price*, counts associated with *disapprove* and *coerce* played a prominent role in separation of event and nonevent points. If a particular location in time of interest has conversation about *government* being carried more than 681 times and a greater negative sentiment than -1.82 persisting in society about *armed conflict* and events related to *coercing* and *disapproval* are occurring more than 3116 and 23078 times respectively, it is being marked suspect to an unrest event.

False Positive Analysis. The following observations are also obtained by reviewing how decision action is being performed in a decision tree. 14% of false positive cases have occurred after a major unrest event's occurrence. Hence the model might be taking into account large discussions happening after unrest event as a measure of discontent in society. 85% of false positive cases have negative sentiments associated with *armed conflict* and *protest*. 48% of false positive cases have large count values of *coerce* and *democracy* which is case with event points. If a certain location has negative sentiment associated with particular factors, administrators of that location can act on those areas.

False Negative Analysis. 61% of false negatives have lesser counts associated with features like *disapprove* and *coerce*. 40% of false negatives don't have necessary negative sentiments associated with features like *price*.

5.4 Prediction of Unrest for One Month into Future

We used GDELT data from January 2015 to November 11th 2017 to train models and predicted top suspect states and counties with the high unrest levels for the following month (Nov 12th 2017–Dec 12th 2017). After completion of the performance period, we went through news articles to find out if unrest events actually happened and the findings are listed below the respective tables (Tables 2 and 3).

Table 2. State level predictions.

State name	Probability score
Florida	0.964
California	0.961
Arizona	0.950
Washington	0.945

Hundreds of people protested against the Haitian ruling on 21st November, 2017 in Florida. Anti-Trump protests also happened in Florida on 26th November 2017. Oakland city in California witnessed a strike by 3000 workers from December 5th to December 11th. A number of protests were observed in Washington against Net Neutrality, the GOP Tax bill, LNG plant. No major protests have been verified in Arizona till December 12th.

Table 3. County level predictions.

County name	State name	Probability score
Philadelphia	Pennsylvania	0.167
San Francisco	California	0.166
Dallas	Texas	0.166
Maricopa	Arizona	0.166
Los Angeles	California	0.163
Cook	Illinois	0.162
Bexar	Texas	0.162
Miami-Dade	Florida	0.160

Hundreds of supporters of the imprisoned rapper Meek Mill rallied outside the Criminal Justice Center, Philadelphia on November 13th which was followed after arrest being discussed more than 36,447 times in Philadelphia during observation period of six months with negative sentiment of −4.1 and 2,191,698 mentions of events related to disapproval being registered. Although no major unrest event happened during the performance window in Maricopa, thousands of people protested against Trump in Phoenix on August 23rd 2017. The model might have sensed tension in public about administration. Hundred people in Cook County, Chicago protested on Black Friday on November 24th. Thousands of protestors participated in a Women's rally in Chicago on January 20th, 2018. These events happened after news articles speaking about human right abuses with a horrific sentiment of -5.2 in Cook County. Several groups like SEIU, FANM on January 12, 2018 in Miami Dade county's little Haiti protested against policies and disrespectful talk of government towards immigrant communities like Haitians. Prior to this negative discussion happened in news with sentiment of −3.5 on Hate speech and −2.1 on government in Miami Dade. No major unrest events occurred in Los Angeles, Bexar and Dallas as per our knowledge.

6 Conclusion

As GDELT collects news every 15 minutes from various news sources, events are recorded continuously across the world. This can essentially be used as a tracker in studying events of interest; especially huge social unrest events. Measuring the emotional change of society as it relates to a specific event, and the events that unfold in their wake, is a great instrument to measure the levels of social unrest in various regions. In our paper, themes and events associated with social unrest have been studied using machine learning techniques in an attempt to identify regions at state and county levels where social unrest might occur in near future. It is difficult to manually track all of the factors that influence social unrest and to understand how they vary in different regions, but this task is easier to achieve with machine learning and large volumes of social recordings (news media). As discussed in the sections above, prior to social unrest events occurring, regions suffering from said events generally see a trend of large volumes of negative news articles being published with respect to certain factors leading up to the time that the event unfolds. In being able to predict the factors that indicate when a social unrest event might occur, government officials could roll out proactive safety measures that could reduce the magnitude of the impact that these events have on local communities. We have also recognized some key social factors and event types that appear to play a more significant role in the buildup of social unrest events. This analysis can be applied to a more granular city level, or to the lesser granularity level of country or continent.

Some of the news articles that were included in the analysis reported different associated locations for the same story, and the analysis of themes discussed in those articles were applied to every location that was identified. This is a less than optimal approach and may not necessarily be the most accurate representation of facts but for this analysis we assumed that both locations were related to the theme in question. Missing data was considered to be zero as it was assumed no data implies no event count or sentiment associated with it. Although an attempt has been made to capture time variation of features using LSTM models in our paper, more time series techniques can be applied to improve performance of model. Also in GDELT's event table we have considered events at root level, we can dig deeper to a more described sub-event level. This will be helpful while analyzing which exact events lead to social unrest. The data that was used from the GDELT database contains themes related to economy, environment, crime and health, and as such can be used to find events related to any of these domains. Alternatively, new features can be extracted for any number of new themes that an analyst would want to evaluate. This research can be combined with existing research on social media using Twitter, Facebook and various other social media sites to capture more information in an attempt to enhance the reliability of social unrest prediction. As news articles reflect what is happening in the world, the analysis of this data aided by the power of machine learning and other advanced techniques of artificial intelligence, can help us to understand trends in society that lead to such detrimental events; our research is a step toward achieving this goal.

References

1. Korolov, R., Lu, D., Wang, J., Zhou, G., Bonial, C., Voss, C., Kaplan, L., Wallace, W., Han, J., Ji, H.: On predicting social unrest using social media. In: IEEE/ACM International Conference on Advances in Social Networks Analysis and Mining (2016)
2. Mishler, A., Wonus, K., Chambers, W., Bloodgood, M.: Filtering tweets for social unrest. In : IEEE 11th International Conference on Semantic Computing (ICSC) (2017)
3. Xu, J., Lu, T.-C., Compton, R., Allen, D.: Civil unrest prediction: a tumblr-based exploration. In: Kennedy, W.G., Agarwal, N., Yang, S.J. (eds.) SBP 2014. LNCS, vol. 8393, pp. 403–411. Springer, Cham (2014). https://doi.org/10.1007/978-3-319-05579-4_49
4. Korkmaz, G., Kuhlman, C.J., Marathe, A., Ramakrishnan, N., Vullikanti, A.: Forecasting social unrest using activity cascade. PLoS ONE **10**(6), e0128879 (2015). https://doi.org/10.1371/journal.pone.0128879
5. Qiao, F., Li, P., Zhang, X., Ding, Z., Cheng, J., Wang, H.: Predicting social unrest events with hidden Markov models using GDELT. Discrete Dyn. Nat. Soc. **2017** (2015). https://www.hindawi.com/journals/ddns/2017/8180272/

Ten Years of Relevance Score for Content Based Image Retrieval

Lorenzo Putzu[2(✉)], Luca Piras[1,2], and Giorgio Giacinto[1,2]

[1] Pluribus One, Cagliari, Italy
[2] Department of Electrical and Electronic Engineering, University of Cagliari, Piazza d'Armi, 09123 Cagliari, Italy
{lorenzo.putzu,luca.piras,giacinto}@diee.unica.it
http://pralab.diee.unica.it

Abstract. After more than 20 years of research on Content-Based Image Retrieval (CBIR), the community is still facing many challenges to improve the retrieval results by filling the semantic gap between the user needs and the automatic image description provided by different image representations. Including the human in the loop through Relevance Feedback (RF) mechanisms turned out to help improving the retrieval results in CBIR. In this paper, we claim that Nearest Neighbour approaches still provide an effective method to assign a Relevance Score to images, after the user labels a small set of images as being relevant or not to a given query. Although many other approaches to relevance feedback have been proposed in the past ten years, we show that the Relevance Score, while simple in its implementation, allows attaining superior results with respect to more complex approaches, can be easily adopted with any feature representations. Reported results on different real-world datasets with a large number of classes, characterised by different degrees of semantic and visual intra- e inter-class variability, clearly show the current challenges faced by CBIR system in reaching acceptable retrieval performances, and the effectiveness of Nearest neighbour approaches to exploit Relevance Feedback.

Keywords: Image retrieval · Image description · Relevance feedback Nearest neighbour

1 Introduction

Content based image retrieval (CBIR) systems include all the approaches to retrieve images from large repositories, by analysing the visual and semantic content of the images. A CBIR system represents each image in the repository as a set of low- and mid-level features such as colour, texture, shape, connected components, etc. and uses a set of distance functions defined over these feature spaces to estimate the similarity between images. The goal of a CBIR system is to retrieve a set of images that is best suited to the user's intention formulated

P. Perner (Ed.): MLDM 2018, LNAI 10935, pp. 117–131, 2018.
https://doi.org/10.1007/978-3-319-96133-0_9

using an image as the query. The performances of CBIR systems are strongly related not only to the employed feature representations, but also to the distance functions used to measure the similarity between images [12,41]. The implicit assumption is that the similarity between images is related to a distance defined over a particular feature space. But since an image may be characterized by a large number of concepts, this leads to the so-called semantic gap between the real user image interpretation and the semantics induced from the low-level features. Furthermore, the availability of personal devices and the possibility for people to capture an unlimited number of photos and videos, rise the amount of multimedia document. Thus, the need for efficient and effective image retrieval systems has become crucial. Indeed, in order to accurately extract information from this vast amount of data, retrieval methods need to quickly discard irrelevant information and focus on the items of interests.

To this end, and to reduce the semantic gap introduced by the automatic image representation, most of the recent CBIR systems has adopted a mechanism of Relevance Feedback (RF) [2,35,43]. RF is a mechanism initially developed to improve text-based information retrieval systems [36], and then extended to CBIR system in [38]. RF techniques involve the user in the process of refining the search, that becomes an iterative process in which the original query is refined interactively, to progressively obtain a more accurate result. In a CBIR system, after the user submits a query image, the system retrieves a series of images and requires user interaction to label the returned images as relevant or non-relevant to the query image. The user's intention is then modified according to this data, producing a new set of images, that is expected to contain a larger number of relevant images. The relationship between the query image and any other image in the database is expressed using a relevance value, that is aimed to directly reflect the user's intention. Many different approaches have been proposed to estimate this relevance value. In this context, nearest neighbour approaches proved to be particularly effective [14]. The aim of these methods is to produce for each image a *Relevance Score* by using the distances to relevant and non-relevant neighbours. Obviously, an image will present a highest relevance score as much as its distance from the nearest relevant image is small compared to the distance of its nearest non relevant image. Although many other approaches for relevance feedback have been proposed in the past ten years, we believe that the Relevance Score can still be considered the state-of-the-art approach for relevance estimation.

Other Relevance Feedback techniques proposed in the literature involve the optimization of one or more CBIR components such as the formulation of a new query or the transformation of the feature space. This kind of approaches extract global properties pertaining to relevant images, and such properties are derived from the set of relevant and non-relevant images retrieved so far. Such global optimization can suffer from the small sample problem, as the number of images displayed to the user for labelling is usually small (e.g., 20 images at a time). In particular, as the size of the database increases, it is very likely that small numbers of relevant images are retrieved in response to the user's query. As a consequence, the modifications of CBIR parameters based on such information may result unreliable or provides poor performance improvement [16]. Nearest Neighbour approaches, instead of trying to estimate global properties pertaining to relevant images, estimated image relevance locally. As a result, such a mechanism is apt to identify classes of images with complex boundaries.

This paper will introduce the reader to the major approaches proposed in the literature to create an effective image retrieval system. Section 2 describes the basic concepts behind the techniques proposed in the content based image retrieval field. Section 3 summarizes the main categories of features used for this task. Section 4 describes some approaches proposed so fat to exploit the relevance feedback, including the Relevance Score. Section 5 proposes a comparison of different RF strategies on real-world challenging datasets, showing the effectiveness of the Relevance Score in different retrieval scenarios, and employing different feature representations. Conclusions and future research perspectives are drawn in Sect. 6.

2 Architecture of CBIR Systems

Many CBIR systems that are based on the automatic analysis of the image content from a computer perspective have been proposed over the years. They present many advantages over traditional image retrieval based on textual queries only. In particular, they avoid either relying on costly professional textual annotations, that are needed to deem the labels and tags as being reliable, or relying on social labels and tags, for which an estimation of their relevance needs to be computed. Moreover, CBIR systems present advantages even in cases where textual annotations are already present, since they could be focused on just some aspects of the image and neglect other important contents [11]. Most of the existing successful CBIR systems are tailored to specific applications, usually referred to as narrow domain systems, and consequently to specific retrieval problems, such as the Computer Aided Diagnosis systems [27], sport events [12], and cultural heritage preservation [7]. In all the previous examples, the objects of interest related to a given query can be either precisely defined or easily modelled due to the knowledge of the semantic content of the images.

Thus, it is easier to manage a CBIR system designed for narrow domain archives, as the semantic content to be searched is clearly defined [41]. Instead the design of a general purpose multimedia retrieval systems is still a challenging task, as such system should be capable of adapting to different semantic contents, and different intents of the users. A huge number of works have been devoted to the special cases of *instance retrieval* or *near duplicate* retrieval [19,45], that consists on retrieving images showing exactly the same object as the query image. But in most of the real cases, the user is not interested in retrieving images with the same subject, but images with *similar* content. As a consequence, this could create some ambiguity, in particular if the image contains multiple objects, or if some objects are just parts of a larger object. This happens because different users, for a given image, may refer to different regions and objects in the image.

Furthermore, an object could even belong to multiple orthogonal classes. Different mechanisms have been proposed to manage these kinds of ambiguity, most of them require the user feedback to improve the retrieval results, while others are mainly based on defining a strong set of features with the most appropriate similarity measure.

3 Features in CBIR

In a CBIR system the description and the representation of the content of a specific image can be provided in multiple ways according to three levels of abstraction [12]. The first level includes the representation of primitive characteristics or low-level features, such as colour, texture and basic geometric shapes. This kind of information although very simple and easy to extract, does not guarantee reliability and accuracy. The second level is devoted to provide a more detailed description of the elements mentioned above, by providing the representation of more complex objects by means of their aggregation and their spatial arrangements which are precisely the logical characteristics or mid-level features. The common perception of an image or a scene is not always related to its description through such low and mid-level features. Indeed, an image can be seen as the representation of different concepts, either related to primitive characteristics but also related to emotions and memories. For these reasons, a third level, called semantic level, is used. It includes the description of the abstract features, such as the meaning of the scenario, or the feelings induced in an observer.

While most of the current CBIR systems employ low-level and mid-level features [12,30], aimed to take into account information like colour, edge and texture [6], CBIR systems that address specific retrieval problems leverage on different kind of features that are specifically designed [40]. Some works exploited low-level image descriptors, such as SIFT [44] originally proposed for object recognition [25], different colour histograms [23] or a fusion of textual and visual information [11] for scene categorization. Also more specific low-level features designed for other applications have been used in CBIR systems, such as the HOG [10] and the LBP [28] descriptors, originally proposed for pedestrian detection and texture analysis respectively. Obviously, these features does not fit for CBIR designed for less specific applications and general purpose retrieval systems [30]. Furthermore, while low-level visual descriptors help improving the retrieval performances for particular purposes, they often fail to provide high-level understanding of the scene. Several CBIR systems that use a combination or fusion of different image descriptors have been proposed in the literature [32]. Fusion approaches are usually categorized into two classes, namely *early*, and *late* fusion approaches. The approaches of early fusion are very common in the image retrieval field, and the simplest and well known solution is based on the concatenation of the feature vectors, such as in [45], where the authors propose two ways of integrating the SIFT and LBP, and the HOG and LBP descriptors, respectively. Instead, the aim of late fusion approaches is to produce a new output by combining either different similarities or distances from the query [13] or different ranks obtained by the classifiers [39].

In this light, *deep learning* approaches implicitly perform feature fusion, modelling high-level features by employing a high number of connected layers composed of multiple non-linear transformation units [32]. Indeed, the Convolutional Neural Network (CNN) model [5] consists of several convolutional layers and pooling layers, where the convolutional layer performs a weighted combination of the input values, while the pooling layer perform a down-sampling operation that reduces the output of the convolutional layer. In [20] it has been shown also that features extracted from the upper layers of the CNN can also serve as good descriptors for image retrieval. It implies that a CNN trained for a given general task has acquired generic representation of objects that will be useful for all sorts of visual recognition tasks [3].

4 Four Approaches to Exploit Relevance Feedback

In the following, we will refer to image retrieval systems where a query image is used as an example to find all the images in the repository that are relevant to that query. Defining which image is relevant or not-relevant to a query is not a trivial problem, in particular if the problem must be addressed using just one image as query. Indeed, there is still a gap between the human perception of the semantic information present in an image, and its computer description, that is typically able to capture just a small subsets of semantic concepts. Moreover, the user that performs the query could not have a specific target in mind, or he could perform the query with an image that is only partially related to the content that he has in mind. In both cases, the retrieval process could not be accomplished in just one step. The mechanism of Relevance Feedback (RF) has been developed to involve the user also in further steps of the process, in particular, to verify if the search results are relevant or not. Indeed, after the user feedback, the system can consider all relevant images as additional examples to better specify the query, and the non-relevant ones as examples of images that the user is not interested in.

A number of RF approaches have been proposed to refine the retrieval results and they can be divided into four main categories. One of the first mechanism of RF used in CBIR tasks and still used in many applications, is based on the so-called *Query Shifting* or *Query-Point Movement* (QPM) paradigm. This technique has been firstly proposed for text retrieval refinement [36] and then adopted in CBIR systems [37]. The assumption behind this approach is that relevant images are clustered in the feature space, but the original query could lie in the region of the feature space that is in some way far from the cluster of relevant images. Accordingly, a new *optimal* query is computed in such a way that it lies near to the euclidean center of the relevant images, and far from the non-relevant images, according to Eq. (1)

$$Q_{opt} = \frac{1}{N_R} \sum_{i \in D_R} D_i - \frac{1}{N_T - N_R} \sum_{i \in D_N} D_i \tag{1}$$

where D_R and D_N are the sets of relevant and non-relevant images, respectively, N_R is the number of images in D_R, N_T the number of the total documents, and

D_i is the representation of an image in the feature space. It is easy to see that this approach can be suited only in cases in which relevant images tend to form a cluster with a small intersection with other images that are not-relevant to the user's interests.

A quite close group of RF approaches are based on *distance or similarity learning*, that instead of optimizing the query with respect to the relevant and not-relevant images retrieved so far, they optimize the distance metric used to compute image similarities. The goal is to have high pair-wise similarity value for images marked as relevant, and a low similarity value between relevant and not-relevant images. In the simplest case, the metric learning may consist in just re-weighting the individual features [31]. A major advantage of these two groups of approaches is that they are relatively fast, but they usually ignore dependencies between features, do not consider local properties in different portions of the feature space, and are only effective if the concept represented by the relevant images consists of a convex region in the feature space.

A third group of approaches are based on the formulation of RF in terms of a *pattern classification* task, by using the relevant and non-relevant image sets to train popular learning algorithms such as SVMs [24], neural networks and self-organizing maps [8,21]. However, in many practical CBIR settings it is usually difficult to produce a high-level generalization as the number of available relevant and non-relevant samples cases may be too small. This kind of problems has been partially mitigated using the *active learning* paradigm [9,34], where the system is trained not only with the most relevant images according to the user judgement, but also with the most informative images that allows driving the search into more promising regions of the feature space [18,33,42].

Finally, RF can be formulated according to a *probabilistic* approach, where the posterior probability distribution of a random variable according to the user feedback [2] is estimated. In particular, the probability densities of the relevant and non-relevant images is used as a similarity measure, as in the case of using a soft classifier [2,15]. This category also includes the nearest neighbour (NN) methods used in this context to estimate the posterior probabilities of an image as being or not relevant to the user's query. NN approaches have been adapted in several forms over the years, but the most used and effective form to compute a *Relevance Score* for each image is through the computation of its distances to its nearest relevant and non-relevant neighbours as follows:

$$rel_{NN}(I) = \frac{||I-NN^{nr}(I)||}{||I-NN^{r}(I)||+||I-NN^{nr}(I)||} \tag{2}$$

where $NN^r(\cdot)$ and $NN^{nr}(\cdot)$ denote the nearest relevant and non relevant image for the image I respectively, and $||\cdot||$ is the metric, typically the Euclidean distance, defined for the feature space. For convenience the equation can be substituted by other ones such as $||I-NN^{nr}(I)||/||I-NN^{r}(I)||$ [1]. In a recent work [2] this ratio has been modified by introducing a smoothed term in order to increase the importance of the images more relevant to the user query with the distance to the closest relevant image. The modified ratio $||I-NN^{nr}(I)||/||(I-NN^{r}(I))^2||$ has been shown to improve the basic one in some cases [2], but in general, it turns out that the original formulation gives surprisingly good results also in comparison to other state-of-the-art techniques [14].

5 Experimental Results

5.1 Datasets

We performed experiments with three real-world, state of the art datasets differing in the number of classes, and in the semantic content of the images. *Caltech* is a well knows image dataset[1] that presents a collection of pictures of objects. In most images, objects are in the centre with fairly similar poses, and with very limited or no occlusion. There are different versions of this dataset, and in this work we used the *Caltech-101* and the *Caltech-256* versions. Caltech-101 is a collection of pictures of objects belonging to 101 categories. It contains a total of 9,144 images and most of the categories have almost 50 images, but the number of images per categories range between 40 and 800. Caltech-256 contains pictures of objects belonging to 256 categories. It contains a total of 30,607 images and the number of images per category ranges greatly between 80 and 827, with an average value of 100 images per category. The *Flower* dataset[2] presents a collection of flower images. This dataset is released in two different versions, and in this work we used the 102 category version *Flowers-102*. Despite this dataset has a similar number of classes as Caltech-101, the two datasets are related to two very different problems. Indeed, Flowers-102 turns out to be a problem of fine retrieval, since it contains the single category object 'Flower' that is subdivided into 102 sub-categories. It consists of 8,189 images, with a number of images per class that ranges between 20 and 238. *SUN-397* is an image dataset for scene categorization. It contains 108,754 images belonging to 397 categories. The number of images varies across categories, but there are at least 100 images per category. In the experimental evaluation these three datasets have been randomly divided into two subsets: the query set, containing a query image for each class, and the search set containing all the remaining images for retrieval.

5.2 Features for Image Representation

In order to assess the performances of the different RF mechanisms on different image representations, we used 8 sets of features: CNN, Colour features, SIFT [25], HOG [10], LBP [28], LLBP [4], Gabor Wavelets [22] and HAAR wavelets [17]. In particular the CNN features have been extracted from the most used CNN architectures that is AlexNet [20]. Each level of a CNN could be used for this purpose. Indeed the first network layers are able to describe just some images characteristics, like point and edges, while the innermost layers can capture high-level features and thus can create a richer image representation. We extracted the features from the second fully connected layer (fc7), that produces

[1] http://www.vision.caltech.edu/Image_Datasets/.
[2] http://www.robots.ox.ac.uk/~vgg/data/flowers/.

a feature vector of size 4096. Colour features instead present a set of features proposed in [26] that includes Colour Histogram, Colour Moments and Colour Auto-correlogram, which concatenated produce a feature vector of size 102. SIFT features instead have been extracted with a grid sampling of 8 pixel size and a window size ranging from 32 to 128. Then the extracted SIFT features have been used to create a BoVW of size 4096. HOG have been computed on HOG blocks composed by 4 cells of 16 pixel size. The blocks are overlapped by one cell per side creating a feature vector of size 6084. Also LBP have been extracted from blocks of 32 pixel size, in order to favour the analysis of small regions of the image, since they have been created for texture analysis. The final feature vector has a size of 2891. In addition to the LLBP, the Gabor and HAAR wavelet features have been also used, and the related feature vectors are sized 768, 5760 and 3456 respectively.

5.3 Experimental Setup

The different RF approaches have been applied after the first retrieval round, where the top k images are returned to the user for labelling as being relevant or not. Values of $k = 20, 50, 100$ have been considered. Then, 4 RF rounds are performed for each query image. Reported results have been computed by automating the RF task, thanks to the availability of labelled datasets. In particular, for a given query image, images are labelled as being relevant or not if it belongs to the same class of the query image. This procedure is useful to obtain objective relevance labels, and to perform repeatable comparisons between various retrieval systems. The underlying assumption is that a user who performs a query using an image belonging to a certain class, should be interested only in images belonging to the same class. The performances attained with the Relevance Score have been compared to other RF approaches following a different paradigm, namely, the QPM paradigm (see Eq. (1)), and a binary Linear SVM classifier [24]. Since the training set that is created to train the SVM is different for each query image, and it is also different at each RF round for each query, the selection of the SVM hyperparameters is not trivial. Thus, the employed SVM uses an error-correcting output code (ECOC) mechanism [29] with a 5-fold cross validation to fit and automatically tune the hyperparameters. The trained SVM is used to classify the images belonging to the repository, and a vector is produced where each component represents the probability that the given image belongs to a certain class. Therefore, given that the SVM has been used to classify relevant and non-relevant images, the provided score could be used as a similarity measure, directly indicating the relevance of an image.

To evaluate the performances of the proposed RF mechanisms, we used three different accuracy metrics: *Precision*, *Recall* and *Average precision*. The Precision (P) (see Eq. (3)) measures the ratio between Relevant Retrieved Images (RRI) within the top k retrieved images. The Recall (R)(see Eq. (4)) measures the ratio between the RRI and the total number of Relevant Images (RI) in the dataset. The Average Precision (AP) (see Eq. (5)) takes into account the position in the results set of the relevant images.

$$P = \frac{RRI}{k\ images} \quad (3) \qquad R = \frac{RRI}{RI} \quad (4) \qquad AP = \frac{1}{Q_i} \sum_{n=1}^{N} \frac{R_i^n}{n} t_n^i \quad (5)$$

where Q_i is the number of relevant images for the i-th query, N is the total number of images of the search set, R_n^i is the number of relevant retrieved images within the n top results, and t_i^n indicates if the n-th retrieved image is relevant (=1) for the i-th query or not (=0).

5.4 Results

In order to give the reader a first idea on the performances attained on one of the considered feature sets, Table 1 shows the AP results after the first retrieval round on each dataset. As it can be observed, the performances obtained are very different for each feature set, although in general the CNN features outperform the other methods. This great imbalance in the initial retrieval performances makes even more interesting the analysis of the approaches on the individual sets of descriptors, in order to verify which one benefits more from that mechanisms. The retrieval performances of the different approaches have been reported in Figs. 1, 2 and 3 in which we compared the various techniques with different features sets on retrieval sets of size 20, 50 and 100 respectively. As a general comment on the attained results, in all the approaches the performances increase after each RF round, but with very different trends. The feature set that take more advantages of an RF approach are the CNN features, as a gain of more than 25% can be obtained for all the datasets when the RF mechanism based on Relevance Score is employed. We can also see that the Relevance Score approach outperforms other approaches in most of the cases. In particular, looking at the trends for CNN features, it can be observed that the Relevance Score allows attaining performance improvements in every RF round, although

Table 1. Comparison of features for image retrieval without relevance feedback.

Features	Caltech101	Caltech256	Flowers	SUN397
CNN	38.53	18.17	29.81	6.14
Colors	2.96	1.05	5.21	0.48
SIFT	9.99	2.43	4.17	0.83
HOG	8.06	2.33	2.72	0.74
LBP	7.50	2.59	4.11	0.85
LLBP	4.05	1.09	4.24	0.62
HAAR	8.79	2.57	5.56	0.78
Gabor	10.26	2.48	4.27	0.59

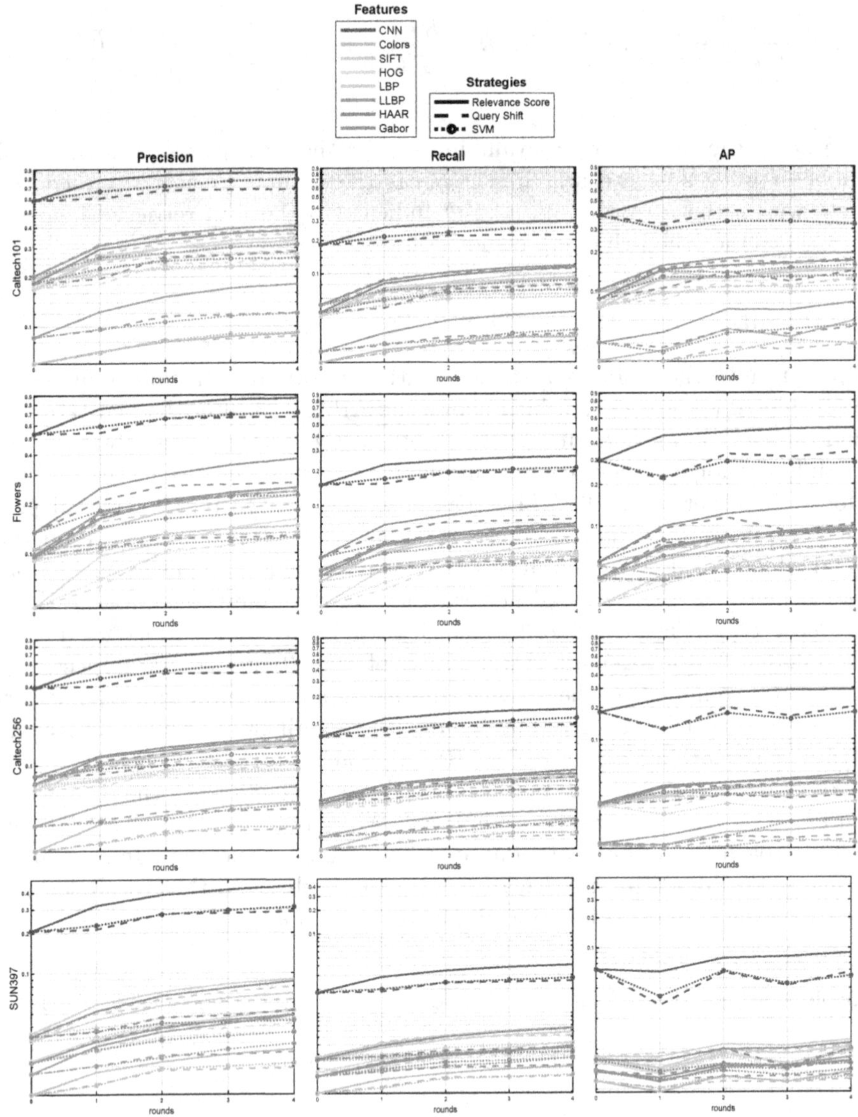

Fig. 1. Performances after 4 RF rounds, when the size of the retrieval set is 20.

it start reducing the amount of the improvement after three rounds, While the other RF approaches shows many fluctuations, as they do not provide increases in every RF round. This trend could be observed in detail in Table 2, where we reported the average results of all datasets and all features, starting from the initial AP value, and then showing the amount of gain in AP after each round, and, finally, the total amount of gain for each RF approach. It is also worth to note that in many cases the SVM outperforms the QS approach for each k

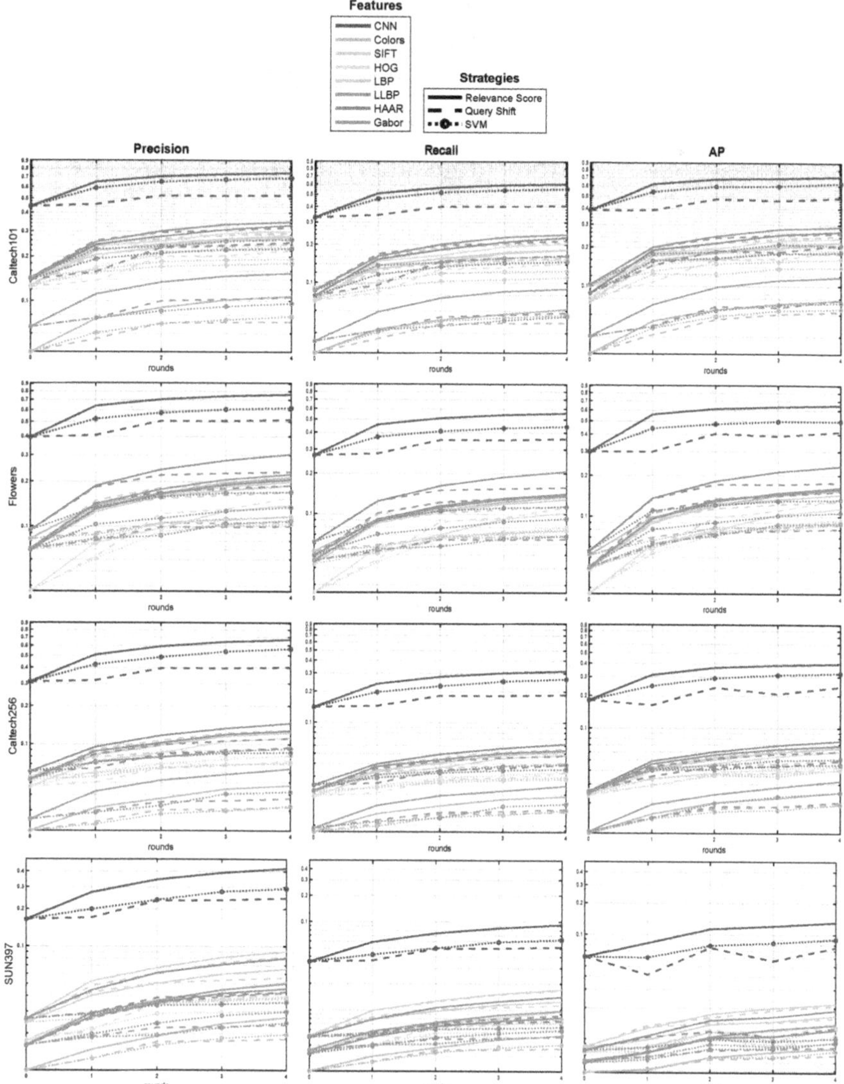

Fig. 2. Performances after 4 RF rounds, when the size of the retrieval set is 50.

value, with the exception of $k = 20$ where, although its precision is higher that that of QS, it presents a lower value for the recall and AP. This is an already known trend of the learning algorithm such as the SVM, that are not able to fit properly a model whit a too small training set. This trend is confirmed also by the fact that the SVM performances arise when k increases, until it reaches the Relevance Score performances on Caltech101 with $k = 100$.

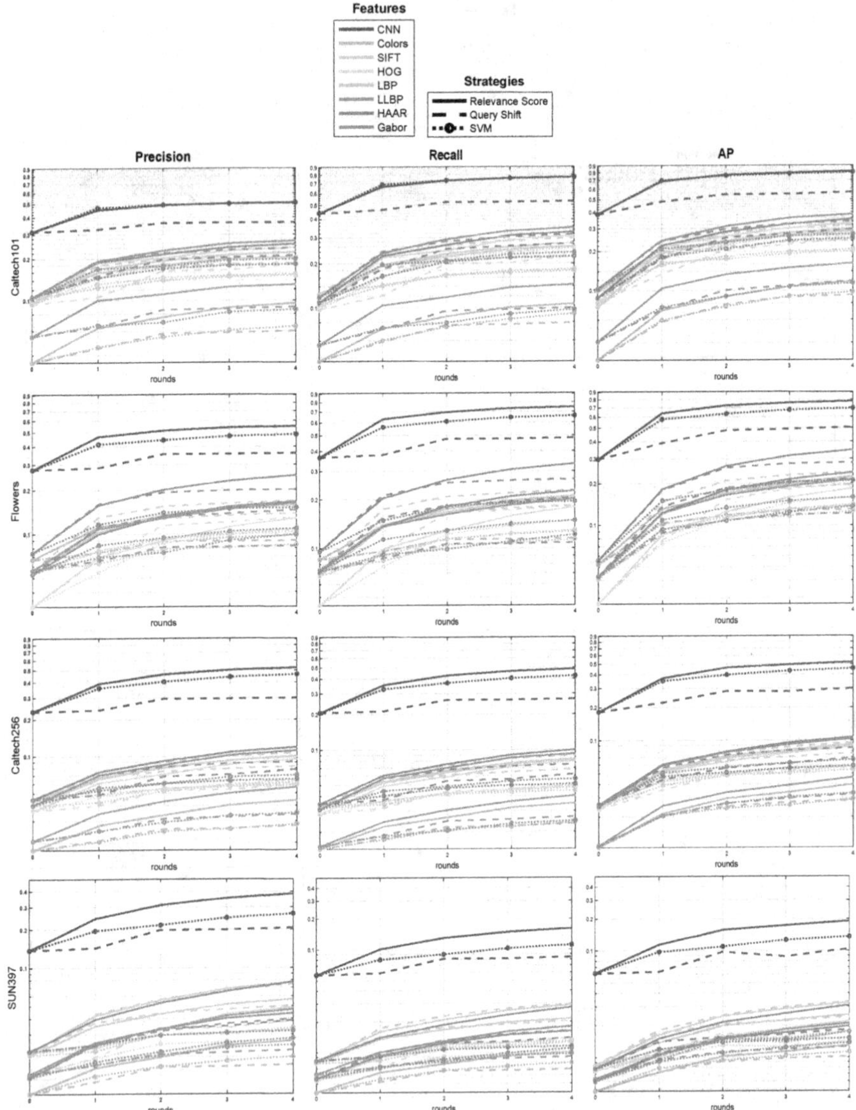

Fig. 3. Performances after 4 RF rounds, when the size of the retrieval set is 100.

6 Conclusion

A Content-Based Image Retrieval (CBIR) system is focused in retrieving images from digital archives comparing their visual content with the semantic content required by the user using a query image as an example. Although during the past years many features sets have been proposed to represent images for CBIR tasks, there is still a semantic gap between the automatic image description and the user needs. The results reported in this paper clearly show the challenges

Table 2. AP gain for the different RF approaches averaged for all the datasets and features set.

RF approach	k	Round				Total gain
		1	2	3	4	
Relevance score	20	2.56	0.97	0.53	0.42	4.47
	50	5.17	2.03	1.24	0.88	9.32
	100	7.57	3.48	2.03	1.52	14.60
Query shift	20	0.15	1.68	−0.45	0.82	2.20
	50	2.36	2.32	0.09	0.89	5.66
	100	4.78	3.32	0.95	0.83	9.87
SVM	20	−0.17	0.95	−0.06	0.27	1.00
	50	2.93	1.17	0.67	0.40	5.16
	100	5.78	1.82	1.36	0.73	9.69

to improve the retrieval performances with different image representations, and thus the needs of approaches to fill the semantic gap. To this end, we tested different Relevance Feedback approaches, and showed that Nearest Neighbour (NN) approaches proved to be able to provide an effective method to exploit user's feedback both on different retrieval problems, and using different feature representations. In particular, although different NN approaches that exploit the relevance feedback paradigm have been proposed in the past ten years, the formulation proposed in [14], still proves to be effective.

Acknowledgements. This work has been supported by the Regional Administration of Sardinia (RAS), Italy, within the project BS2R - Beyond Social Semantic Recommendation (POR FESR 2007/2013 - PIA 2013).

References

1. Arevalillo-Herráez, M., Domingo, J., Ferri, F.J.: Combining similarity measures in content-based image retrieval. Pattern Recogn. Lett. **29**(16), 2174–2181 (2008)
2. Arevalillo-Herráez, M., Ferri, F.J., Domingo, J.: A naive relevance feedback model for content-based image retrieval using multiple similarity measures. Pattern Recogn. **43**(3), 619–629 (2010)
3. Babenko, A., Slesarev, A., Chigorin, A., Lempitsky, V.: Neural codes for image retrieval. In: Fleet, D., Pajdla, T., Schiele, B., Tuytelaars, T. (eds.) ECCV 2014. LNCS, vol. 8689, pp. 584–599. Springer, Cham (2014). https://doi.org/10.1007/978-3-319-10590-1_38
4. Rosdi, B.A., Shing, C.W., Suandi, S.A.: Finger vein recognition using local line binary pattern. Sensors **11**, 11357–11371 (2011)
5. Bengio, Y., Courville, A., Vincent, P.: Representation learning: a review and new perspectives. IEEE Trans. Pattern Anal. Mach. Intell. **35**(8), 1798–1828 (2013)
6. Chatzichristofis, S.A., Boutalis, Y.S.: CEDD: color and edge directivity descriptor: a compact descriptor for image indexing and retrieval. In: Gasteratos, A., Vincze, M., Tsotsos, J.K. (eds.) ICVS 2008. LNCS, vol. 5008, pp. 312–322. Springer, Heidelberg (2008). https://doi.org/10.1007/978-3-540-79547-6_30

7. Chen, H.: A socio-technical perspective of museum practitioners' image-using behaviors. The Electron. Libr. **25**(1), 18–35 (2007)
8. Chen, Y., Zhou, X.S., Huang, T.: One-class SVM for learning in image retrieval. ICIP **1**, 34–37 (2001)
9. Cohn, D.A., Atlas, L.E., Ladner, R.E.: Improving generalization with active learning. Mach. Learn. **15**(2), 201–221 (1994)
10. Dalal, N., Triggs, B.: Histograms of oriented gradients for human detection. In: Proceedings of CVPR, pp. 886–893 (2005)
11. Dang-Nguyen, D.T., Piras, L., Giacinto, G., Boato, G., De Natale, F.G.B.: Multimodal retrieval with diversification and relevance feedback for tourist attraction images. ACM Trans. Multimedia Comput. Commun. Appl. **13**(4), 49:1–49:24 (2017)
12. Datta, R., Joshi, D., Li, J., Wang, J.Z.: Image retrieval: ideas, influences, and trends of the new age. ACM Comput. Surv. **40**(2), 1–60 (2008)
13. Escalante, H.J., Hérnadez, C.A., Sucar, L.E., Montes, M.: Late fusion of heterogeneous methods for multimedia image retrieval. In: Proceedings of the 1st ACM International Conference on Multimedia Information Retrieval, pp. 172–179 (2008)
14. Giacinto, G.: A nearest-neighbor approach to relevance feedback in content based image retrieval. In: Proceedings of the 6th ACM International Conference on Image and Video Retrieval, CIVR 2007, pp. 456–463. ACM, New York (2007)
15. Giacinto, G., Roli, F.: Bayesian relevance feedback for content-based image retrieval. Pattern Recogn. **37**(7), 1499–1508 (2004)
16. Giacinto, G., Roli, F.: Nearest-prototype relevance feedback for content based image retrieval. In: ICPR, vol. 2, pp. 989–992 (2004)
17. Graps, A.: An introduction to wavelets. IEEE Comput. Sci. Eng. **2**(2), 50–61 (1995)
18. Hoi, S.C.H., Jin, R., Zhu, J., Lyu, M.R.: Semisupervised SVM batch mode active learning with applications to image retrieval. ACM Trans. Inf. Syst. **27**(3), 16:1–16:29 (2009)
19. Jégou, H., Zisserman, A.: Triangulation embedding and democratic aggregation for image search. In: Proceedings of CVPR, pp. 3310–3317. IEEE Computer Society (2014)
20. Krizhevsky, A., Sutskever, I., Hinton, G.E.: Imagenet classification with deep convolutional neural networks. In: 26th Annual Conference on Advances in Neural Information Processing Systems, pp. 1106–1114 (2012)
21. Laaksonen, J., Koskela, M., Oja, E.: PicSOM-self-organizing image retrieval with MPEG-7 content descriptors. IEEE Trans. Neural Netw. **13**(4), 841–853 (2002)
22. Lee, T.S.: Image representation using 2D Gabor wavelets. IEEE Trans. Pattern Anal. Mach. Intell. **18**(10), 959–971 (1996)
23. van Leuken, R.H., Garcia, L., Olivares, X., van Zwol, R.: Visual diversification of image search results. In: ACM International Conference on World Wide Web, pp. 341–350 (2009)
24. Liang, S., Sun, Z.: Sketch retrieval and relevance feedback with biased SVM classification. Pattern Recogn. Lett. **29**(12), 1733–1741 (2008)
25. Lowe, D.G.: Distinctive image features from scale-invariant keypoints. Int. J. Comput. Vis. **60**(2), 91–110 (2004)
26. Mitro, J.: Content-based image retrieval tutorial. ArXiv e-prints (2016)
27. Müller, H., Clough, P.D., Deselaers, T., Caputo, B. (eds.): ImageCLEF: Experimental Evaluation in Visual Information Retrieval. Springer, Heidelberg (2010). https://doi.org/10.1007/978-3-642-15181-1

28. Ojala, T., Pietikäinen, M., Mäenpää, T.: Multiresolution gray-scale and rotation invariant texture classification with local binary patterns. IEEE Trans. Pattern Anal. Mach. Intell. **24**(7), 971–987 (2002)
29. Passerini, A., Pontil, M., Frasconi, P.: New results on error correcting output codes of kernel machines. IEEE Trans. Neural Netw. **15**(1), 45–54 (2004)
30. Pavlidis, T.: Limitations of content-based image retrieval. Technical report, Stony Brook University (2008)
31. Piras, L., Giacinto, G.: Neighborhood-based feature weighting for relevance feedback in content-based retrieval. In: WIAMIS, pp. 238–241. IEEE Computer Society (2009)
32. Piras, L., Giacinto, G.: Information fusion in content based image retrieval: a comprehensive overview. Inf. Fusion **37**, 50–60 (2017)
33. Piras, L., Giacinto, G., Paredes, R.: Enhancing image retrieval by an exploration-exploitation approach. In: Perner, P. (ed.) MLDM 2012. LNCS (LNAI), vol. 7376, pp. 355–365. Springer, Heidelberg (2012). https://doi.org/10.1007/978-3-642-31537-4_28
34. Piras, L., Giacinto, G., Paredes, R.: Passive-aggressive online learning for relevance feedback in content based image retrieval. In: Proceedings of the 2nd International Conference on Pattern Recognition Applications and Methods, pp. 182–187 (2013)
35. Piras, L., Tronci, R., Giacinto, G.: Diversity in ensembles of codebooks for visual concept detection. In: Petrosino, A. (ed.) ICIAP 2013. LNCS, vol. 8157, pp. 399–408. Springer, Heidelberg (2013). https://doi.org/10.1007/978-3-642-41184-7_41
36. Rocchio, J.J.: Relevance feedback in information retrieval, pp. 313–323. Prentice Hall, Englewood Cliffs (1971)
37. Rui, Y., Huang, T.S., Mehrotra, S.: Content-based image retrieval with relevance feedback in MARS. In: International Conference on Image Processing Proceedings, pp. 815–818, October 1997
38. Rui, Y., Huang, T.S., Mehrotra, S.: Relevance feedback: a power tool in interactive content-based image retrieval. IEEE Trans. Circuits Syst. Video Technol. **8**(5), 644–655 (1998)
39. da Torres, R.S., Falcão, A.X., Gonçalves, M.A., Papa, J.P., Zhang, B., Fan, W., Fox, E.A.: A genetic programming framework for content-based image retrieval. Pattern Recogn. **42**(2), 283–292 (2009)
40. Sivic, J., Zisserman, A.: Efficient visual search for objects in videos. Proc. IEEE **96**(4), 548–566 (2008)
41. Smeulders, A.W.M., Worring, M., Santini, S., Gupta, A., Jain, R.: Content-based image retrieval at the end of the early years. IEEE Trans. Pattern Anal. Mach. Intell. **22**(12), 1349–1380 (2000)
42. Tong, S., Chang, E.Y.: Support vector machine active learning for image retrieval. In: ACM Multimedia, pp. 107–118 (2001)
43. Tronci, R., Murgia, G., Pili, M., Piras, L., Giacinto, G.: ImageHunter: a novel tool for relevance feedback in content based image retrieval. In: Lai, C., Semeraro, G., Vargiu, E. (eds.) New Challenges in Distributed Information Filtering and Retrieval. SCI, vol. 439, pp. 53–70. Springer, Heidelberg (2013). https://doi.org/10.1007/978-3-642-31546-6_4
44. Tsai, C.M., Qamra, A., Chang, E., Wang, Y.F.: Extent: interring image metadata from context and content. In: IEEE International Conference on Multimedia and Expo, pp. 1270–1273 (2006)
45. Yu, J., Qin, Z., Wan, T., Zhang, X.: Feature integration analysis of bag-of-features model for image retrieval. Neurocomputing **120**, 355–364 (2013)

When Different Is Wrong: Visual Unsupervised Validation for Web Information Extraction

Benoit Potvin(✉) and Roger Villemaire

Department of Computer Science, Université du Québec à Montréal, Montréal H3C 3P8, Canada
Potvin.benoit.2@courrier.uqam.ca

Abstract. This paper shows how visual information can be used to identify false positive entities from those returned by a state-of-the-art web information extraction algorithm and hence further improve extraction results. The proposed validation method is unsupervised and can be integrated into most web information extraction systems effortlessly without any impact on existing processes, system's robustness or maintenance. Instead of relying on visual patterns, we focus on identifying visual outliers, i.e. entities that visually differ from the norm. In the context of web information extraction, we show that visual outliers tend to be erroneous extracted entities. In order to validate our method, we post-processed the entities obtained by Boilerpipe, which is known as the best overall main content extraction algorithm for web documents. We show that our validation method improves Boilerpipe's initial precision by more than 10% while F_1 score is increased by at least 3% in all relevant cases.

1 Introduction

Visual information plays an important role in web pages that have been designed for humans. Web technologies have evolved to operate on powerful devices that can execute considerable front end calculation and enhance user experience. Visually rich documents such as web pages and PDF files contain visual information that supplements the meaning of textual information and facilitates comprehension. Without visual formatting, a website would be much more difficult to understand and navigate, perhaps even incomprehensible to the user. Accordingly, visual features that help human users to understand a document can also help data extraction.

Recent works have emphasized the importance of visual elements in web mining tasks [4,5,11,13,20]. Visually oriented web information extraction (WIE) methods use discriminant regularities (patterns) across visual information in order to improve extraction results. Visual characteristics can obviously be used with different motivations by web designers, sometimes in very creative ways. Consequently, most visual patterns are not expected to be consistent across the

P. Perner (Ed.): MLDM 2018, LNAI 10935, pp. 132–146, 2018.
https://doi.org/10.1007/978-3-319-96133-0_10

World Wide Web. In order to rely on consistent patterns, visually oriented WIE methods have a limited range of application. Accordingly, these methods can be classified in two broad categories, i.e. whether the visual regularities rely on a limited set of documents (a corpus) or on an object with recurrent visual cues (e.g. a table or a content block). In the former case, good performances are obtained at the expense of generality (i.e. the possibility to process unseen documents) and robustness (i.e. the stability of the method following template modifications). In the latter case, most visual information is ignored in order to rely on generic regularities.

Many disadvantages are associated with the use of currently available visually oriented WIE methods:

1. Extraction time is often compromised. Instead of relying on the HTML response of the HTTP request, CSS properties must be computed for all document object model (DOM) nodes. On large sets of documents, extraction time can become impractical.
2. Visually oriented WIE methods are laborious to develop. Visual information is managed by means of ad hoc knowledge, i.e. rules or patterns that have been defined to fit a specific need. These extraction rules have to be crafted by experts or learned through (semi-)supervised algorithms on an annotated corpus.
3. Integration to existing systems can be arduous. In most cases, visually oriented WIE methods have to be combined with other extraction methods based on different types of patterns (e.g. across HTML tags). This can necessitate considerable efforts and/or reengineering of existing systems.
4. Visual patterns can compromise system's robustness. Template modifications can have consequences on exploited regulaties. In the industry, robustness (and subsequently maintenance) is a key issue of WIE systems.

Although visual information plays a key role in the meaning of web documents, the use of visually oriented WIE methods involves significant drawbacks. This gap between the importance of visual information in web documents and the possibility to optimally exploit this type of information in web mining tasks has motivated our research.

This paper shows how visual information from a set of formerly extracted entities can be used following a WIE task to identify false positive entities and hence further improve extraction results. The method is unsupervised and does not rely on any visual pattern. Instead, we focus on identifying visual outliers, i.e. entities that visually differ from the norm. In the context of WIE, we show that visual outliers tend to be erroneous extracted entities. We assume that (1) state-of-the-art WIE systems extract more true positive entities than false positive entities and (2) extracted entities are visually similar.

The advantages of the proposed validation method are the following:

1. The method is unsupervised and do not require any annotation, learning, or rule definition.

2. The method can be integrated into most state-of-the-art WIE systems effortlessly, as it is a validation process based on formerly extracted entities.
3. The method has no impact on the robustness or maintenance of the system because it does not rely on visual patterns and is used for validation.
4. Only computed visual properties of extracted entities are required, which can represent a substantial saving in computation time compared to extraction methods that rely on visual patterns.

In order to validate our method, we post-processed the entities obtained by the top state-of-the-art extraction algorithm Boilerpipe, which is the best overall main content extraction algorithm for web documents [24]. Main content extraction is an important WIE task for both research and industry, as it allows to remove the surplus "clutter" (boilerplate, templates) around the main textual content of a web page and improve subsequent extraction tasks. Our method improves Boilerpipe's initial precision by more than 10% while F_1 score is increased by at least 3% in all relevant cases.

The contributions of this paper are:

1. We show that in a set of formerly extracted entities obtained by a state-of-the-art data extraction algorithm, visual outliers tend to be false positive entities.
2. We also show that visual outliers can be eliminated in order to improve precision and F_1 score (i.e. with minimal impact on recall).
3. We show that two established anomaly detection algorithms (k-NN and HBOS) can be used to identify relevant visual outliers.
4. We improve Boilerpipe main content extraction algorithm, which is the best overall main content extraction algorithm.
5. More generally, we show how visual information of web documents can be exploited in an unsupervised manner in web data extraction tasks without impacting on system's flexibility and robustness.
6. To our knowledge, we are the first authors to introduce *visual outliers*, i.e. point anomalies based on visual information, in WIE.

2 Background

Web wrappers (also called *web extractors*) are algorithms that extract data from unstructured or semi-structured web sources and map them to a suitable structured format for further processing [7]. However, wrappers rarely embed the full process executed by browsers [11,24].

When a human user visits a website the web browser creates a well-formed DOM tree from the (possibly broken) HTML code contained in the HTTP response. We will refer to this tree as the *first DOM tree*. The DOM is a W3C recommendation that allows programs and scripts to dynamically access and update the content, structure and style of documents. The browser then parses stylesheets and generates style boxes for all elements of the DOM tree according to the CSS box model and CSS visual formatting model [5]. JavaScript code is parsed and executed in order to update HTML elements, attributes, CSS style properties and events, yielding what we will call the *rendered DOM tree*. Finally, the browser renders the page on the user's screen.

Most web wrappers rely on the first DOM tree in order to extract information [11,24] despite the fact that it represents an approximation of the resulting page, i.e. a lone well-formed HTML document [5] where the inherent complexity of visually rich websites is mostly ignored [20,24]. Wrappers solely based on the first DOM tree are limited, particularly when:

1. HTML structure is highly variable or complex [9].
2. HTML code is not written properly (e.g. when a table is defined with *div* elements with absolute positioning) [11,12].
3. JavaScript code is present [9,20,24].
4. Web pages are visually rich [4,5,8–11,13].
5. Source code is hardly accessible [9].

Visually oriented WIE methods[1] usually overcome such limitations by dealing with the rendered representation of web documents.

3 Anomaly Detection

Anomaly detection is the process of identifying unexpected items or events in a dataset [14]. Unsupervised methods mostly deal with point anomalies, i.e. single anomalous instances that differ from the norm. These methods typically return, for each instance, a score based on the intrinsic properties of the dataset. This score is interpreted as a degree of abnormality [14]. For the detection of visual outliers in a set of extracted entities, we will focus on point anomalies and therefore use unsupervised methods.

Point anomalies can furthermore differ locally or globally from the norm, i.e. depending if each item is compared to the whole dataset or only to its closest neighborhood. In our case, visual outliers differ from all other elements, so we restrict ourselves to global anomaly detection methods.

Goldstein and Uchida [14] present a comparative evaluation of 19 unsupervised anomaly detection algorithms on ten different datasets from multiple application domains. For global anomaly detection, as in our case, they recommend k-nearest-neighbors based algorithms if computation time is not an issue and histogram-based algorithms when computation time is essential, especially for large datasets.

K-nearest-neighbors techniques assume that normal items occur in dense neighborhoods while anomalous instances are far from their closest neighbours [6]. The anomaly score of each item is computed relatively to the distance of its k-nearest-neighbors. This distance can be measured relatively to the kth-nearest-neighbors or to the average distance of all of the k-nearest-neighbors. The first

[1] In the literature, visual web information extraction may refer to the use of a graphical user interface (GUI) that allows the user to generate wrappers. This is not the intended meaning here as we refer to the visual formatting of documents.

method is referred to as k^{th}-NN and the latter as k-NN. In practical applications, k-NN is often preferred [14].

Histogram-based anomaly detection algorithms use histograms to maintain a profile of normal instances. Such approach is also referred to as frequency-based or counting-based [6]. For each feature of the dataset, a histogram is created based on the different values taken by that feature. Then each instance is evaluated according to the profile of its features. The most common approach for unsupervised histogram-based algorithms is to compute an anomaly score based on the height of the histogram in which each feature falls. The histogram-based outlier score (HBOS) algorithm proposed by Goldstein and Uchida [14] is obtained by multiplying the inverse heights of each histogram—representing the density estimation—in which the features resides. It is worth noting that HBOS assumes the independence of the features. This allows a fast processing speed. In some cases HBOS can process a dataset under a minute, whereas nearest-neighbors based algorithms take over 23 h [15].

We will hence use both k-NN and HBOS algorithms for visual outlier detection.

4 Proposed Method

The proposed method consists of three steps, as shown in Fig. 1:

1. *Information gathering*: Given a set of extracted DOM nodes by some WIE system, we use the XML Path Language (XPath) expressions of each node and their related URLs in order to obtain their visual characteristics from the rendered DOM tree, i.e., the computed style properties, as their are shown to the user.
2. *Anomaly Detection*: Given the set of CSS characteristics of all extracted nodes, we use an unsupervised anomaly detection algorithm to detect visual outliers. For each node, we hence obtain an *anomaly score* for which high scores denote entities that differ the most from the norm.
3. *Cleaning task*: Based on the anomaly scores, we rank the nodes in descending order and we successively delete the most visually abnormal ones. Visual outliers are eliminated from the initial set of extracted nodes and an improved set of nodes is returned.

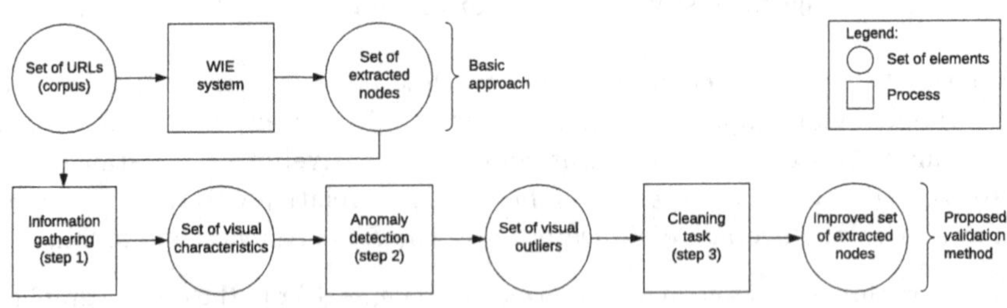

Fig. 1. The successive steps of the proposed method

4.1 Rationale

The rationale of our approach is that WIE systems aim at extracting entities of a specific kind, such as titles, prices, product descriptions, the main content of news articles, etc. These entities can either have a simple structure like titles (strings) and prices (numbers), or a complex one such as product descriptions (formed of specific fields) or main content of news articles (containing a title, a publication date, author's names, a location, subtitles, paragraphs and correction notes).

Assuming a WIE system with fair performance, most extracted DOM nodes should belong to the same type of entity or, in the case of complex entities, should be composed of a fixed set of types (e.g., all news articles should contain a title, an author, a publication date, etc.). We should also expect that similar entities share similar visual characteristics. For example, a set of titles extracted from scientific articles should share some common visual patterns. However, we do not try to identify these visual similarities, but only leverage their existence in order to detect and remove visual outliers.

4.2 Data Normalization

CSS values can be quantitative or qualitative. We normalize all data in order to deal exclusively with numeric values. Moreover, k-NN algorithm assumes that attributes are normalized and are of equal importance. We use standard normalization techniques for data mining as shown in [25]:

1. If value is numeric then we keep it as it is.
2. If value is a RGB color then R, G and B are separated in three different columns, each containing a numeric value.
3. If value is numeric but followed by some unit of measurement then we use the same unit and only keep the numerical part.
4. If value is qualitative then we create a new column for each categorial possibility (according to the W3C CSS specification) and store the value as binary numbers (i.e. "1" in the corresponding column, else "0").

Consequently, we obtain a set of multivariate tabular data where all CSS values are numeric. There are hundreds of CSS properties and most of them are in practice rarely used. Therefore, we retain the 32 CSS properties where variations are the most frequent.[2]

[2] Retained properties are the following: background-color; border-bottom-color; border-bottom-style; border-bottom-width; border-left-color; border-left-style; border-left-width; border-right-color; border-right-style; border-right-width; border-top-color; border-top-left-radius; border-top-right-radius; border-top-style; border-top-width; color; font-size; font-style; font-weight; margin-bottom; margin-left; margin-right; margin-top; outline-color; padding-bottom; padding-left; padding-right; padding-top; position; text-align; text-decoration; visibility;.

5 Evaluation

In this section we evaluate our method based on the extraction results of Boilerpipe[3], a well-known main content extraction algorithm [18,24]. Boilerpipe aims to remove the surplus "clutter" (boilerplate, templates) around the main textual content of a web page (i.e. all content that is not related to the main content, e.g. navigational elements, advertisements, footers, etc.). It uses a set of shallow text features—such as text density and link density—to classify the individual text elements in web pages. Boilerpipe is the overall best algorithm for main content extraction [19,24] and has a specific strategy that is tuned towards news articles.

5.1 Performance

Standard measures of performance for wrappers are precision, recall and F_1 score. Precision (P) is the quotient $\frac{TP}{TP+FP}$ of the number of true positive elements on the number of retrieved elements (true and false positives). Recall (R) is the quotient $\frac{TP}{TP+FN}$ of the number of true positive elements on the number of relevant elements (true positives and false negatives). F_1 score is the weighted harmonic mean $\frac{2P \cdot R}{P+R}$ of precision and recall.

Since our method consists in filtering out some elements extracted by an existing WIE system, the number of retrieved true positives can only decrease while the absolute number of true positives and false negatives remains constant. Recall can hence only decrease. However, retaining very few elements (low recall) could dramatically improve precision. We will hence use the F_1 score to evaluate our method.

Finally, in order to compute precision, recall and F_1 score, the number of elements can be calculated either from the number of bytes or from the number of DOM nodes. However, in content extraction the number of bytes is more relevant, hence we use this measure. One could also argue that there could be a large number of irrelevant nodes containing few bytes. Counting nodes would then artificially boost the impact of our method.

5.2 Dataset

There are two fairly well known datasets for the evaluation of content extraction algorithms. Cleaneval evaluation dataset[4], whose documents mostly date to 2006 and L3S-GN1 from the authors of Boilerpipe, which contains Google news articles dating mostly from 2008.

However, documents in these datasets consist of basic HTML files without CSS and Javascript files. Consequently, we created our own dataset. Similarly to L3S-GN1, we used Google News to obtain the first one hundred news articles

[3] https://boilerpipe-web.appspot.com/.

[4] https://cleaneval.sigwac.org.uk/.

from three different sources (300 news articles in total): The New York Times[5], The Guardian[6], and Le Devoir[7]. The size of our dataset is comparable to similar datasets as no learning task is required and all documents are used for evaluation (L3S-GN1 has 740 documents and L3S-GN1 has 621 documents).

Since our goal is not to re-evaluate Boilerpipe, we did not proceed to the annotation of the whole corpus but rather identified all false positives resulting from the Boilerpipe tool on our corpus. For annotation we used the CleanEval guidelines[8] for boilerplate removal.

On the corpus on which it has been trained (L3S-GN1), Boilerpipe's articles extractor obtains a precision of 0.9312, a recall of 0.9550, and a F_1 score of 0.9388 [19]. On another well-known corpus (CleanEval), it obtains a precision of 0.9485, a recall of 0.7643, and a F_1 score of 0.8041. Weninger et al. [24] evaluated the precision of Boilerpipe's articles extractor on a corpus of recent websites (2015) of all kinds (i.e. not just news articles). They obtained a precision of 0.8579, a recall of 0.6321, and a F_1 score of 0.7279.

On our corpus, Boilerpipe's articles extractor obtains a precision of 0.8442 on 10696 extracted entities. Boilerpipe's recall is however unknown because we only annotated false positives from retrieved entities. In order to tackle this uncertainty of recall we will compute F_1 values from estimated recall values of 0.65, 0.75, 0.85, and 0.95 and show a consistent improvement of F_1 scores across all these values.

5.3 Experimental Setup

We used the Boilerpipe Java Library[9] with the "ArticleExtractor" strategy, i.e. the best strategy for extracting the main content of news articles. Boilerpipe relies on the first DOM tree instance of web pages and it is therefore necessary to map the nodes of the first DOM tree to the nodes of the rendered DOM tree.

We use XPath expressions in order to localize the nodes and obtain their computed style properties. XPath is the W3C recommended and preferred tool to address nodes of the DOM and has been largely used in WIE systems [11]. Although it is not a difficulty for most WIE systems to associate extracted nodes to XPath expressions, computed style properties are generated according to the nodes of the rendered DOM tree. Consequently, XPath expressions must be valid on the rendered DOM tree. When extracted nodes rely on the first DOM tree, our method requires to map the XPath expressions of extracted nodes (of the first DOM tree) to XPath expressions of the corresponding nodes in the rendered DOM tree.

Instead of modifying Boilerpipe's library in order to obtain the XPath expressions of extracted nodes, we used an already implemented debug function that

[5] https://www.nytimes.com/.
[6] https://www.theguardian.com/.
[7] http://www.ledevoir.com/.
[8] https://cleaneval.sigwac.org.uk/annotation_guidelines.html.
[9] https://github.com/kohlschutter/boilerpipe.

shows how Boilerpipe segments the page in different sections. Each section is numbered by Boilerpipe according to its order of appearance in the HTML document. Consequently, it is possible to associate each section to a set of nodes in the rendered DOM tree. After the extraction process the same debug function can be used to get the list of all sections that have been identified as a part of the main content. Consequently, we obtain the set of all XPath expression of extracted nodes by Boilerpipe.

In order to access the rendered DOM tree, we use PhantomJs[10], a well-known headless browser [24]. PhantomJs allows manipulation to rendered web pages through a JavaScript API. Computed style properties are obtained with the *getComputedStyle()* javascript function[11]. As a result, PhantomJs returns the CSS properties of requested DOM nodes and saves them in a CSV file. Then a simple script normalizes all values in the CSV file according to the previously discussed normalization strategy.

Anomaly detection algorithms take this normalized CSV file as an input. For HBOS and k-NN we used Goldstein et al.'s RapidMiner[12] library [14]. An anomaly score is computed for each node and Rapidminer adds this score to the original CSV file.

Finally, extracted nodes are ranked in descending order according to their anomaly score and are successively deleted one by one. Through the deletion process we compute precision, recall, and F_1 scores.

5.4 Results

Figure 2 shows the distribution of Boilerpipes's retrieved elements (true positives in blue and false positives in red) according to their computed visual anomaly scores. The distribution of false positives in function of their anomaly scores for 100-NN, 200-NN, 500-NN, and HBOS is similar, which confirms our hypothesis that visual outliers—i.e. entities with high visual anomaly scores—tend to be false positives (red). It is worth noting that there is a surprisingly high quantity of false positives for the main content extraction task despite a precision of 0.8442. Accordingly, these nodes contain significantly less bytes of information. For example, a subscription form in the core text of an article can generate dozens of nodes while a relevant paragraph is only associated to one node. It remains to be shown to which extent filtering out elements with the HBOS or k-NN algorithms can improve F_1 score.

In order to evaluate our method we remove nodes in descending visual anomaly scores stopping when the F_1 score is maximal.

As shown on the right of Table 1, the number of removed bytes ranges between 2% and 20%, while the number of removed nodes ranges between 5% and 50%. With the exception of 10-NN and 50-NN, which improve precision by less than

[10] http://phantomjs.org/.

[11] Most developer tools included in browsers, such as Firebug for Firefox or Chrome DevTools, allow to access computed style properties of DOM nodes.

[12] https://rapidminer.com/.

10% (see Table 2), the number of removed bytes ranges between 14%–20% and the number of removed nodes ranges between 35%–50%. The extent to which these values would be adequate for other corpora fall outside the scope of this paper and is left to further investigations.

As for the appropriate value of k for the k-NN algorithms, we experimented with different values, showing that precision and F_1 score improve up to around $k = 200$, with more minor gains afterwards (see Table 2).

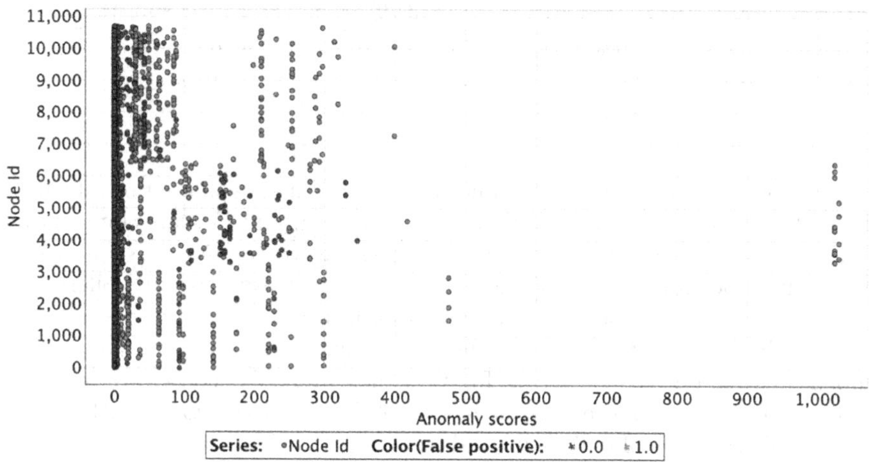

Fig. 2. Boilerpipe's retrieved entities; k-NN anomaly scores for $k = 200$. (Color figure online)

As said before, we identified all true and false positives hence we are able to compute exact values for precision (as shown on the left of Table 1). Furthermore, Table 2 shows that our method offers substantial precision improvement (more than 10%) in all cases except 10-NN and 50-NN.

Table 1. F_1 scores and related results for k-NN with different values of k and HBOS.

	Precision	Recall measures								% deleted nodes	% deleted bytes
		0.65		0.75		0.85		0.95			
		R	F_1	R	F_1	R	F_1	R	F_1		
Boilerpipe	0.8442	0.65	0.7345	0.75	0.7943	0.85	0.8471	0.95	0.8940	0	0
10-NN	0.8575	0.6449	0.7361	0.7441	0.7968	0.8433	0.8503	0.9425	08980	4.78%	2.32%
50-NN	0.8786	0.6407	0.7410	0.7392	0.8029	0.8378	0.8577	0.9364	0.9066	10.79%	5.29%
100-NN	0.9389	0.6213	0.7478	0.7169	0.8130	0.8126	0.8712	0.9081	0.9233	34.92%	14.05%
200-NN	0.9883	0.6129	0.7566	0.7072	0.8245	0.8015	0.8852	0.8958	0.9398	48.18%	19.46%
500-NN	0.9963	0.6129	0.7589	0.7072	0.8272	0.8015	0.8883	0.8958	0.9434	52.62%	20.10%
HBOS	0.9450	0.6345	0.7592	0.7321	0.8250	0.8297	0.8836	0.9273	0.9361	30.40%	12.80%

Table 2. Increase in % compared to Boilerpipe.

	Precision	Recall measures							
		0.65		0.75		0.85		0.95	
		R	F_1	R	F_1	R	F_1	R	F_1
10-NN	1.57%	−0.79%	0.22%	−0.79%	0.31%	−0.79%	0.38%	−0.79%	0.45%
50-NN	4.07%	−1.43%	0.89%	−1.43%	1.08%	−1.43%	1.25%	−1.43%	1.41%
100-NN	11.22%	−4.41%	1.81%	−4.41%	2.36%	−4.41%	2.84%	−4.41%	3.27%
200-NN	17.07%	−5.71%	3.01%	−5.71%	3.79%	−5.71%	4.49%	−5.71%	5.12%
500-NN	18.00%	−5.70%	3.33%	−5.70%	4.14%	−5.71%	4.86%	−5.71%	5.52%
HBOS	11.94%	−2.39%	3.37%	−2.39%	3.87%	−2.39%	4.31%	−2.39%	4.71%

However, we don't have exact values for recall. Using estimated recall values of 0.65, 0.75, 0.85, and 0.95 for Boilerpipe, we can compute F_1 scores for our own method. This allows us to show that in all cases except 10-NN, 50-NN and 100-NN, our method improves F_1 score by at least 3%, irrespective of Boilerpipe's estimated recall (see Table 2). While our approach cannot improve recall, this significant improvement in F_1 score shows that the increase in precision largely compensates the decrease of recall. Figure 3 shows how F_1 scores variate across the deletion of visual outliers for 200-NN, 500-NN, and HBOS algorithms and initial recall measures of 0.65 (left) and 0.95 (right).

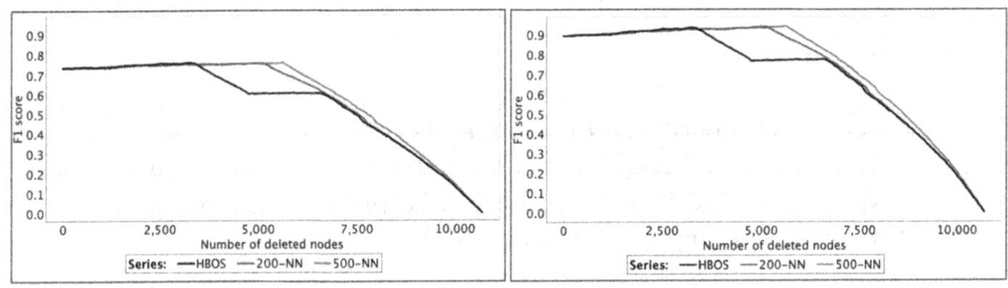

Fig. 3. Variation of F_1 scores across node deletion for 200-NN, 500NN, and HBOS for recall values of 0.65 (left) and 0.95 (right).

As shown in Table 2, the best improvements for precision are obtained with the 200-NN and 500-NN algorithms. While HBOS is outperformed by k-NN, this is at the expense of setting k to considerably high values. This tends to confirm Goldstein and Uchida's recommendation to use HBOS on large datasets [14].

Another interesting fact is that 200-NN and 500-NN must delete considerably more bytes (and nodes) than HBOS in order to reach similar F_1 scores. This is related to the distribution of false positives according to their anomaly score computed by each algorithm (see Fig. 2). Although most true positives have a low anomaly score, HBOS gives higher scores to more values than k-NN. From this point of view, HBOS is more *efficient* as it achieves similar results with less actions.

On our dataset 500-NN has a running time of approximately 10 s while HBOS running time is below a second. However, one would expect that on very large datasets HBOS would outperform 500-NN in terms of execution time.

6 Related Works

To our knowledge, we are the first authors to use *visual outliers*, i.e. point anomalies based on visual information for web data extraction. Agyemang et al. [1–3] introduced the concept of *web outlier* in data mining. While data mining is the process of discovering patterns in data [25], web outlier mining is defined as "the discovery and analysis of rare and interesting patterns from the web" [2]. Agyemang [1] presents a web content outlier mining framework that uses textual data of web pages in order to find web documents with varying contents from a set of similar documents. The discovery of web outliers from Agyemang's definition still relies on document patterns. Agyemang et al.'s original ideas have been extended in recent web text outlier mining applications [16,17]. Visual information of web documents is however not considered.

Recent works demonstrate the importance of visual information for data extraction. Apostolova and Tomuro [4] evaluated the performance of SVM classifiers on the task of identifying 12 types of named entities in online commercial real estate flyers. They show that the addition of visual features increase their overall F_1 score from 0.83 to 0.87 (4%), and up to 19% for visually salient features. Apostolova et al. give an excellent example of how corpus-specific visual patterns can improve extraction results.

Gogar et al. [13] uses convolutional neural networks in order to create web wrappers that can extract information in non-trivial cases. They propose a method for combining textual and visual information into a single neural net (called *Text Maps*). They evaluate their method on a task of product information extraction and show that the resulting wrapper can extract information on previously unseen websites with an overall accuracy of 93.7%. This method shows how important visual information is for web information extraction. Their results suggest that visual data itself can outperform textual data. Moreover, they show that the combination of both inputs (textual data and visual data) do not have a significant difference from the results achieved with only visual data (at least using convolutional neural networks [21]).

Visual information is also of great interest for anomaly detection, especially when dealing with media resources such as images and videos. Li et al. [22] show that burn injury diagnostic imaging devices can be improved by outlier detection. The deletion of outliers allows to reduce the variance of training data and improve device's accuracy from 63% to 76%. Vu et al. [23] present a unified framework for anomaly detection in video surveillance based on restricted Boltzmann machines (RBM). Their system works directly on image pixels rather than handcrafted features and is unsupervised as it does not require labels. Other examples include satellite imagery, spectroscopy, medical imagery and video surveillance [6].

There are also works in WIE that use anomaly detection. For example, FluxFlow [26] is a system designed for detecting, exploring, and interpreting anomalous conversational threads in Twitter. It integrates a visualization module that displays anomalous threads and their contextual information with various views in order to facilitate deeper analysis. 239 features are used to compute anomalous threads, which include user profile and user network features (individual level) and temporal and content features (thread level). FluxFlow does not include visual features.

Our motivation to improve state-of-the-art algorithm Boilerpipe has been influenced by a recent meta-analysis on web content extraction algorithms. Weninger et al. [24] evaluate 11 different algorithms for web content extraction on 4 news corpora from 2000 to 2015, each corpus containing 250 websites from 10 news sources spread over 5 year periods (i.e. 2000–2004, 2005–2009, 2010–2014, 2015). The study aims to evaluate how web content extractors are impacted by the changing web in order to make recommendations for future content extraction algorithms. Their results show that Boilerpipe has the best performance on most sets of documents but also underscore a robustness problem in web content extraction. In fact, most of the worst extraction results are obtained on the 2015 corpus. Weninger et al. argue that this is due to web's increasing reliance on external sources for content and data via JavaScript, iframes, etc. They give as an example the fact that the most frequent last-word found by many content extractors on New York Times articles is "loading...". Consequently, they recommend to perform content extraction on the rendered DOM tree (with tools like PhantomJs) in order to manage external scripts. They also suggest that visual information may improve extraction effectiveness, notwithstanding the fact that visual patterns can impact system's robustness. However, for most algorithms, these recommendations would require substantial modifications and may even be inconsistent with existing approaches.

In this paper, we showed that our validation method fulfills Weninger et al.'s recommendations for web content extraction, while maintaining robustness and minimizing integration efforts.

7 Conclusion

In this paper we presented a novel unsupervised validation method that uses visual information of formerly extracted entities in order to eliminate false positive entities and improve extraction results. We introduced the concept of *visual outliers*, i.e. point anomalies based on visual information in web information extraction. We showed that two established anomaly detection algorithms (k-NN and HBOS) can be used in order to identify relevant visual outliers. We applied our method to top state-of-the-art main content extraction algorithm Boilerpipe and showed that visual outliers can be eliminated in order to improve precision and F_1 score. The proposed validation method can be integrated effortlessly into most WIE systems without impacting on system's flexibility, robustness, and maintenance. Moreover, only computed visual properties of extracted entities are required, which can represent a substantial economy in computation time compared to extraction methods that rely on visual patterns. Future research projects will extend the scope of the proposed method in order to validate its application on large corpora, different extraction tasks, and other WIE methods.

Acknowledgements. The authors gratefully acknowledge the financial support of the Natural Sciences and Engineering Research Council of Canada (NSERC).

References

1. Agyemang, M.: Web content outlier mining: motivation, framework, and algorithms. University of Calgary (2006)
2. Agyemang, M., Barker, K., Alhajj, R.: Framework for mining web content outliers. In: Proceedings of the 2004 ACM Symposium on Applied Computing, pp. 590–594. ACM (2004)
3. Agyemang, M., Barker, K., Alhajj, R.: Web outlier mining: discovering outliers from web datasets. Intell. Data Anal. **9**(5), 473–486 (2005)
4. Apostolova, E., Tomuro, N.: Combining visual and textual features for information extraction from online flyers. In: EMNLP, pp. 1924–1929 (2014)
5. Burget, R., Rudolfova, I.: Web page element classification based on visual features. In: 2009 First Asian Conference on Intelligent Information and Database Systems, ACIIDS 2009, pp. 67–72. IEEE (2009)
6. Chandola, V., Banerjee, A., Kumar, V.: Anomaly detection: a survey. ACM Comput. Surv. (CSUR) **41**(3), 15 (2009)
7. Chang, C.H., Kayed, M., Girgis, M.R., Shaalan, K.F.: A survey of web information extraction systems. IEEE Trans. Knowl. Data Eng. **18**(10), 1411–1428 (2006)
8. Chenthamarakshan, V., Varadarajan, R., Deshpande, P.M., Krishnapuram, R., Stolze, K.: WYSIWYE: an algebra for expressing spatial and textual rules for information extraction. In: Gao, H., Lim, L., Wang, W., Li, C., Chen, L. (eds.) WAIM 2012. LNCS, vol. 7418, pp. 419–433. Springer, Heidelberg (2012). https://doi.org/10.1007/978-3-642-32281-5_41
9. Della Penna, G., Magazzeni, D., Orefice, S.: Visual extraction of information from web pages. J. Vis. Lang. Comput. **21**(1), 23–32 (2010)
10. Della Penna, G., Magazzeni, D., Orefice, S.: A spatial relation-based framework to perform visual information extraction. Knowl. Inf. Syst. **30**(3), 667 (2012)
11. Ferrara, E., De Meo, P., Fiumara, G., Baumgartner, R.: Web data extraction, applications and techniques: a survey. Knowl.-Based Syst. **70**, 301–323 (2014)
12. Gatterbauer, W., Bohunsky, P.: Table extraction using spatial reasoning on the CSS2 visual box model. In: Proceedings of the 21st National Conference on Artificial Intelligence (2006)
13. Gogar, T., Hubacek, O., Sedivy, J.: Deep neural networks for web page information extraction. In: Iliadis, L., Maglogiannis, I. (eds.) AIAI 2016. IAICT, vol. 475, pp. 154–163. Springer, Cham (2016). https://doi.org/10.1007/978-3-319-44944-9_14
14. Goldstein, M., Uchida, S.: A comparative evaluation of unsupervised anomaly detection algorithms for multivariate data. PLoS ONE **11**(4), e0152173 (2016)
15. Goldstein, M.B.: Anomaly Detection in Large Datasets. Verlag Dr. Hut, Munich (2014)

16. Huosong, X., Zhaoyan, F., Liuyan, P.: Chinese web text outlier mining based on domain knowledge. In: 2010 Second WRI Global Congress on Intelligent Systems (GCIS), vol. 2, pp. 73–77. IEEE (2010)
17. Khan, M.R.R., Ahmed, M.I., Riyad, M.A.: A novel analytical approach for identifying outliers from web documents. Int. J. Appl. Eng. Res. **12**(22), 12156–12161 (2017)
18. Kohlschütter, C., Fankhauser, P., Nejdl, W.: Boilerplate detection using shallow text features. In: Proceedings of the Third ACM International Conference on Web Search and Data Mining, pp. 441–450. ACM (2010)
19. Kovacic, T.: Evaluating Web Content Extraction Algorithms. University of Ljubljana, Ljubljana (2012)
20. Krüpl-Sypien, B., Fayzrakhmanov, R.R., Holzinger, W., Panzenböck, M., Baumgartner, R.: A versatile model for web page representation, information extraction and content re-packaging. In: Proceedings of the 11th ACM Symposium on Document Engineering, pp. 129–138. ACM (2011)
21. LeCun, Y., Bengio, Y., Hinton, G.: Deep learning. Nature **521**(7553), 436 (2015)
22. Li, W., Mo, W., Zhang, X., Lu, Y., Squiers, J.J., Sellke, E.W., Fan, W., DiMaio, J.M., Thatcher, J.E.: Burn injury diagnostic imaging device's accuracy improved by outlier detection and removal. In: SPIE Defense+ Security, p. 947206. International Society for Optics and Photonics (2015)
23. Vu, H., Nguyen, T.D., Travers, A., Venkatesh, S., Phung, D.: Energy-based localized anomaly detection in video surveillance. In: Kim, J., et al. (eds.) PAKDD 2017. LNCS (LNAI), vol. 10234, pp. 641–653. Springer, Cham (2017). https://doi.org/10.1007/978-3-319-57454-7_50
24. Weninger, T., Palacios, R., Crescenzi, V., Gottron, T., Merialdo, P.: Web content extraction: a meta-analysis of its past and thoughts on its future. ACM SIGKDD Explor. Newsl. **17**(2), 17–23 (2016)
25. Witten, I.H., Frank, E., Hall, M.A., Pal, C.J.: Data Mining: Practical Machine Learning Tools and Techniques. Morgan Kaufmann, Los Altos (2016)
26. Zhao, J., Cao, N., Wen, Z., Song, Y., Lin, Y.R., Collins, C.: # FluxFlow: visual analysis of anomalous information spreading on social media. IEEE Trans. Vis. Comput. Graph. **20**(12), 1773–1782 (2014)

An Efficient Approximate EMST Algorithm for Color Image Segmentation

Xia Li Wang[1(✉)] and Xiaochun Wang[2]

[1] School of Information Engineering, Changan University, Xi'an 710049, China
xlwang@chd.edu.cn

[2] School of Software Engineering, Xi'an Jiaotong University, Xi'an 710061, China
xiaocchunwang@mail.xjtu.edu.cn

Abstract. Efficient Euclidean minimum spanning tree algorithms have been proposed for large scale datasets which run typically in time near linear in the size of the data but may not usually be feasible for high-dimensional data. For data consisting of sparse vectors in high-dimensional feature spaces, however, the calculations of an approximate EMST can be largely independent of the feature space dimension. Taking this observation into consideration, in this paper, we propose a new two- stage approximate Euclidean minimum spanning tree algorithm. In the first stage, we perform the standard Prim's MST algorithm using Cosine similarity measure for high-dimensional sparse datasets to reduce the computation expense. In the second stage, we use the MST obtained in the first stage to complete an approximate Euclidean Minimum Spanning Tree construction process. Experimental results for color image segmentation demonstrate the efficiency of the proposed method, while keeping high approximate precision.

Keywords: Minimum spanning tree · Approximate minimum spanning tree
Euclidean minimum spanning tree · Euclidean distance
Cosine similarity measure

1 Introduction

Given N data points in d-dimensional space R^d, a Euclidean minimum spanning tree or EMST is a minimum spanning tree in which the weight of the edge between every pair of points is their Euclidean distance. In a simpler term, an EMST is a tree that connects a set of data points using lines such that the total sum of all the line lengths is minimized and any data points can be reached from others by following the lines. However, in today's EMST tasks for a modern large-scale dataset of size N, there are $V = N$ vertices and $E = N(N - 1)/2$ edges in the complete graph and standard EMST algorithms have a time complexity roughly equal to $O(dN^2)$ [1, 2]. For standard EMST algorithms to be efficient, some index structures, such as kd-tree [3], come in handy. However, for high-dimensional data, kd-tree based approaches will lose their efficiency and perform even worse than the naïve brute force methods (e.g., the Prim's algorithm

P. Perner (Ed.): MLDM 2018, LNAI 10935, pp. 147–159, 2018.
https://doi.org/10.1007/978-3-319-96133-0_11

[2]). Fortunately, in many practical applications, an exact EMST can be generally replaced by an approximate one without degrading the quality of the final application.

Due to its extensive applications in image segmentation [4, 5], cluster analysis [6–8], classification [9], and manifold learning [10], many EMST algorithms have been developed. In particular, we are interested in EMST applications in data clustering, often referred to as a single-linkage clustering in the clustering literatures [4, 11, 12]. When the weight associated with each edge denotes a distance between two end points, any edge in an EMST will be the shortest distance between two subtrees that are connected by that edge. Therefore, removing the longest edge will theoretically result in a two-cluster grouping. Removing the next longest edge will result in a three-cluster grouping, and so on. This corresponds to choosing the breaks where the maximum weights occur in the sorted edges. Though simple and elegant, EMST-based clustering algorithms usually begin by constructing an EMST over a given dataset, which is often the bottle neck for fast EMST-based applications.

Being an important data mining technique, clustering algorithm aims to partition a given dataset into groups such that intra-cluster similarity is high while inter-cluster similarity is low. For this purpose, a wealth of similarity criteria (e.g., Euclidean distances, Jaccard index, etc.) have been developed. In statistics and related fields, a similarity measure or similarity function is a real-valued function that quantifies the similarity between two objects and takes on large values for similar objects and either zero or a negative value for very dissimilar objects. Although no single definition of a similarity measure exists, usually such measures are in some sense the inverse of distance metrics. Cosine similarity and Euclidean distance are example of this observation. Cosine similarity is a measure of similarity between two non-zero vectors that measures the cosine of the angle between them. The cosine of 0 is 1, and it is less than 1 for any other angle in the interval $[0, 2\pi)$. It is thus a judgment of orientation and not magnitude. Two vectors with the same orientation have a cosine similarity of 1, two vectors at 90° have a cosine similarity of 0, and two vectors diametrically opposed have a similarity of -1, independent of their magnitude. Cosine similarity is particularly used in positive space, where the outcome is neatly bounded in [0, 1]. Note that these bounds apply for any number of dimensions, and cosine similarity is most commonly used in high-dimensional positive spaces. In mathematics, a metric or distance function is a function that defines a distance between each pair of elements of a set. A set with a metric is called a metric space. The most familiar metric space is d-dimensional Euclidean space. In fact, a "metric" is the generalization of the Euclidean metric arising from the four long-known properties of the Euclidean distance. The Euclidean metric defines the distance between two points as the length of the straight line segment connecting them.

In this paper, we propose a two-stage approximate EMST algorithm that is both computationally efficient and competent with the state-of-the-art EMST algorithms. In the first stage, Cosine similarity is used in the computation of an exact MST. Since the Cosine similarity measures the similarity between two non-zero vectors in an inner product fashion, the construction of an MST based on this similarity measure can be very efficient, especially for high-dimensional highly sparse data. In the second stage, an approximate EMST is constructed by replacing the Cosine dissimilarity value with the Euclidean distance between the corresponding two nodes of the MST produced in the first stage. Experiments conducted on two image data manifest its effectiveness and efficiency in comparison with state-of-the-art EMST algorithms.

The rest of this paper is organized as follows. Section 2 gives a review of some related work in MST. Section 3 presents the proposed approximate EMST algorithm. In Sect. 4, we present the results of experiments conducted to evaluate the performance of the proposed algorithm. Finally, conclusions are given in Sect. 5.

2 Related Work

Being the first MST algorithm, Borůvka's algorithm begins with each vertex of a given connected and weighted graph G = (E, V) being a tree. For each consecutive iteration, it selects the shortest edge from a tree to another tree and combines them until all the trees are combined into one tree [13]. Being another popular MST algorithm, Kruskal's algorithm starts with sorting all the edges by their weights in a non-decreasing order, treats each vertex as a tree, and iteratively combines the trees by adding edges in the sorted order excluding those leading to a cycle until all the trees are combined into one tree [1]. Proposed independently by Jarník [14], Prim [2] and Dijkstra [15], the famous Prim's algorithm first arbitrarily selects a vertex as a tree and then repeatedly adds the shortest edge that connects a new vertex to the tree until all the vertices are included. The time complexity of these classic MST algorithms is O(ElogV).

Unfortunately, these classical MST algorithms require a quadratic running time to construct an EMST for modern large datasets. To improve, in 1978, Bentley and Friedman proposed to augment Prim's algorithm with a kd-tree to enhance the search for the next edge to add to the tree, which can result in an O(NlogN) running time for most data distributions [3]. To provide a theoretical analysis, in 1985, Preparata and Shamos gave a lower bound for the EMST problem of (NlogN), which has been the tightest known lower bound [16]. Being the basis of most recent EMST algorithms, the concept of Well-Separated Pair Decomposition (WSPD) was proposed by Callahan and Kosaraju's in 1993 [17]. Following this line, in 2000, Narasimhan and Zachariasen introduced WSPD to Boruvka's algorithm to find edges of an MST [18]. Realizing that the constant in the O(N) size of the WSPD grows exponentially with the data dimension and is often very large in practice, in 2010, March et al. presented a new dual-tree algorithm for efficiently computing the EMST, which is superficially similar to the method in [17] except that the WSPD is replaced by a new dual-tree data structure and referred to in the following as FEMST algorithm [19]. Experiments conducted on large scale astronomical data sets demonstrated the scalability of their method.

Being an alternative, in addition to exact EMST problem, approximate EMST algorithms have been also developed since, in many practical applications, an exact EMST can be generally replaced by an approximate one without degrading the quality of the final application. In 1988, Vaidya employed a group of grids to partition a data set into cubical boxes of identical size [20]. For each box, a representative point was determined to which points within the cubical box were connected. Any two representatives of two cubical boxes were connected if corresponding edge length was between two specific thresholds. In 1993, Callahan and Kosaraju utilized WSPD of a data set to extract a sparse graph from the complete graph and then applied an exact MST algorithm to it [17]. In 2009, Wang et al. proposed to detect longest edges in an EMST at an early stage by employing a divide-and-conquer scheme to construct an approximate EMST for clustering [21]. At the same year, Lai et al. proposed a two-stage Hilbert curve based approximate EMST algorithm for clustering as well [23]. In 2014, Wang et al. proposed a fast EMST algorithm, which is superficially similar to the method in [3] except that the kd-tree is replaced by the iDistance indexing structure for fast kNN search in high-dimensional datasets [24]. The authors argued that the algorithm had an expected O(NlogN) running time, but did not prove this rigorously. In 2015, Zhong et al. proposed a fast two-stage Euclidean minimum spanning tree (FEMST) algorithm which employs a divide-and-conquer scheme to produce an approximate EMST with theoretical time complexity of $O(N^{1.5})$ [24]. In the first stage, K-means is employed to partition a dataset into $N^{1/2}$ clusters, to each of which an exact EMST algorithm is applied. The produced $N^{1/2}$ EMSTs are connected to form an approximate EMST. In the second stage, the dataset is repartitioned so that the neighboring boundaries of a neighboring pair produced in the first stage are put into a cluster. With these $N^{1/2} - 1$ clusters, another approximate EMST is constructed. Finally, the two approximate EMSTs are combined into a graph and a more accurate EMST is generated from it.

3 The Proposed Approximate EMST Algorithm

In this section, we propose a two-stage approximate EMST algorithm based on a more analytical examination on Euclidean distance definition and Cosine similarity definition. Based on this observation, a new fast approximate EMST algorithm is then developed for high-dimensional highly sparse image data.

3.1 A Simple Idea

Given two data points in d-dimensional space, $x = (x_1, x_2, \ldots, x_d)$ and $y = (y_1, y_2, \ldots, y_d)$, the standard Euclidean distance (dist) from x to y, or from y to x is given by the following formula,

$$dist_{\text{Euclidean}}(x, y) = dist_{\text{Euclidean}}(y, x) = \sqrt{\sum_{i=1}^{d} (x_i - y_i)^2} \quad (1)$$

The position of a point in a Euclidean d-space is a Euclidean vector. So, x and y are Euclidean vectors, starting from the origin of the space, and the Euclidean norm, Euclidean length, or magnitude of a vector measures the length of the vector,

$$\|x\| = \sqrt{x_1^2 + x_2^2 + \cdots + x_d^2} = \sqrt{x \bullet x} \quad (2)$$

The position of a point in a Euclidean d-space is a Euclidean vector. The normalized Euclidean distance between two vectors, x and y, is defined to be,

$$dist_{\text{Euclidean-normalized}}(x, y) = \sqrt{\sum_{i=1}^{d} \left(\frac{x_i}{\|x\|} - \frac{y_i}{\|y\|}\right)^2} \tag{3}$$

Being most commonly used in high-dimensional positive spaces, as illustrated in Fig. 1, Cosine similarity between x and y is defined to be,

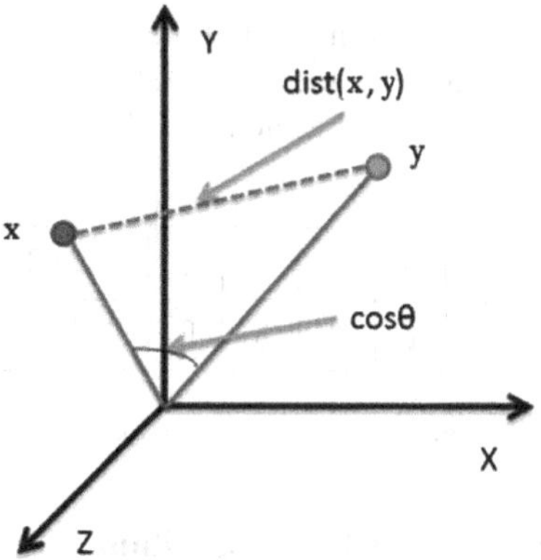

Fig. 1. An illustration of Euclidean distance and Cosine similarity.

$$sim_{consine}(x, y) = \cos(\theta) = \frac{x \bullet y}{\|x\| \cdot \|y\|} \tag{4}$$

Cosine distance is a term often used for the complement in positive space and defined to be,

$$dist_{Co\sin e}(x, y) = 1 - sim_{con\sin e}(x, y) = 1 - \cos(\theta) = 1 - \frac{x \bullet y}{\|x\| \cdot \|y\|} \tag{5}$$

According to the definitions of the normalized Euclidean distance and Cosine distance, we have,

$$dist_{\text{Euclidean-normalized}}(x, y) = \sqrt{\sum_{i=1}^{d} \left(\frac{x_i}{\|x\|} - \frac{y_i}{\|y\|}\right)^2} == \sqrt{2 - 2\cos(\theta)} = \sqrt{2dist_{\text{cosine}}(x, y)} \tag{6}$$

From Eq. (6), it can be seen that, if the Euclidean norm for each data vector is pre-computed, the calculation of the normalized Euclidean distance between two vectors can be reduced to the computation of the inner product of two vectors. Further, if data vectors in high-dimensional data are highly sparse as shown in the following figure, a considerate amount of distance computations can be saved.

Therefore, for highly sparse vectors where only those dimensions for non-zero values need to be considered, not only pairwise distance computations can be reduced to the inner product among a few dimensions, but also the vectors can be represented more efficiently by only remembering the dimensions for which the attribute values are not zeros, which is especially good for highly sparse high-dimensional datasets. For example, for vector y in Fig. 2, only two integers (for the nonzero dimensions) and two float values can be used to represent it in stead of the original eleven float values. Then for inner product computation, only the second and the eleventh dimensions of vector x need to be checked since vector y only has nonzero elements in these two dimensions.

x	7	0	5	0	0	0	0	0	0	3	0
y	0	1	0	0	0	0	0	0	0	0	9

Fig. 2. Two sample vectors.

3.2 Our Proposed Approximate EMST Algorithm

Based on the above discussion, our proposed approximate EMST (AEMST) algorithm can be summarized in the following Table 1.

Table 1. An AEMST Based Clustering Algorithm

Input:	S: a set of N data objects; p: a loosely estimated minimum number of data points in a cluster; q: a loosely estimated maximum number of data points in a cluster
Output:	C: a desired set of clusters
Begin:	
1:	Perform standard Prim's algorithm to construct a MST using cosine distance, that is, Eq. (6). The resulted MST is contained in two arrays, one for storing the edge weight and the other for remembering the parent index of each node
2:	Update the obtained MST by replacing the edge weights with the exact Euclidean distances, that is, Eq. (1), between each node and its parent node calculated using the Euclidean distance definition to obtain the approximate EMST
3:	To obtain a cluster label for each data point, sort all the edges of the produced EMST in a non-increasing order, always cut the largest edges and cut an edge only when the sizes of both clusters resulted by cutting that edge are larger than a loosely estimated minimum number of data points, until the size of the largest cluster becomes smaller than an estimated maximum number of data points
5:	Return the obtained set of clusters, C
End	

To summarize, the numerical parameters the algorithm needs from the user include the data set, S, the loosely estimated maximum number of data points in a group and the loosely estimated minimum number of data points in a group.

3.3 Color Image Segmentation Model

EMST-based clustering receives much attention as a competitive clustering algorithms emerging in recent years and is mainly applied to image segmentation. To allow things to be recognized through the process of assembling sensory information into a useful and reliable representation of the world, there are many methods available for feature binding and sensory segmentation in practice. Currently, the most competitive approaches for image segmentation are formulated as clustering models for probabilistic grouping of distributional feature vectors. Based on the feature vectors extracted from image patches, image segmentation can be realized by partitioning the set of image patches into a number of disjoint clusters or segments.

Natural images contain statistical regularities which distinguish objects from each other and from random noise. For successful visual recognition, each object must have attributes that can be used to differentiate it from others and to segment the whole image into meaningful objects. For image segmentation and object identification, instead of using the color information of a single pixel, color pixels in a small local region of an image are considered to form a color histogram-based feature vector. As illustrated in Fig. 3, to obtain feature vectors, for a given image, a moving window of size $N \times N$ hops by M pixels in the row and column directions but not to exceed the border of the image. The moving windows are overlapping to allow a certain amount of fuzziness to be incorporated so as to obtain a better segmentation performance. The window size controls the spatial locality of the result and the window hopping step controls the resolution of the result. A decrease in the step gives rise to an increased resolution but an increased processing time.

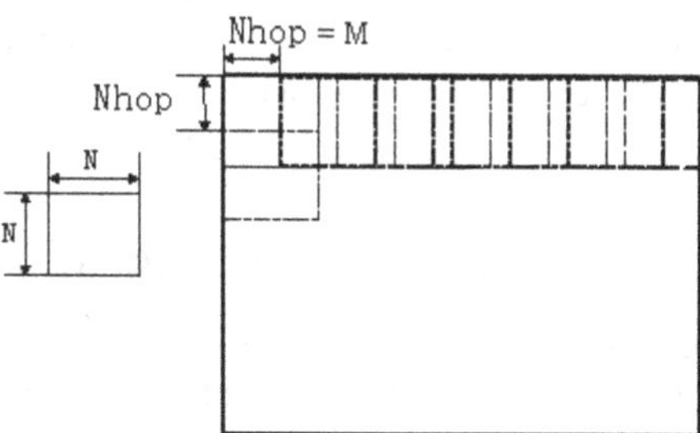

Fig. 3. Image segmentation model.

To obtain the color features, the images from RGB color space are first converted into HSV color space. Next, the chromaticity H is evenly divided into r intervals, while the saturation S and the brightness V are evenly divided into s and t intervals, respectively. Thus, each of them is divided into three parts: r color values, s saturation values, and t brightness values. All the interval combination can generate $r \times s \times t$

different color feature dimensions. Then, all the pixels in the moving window are assigned to the corresponding possibility, and a highly sparse feature vector (color histogram) is obtained.

Then the set of color-histogram-based feature vectors for each image are taken as the input to the EMST-based clustering algorithm to obtain a cluster label for each corresponding color feature vector.

4 Experimental Results

In this section, experiments are conducted to evaluate the performance of our proposed algorithms in comparison to the standard EMST algorithm in the task of EMST based clustering for color image segmentation on two different image datasets. First, to check the technical soundness of this study, we would like to show that our proposed approximate EMST algorithm can be competent with classic EMST algorithm in the clustering accuracy. And then we evaluate the runtime performance of our algorithm with respect to the standard EMST algorithm. All the data sets are briefly summarized in Table 2.

Table 2. Descriptions of all datasets

Data name	Data size	Dimension	Image size	N	M
image_Data1	13,650	1000	1280 × 720	10	6
image_Data2	11,505	1000	481 × 321	10	6

We implement all the algorithms in C++ and perform all the experiments on a computer with AMD A6-4400 M Processor 2.70 GHz CPU and 4.00 G RAM. The operating system running on this computer is Windows 7. In our evaluation, we focus on the classification accuracy of these EMST-based clustering algorithms on different data sets. The results show that, overall, our proposed approximate EMST-based clustering algorithm can be competent with standard EMST based clustering algorithms if not better.

4.1 Performance of Our Algorithm on Color Image Segmentation

In this subsection, we use two image datasets to show that the proposed approximate EMST based clustering method can effectively identify major percepts in several outdoor environments. All two datasets, image_Data1 and image_Data2, are shown in the left plots of Figs. 4 and 5, respectively. The first image data, image_Data1, consists of 5 images taken around an outdoor environment, using SONY Digital Handycam DCR VX2000 digital video camera recorder. The second image data, image_-Data2, consists of 5 natural images from the Berkeley Segmentation Dataset that contain complex layouts of distinct textures, thin and elongated shapes and relatively large illumination changes and is therefore a challenge for segmentation task.

Fig. 4. Performance on image_Data1 (left) original data, (middle) experiment results obtained by standard EMST algorithm, and (right) results obtained by our proposed approximate EMST. (Color figure online)

The images are first converted from RGB color space into HSV color space. To obtain feature vectors, for each image, a moving window of size 10×10 is shifted by 6 pixels in the row and column directions but not to exceed the border of the image. The chromaticity H is evenly divided into 10 intervals, while the saturation S and the brightness V are evenly divided into 10 intervals. All the interval combinations can generate 1000 different color bins for the generation of color histogram for each image patch. Then, all the pixels in the moving window are assigned to the corresponding possibility, and a sparse feature vector (color histogram) is obtained. There are three columns in each figure. The left column displays the five testing images used. The middle column displays the clustering results based on standard EMST. The right column displays the clustering results based on our proposed approximate EMST algorithm.

Fig. 5. Performance on image_Data2 (left) original data, (middle) experiment results obtained by standard EMST algorithm, (right) results obtained by our proposed approximate EMST. (Color figure online)

From the original images of the first image_Data1 shown in Fig. 4, it is easy to identify that the major percepts include dark green trees, tree shadows, wide grey concrete road and manhole covers on the concrete road. To ease visual examination, in the segmented images, a color close to the real one is chosen to denote the percepts.

There are 7 clusters obtained from this dataset by our method. Typical stable percepts detected (and the corresponding colors representing them) include road percepts, several different grass and tree percepts, and the tree and wall shadows. For this outdoor environment, it is a little harder to segregate the more detailed percepts bundled into green. From the segmented images as shown on the first row of Fig. 4, our method performs a little better than standard EMST based clustering method which segments the grey road in foreground into more percepts. From the segmented images as shown on the second and the fourth rows of Fig. 4, both methods perform similarly well. On the third and fifth rows of Fig. 4, it can be seen that standard EMST based clustering method performs better while our method can not correctly discover the manhole covers. To summarize, overall, our algorithm can discover all the major percepts most of time.

The second image dataset, imageData2, has 5 different scenes which are shown in Fig. 5. The first scene is composed of a church building with white wall, three white crosses on the top, a dark brown wood railing entrance in the front, some dark building on the lower-right corner. The second scene is composed of a deer standing on brown to yellowish green grass with a dark green forest behind. The third scene has several undulating mountains in different colors under a setting sun. The fourth scene has two different kinds of rocks, a smaller one on top of a larger one, with the lower one engraved with a pair of eyebrows, a pair of eyes and a nose. The last scene contains a golden pyramid on a desert in front of a light grayish sky.

There are 17 clusters obtained from this dataset by our method. From the segmented figures, it can be seen that both methods can separate the sky from the church wall and thus performs reasonably well for the first and the third scenes. Standard EMST based clustering method performs better for the second and the last scenes than our method. Our method performs better for the fourth scene. To summarize, overall, our algorithm can discover all the major percepts most of time and performs competently.

4.2 Runtime Performance of Our Proposed Algorithm

The running time performances of standard Prim's algorithm (referred to as BF), the FEMST algorithm and our proposed approximate EMST algorithm are summarized in Table 3 for the two image data, respectively. It can be seen from the table that our method is significantly faster than the standard Prime's algorithm.

Table 3. The running time performance

Dataset	BF (s)	FEMST (s)	Our method (s)
image_Data1	9 771	12 161	1 850
image_Data2	7 121	9 755	1 846

We note that since the Prim's algorithm is used for the construction of approximate EMST in our algorithm, the running time complexity of the proposed method is not proportional to the size of the data set (that is, N). On the other hand, our proposed AEMST algorithm first uses the Cosine distance which relies on the number of inner products to be computed and thus is proportional to the size of nonzero dimensions. For this reason, the scalability of the algorithm is proportional to the dimension size of the data and the bottleneck operation of our proposed method is in the compression of the major information which is contained in the original data into a few dimensions by way of some mathematic transformation. Therefore, our proposed AEMST is essentially data dependent over the data space, and it is not possible to propose a reasonably tight estimation of the complexity behavior in closed form.

5 Conclusion

In this paper, we develop a new technique for approximate EMST problems that is especially suited to highly sparse high-dimensional data sets. The method utilizes the relationship between the Euclidean distance and the Cosine distance and works by removing distance computations involved in the Euclidean distances. This technique for approximate EMST based on the inner product calculations has advantages over simple index based FEMST that works remarkably well for large low dimensional datasets but cannot overcome the effects of the dimensionality curse. Experiments conducted on image datasets demonstrate the efficiency of the proposed method. In our future work, we would like to extend our work to larger-sized high dimensional data using dimensionality reduction techniques to further reduce the computation cost.

Acknowledgment. The authors would like to thank the Chinese National Science Foundation for its valuable support of this work under award 61473220 and all the anonymous reviewers for their valuable comments.

References

1. Kruskal, J.B.: On the shortest spanning subtree of a graph and the traveling salesman problem. Proc. Am. Math. Soc. **7**, 48–50 (1956)
2. Prim, R.C.: Shortest connection networks and some generalizations. Bell Syst. Tech. J. **36**, 567–574 (1957)
3. Bentley, J., Friedman, J.: Fast algorithms for constructing minimal spanning trees in coordinate spaces. IEEE Trans. Comput. **27**, 97–105 (1978)
4. An, L., Xiang, Q.S., Chavez, S.: A fast implementation of the minimum spanning tree method for phase unwrapping. IEEE Trans. Med. Imaging **19**(8), 805–808 (2000)
5. Xu, Y., Uberbacher, E.C.: 2D image segmentation using minimum spanning trees. Image Vis. Comput. **15**, 47–57 (1997)
6. Zahn, C.T.: Graph-theoretical methods for detecting and describing gestalt clusters. IEEE Trans. Comput. **C20**, 68–86 (1971)
7. Xu, Y., Olman, V., Xu, D.: Clustering gene expression data using a graph-theoretic approach: an application of minimum spanning trees. Bioinformatics **18**(4), 536–545 (2002)
8. Zhong, C., Miao, D., Wang, R.: A graph-theoretical clustering method based on two rounds of minimum spanning trees. Pattern Recogn. **43**(3), 752–766 (2010)
9. Juszczak, P., Tax, D.M.J., Pękalska, E., Duin, R.P.W.: Minimum spanning tree based one-class classifier. Neurocomputing **72**, 1859–1869 (2009)
10. Yang, L.: Building k edge disjoint spanning trees of minimum total length for isometric data embedding. IEEE Trans. Pattern Anal. Mach. Intell. **27**(10), 1680–1683 (2005)
11. Malik, J., Belongie, S., Leung, T., et al.: Contour and texture analysis for image segmentation. Int. J. Comput. Vis. **43**(1), 7–27 (2001)
12. Bach, F.R., Jordan, M.I.: Blind one-microphone speech separation: a spectral learning approach. In: Proceedings of NIPS 2004, Vancouver, B.C., pp. 65–72 (2004)

13. Borůvka, O.: O jistém problému minimálním (About a certain minimal problem). Práce moravské přírodovědecké společnosti v Brně, III, pp. 37–58 (1926). (in Czech with German summary)
14. Jarník, V.: O jistém problému minimálním (About a certain minimal problem). Práce moravské přírodovědecké společnosti v Brně, VI, pp. 57–63 (1930). (in Czech)
15. Dijkstra, E.W.: A note on two problems in connexion with graphs. Numer. Math. **1**(1), 269–271 (1959)
16. Preparata, F.P., Shamos, M.I.: Computational Geometry. Springer, New York (1985). https://doi.org/10.1007/978-1-4612-1098-6
17. Callahan, P., Kosaraju, S.: Faster algorithms for some geometric graph problems in higher dimensions. In: Proceedings of 4th Annual ACM-SIAM Symposium on Discrete Algorithms, pp. 291–300 (1993)
18. Narasimhan, G., Zachariasen, M., Zhu, J.: Experiments with computing geometric minimum spanning trees. In: Proceedings of ALENEX 2000, pp. 183–196 (2000)
19. March, W.B., Ram, P., Gray, A.G.: Fast Euclidean minimum spanning tree: algorithm, analysis, and applications. In: Proceedings of 16th ACM SIGKDD International Conference on Knowledge Discovery and Data Mining (KDD), Washington, pp. 603–612 (2010)
20. Vaidya, P.M.: Minimum spanning trees in k-dimensional space. SIAM J. Comput. **17**(3), 572–582 (1988)
21. Wang, X., Wang, X., Wilkes, D.M.: A divide-and-conquer approach for minimum spanning tree-based clustering. IEEE Trans. Knowl. Data Eng. **21**(7), 945–958 (2009)
22. Lai, C., Rafa, T., Nelson, D.E.: Approximate minimum spanning tree clustering in high-dimensional space. Intell. Data Anal. **13**, 575–597 (2009)
23. Wang, X., Wang, X.L., Zhu, J.: A new fast minimum spanning tree based clustering technique. In: Proceedings of the 2014 IEEE International Workshop on Scalable Data Analytics, 14–17 December, Shenzhen, China (2014)
24. Zhong, C., Malinen, M., Miao, D., Fränti, P.: A fast minimum spanning tree algorithm based on K-means. Inf. Sci. **295**(C), 1–17 (2015)

Personalized Blended E-learning System Using Knowledge Base Approach Based on Information Processing Speed Cognitive

Qumar Ibrahim(✉) and Md. Tanwir Uddin Haider

Computer Science and Engineering Department,
National Institute of Technology Patna, Patna, India
qumar.ibrahim@gmail.com, tanwir99@yahoo.com

Abstract. In recent years we have been observed that the demand of e-learning system increased due to vast growth in the field of electronic media. The e-learning system facilitates the learning from any parts of the world. The system of e-learning can also be made personalized blended e-learning system based on certain cognitive traits of the learner. Using blended e-learning system the performance of learners will be improved as well as the interest in that particular subject will also be developed. This paper discusses a blended e-learning system based upon the ability of information processing speed cognitive of a learner. The blended is based on combination of traditional video lecture and practical visualize videos. The students are grouped into three clusters on the basis of their cognitive ability. The machine will recognize each cluster and accordingly it will provide the learning contents. The inference rules and knowledge based approach have been used to make the machine under stable. The features are extracted for each student learning behavior. The system decides about the learning contents for each cluster based upon the observed behavior of that particular cluster and also monitors the growth of all the students.

Keywords: Information processing speed · Inference rule · Blended learning

1 Introduction

An e-learning system is a technique of teaching with the help of electronic resources like computer systems, mobile phones, etc. The e-learning system is very common now a day due to easy accessibility of learning resources through internet. The e-learning approach provides a convenient way of teaching methodology where the learner and teacher both may chose the time and place according to their convenience. The e-learning system is being improved to make the teaching in a personalized way so that any individual would be able to learn according to their wish. The e-learning system can also be made hybrid or blended. In hybrid approach two or more teaching techniques are combined together to understand the course easily and this technique is called as Blended e-learning system.

The Blended e-learning system [3] is an online learning system in which the topics are simplified by clarifying certain terms used within that topic. In this paper the term

P. Perner (Ed.): MLDM 2018, LNAI 10935, pp. 160–167, 2018.
https://doi.org/10.1007/978-3-319-96133-0_12

blended implies the grouping of traditional video lecture with practical visualize videos. In this way students can understand it in a better way. The blended e-learning system can be personalized. The personalization can be brought by using various factors like learning style, cognitive skills, and etc. In our system the personalization is based upon the information processing speed cognitive. The information processing speed cognitive [4] is the ability of students to grasp the things quickly and response the questions related to that in a correct way. The main motive behind using this cognitive is to overcome the most observed differences found among the students while teaching in a traditional classroom which is the processing capability of each student. It is observed that some students can learn very fast and some students learn slowly that is why the teaching should be done according to their learning speed otherwise these students will have lack of interest in that subject.

The huge growth of digitalization is motivating to develop an e-learning environment where the system itself acts as an online tutor. The system should have intelligence like a traditional tutor so that it could judge the learning pattern and capability of each individual student. The teaching pattern has a major impact on the learning style of an individual. It helps in developing the interest in a particular topic and also makes the topic easily understandable. The teaching pattern for each student should be according to their learning pattern. The machine should be able to identify the pattern which is being followed by a group of a particular type of students. The students should be categorized based on certain parameters like learning styles or cognitive traits and after categorizing they should be monitored to get the idea about learning skills of each category of students. The learning skills of students are stored in the database and the learning contents are delivered by the system according to the stored behavior of the cluster to which that student belongs. The learning behavior of a student is analyzed by clustering them into three different clusters using a k-means clustering algorithm. For this a tool has been developed to take a test and calculates the data of response time, correct response and reading time. These data help in clustering of students on the basis of information processing speed cognitive. After the clustering of students a common video lecture is delivered to all the students and a dictionary of all the terms used within that particular video lecture is created. The student having any doubt may click on that term within the dictionary. The system analyses the clicking behavior of each student and then a common behavior for each cluster is identified.

2 Related Work

A number of research works have been done in this field. Rani et al. [1] have work on an adaptive e-learning system in which personalization is achieved with the help of monitoring agents. It is an ontology-based e-learning system where software agents are used to monitor the learners' activities and store the information in a Knowledge-Based System on cloud storage. The Felder-Silverman model is used to find the learning style in this model. The Personalize e-learning content is provided to the learners with the help of Ontology which stores the semantic rule for an individual's learning style as the learners' requirements change dynamically.

Djenic et al. [2] have developed a fundamental programming course using Blended Learning system where the development is based upon lecturer's experience. The lecture notes are in the text format and blended videos are used to elaborate and explain certain terms used within the lecture notes.

Prez Sanagustn [6] describes that in formal learning the learning objectives are decided by the members who teach you while in informal learning system, the topic, subject, etc. is decided by the learners itself but they do not control the objectives to accomplish this learning. Whereas, in informal learning the aims are not defined because learning is totally depend upon learner, but the learner controls the means that can result into learning.

FitzGerald [7] and Frohberg [8] say that new technologies like smartphones and Augmented Reality technologies are being grabbed by the researchers to develop new opportunities for learning with informal and non-formal approach.

Phobuna and Vicheanpanya [9] have used intelligent tutoring system and adaptive hypermedia to develop knowledge base to create an adaptive e-learning system.

Duo and Ying [10] has worked on a personalized e-learning system in which there are three types of intelligent agents are active learner agent, teacher agent and system agent. These agents monitor the students browsing activities over internet and by compiling all the information the system suggest the most appropriate learning content to the learner. The learning contents are analyzed and then kept in an organized manner by the system.

3 Architecture and Methodology

The below architecture shown in Fig. 1 consists of three modules the first module describes that students are grouped into three clusters. The clustering of students is done based on their information processing speed cognitive which are classified into slow processing speed, medium processing speed and fast processing speed.

The students can start learning through the traditional videos on any topic and as the video lecture is completed a dictionary of terms is created. The dictionary shows only those terms which are used within the topic and it is done by using tokenization technique. In this technique the stop words are removed and the keywords related to the specific topic will retain.

The Fig. 2 shows that the blended videos within the repository are linked with the associated terms within the dictionary so that any student can access those blended videos by clicking on the term. The clicking behavior of each student is stored by the system to be further analyzed. The behaviors of the clusters are analyzed by taking only those features which are having the maximum number of occurrences.

The second module of the architecture is to make the machine understandable about the behavior of students according to their cluster by creating a knowledge base and applying inference rule. The knowledge base [5] is a technique to store structured and unstructured information. It stores the large data in a structured way for a long period of time. The data can be stored in the form of object model that is classes, subclasses and instances which can be reused and analyzed. Whereas, the inference rule is used to infer the information through the knowledge base and then enables the system to take a

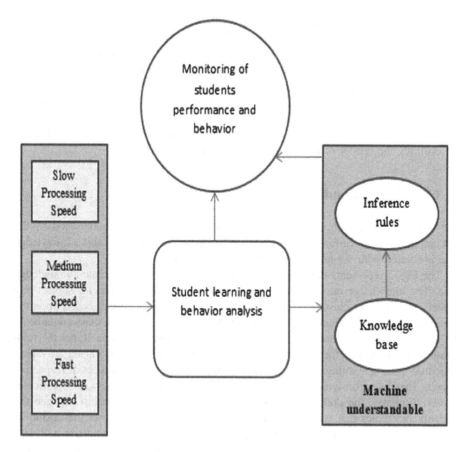

Fig. 1. Framework of the system.

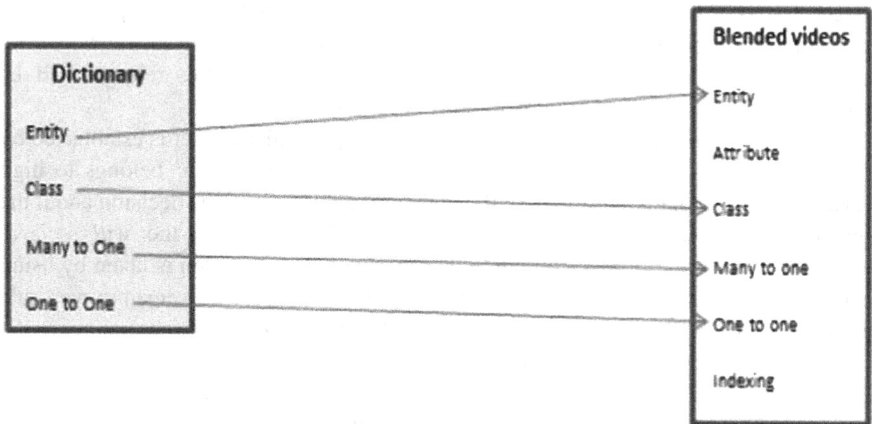

Fig. 2. Linking of terms with blended videos within the repository.

decision. Here the machine performs the task to deliver the learning contents in a personalized way that is elaborate those things in more detail having most of the students of a particular cluster are facing difficulties in understanding. The machine uses some practical videos to make these terms understandable by the students which are delivered to a student just after the completion of a traditional video lecture.

The third module is to monitor students' performance by taking the cognitive test related to the topics taught. Further, the cognitive i.e. their processing speed of a student is observed that whether it has been improved or fallen up to a threshold value. Accordingly the cluster of that student is changed by the machine.

4 Contribution and Operation

The behavior of students with respect to their clusters should be understood by the machine so that the learning system could be made personalized. For this a knowledge base is created with the help of ontology by using Protégé tool.

The Fig. 3 represents the graph of the ontology that is used to create a knowledge base. The knowledge base contains three types of class that is student, subject and clusters. The student class is further categorized into three subclasses based upon the processing speed of students that is high, medium and low. The topic class contains all the topics that is contained within the subject. This knowledge base stores the relationship of a topic with the cluster having most of the students of that cluster are facing difficulties during learning. The solid violet color lines show the instances of respective classes. The dotted brown color line represents the relation of a topic with a particular cluster. Similarly, the dotted green color lines represent the relationship between the students and topics that is established by making a rule for that shows that a student belonging to a particular cluster will receive the videos for the topics associated with that cluster. This ontology is created over Protégé where certain rules are developed to make the machine understandable.

The Fig. 4 shows the different clusters and also the topics associated with these clusters. For example the topic T1, T2 is associated with high processing speed cluster. The *'relates_to'* object property is used to make a generalization for each cluster with the associated topics which are needed to be elaborated and this relation will be implemented over all the students belonging to that cluster.

The Fig. 5 shows that the students relation with a particular cluster is established by using *'IsIn'* object property like in above figure the student 'Arun' belongs to high cluster so based upon this information the machine is able to take the decision about the topics which may be difficult to understand by 'Arun' and by using the *'will_recieve'* object property these topics are shown by the machine. This decision is taken by using the information through the knowledge base, applying rules using semantic web rule language (SWRL) over this knowledge base and synchronizing the reasoner.

The Fig. 6 shows the SWRL rule that is developed to enable the decision making capability within the machine. The rule shows that a student 's' belongs to cluster 'c' represented as *IsIn(?s, ?c)* and the topic *'t'* related to cluster *'c' relates_to(?c, ?t),* will receive a practically elaborated video according to his respective cluster.

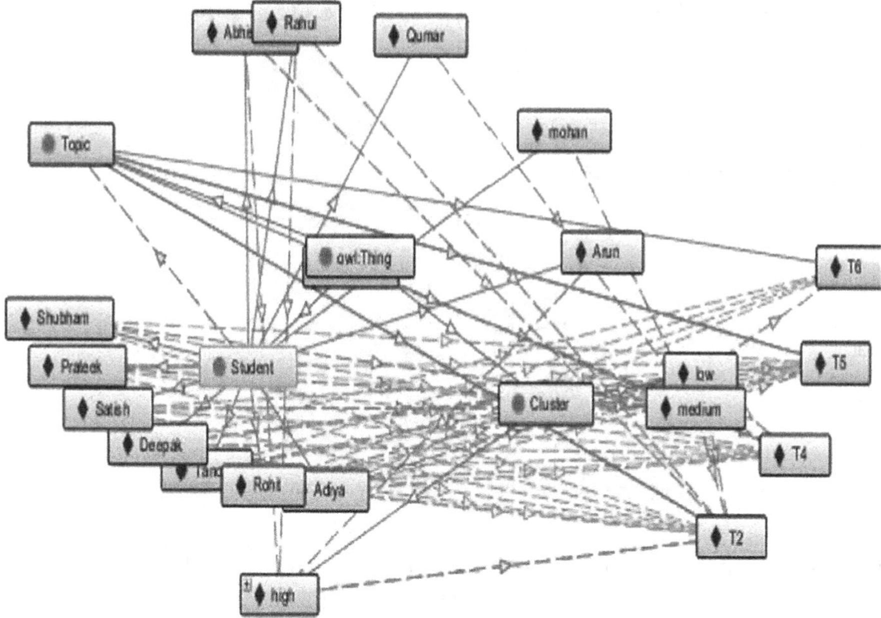

Fig. 3. Ontology graph. (Color figure online)

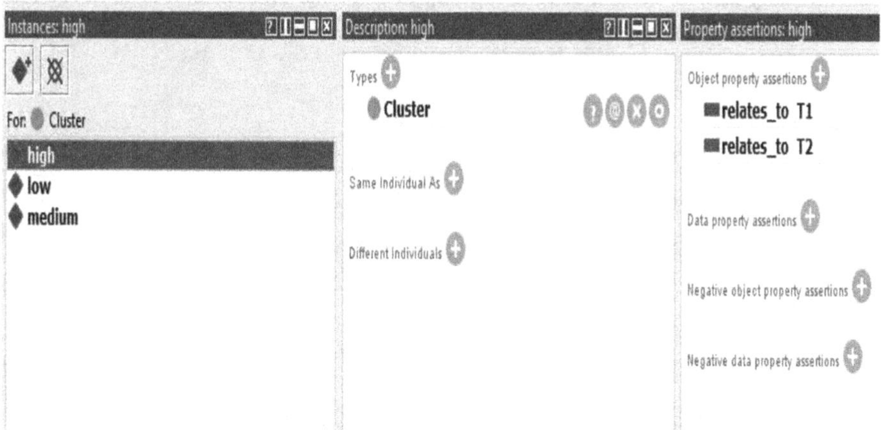

Fig. 4. Snapshot of students cluster developed on Protégé.

5 Results and Conclusions

The advantage of blended e-learning system is tested by performing a survey over 100 students. The students were divided into 2 groups of 50 students each where first group of 50 students were taught ER-diagram topic of DBMS with the help of traditional

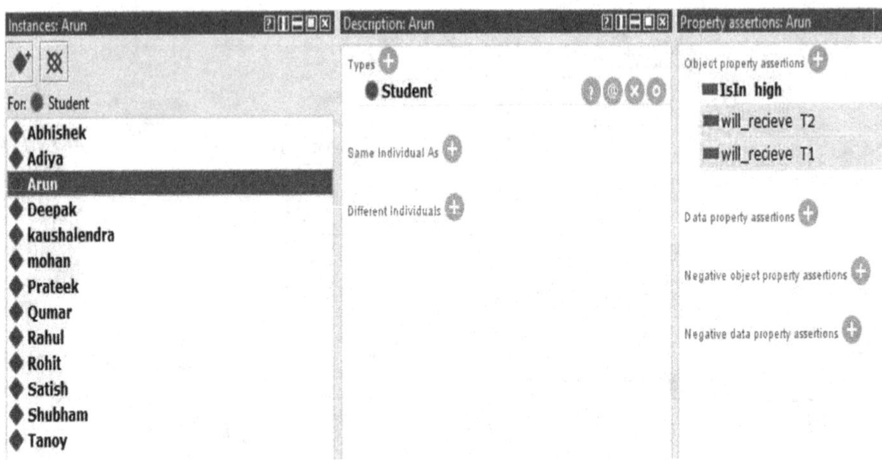

Fig. 5. Snapshot of the output obtained by executing the reasoner.

video lecture and the second group of 50 students were taught with the combination of both the traditional videos and the blended videos. Now a test on that topic is conducted over these 100 students with the similar questions and it is observed that there was a drastic change in the average marks of first group of students with the second group of students. The average score of second group of students was increased by approximately 20% as compared to first group of students. Therefore, it is observed that blended e-learning system improves the performance of a learner.

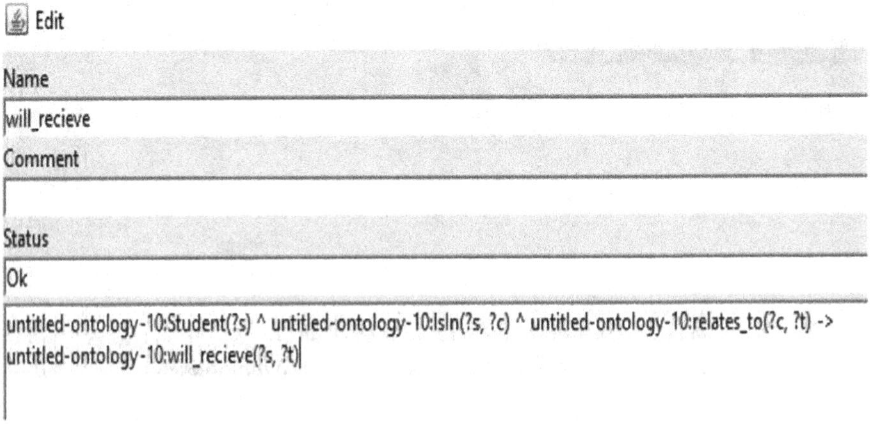

Edit

Name

will_recieve

Comment

Status

Ok

untitled-ontology-10:Student(?s) ^ untitled-ontology-10:IsIn(?s, ?c) ^ untitled-ontology-10:relates_to(?c, ?t) -> untitled-ontology-10:will_recieve(?s, ?t)

Fig. 6. Snapshot of the rule developed for executing the hermit reasoner.

The blended learning system brings about the flexibility, personalization, efficiency in the learning environment. It improves the performance of a student by making the course easier to understand. The system also keeps on monitoring the students' performance and based on that their clusters are updated. The other cognitive skills like working memory capacity may be used to make the learning system more personalized. The accuracy of the cognitive based clustering may be improved by applying some other clustering techniques. The students monitoring techniques could also be further improved. The intelligence could be developed to examine the learning contents and unifying them in a sequential way repeatedly by the system.

References

1. Rani, M., Nayak, R., Vyas, O.P.: An ontology-based adaptive personalized e-learning system assisted by software agents on cloud storage. Knowl.-Based Syst. **90**, 33–48 (2015)
2. Djenic, S., Krneta, R., Mitic, J.: Blended learning of programming in the internet age. IEEE Trans. Educ. **54**(2), 247–254 (2011)
3. Khlaisang, J., Likhitdamrongkiat, M.: E-learning system in blended learning environment to enhance cognitive skills for learners in higher education. Procedia – Soc. Behav. Sci. **174**, 759–767 (2015)
4. Kyllonen, P.C., Zu, J.: Use of response time for measuring cognitive ability. J. Intell. **4**(4), 14 (2016)
5. https://en.wikipedia.org/wiki/Feature_selection
6. Prez-Sanagustn, M., Hernndez-Leo, D., Santos, P., Delgado Kloos, C., Blat, J.: Augmenting reality and formality of informal and non-formal settings to enhance blended learning. IEEE Trans. Learn. Technol. **7**(2), 118–131 (2014)
7. FitzGerald, E., Adams, A., Ferguson, R., Gaved, M., Mor, Y., Rhodri, T.: Augmented reality and mobile learning: the state of the art. In: Proceedings of the World Conference on Mobile and Contextual Learning, pp. 62–69, October 2012
8. Frohberg, D., Goth, C., Schwabe, G.: Mobile learning projects: a critical analysis of the state of the art. J. Comput. Assist. Learn. **25**(4), 307–331 (2009)
9. Phobuna, P., Vicheanpanya, J.: Adaptive intelligent tutoring systems for e-learning systems. Procedia Soc. Behav. Sci. **2**, 4064–4069 (2010)
10. Duo, S., Ying, Z.C.: Personalized e- learning system based on intelligent agent. Phys. Procedia **24**, 1899–1902 (2011)

A Tag2Vec Approach for Questions Tag Suggestion on Community Question Answering Sites

Pradeep Kumar Roy and Jyoti Prakash Singh(✉)

National Institute of Technology Patna, Patna, Bihar, India
pkroynitp@gmail.com, jps@nitp.ac.in

Abstract. There are several reasons behind a question do not receive an answer. One of them is user do not provide the proper keyword called *Tag* to their question that summarizes their question domain and topic. Tag plays an important role in questions asked by the users in Community Question Answering (CQA) sites. They are used for grouping questions and finding relevant answerers in these sites. Users of these sites can select a tag from the existing tag list or contribute a new tag to their questions. The process of tagging is manual, which results in inconsistent and sometimes even incorrect or incomplete tagging. To overcome this issue, we design an automatic tag suggestion technique which can suggest tags to the users based on their question text. It serves to minimize the error of the manual tagging system by providing more relevant tags to questions. The performance of the proposed system is evaluated using Precision, Recall, and F1-score.

Keywords: Community Question Answer · Machine learning Classification · Tag suggestion · Stack Exchange

1 Introduction

Community Question Answering (CQA) sites are rapidly becoming a benchmark in user generated-content. The CQA site has been very influential in the way people share knowledge and discuss complicated topics online. It is being used by millions of users to find answers on complex, subjective or content dependent questions [9,15,20]. It optimally harnesses the collective intelligence of the whole online community. Every CQA sites have its own user interface through which a user can post the questions and answers. CQA sites such as Yahoo! Answers have a pre-defined topical hierarchy, where every question posted by the user will automatically be assigned to the related topics. If the relevant topic hierarchy is not present, then it goes to *Others* category [16]. Stack Exchange has a tag based user interface for posting the question which is different from Yahoo! Answers. In Stack Exchange, the user has to give the question title, question body, and the question tag. Question tags are used to assign a set of topics to a particular

P. Perner (Ed.): MLDM 2018, LNAI 10935, pp. 168–182, 2018.
https://doi.org/10.1007/978-3-319-96133-0_13

question. Some question tags are already present in a list on the site. However, if a user feels the question needs a different tag, they are free to assign a new tag to their question. Most of the community members search question using these tags [2].

Stack Exchange allows users to add tags to the question they have posted which helps in structuring and organizing the questions, thus making it easier for the user of common interest to find it. Moreover, the experts of related field can subscribe to the tags to get an update on new questions or any new digest related to that tag. This facilitates both the questioner and people who are interested in answering the questions. Stack Exchange allows the user to attach upto five tags to a question. To encourage users, Stack Exchange also allow users to create new tags, this makes the number of tags on Stack Exchange infinite. The manual creation of tags can be of great help to the experienced user, but it may be quite cumbersome for naive and inexperienced users as tagging a post incorrectly can potentially defer and delay the probable answers to the questions. The inexperienced users sometimes strugles to attach proper tags to the questions (incorrect tagging and incomplete tagging) [17,23]. The incorrect tag in the question can route the question to non-related experts, whereas incomplete tag may not route question to an expert at all.

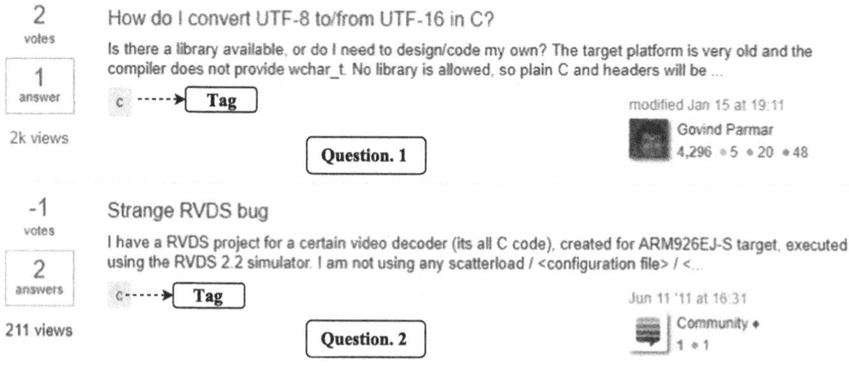

Fig. 1. Questions having single tag. Retrieve from https://stackoverflow.com/questions/674575and/310558 on 20/08/2017

As shown in Fig. 1, there are two questions present and both have a single tag. Question. 1 received 2000 community users views and got a single answer. This suggests that question is not seen by the right community users or not framed well. In Question. 2, total 211 community users have seen the questions, and got only 2 answers. The question also received a negative vote, that indicates the question doesn't fit well for the respective site or might be wrongly tagged by the questioner. User generated tags also suffers from the problem of tag idiosyncrasy i.e., the wording of tag may differ from person to person. For example, a tag on PHP can be created are PHP5 and PHP-5 by two different users, also there are number of tags are present with their synonym as shown in Fig. 2, we can see from the Fig. 2, the master tags are more frequently used by the users compared to their synonym, the master tag such as *shiny* was used 10105 times to tag the questions whereas their synonym *shinyapp.io* used only 39 times. However, both the tags present on the CQA sites. This was happening due to the freedom of tag creation by the users, and hence the number of tags increase enormously. In order to have better categorization and homogenization of tags, we introduce a tag suggestion system. When a user is entering the question, the system reads the question title and body and suggests some tags. The tag suggestion system is machine learning based system. This is trained on existing tags of Stack Exchange. We used Word2Vec model [8] to combine the similar tags together. It helps the user in correctly selecting the appropriate tags which will facilitate routing of the question to proper experts or users in that domain. Furthermore, it will help in reducing the problem of the idiosyncrasy of tagging which will encourage a more friendly CQA forum.

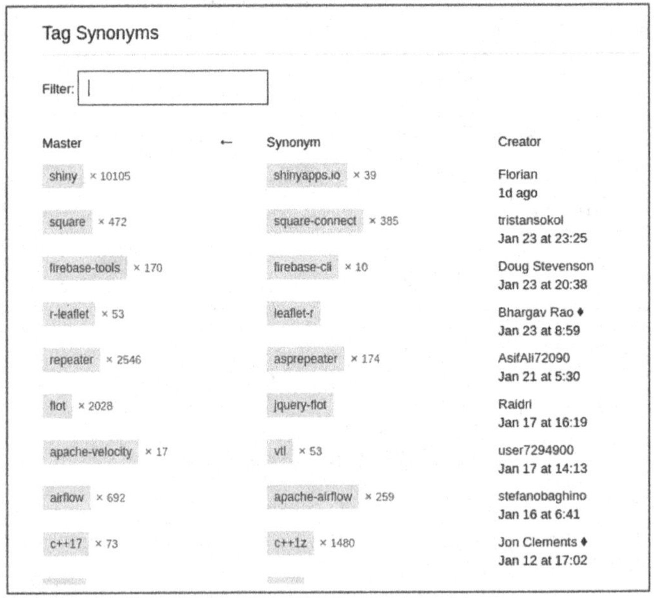

Fig. 2. A sample of tag synonym present on stack overflow

The rest of the paper is organized as follows: In Sect. 2, related works have been discussed. Section 3 consists of our proposed methodology. In Sect. 7, the results of the experiment have been discussed, and Sect. 8 concludes our work.

2 Literature Review

Community Question Answering (CQA) sites are popular due to the availability of good content in the form of question and answers. The community member expects the best answer to their question in less time. For this, it is very important that the posted question reaches to the right answerer. But, due to the mistagging, or incomplete tagging, the questions are not routed to the right experts. This leads to the late answer as well as low-quality answers. To address this problem some works already have been proposed by the researchers for automatic tagging system. An automatic tagging system is of either content-based [12,14] or collaborative methods [25]. Miao et al. [7] address the issue of new category identification problem in CQA site. They find out the potential categories which are not included in the current topic hierarchy. To achieve this issue authors used two different methods (i) Probabilistic Latent Semantic Analysis (PLSA) and (ii) Semi-supervised topic modelling. PLSA is a uni-gram language-based model which is used for grouping. Whereas in semi-supervised topic modelling: a prior knowledge about the specific topic is modelled in a probabilistic manner. The shortcoming of their work was that it is practically infeasible to build prior distribution manually. Singh and Visweswariah [18] proposed a model to allocate a pre-defined category to a new question on Yahoo! Answers. Yahoo! Answers have a three-level hierarchical structure with around twenty-six category at top-level and almost 1065 leaf level category. In Yahoo! Answers it was found that more than 17% of the top categories questions were of type Others while the count for the overall leaf category was only 10%.

Wang et al. [22] extracted the relevant term from the title and the description of the post using the *tf-idf* technique and rank them and achieve the recall value in the range of 0.08 to 0.22 for three different datasets. The recall value of their proposed system is very less that needs improvements using other approaches. Lipczak et al. [5] extracted the relevant term from the text and the relevancy of the extracted term was checked by their usage as a tag in the training set of the data. They divide the dataset into two parts (i) content-based recommendation and (ii) graph-based recommendation. However, their system only achieved 0.18 F1-score for content-based recommendation system and 0.32 F1-score for graph-based recommendation system which again not satisfactory. Zhou et al. [26] proposed a system which assigns a topic to the new post automatically. Yahoo! Answers have their own topic hierarchy, where every post is labelled with a pre-existing category, but if no topic was not assigned to a post, then by default it is labelled as *Others*. To check the efficiency of the proposed approach they used the concept of matchk i.e., the relative number of labels for which at least one of the top k label is correct. The proposed method is up to 11% more accurate than the best baseline method, i.e., Maximum Concept Weight Labelling (MCWL) with match10.

Nie et al. [11] proposed an automatic tagging approach, this approach is effective to solve the mistakes of the manual tagging system, such as incomplete tagging, wrong tagging. Their proposed methodology work in two phases: First, the off-line process in which they collected a set of similar questions and tags associated with them and make a hyper-graph using question tags and user information. Second, the relevant tags were selected based on the semantically similar score. The ranking of tags is done by considering informativeness (users tagging behaviour), stability (voting method) and closeness (by initial component) of a question and based on these terms, the relevant tags are selected. The major issue with the existing literature is about the overall accuracy of their system. As per our knowledge, none of the existing models achieve more than 50% accuracy to predict the tag for the questions. To overcome these issues, we proposed a system that used only the question data such as *Title*, *Body* and *Tags* to learn and suggest a tag for new questions.

3 Proposed Methodology

There are number of tags already available on different topics of CQA sites such as Stack Exchange, still, it has seen that users are not able to tag their question accurately. Mistagging leads to wrong question routing, and hence, number of questions unable to get a single answer. To resolve such issues we proposed a text-based tag suggestion system. We choose two diverse subjects such as *Biology* and *Robotics* from Stack Exchange dataset. The reason for selecting these subjects dataset is that they contain a huge number of tags and a number of tags with similar meaning. The complete architecture of the proposed methodology is shown in Fig. 3, where *FI* is the abbreviation of Feature Importance. The proposed system works in two parallel phases, the first phase is about clustering the tags that used for labelling, whereas in the second phase we identify the important features from question's title and question's body. The details of each phase are presented in following sections.

3.1 Dataset Description

The dataset were downloaded from https://archive.org/details/stackexchange [21], which were in Extensible Markup Language (XML) format. The dataset contains a number of different XML files such as Post.xml, Votes.xml, Comments.xml and etc. The questions are extracted from Post.xml and store in a separate comma-separated value (CSV) file. The question data consists of three attributes, namely *Title, Body*, and *Tags.*

Title: It is the short description of the question. A good title helps the question get the attention it deserves.
Body: It is the detailed description of the question asked by the users.
Tags: Tags are a set of keywords, that are used to associate topic or area to the question.

3.2 Data Pre-processing

The downloaded dataset contains several invalid characters as well HTML links that were not relevant to our work. we transform the dataset by removing such characters. The data preprocessing involve the following:

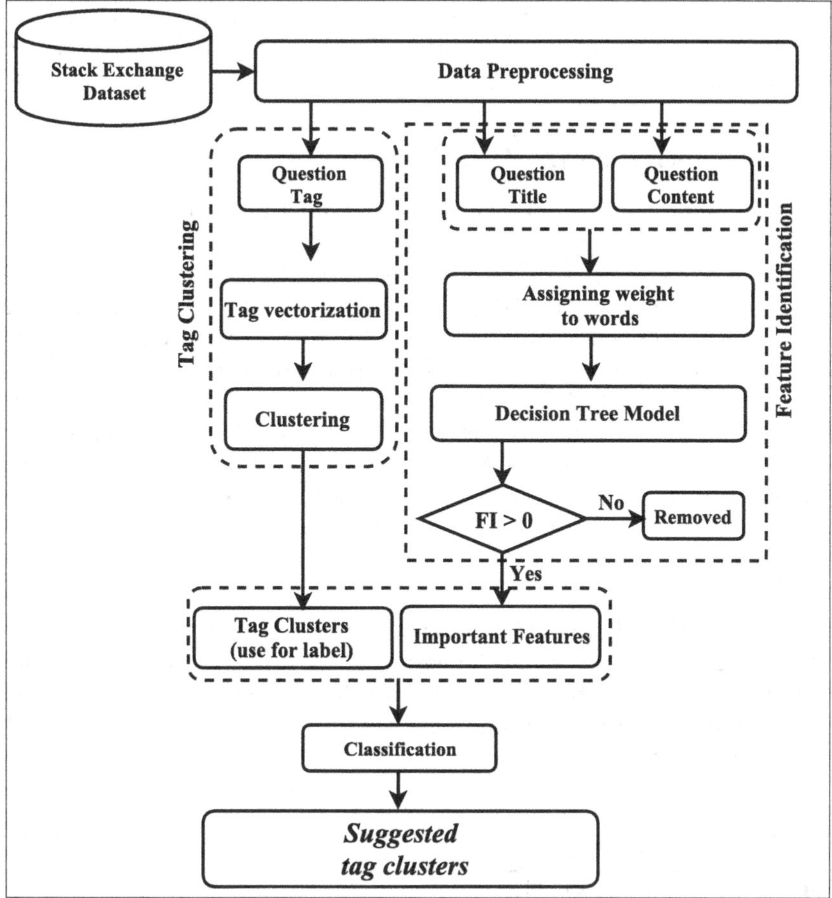

Fig. 3. The architecture of proposed system

- Removal of HTML tag and URI (Uniform Resource Identifier) content.
- Removal of punctuation from *Title* and *body.*
- Removal of stop words from *Title* and *Body.*
- Nouns were extracted from the *Title* and *Content.* The remaining words were removed from the questions. The reason for doing so, is that the tags of a question depends on the nouns used in the question, and not on other parts of speech.
- Stemming is done on extracted Nouns.

4 Tag Clustering

The input to this phase was questions tags and the output was tag clusters. The question tags in the dataset were unlabelled. After analysis of the tags, it was found that a number of different tags appear only once but they are similar to other tags, they could be grouped together. To group these tags we vectorize the words and take clusters.

4.1 Tag Vectorization (A Word2Vec Approach)

There was a number of tags present in the dataset having similar meaning, we needed to place them together. But, due to a large number of tags, the manual process was not efficient. So, we used tag vectorization to represent the similar tags into the same cluster. Word2Vec [8] is used to transform a tags into a vector space. The tag vector consists of the probability of other tags, with the root tag. The size of the tag vector is user-defined where the output size varies from 1 to maximum (user-defined). We set 300 as an output parameter. Hence, for every word we have the probability of 300 different words (the size of the vector is $M * N, where\ M\text{: } Input\ words, N\text{: } Output\ words$). The word with the high probability value is the most similar word to the root word.

4.2 Clustering

The number of tags in both datasets was very high and most of the tags preserve same meaning, so to reduce the dimension of the tags, we merge the similar tag together by applying the clustering algorithm, the detail steps of clustering process was explained in Algorithm 1. The obtained clusters were used as a label for the classification.

Algorithm 1. Tag Clustering

```
Input: Questions, and Tag file
Output: List of tagClusters
ques[]:=All questions of selected topic;
tag[]:=All tags of selected topic;
tagVector[];
k=0;
wv[]=Word2Vec(input, output);
/* input: ques; output: No. of words/*;
for i=0 to length(wv);
   if wv['word'] == tag['word'] then   tagVector[k] = wv['word'];
      k=k+1;
K-Means(tagVector[k]) /* tagVector[k] given as input to K-Means
 algorithm/*;
return(tagClusters)
```

To cluster the tag we used K-Means algorithm [3] as explained in Algorithm 1. The Algorithm 1, works as follows: We first, appended all the question and stored in variable *ques*[ ] and tags stored in variable *tag*[ ]. Then, *ques*[ ] sent as input to the neural network training algorithm word2vec, setting the output vector length of 300. We stored the list of words with the associated vectors in the variable $wv[\ ]$. From the complete list of words, store the vector values of only the words appearing as tags of questions, in variable $tagVector[\ ]$. We applied K-Means clustering algorithm to the set of vectors $tagVector[\ ]$. Finally, we assigned unique Ids to the obtained clusters. These unique ids will act as class labels for training and testing of the proposed system.

The number of tag clusters obtained in the topic biology was 107 while the number of tag clusters obtained in robotics was 30. Each cluster contains on an average 5 tags. The maximum and the minimum number of tags in a cluster was 33 and 1 for *Biology*, and for the *Robotics*, the maximum and the minimum number of tags in a cluster was 35 and 1. For each topic clustering of tags needed to be done only once. After clustering of the tags, we used the clustered labels of the corresponding tags for training the classifier. After this step, we got a cluster label for every tag that used as a label to train the classifier.

5 Feature Identification

Assignment of Weight to Words: We got the tag cluster that we used for data labelling from Sect. 4. In this section, our target was to find the relevant features from the question's title and body. On manual analysis of questions of both the datasets, it was found that the proper nouns in the question text are the most critical in determining tags of the questions. Therefore, we needed proper nouns as the feature of the training set. The proper noun was extracted using NLTK tool [1]. We gave more weights to the noun that appeared in both titles as well as in the body of the question as compared to other nouns. The steps of the weight assigning process are discussed in Algorithm 2.

The Algorithm 2 works as follows: we first perform *Part of speech* tagging on the question's title and body, and extract the proper noun (NP) from them, then, *Stemming* [19] is done. The extracted proper nouns along with their frequency of appearance in all the questions are saved as a bag of words. Thus the bag of words contains two attributes, i.e., the nouns and its frequency (count). If the frequency of noun is greater than 10 then such nouns are kept in the bag of words and all the remaining nouns are removed. After this, we were left with 6,754 nouns in the topic *Biology* and 2,441 nouns in *Robotics*. Weights are assigned to the remaining nouns using the following rule.

- Initially assign the weights of all the words to 0.
- Count the number of times a the noun appears in the title and store the count in t.

$$weight = weight + a * t \tag{1}$$

- Count the number of times the nouns appears in the body and store the count in c.

$$weight = weight + c \tag{2}$$

Algorithm 2. Weight Assignment

```
Input: Question's Title and Body
Output: Word with their weight
ques[][]:=All questions with their title of selected topic;
POS(ques) /* Nouns are extracted/*;
Stemming(ques);
BagofWord= words with their frequency;
bow=BagofWords;
bog1[];
for(i=0 to len(bow));
   if frequecy of bog['word'] > 10 then      bog1[k] = bog['word'];
      k=k+1;
/*Weight (w) assignment/* ;
w=0 /* Initial weight of all words=0/*;
t=0 /* Counter (t)=0/*;
for (k=0 to len(ques[Title])):;
         if bog1[k] in ques[Ttile] then t=t+1;
         k=k+1;
          weight (w)=w+a*t;
         else if bog1[k] in ques[Body] then
;                 t=t+1;
          k=k+1;
         weight(w)=w+b*t;
         else ;
         weight (w)=w+t;
return(w)
```

If the noun present in body has appeared in some title also, then:

$$weight = weight + b * c \tag{3}$$

Here a and b are constants such that $a > b$. From our experimental analysis, we found the value of a and b to be 0.625 and 0.375 respectively.

After extracting the *Nouns* using the NLTK tool from the question posts, we found that a lot of irrelevant parts of speech still exist in our dataset. To minimize the effect of these irrelevant words in our features, we devise a new weighing strategy. This weighing strategy intends to assign higher weights to relevant noun words and lower weights to irrelevant words. The theory behind this weighting scheme is that the proper noun words have a lower number of synonyms as opposed to words belonging to other parts of speech (e.g., verbs, common nouns, adverbs etc.). Thus we find the number of synonyms for each word. The words with the low number of synonyms are given higher weight and those with a high number of synonyms are given less weight. For finding synonyms of extracted words we use scrapped data from www.thesaurus.com. The new weighing scheme follows the following steps:

- For each word find the number of synonyms found in www.thesaurus.com and store it in s.
- New weight ($weight_1$) of the word is:

$$weight_1 = \frac{weight}{s} \tag{4}$$

Feature Extraction: The number of features in the dataset of biology was 6,754 and the number of features in robotics was 2,441. This is a large number of feature set and a further reduction in the number of features was required. To reduce the number of features the list of words in each topic are transferred to the Decision Tree model [13] based on their new weights ($weight_1$). In the Decision Tree model, the feature importance (FI) of each word is calculated. If $FI \geq 0$ then the required words are considered to be features for classification. Thus, the final features used for classification are the proper noun words that have feature importance greater than zero and the class label for these features is their respective tag cluster index. After this phase, the number of words remaining for features in biology is 2,340 and in robotics is 674.

6 Classification

The goal of the proposed system is to suggest some tags to the users based on their question text to improve the tagging accuracy as well as question routing process. Using the features extracted, and their corresponding tag cluster index as the class label, we trained the system with five different classifiers (i) Random Forest, (ii) Gradient Boosting, (iii) Naive Bayes, (iv) Support Vector Machine (SVM), and (v) Logistic Regression.

7 Results and Discussion

In this section, we presents the detail results of the proposed system. To measure the performance of our proposed model we chose widely used metrics such as precision (5), recall (6) and F1-Score (7).

$$Precision = \frac{True\ Positive}{True\ Positive + False\ Positive} \tag{5}$$

$$Recall = \frac{True\ Positive}{True\ Positive + False\ Negative} \tag{6}$$

$$F1\text{-}Score = 2 * \frac{precision * recall}{precision + recall} \tag{7}$$

Precision is defined as the part of relevant instances among the retrieved instances. Whereas, recall is the fraction of relevant instances that have been retrieved over the total amount of relevant instances. F1-Score is the harmonic mean of precision and recall value.

The complete dataset was divided into two parts called training and testing: The training set for *Biology* contains 9,858 questions each having 2,340 features. While the training set for *Robotics* contains 2,076 questions each having 674 features. All Five classifiers were tested with the remaining 3,286 questions in biology and the remaining 693 questions in Robotics. The statistics of training and testing dataset presented in Tables 1 and 2 respectively.

Table 1. Statistics of training dataset

Topic	Total questions	Number of features	Number of label
Biology	9,858	2,340	107
Robotics	2074	674	30

Table 2. Statistics of test dataset

Topic	Total questions
Biology	3286
Robotics	693

We trained and tested five different classifiers with *Robotics* dataset to check the performance of the proposed model, namely (i) Random Forest (RF), (ii) Gradient Boosting (GB), (iii) Naive Bayes (NB), (iv) Support Vector Machine (SVM), and (v) Logistic Regression (Logistic). First, Random Forest classifier is used which is an ensemble learning-based classifier and achieved the recall value 0.36 and the precision was 0.61, after that the remaining four classifiers were used. The experimental results confirmed that the performance using the *Random Forest* classifier is the best one whereas *Naive Bayes* giving worst result. The recall was 0.21 and precision was 0.39 obtained using the Naive Bayes classifier as can be seen from the Table 3.

The similar process was done with the *Biology* dataset, here also we got the best performance using the *Random Forest* classifier with recall value 0.26 and precision value is 0.55 whereas the *Naive Bayes* performing worst where the recall was 0.16 and precision value was 0.32 as can be seen from Table 4.

The average F1-Score with the best performing classifier (Random Forest) across the topics came out to 0.40 as shown in Table 3. Which seems to be low, but this is due to the number of classes being too high with respect to the number of training data. The total number of classes in the *Biology* dataset is 107 while the training data contains only 9,858 questions. For Robotics dataset the number of *Classes* was 30 while the training data contains only 2,076 questions. This means that the classifiers got on an average only 92 questions per *Class* in *Biology* and 69 questions in *Robotics* to learn a new *Class*. This low number

Table 3. Precision, recall and F1-scores for the topic 'Robotics'

Topic	Classifier	Precision	Recall	F1-score
Robotics	**Random Forest**	**0.61**	**0.36**	**0.46**
	Gradient Boosting	0.56	0.34	0.43
	Naive Bayes	0.39	0.21	0.28
	SVM	0.50	0.32	0.39
	Logistic	0.59	0.34	0.43

Table 4. Precision, recall and F1-scores for the topic 'Biology'

Topic	Classifier	Precision	Recall	F1-score
Biology	**Random Forest**	**0.55**	**0.26**	**0.35**
	Gradient Boosting	0.51	0.24	0.33
	Naive Bayes	0.32	0.16	0.21
	SVM	0.43	0.22	0.31
	Logistic	0.53	0.25	0.34

of records is not sufficient for proper training of a classifier. Also, the class distribution of the data is uneven. Some classes occur frequently while some classes occur only once. The classes with low occurrence drag down the accuracy of the classifier.

The average F1-Score of 0.40 of our proposed system is good for a suggestion system, this result captures the accuracy of the classifier. On an average 58% (Precision) of the time, people are choosing correct tags based on our suggested model. This result is bound to increase once this suggestion system is implemented by the site, as people are likely to choose tags more conveniently and accurately from the tags suggested by the system instead of giving inaccurate or incomplete tags. The proposed model was performed better compared to other existing tag recommendation models. A detailed comparison of F1-Score of our proposed model with other models are presented in Table 5. However, the dataset used by them are different that we used. A model proposed by Wang et al. [22] achieved the recall value in the range of 0.08 to 0.22 using the *tf-idf*. Similarly, the model proposed by [5] achieved 0.32 F1-score for the best case. Our proposed system performed 63% and 18% better than the model proposed by Wang et al. [22] whereas 43% and 9% better than the model proposed by the Lipczak et al. [5] with the topic *Robotics* and *Biology* respectively. The other results presented in Table 5 is based on recommending tags for bookmarks which were similar to recommending the tags for the question. But, We are referencing these results here because of the similarity of the nature of the problem. Our proposed system not only perform better than the above-stated system but also performed better than the system proposed by the researchers as listed in Table 5.

Table 5. Comparison of F1-score with existing tag recommendation models

Proposed by	F1-score
Lipczak et al. [5]	32.46
Mrosek et al. [10]	26.47
Ju and Hwang [4]	21.11
Lops et al. [6]	20.18
Wang et al. [22]	17.94
Zhang et al. [24]	17.35
Our system (average F1-score)	**40.00**

8 Conclusion

Due to the incomplete tagging or incorrect tagging of the questions, CQA sites suffer from a number of issues such as wrong question routing, late answers, even sometimes question doesn't receive a single answer. To overcome these issues, a tag suggestion technique has been proposed in this paper. This technique is completely based on textual data of the questions. The input given to the algorithm is the question's *Title*, *Body* and *Tags* for training purpose. While testing, we have given only the question's *Title* and *Body* to the system, and the system suggests a set of tags for the input question. The questioner can choose the relevant tags from the suggested list of tags. The proposed technique is helpful for naive users, as most of the time, they give a generic or incomplete tag to their question, due to this their question does not get expected responses from the peer users. Hence, this technique may also potentially improve the performance of other CQA systems depend on tags. The proposed system has some limitations such as the system was tested with only two topics also some topics. Also, some major topics like Science, Programming and other are not yet tested. The future work may include the testing of the proposed system with different topics of Stack Exchange and the dataset of different CQA site. The same system may be performed better if the deep learning approach can be used to suggest the tags to the question.

References

1. Bird, S.: NLTK: the natural language toolkit. In: Proceedings of the COLING/ACL on Interactive Presentation Sessions, pp. 69–72. Association for Computational Linguistics (2006)
2. Diakopoulos, N.A., Shamma, D.A.: Characterizing debate performance via aggregated Twitter sentiment. In: Proceedings of the SIGCHI Conference on Human Factors in Computing Systems, CHI 2010, pp. 1195–1198. ACM (2010)
3. Hartigan, J.A., Wong, M.A.: Algorithm as 136: a k-means clustering algorithm. J. Roy. Stat. Soc.: Ser. C (Appl. Stat.) **28**(1), 100–108 (1979)

4. Ju, S., Hwang, K.B.: A weighting scheme for tag recommendation in social bookmarking systems. In: Proceedings of the ECML/PKDD 2009 Discovery Challenge Workshop, pp. 109–118 (2009)
5. Lipczak, M., Hu, Y., Kollet, Y., Milios, E.: Tag sources for recommendation in collaborative tagging systems. ECML PKDD Discov. Chall. **497**, 157–172 (2009)
6. Lops, P., De Gemmis, M., Semeraro, G., Musto, C., Narducci, F.: Content-based and collaborative techniques for tag recommendation: an empirical evaluation. J. Intell. Inf. Syst. **40**(1), 41–61 (2013)
7. Miao, Y., Li, C., Tang, J., Zhao, L.: Identifying new categories in community question answering archives: a topic modeling approach. In: Proceedings of the 19th ACM International Conference on Information and Knowledge Management, pp. 1673–1676. ACM (2010)
8. Mikolov, T., Sutskever, I., Chen, K., Corrado, G.S., Dean, J.: Distributed representations of words and phrases and their compositionality. In: Advances in Neural Information Processing Systems, pp. 3111–3119 (2013)
9. Molino, P., Aiello, L.M., Lops, P.: Social question answering: textual, user, and network features for best answer prediction. ACM Trans. Inf. Syst. (TOIS) **35**(1), 4 (2016)
10. Mrosek, J., Bussmann, S., Albers, H., Posdziech, K., Hengefeld, B., Opperman, N., Robert, S., Spira, G.: Content-and graph-based tag recommendation: two variations. In: ECML PKDD Discovery Challenge, pp. 189–199 (2009)
11. Nie, L., Zhao, Y.L., Wang, X., Shen, J., Chua, T.S.: Learning to recommend descriptive tags for questions in social forums. ACM Trans. Inf. Syst. (TOIS) **32**(1), 5 (2014)
12. Nishida, K., Fujimura, K.: Hierarchical auto-tagging: organizing Q&A knowledge for everyone. In: Proceedings of the 19th ACM International Conference on Information and Knowledge Management, pp. 1657–1660. ACM (2010)
13. Quinlan, J.R.: Induction of decision trees. Mach. Learn. **1**(1), 81–106 (1986)
14. Rekha, V.S., Divya, N., Bagavathi, P.S.: A hybrid auto-tagging system for stackoverflow forum questions. In: Proceedings of the 2014 International Conference on Interdisciplinary Advances in Applied Computing, p. 56. ACM (2014)
15. Roy, P.K., Ahmad, Z., Singh, J.P., Alryalat, M.A.A., Rana, N.P., Dwivedi, Y.K.: Finding and ranking high-quality answers in community question answering sites. Glob. J. Flex. Syst. Manag. **19**(1), 53–68 (2018)
16. Shah, C.: Measuring effectiveness and user satisfaction in Yahoo! answers. First Monday **16**(2) (2011)
17. Shah, R.R., Samanta, A., Gupta, D., Yu, Y., Tang, S., Zimmermann, R.: PROMPT: personalized user tag recommendation for social media photos leveraging personal and social contexts. In: 2016 IEEE International Symposium on Multimedia (ISM), pp. 486–492. IEEE (2016)
18. Singh, A., Visweswariah, K.: CQC: classifying questions in CQA websites. In: Proceedings of the 20th ACM International Conference on Information and Knowledge Management, pp. 2033–2036. ACM (2011)
19. Singh, J., Gupta, V.: Text stemming: approaches, applications, and challenges. ACM Comput. Surv. (CSUR) **49**(3), 45 (2016)
20. Singh, J.P., Irani, S., Rana, N.P., Dwivedi, Y.K., Saumya, S., Roy, P.K.: Predicting the helpfulness of online consumer reviews. J. Bus. Res. **70**, 346–355 (2017)
21. stackexchange.com, January 2017. https://archive.org/details/stackexchange
22. Wang, J., Hong, L., Davison, B.D.: Tag recommendation using keywords and association rules. In: Proceedings of the ACM SIGKDD Conference on Knowledge Discovery and Data Mining, Bled, Slovenia (2009)

23. Wang, J., Luo, N.: A new hybrid popular model for personalized tag recommendation. JCP **11**(2), 116–123 (2016)
24. Zhang, Y., Zhang, N., Tang, J.: A collaborative filtering tag recommendation system based on graph. In: ECML PKDD Discovery Challenge, pp. 297–306 (2009)
25. Zhao, S., Du, N., Nauerz, A., Zhang, X., Yuan, Q., Fu, R.: Improved recommendation based on collaborative tagging behaviors. In: Proceedings of the 13th International Conference on Intelligent User Interfaces, pp. 413–416. ACM (2008)
26. Zhou, G., Cai, L., Liu, K., Zhao, J.: Exploring the existing category hierarchy to automatically label the newly-arising topics in CQA. In: Proceedings of the 21st ACM International Conference on Information and Knowledge Management, pp. 1647–1651. ACM (2012)

A Crowd Sensing Approach to Video Classification of Traffic Accident Hotspots

Bernhard Gahr[1(✉)], Benjamin Ryder[2], André Dahlinger[1], and Felix Wortmann[1]

[1] University of St. Gallen, 9000 St. Gallen, Switzerland
bernhard.gahr@unisg.ch
[2] ETH Zurich, 8000 Zurich, Switzerland

Abstract. Despite various initiatives over the recent years, the number of traffic accidents has been steadily increasing and has reached over 1.2 million fatalities per year world wide. Recent research has highlighted the positive effects that come from educating drivers about accident hotspots, for example, through in-vehicle warnings of upcoming dangerous areas. Further, it has been shown that there exists a spatial correlation between to locations of heavy braking events and historical accidents. This indicates that emerging accident hotspots can be identified from a high rate of heavy braking, and countermeasures deployed in order to prevent accidents before they appear. In order to contextualize and classify historic accident hotspots and locations of current dangerous driving maneuvers, the research at hand introduces a crowd sensing system collecting vehicle and video data. This system was tested in a naturalistic driving study of 40 vehicles for two months, collecting over 140,000 km of driving data and 36,000 videos of various traffic situations. The exploratory results show that through applying data mining approaches it is possible to describe these situations and determine information regarding the involved traffic participants, main causes and location features. This enables accurate insights into the road network, and can help inform both drivers and authorities.

1 Introduction

In recent years the number of road fatalities have been steadily increasing to over 1.2 million deaths per year [35]. This increase is not limited to developing countries with poor infrastructure, for example, the USA saw a 5.6% increase in the number of deaths from traffic accident from 2015 to 2016, and 8.4% from 2014 to 2015 [27]. Similarly, the number of road deaths in the European zone has increased in 2015 and 2016, widening the gap to the goal of reducing road fatalities from 2010 by 50% until 2020 [1]. Besides the well known reasons of driver distraction, i.e., driving under influence of alcohol or drugs, speeding, and other reckless driving behavior, many external factors can lead to traffic accidents. These external factors, such as poor road management, infrastructure challenges and weather conditions, can develop into locations with a high

P. Perner (Ed.): MLDM 2018, LNAI 10935, pp. 183–197, 2018.
https://doi.org/10.1007/978-3-319-96133-0_14

likelihood of traffic accidents, otherwise known as 'accident hotspots' [15,30]. Further, vehicle collisions with wildlife might not have such a high impact on human mortality, however damages related to these accidents are estimated to have an annual cost of over $8.3 Billion in the USA alone [9].

Since incorrect driver behavior has been identified as one of the main reasons for traffic accidents [5], many of the road safety increasing programs focus on educating drivers and finding countermeasures for wrong behavior. Authorities within the EU, or the NHTSA in the USA, hence introduced guidelines on driver education to improve the traffic safety situation [13,16]. Moreover, several research groups have shown the positive effects that in-vehicle warnings of upcoming accident hotspots have on driver behavior [3,31,33]. In these situations, people preferred contextualized warnings of danger rather than simple unspecific warnings [26]. A second approach to reduce traffic accidents involves programs focused on improvements to the road infrastructure, examples of this include EuroRAP [20], RANKERS [11], or EUROTAP [2]. For example, recent research on road network characteristics shows that the highest risk of fatal or severe crashes occur on road networks with a very low intersection density [22], and it was found that safety outcomes improve as the intersection density increases. Additionally, the reorganization of road structures implementing 'shared spaces' have shown potential in increasing safety for pedestrians and reducing the number of accidents [24]. Other projects, such as in the Banff National Park, found that the building of animal crossings led to a severe reduction of animal related accidents [8]. Independent of the countermeasure employed at a certain location, a concrete knowledge of the actual cause for the accidents can aid in improving the road infrastructure.

One generalized approach to develop this vital understanding of our road networks includes programs such as the DaCoTa project. This project led to the combination and analysis of accident investigations from 30 European countries [19], and resulted in the Road Safety Knowledge System [12]. However, since this approach relies on historic accidents, it falls short in preventing new upcoming accident hotspots and contextualize them in real-time. Therefore we investigate, whether recent content deprived accident hotspots and current dangerous situations can be detected and classified in a immediate manner. As such, in the paper at hand we introduce a crowd, or community, sensing approach which collects data from the Controller Area Network (CAN-Bus) of vehicle's on board computer, and enriches this with video data from an on-board smartphone camera. Videos are recorded when these vehicles cross a historically known dangerous location, or performs a dangerous driving maneuver. This way, we tackle two hotspot classification issues at the same time: First, locations which are historically known to be dangerous, but lack contextual information about the main accident cause can be analyzed in a retrospective manner. This community sensing approach means that expensive and high maintenance equipment, like stationary surveillance cameras, do not have to be installed. Second, by recording video sequences from dangerous driving maneuvers, developing dangerous locations can be identified and classified before an accident occurs. The increasing number of dash cameras in recent years shows the potential of identifying accident causes out of video sequences. The approach is hereby based on the idea that only a subset of all accidents (A) are reported (A_r). Further, we assume that all accidents are a subset of critical brake events (B_c), which are a subset of all brake events (B).

$$A_r \subset A \subset B_c \subset B. \tag{1}$$

In the following chapters, the paper at hand introduces a crowd sensing system, and the associated field study, which resulted in a dataset enabling the automated classification of traffic accident hotspots. The remainder of the paper is structured as follows. In Sect. 2 related work is given and in Sect. 3 the system is presented. Further, Sect. 4 provides a description of a field study setup and the collected data is presented in Sect. 5. Preliminary results are discussed in Sect. 6 and the paper concludes with an outlook to the remaining work in Section 7.

2 Related Work

Crowd sensing, sometimes referred to as crowd sourcing, is a relatively new paradigm which typically utilizes mobile devices to passively gather data and provide wide-ranging ubiquitous applications [21]. In previous research it was shown that sensors in smart phones, such as an accelerometer, gyroscope and GPS, are well suited for real-time community sensing tasks, e.g., predicting thermal hotspots by clustering sink- and climb rates of paragliders [34]. Crowd sensing of insights related to the road network has previously been used for detecting traffic anomalies, such as disasters or protests [28], along with identifying infrastructure, e.g., traffic lights and stop signs [18]. More recently, a crowd sourcing approach was analyzed for the detection road anomalies from driving video records [7].

The aim of the research at hand is to classify both accident hotspots and dangerous driving behavior on the road network following a crowd sensing approach. Prior research has demonstrated a similar contextualization task, focusing on traffic scenes using video data from static surveillance cameras [23]. The need for such classification is motivated by recent findings which demonstrate that users prefer contextualized in-vehicle warnings over an 'all purpose' general warning sign [26]. A prerequisite of the first task, accident hotspots classification, is identifying the locations of existing accident hotspots where driving videos should be recorded. To achieve this, we build upon the results of previous research in Switzerland, where accident hotspots were identified using police accident reports and the DBSCAN clustering technique [31]. This hotspot dataset provided GPS locations of over 1,600 accident hotspots, along with the three descriptive layers of the involved traffic participating objects ("What"), the predominant cause ("Why"), and the location ("Where").

The second task, contextualizing video captured during dangerous driving situations, builds upon a large body of existing research on driving event detection. Since the system should collect the dangerous driving maneuvers out of a real-world study, we cannot rely on Time-to-collision (TTC) values found in previous research [17], but rather velocity, deceleration, and jerk (change of acceleration, $j = \frac{da}{dt}$) measurements. Rather high deceleration values of 0.75 g (7.35 m/s^2) to cover 4% to 0.5% of all braking events in an urban area were found in [14]. More practical, but not rigorously tested, values to classify braking events were given in [4] with decelerations of below 2.0 m/s^2 as "low danger", above 3.9 m/s^2 as "high danger", and everything in between as "medium danger". These same thresholds were taken for both acceleration and turning. Recent research also considered the jerk-rate as a good indicator for critical driving events [29] and values of -2 m/s^3 were found to be good classifier thresholds when brake event locations were correlated with historical accident locations [32]. Moreover, a sophisticated categorization of dangerous driving has been given by the "jerk feature", calculated via the standard deviation of the jerk rate within a specific time window over the average jerk rate of the current road type [25]. The thresholds from this study were suggested to be between 0.5 and 1.0.

Summing up, we find that smartphones and driving data from vehicles have previously been utilized, and are well suited, for community sensing applications. Driver behavior is improved when users are provided with in-vehicle warnings of upcoming dangerous locations, and contextualized information in these warnings is preferred. Finally, thresholds for identifying brake events vary significantly across research studies, however, medium and high danger events typically lie around 2 m/s^2 for deceleration and -2 m/s^3 for the jerk rate.

3 System Description

In the following section we present the crowd sensing system which was developed to address the challenge of classifying dangerous locations on the road network. The system is mainly comprised of three parts: the car and its associated sensors, the smartphone which relays all data from the vehicle to the server and contributes recoded video sequences and its own sensor information, and the server which stores and processes the data and provides accident hotspot information back to the driver's smartphone. An overview of the system is shown in Fig. 1.

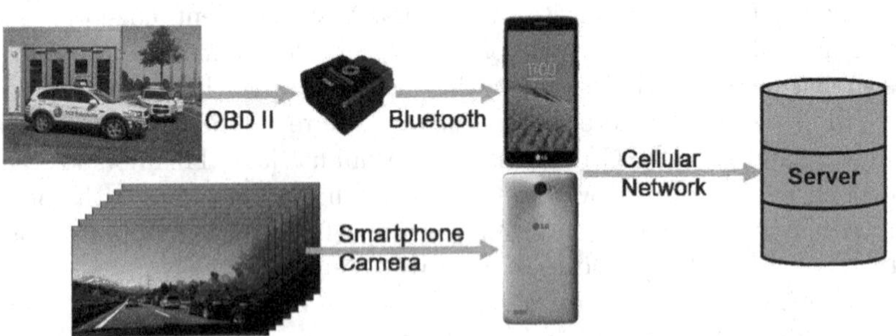

Fig. 1. Overview of the system setup.

The car's CAN-Bus was accessed via a dongle mounted at the OBD-II port and data collected was transfered via Bluetooth to the smartphone. The signals which were queried from the CAN-Bus by the dongle and transmitted to the server are shown in Table 1. These signals are not typically available with a constant frequency on the CAN-Bus, rather they are triggered on a change of the associated sensor. However, many change frequently enough to consider them a continuous signal, and for these we present an approximate frequency.

Table 1. Collected CAN-bus Signals

Signal name	Range/Measure	Approx. frequency
Wheel speed {fr, fl, br, bl}[a]	$\frac{km}{h}$	10 Hz
Wheel slip status	p-/ n-/ no- slip[b]	onChange
ABS active	on/off	onChange
ABS failed	on/off	onChange
ESP	on/off	onChange
Engine speed	rpm	10 Hz
Odometer	km	onChange
Brake pressure	Pa	30 Hz
Brake Pedal & Throttle position	%	30 Hz
Brake pressure detected	on/off	onChange
Steering wheel angle	deg	10 Hz
Gear	p/r/n/d[c]	onChange
Vehicle acceleration	$\frac{m}{s^2}$	10 Hz
Long. & Lat. acceleration	$\frac{m}{s^2}$	10 Hz
Lights	o/p/d/f[d]	onChange
Head lights	on/off	onChange
Fuel level	%	onChange
Fuel consumption	$\frac{l}{h}$	10 Hz
Fuel consumption/100 km	$\frac{l}{100\,km}$	10 Hz
Outside temperature	Celsius	onChange
Yaw rate	deg	10 Hz

[a]f: front, b: back, l: left, r: right
[b]p: positive, n: negative
[c]p: park, r: reverse, n: neutral, d: drive
[d]o: off, p: park, d: dimmed, f: full

On the smartphone, the received CAN-Bus signals were further enriched with video sequences, location information and additional sensor data from phone itself. The additional signals generated from the smartphone are presented in Table 2. In order to get an optimum view of the street scene ahead of the car, users of the system were asked to mount the smartphone so that the rear camera

Table 2. Collected smart phone signals

Signal name	Range/Measure	Approx. frequency
Bearing	deg	1 Hz
GPS altitude	Masl[e]	1 Hz
GPS latitude	deg	1 Hz
GPS longitude	deg	1 Hz
GPS speed	$\frac{m}{s}$	1 Hz
GPS accuracy	m	1 Hz
Accelerometer	{x,y,z}-$\frac{m}{s^2}$	1 Hz
Magnetometer	{x,y,z}-μT	onCalib[f]
Cellular network signal strength	Type[g], dBm	1 Hz
Video recordings	Video sequences	onTrigger

[e]Meters above sea level
[f] Only for calibration, to get the phones orientation
[g]2G, 3G, LTE

was facing out of the windshield onto the street. To better aid in the calibration of the smartphone's field of view, a live camera feed was shown to the user for six seconds on the smartphone's screen whenever a start of the engine was detected from the CAN-Bus. After the calibration screen the system provided eco-driving feedback, as introduced in [10], together with accident hotspot warnings [31] to the users. The collected data from the CAN-Bus and the smartphone were transmitted via the cellular network to a server, where they were stored in a database. With the latency of the cellular network as the main delay contribution, the system can provide insights in a real-time manner to interested road authorities and potentially other traffic participants.

3.1 Video Recordings

As introduced in Sect. 1, the aim of the system is to enrich historic accident hotspots with context from captured video sequences, and in addition classify current dangerous driving maneuvers. In order to accurately capture relevant video footage, live data from the smartphone camera was constantly written to and stored in a ring buffer of four seconds length. Whenever a user crossed a historic hotspot location, the buffer together with a further video sequence of three seconds was extracted and sent to the server. With this implementation, we received a video sequence that typically showed the entry and exit each time a historic hotspot was encountered.

To enable to video classification of dangerous driving maneuvers, the video extract was triggered on the detection of these events. Due to the many different definitions of a "dangerous driving behavior" we took the simple, yet practical definition of such events as a magnitude acceleration of greater than 2 m/s^2 in the horizontal plane [4], i.e.

$$\sqrt{\left(\frac{d^2x(t)}{dt^2}\right)^2 + \left(\frac{d^2y(t)}{dt^2}\right)^2} > 2\left[\frac{m}{s^2}\right], \tag{2}$$

in other words the L2-norm, whereby $x(t)$ and $y(t)$ denoted the coordinates in the horizontal plane dependent on the time t. Hence, the videos are triggered on heavy deceleration, acceleration, cornering, or a combination of these. Once the dangerous event is detected, the ring buffer and an additional sequence of three seconds is extracted and transmitted to the server. Similar to the historical hotspots, this results in video sequences which show how the user enters and exits each dangerous situation.

4 Study Description

In order to evaluate the proposed crowd sensing system, a field study was conducted with the Swiss road assistance service "Touring Club Suisse" (TCS). In total, 41 TCS drivers, or patrollers, were recruited to take part in a two month naturalistic driving study. All drivers were male, between 23 and 64 years old, and 37 years on average. We installed our system into each of the patrollers' cars, all same make and model of Chevrolet Captiva 2.2, without assigning them with any specific driving tasks. As such, they were instructed to simply follow their daily work and so did not follow any specific routes nor did they perform any predefined driving maneuvers. These drivers were situated at nine different base locations in the German speaking area of Switzerland with at least three drivers per base location, as shown in Fig. 2. The locations covered mountainous (southern area, base locations 3 to 5) as well as hilly to almost flat (northern area) regions. Part of the general employment conditions of the participants are taking driver safety training courses, as well as fuel efficient driving lessons. As such, a safe and "calm" driving behavior was typically expected of the drivers.

Fig. 2. The base locations of the drivers distributed in the German speaking area of Switzerland.

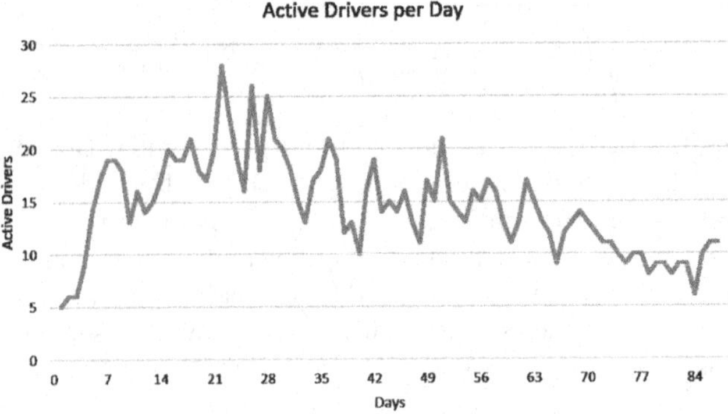

Fig. 3. Number of active drivers per day.

The study ran for over two months, including the staging and dismantling phases at the beginning and the end of the study respectively. The number of active drivers per day is given in Fig. 3. Disregarding the first and last days, we can see that at least 10 drivers were actively using our system on most days, with peak days reaching over 25 active drivers. The high fluctuations are a result of week days and the demand of call-outs per day[1].

5 Collected Dataset

Within the two months of the field study over 140,000 km of driving data were collected in total. The accumulated kilometers over the whole fleet is shown in Fig. 4. A constant increase in the accumulated kilometers can be seen. In Fig. 5 the blue line shows the total amount of driven kilometers per driver. The driver IDs are ordered by the total distance driven during the field study. Out of the 40

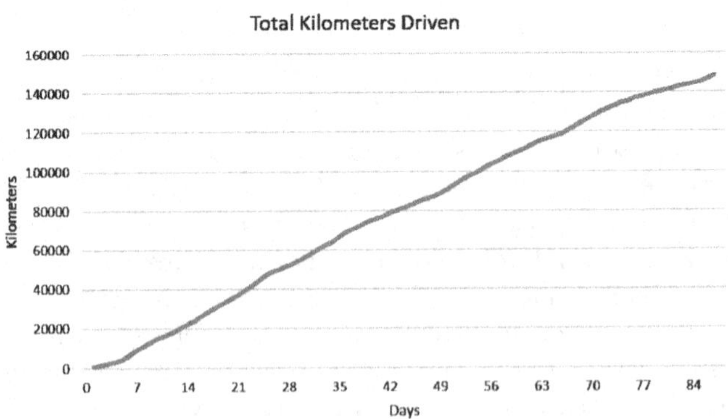

Fig. 4. Total driven kilometers of the whole fleet.

[1] Beside the normal road rescue tasks, patrollers had stand-by duty.

drivers that signed up for the field study, three never used the system. Another three drivers only marginally used the system, driving less than 1,000 km each. The three most active drivers reached almost 8,000 km within the two and a half months.

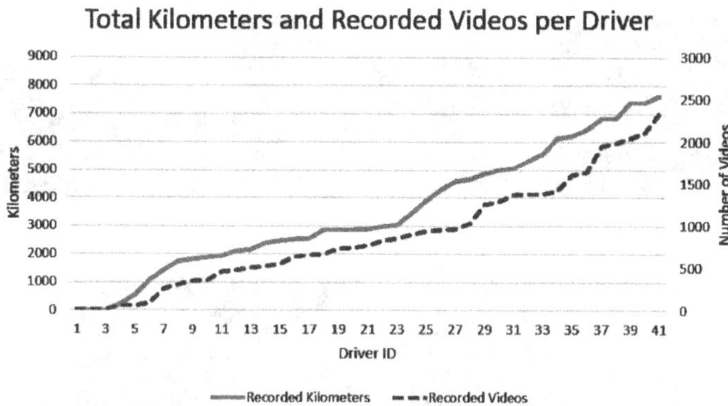

Fig. 5. Total kilometers driven and videos recorded per driver.

Furthermore, Fig. 5 additionally shows the number of videos recorded per driver, denoted by the orange dashed graph. Within the whole field study over 36,000 video sequences were collected, resulting in over 36 GB of video data. As described in Sect. 3.1 the videos were triggered during an accident hotspot crossing and on the detection of dangerous driving maneuvers. The number of videos increased more or less proportional to the total kilometers recorded, reaching a maximum of over 2,300 videos for one driver. Importantly, since the patrollers ultimately chose freely the final mounting of the smartphone within the car, two sets of videos were deemed to be unusable. These videos were generated by Driver 19, who placed the phone in such a way, that only the motor toll sticker was recorded, and Driver 31, who left the phone somewhere in the side door without any sight onto the road. Therefore of the 36,341 videos, approximately 34,000 videos are labeled as usable.

6 Preliminary Results

In the following section we present the results of two exploratory classification analyses conducted for the collected historic accident hotspot and dangerous event recorded videos respectively. Hereby, for each task we attempt to generate warning feedback similar to the guidelines introduced by the NHTSA, so that drivers are provided with information about an accident hotspot in form of a warning sign and non-critical supporting text [6]. The hotspot information will be described—whenever possible—by the three categories of location description, predominant cause, and involved traffic objects similar to the approach in [31].

The available information is hereby ranked by the following descending order of importance: objects, cause, and location. If no specific attribute in any of the category can be found a general warning sign with the text "Attention: Dangerous Location" is assigned to the hotspot.

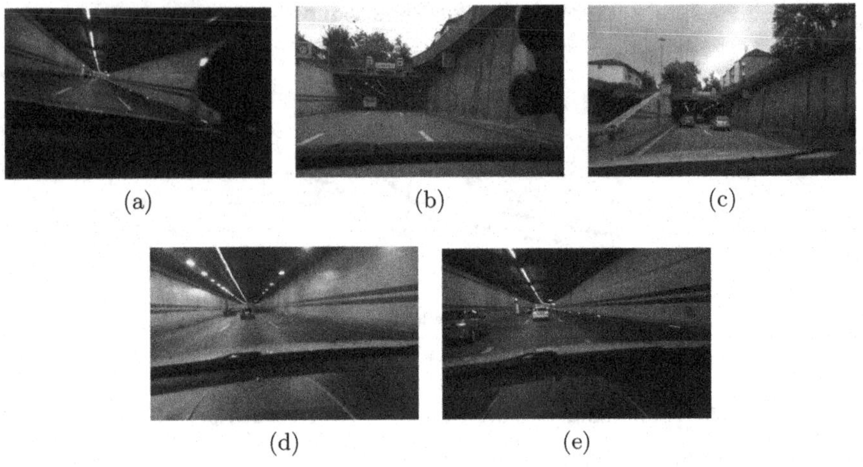

Fig. 6. Frames of five videos at a historic tunnel accident hotspot.

6.1 Historic Hotspot Contextualization

The videos recorded from existing hotspots can serve as a good indicator for the location contextualization layer describing the scenery. Out of the video sequences, large static objects can be easily identified, for example, a tunnel as shown in Fig. 6. Further, as it can be seen in Fig. 7, road junctions, crosswalks, bike-lanes and other infrastructure can also be identified.

In addition, the traffic participants, such as pedestrians (c.f. Fig. 7(e)) or cyclists (Fig. 7(f)), can be determined through investigating multiple videos of the same hazardous location, and a dominant object class assigned to the hotspot. In the same way, if the majority of video sequences of a hotspot show a particular traffic situation which could cause accidents, then this can be identified. For example, videos in Fig. 7 might lead to a "rear-end collision" categorization due to the repeatedly observed close proximity of vehicles.

6.2 Dangerous Driving Contextualization

As described in Sect. 3.1, we used utilized a simple yet practical threshold of acceleration magnitude of 2 m/s^2 for dangerous event detection. However, since the system collected the vehicle accelerometer measurements in addition to the

Fig. 7. Frames of six videos at a historic rear-end collision accident hotspot.

video sequences, we can assign the maximum magnitude acceleration and maximum jerk rates a-posteriori to each video. The following Figures will show situations, which are characterized either by a high acceleration value following the definition of [4], or by a high jerk rate, following the definition of [29, 32] as a dangerous situation. Similar to the classification of historic hotspots, the classification of a locations where dangerous driving maneuvers occurred is demonstrable from the recorded video sequences.

Fig. 8. Two frames of a video labeled as a "dangerous" driving maneuver by a high acceleration rate.

The CAN bus recordings for the videos shown in Fig. 8 show medium jerk values and high acceleration values. For the sequences in Fig. 8, the location attribute can be described as a sharp curve. In Fig. 9 a sequence characterized by high jerk rates is shown. Similar to Fig. 8, the location is characterized by a sharp curve. However, the driver enters this second curve with an excessive speed, so much so that we observe objects sliding through the driver's cabin. Where in the first sequence, a "sporty", but still controlled type of driving was

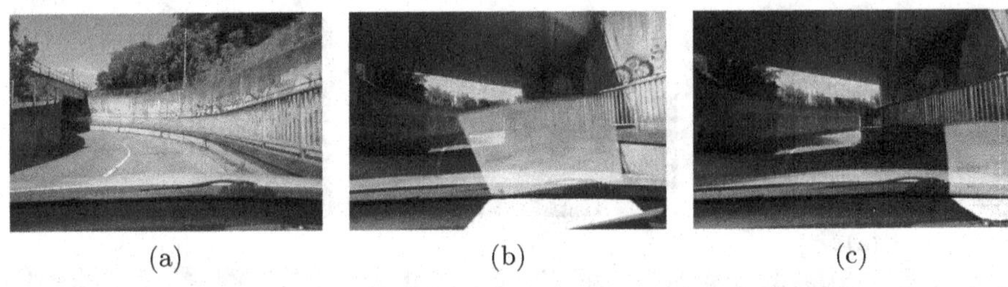

(a) (b) (c)

Fig. 9. Three frames of a video labeled as a "dangerous" driving maneuver by a high jerk rate.

performed, the cause attribute of the later one can be characterized as heedless driving. As such, the classification of the predominant cause in this situation would lead to a "swerving" or "control speed" warning. Further, weather and light conditions, i.e. entering from the bright street into the dark underpass can lead to uncontrolled maneuvers.

The location shown in Fig. 10 is classified as a crossing, where the patroller had the right of way. In this sequence we find that in the object layer another vehicle is identified, and later observe that this car disregards the patroller's right of way causing him to brake sharply. Hence, on the predominant cause layer, this situation is described as "disregarding the right of way".

(a) (b)

Fig. 10. Frames of a video characterized by disregarding the right of way.

Finally, in the sequence in Fig. 11 the driver had a delayed reaction to the vehicle ahead braking, and had to come to an emergency stop in order avoid a collision. In addition to the classification of the cause, in this case "heedless diving", the vehicle ahead can by identified in the object layer and leads to a "rear-end collision" contextualization.

(a) (b)

Fig. 11. Two frames of a video characterized by a rear-end collision situation.

7 Conclusion

In this work we introduced a crowd sensing system to tackle the problem of classifying traffic accidents and hotspots, and presented initial exploratory results demonstrating the potential of this approach. In previous work we saw that concrete countermeasures improve the road safety if the correct, i.e. the cause of accident matching counteraction, are taken. In order to identify the correct cause of a hotspot the system was taken into the field and collected a large dataset of CAN-Bus and video recordings with over 140,000 km and over 34,000 video sequences. Our first results indicate that with this data set we are able to identify dangerous hotspots within a large part of Switzerland and classify the main accident causes in order to warn drivers and road authorities with concrete information of these areas.

The preliminary results of this work should be assessed in light of its limitations. Despite the size and uniqueness of the collected dataset, we have to acknowledge that the first results are based on a homogeneous group of experienced drivers. As such, a generalization of the field study data to the normal population of traffic participants should be undertaken with caution, and future field studies should consist of more regular drivers. However, due to the experience and risk averse driving style of the patrollers, we assume that dangerous driving maneuvers occurred less frequently than a more representative group of study participants.

Further, a promising next step to this research, omitted here for brevity, are the first insights from using publicly available image classification models for object identification. This approach shows very promising results, and indicates that automated models can already be used to identify objects and reduce the laborious task of labeling all videos by hand. In future work, the object identification of such videos should be extended towards automated classification of traffic ancient hotspots and dangerous driving maneuvers. Implications of video sequences from the three layers used in this work to contextualize a hotspot should be drawn automatically to build an actual real-time system.

References

1. Adminaite, D., Jost, G., Stipdonk, H., Ward, H.: Ranking EU progress on road safety. Technical report (2016). http://etsc.eu/10th-annual-road-safety-performance-index-pin-report
2. Allgemeinder Deutscher Automobil Club: EUROTAP (2007). https://ec.europa.eu/transport/road_safety/sites/roadsafety/files/pdf/projects/eurotap.pdf
3. An, P.E., Harris, C.J.: An intelligent driver warning system for vehicle collision avoidance. IEEE Trans. Syst. Man Cybern. Part A: Syst. Hum. **26**(2), 254–261 (1996)
4. Bergasa, L.M., Almeria, D., Almazan, J., Yebes, J.J., Arroyo, R.: DriveSafe: an app for alerting inattentive drivers and scoring driving behaviors. In: IEEE Intelligent Vehicles Symposium, Proceedings, pp. 240–245 (2014)
5. Bohnenblust, D., Pool, M.: Verkehrsunfälle in der schweiz 2016. Bundesamt für Statistik, BFS (2017). https://www.bfs.admin.ch/bfs/de/home/statistiken/mobilitaet-verkehr/unfaelle-umweltauswirkungen/verkehrsunfaelle.assetdetail.3103126.html
6. Campbell, J.L., Richard, C.M., Brown, J.L., McCallum, M.: Crash Warning System Interfaces: Human Factors insights and lessons learned. Technical report (2007)
7. Chen, H.T., Lai, C.Y., Shih, C.A.: Toward community sensing of road anomalies using monocular vision. IEEE Sens. J. **16**(8), 2380–2388 (2016)
8. Clevenger, A.P., Ford, A.T., Sawaya, M.A.: Banff wildlife crossings project: Integrating science and education in restoring population connectivity across transportation corridors. Final report to Parks Canada Agency, Radium Hot Springs, British Columbia, Canada, p. 165, June 2009
9. Clevenger, T., Cypher, B.L., Ford, A., Huijser, M., Leeson, B.F., Walder, B., Walters, C.: Wildlife-Vehicle Collision Reduction Study. Technical report (2008)
10. Dahlinger, A., Wortmann, F., Tiefenbeck, V., Ryder, B., Gahr, B.: Feldexperiment zur wirksamkeit von konkretem vs. abstraktem eco-driving feedback. In: Wirtschaftsinformatik Konferenz (WI), March 2017. https://www.alexandria.unisg.ch/250432/
11. European Commision: Ranking for European Road Safety (2008). https://ec.europa.eu/transport/road_safety/sites/roadsafety/files/pdf/projects/rankers.pdf
12. European Commision: Road safety knowledge system (2008). http://safetyknowsys.swov.nl
13. European Commission: A strategic approach to implementing countermeasures (2018). https://ec.europa.eu/transport/road_safety/specialist/knowledge/young/implementation_process/a_strategic_approach_to_implementing_countermeasures_en
14. Glassco, R.A., Cohen, D.S.: Collision avoidance warnings approaching stopped or stopping vehicles. In: America, I.T.S. (ed.) 8th World Congress on Intelligent Transport Systems, Australia, Sydney (2001)
15. Goniewicz, K., Goniewicz, M., Pawłowski, W., Fiedor, P., Lasota, D.: Road safety in poland: magnitude, causes and injuries. Wiadomosci lekarskie (Warsaw, Poland: 1960), **70**(2 pt 2), 352–356 (2017)
16. Goodwin, A., Thomas, L., Kirley, B., Hall, W., O'Brien, N., Hill, K.: Countermeasures That Work: A highway safety countermeasure guide for State highway safety offices. Eighth edition. (Report No. DOT HS 812 202). National Highway Traffic Safety Administration, Washington, D.C., January 2015

17. Hogema, J., Janssen, W.: Effects of intelligent cruise control on driving behaviour: a simulator study, January 1996
18. Hu, S., Su, L., Liu, H., Wang, H., Abdelzaher, T.F.: Smartroad: smartphone-based crowd sensing for traffic regulator detection and identification. ACM Trans. Sen. Netw. **11**(4), 55:1–55:27 (2015)
19. Jähi, H., Muhlrad, N., Buttler, I., Gitelman, V., Bax, C., Dupont, E., Giustiniani, G., Machata, K., Martensen, H., Papadimitriou, E., Persia, L., Talbot, R., Vallet, G., Yannis, G.: Investigating road safety management processes in Europe. Procedia - Soc. Behav. Sci. **48**, 2130–2139 (2012)
20. Lynam, D., Castle, J., Martin, J., Lawson, S.D., Hill, J., Charman, S.: EuroRAP 2005–06 technical update. Traffic Eng. Control **48**(11), 477–484 (2007)
21. Ma, H., Zhao, D., Yuan, P.: Opportunities in mobile crowd sensing. IEEE Commun. Mag. **52**(8), 29–35 (2014)
22. Marshall, W.E., Garrick, N.W.: Street network types and road safety: a study of 24 California cities. Urban Des. Int. **15**(3), 133–147 (2010)
23. Misra, I., Shrivastava, A., Hebert, M.: Watch and learn: Semi-supervised learning of object detectors from videos. CoRR abs/1505.05769 (2015)
24. Moody, S., Melia, S.: Shared space research, policy and problems. In: Proceedings of the Institution of Civil Engineers - Transport, vol. 167, no. 6, pp. 384–392 (2014)
25. Murphey, Y.L., Milton, R., Kiliaris, L.: Driver's style classification using jerk analysis. In: 2009 IEEE Workshop on Computational Intelligence in Vehicles and Vehicular Systems, CIVVS 2009 - Proceedings, pp. 23–28 (2009)
26. Naujoks, F., Neukum, A.: Specificity and timing of advisory warnings based on cooperative perception. In: Mensch und Computer Workshopband, pp. 229–238 (2014)
27. NHTSA, N.H.T.S.: 2016 Motor Vehicle Crashes: Overview. Traffic safety facts research, pp. 1–9 (2017). https://crashstats.nhtsa.dot.gov/Api/Public/Publication/812456
28. Pan, B., Zheng, Y., Wilkie, D., Shahabi, C.: Crowd sensing of traffic anomalies based on human mobility and social media. In: Proceedings of the 21st ACM SIGSPATIAL International Conference on Advances in Geographic Information Systems, SIGSPATIAL 2013, pp. 344–353. ACM, New York (2013)
29. Pande, A., Chand, S., Saxena, N., Dixit, V., Loy, J., Wolshon, B., Kent, J.D.: A preliminary investigation of the relationships between historical crash and naturalistic driving. Accid. Anal. Prev. **101**, 107–116 (2017)
30. Ruikar, M.: National statistics of road traffic accidents in India. J. Orthop. Traumatol. Rehabil. **6**(1), 1–6 (2013)
31. Ryder, B., Gahr, B., Dahlinger, A.: An in-vehicle information system providing accident hotspot warnings. In: ECIS 2016 Proceedings. Prototypes, AIS Electronic Library (AISeL) (2016)
32. Ryder, B., Gahr, B., Dahlinger, A., Zundritsch, P., Wortmann, F., Fleisch, E.: Spatial prediction of traffic accidents with critical driving events – insigths from a nationwide field study. Transp. Res.: Part A (2017, submitted)
33. Werneke, J., Vollrath, M.: How to present collision warnings at intersections?- A comparison of different approaches. Accid. Anal. Prev. **52**, 91–99 (2013)
34. Wirz, M., Strohrmann, C., Patscheider, R., Hilti, F., Gahr, B., Hess, F., Roggen, D., Tröster, G.: Real-time detection and recommendation of thermal spots by sensing collective behaviors in paragliding. In: Proceedings of 1st International Symposium on From Digital Footprints to Social and Community Intelligence, SCI 2011, pp. 7–12. ACM, New York (2011)
35. World Health Organization: Global Status Report on Road Safety 2015. Technical report (2015)

Spam Review Detection Using Ensemble Machine Learning

Shwet Mani, Sneha Kumari, Ayushi Jain(✉), and Prabhat Kumar

Computer Science and Engineering Department, National Institute of Technology Patna, Patna, India
shwetmani100@gmail.com, sneha.nitp.gaya@gmail.com, {ayushi.cspg16, prabhat}@nitp.ac.in

Abstract. The importance of consumer reviews has evolved significantly with increasing inclination towards e-Commerce. Potential consumers exhibit sincere intents in seeking opinions of other consumers. These consumers have had a usage experience of the products they are intending to make a purchase decision on. The underlying businesses also deem it fit to ascertain common public opinions regarding the quality of their products as well as services. However, the consumer reviews have bulked over time to such an extent that it has become a highly challenging task to read all the reviews and detect their genuineness. Hence, it is crucial to manage reviews since spammers can manipulate the reviews to demote or promote wrong product. The paper proposes an algorithm for detecting the fake reviews. Since the proposed work concentrates only on text. So, n-gram (unigram + bigram) features are used. Supervised learning technique is used for reviews filtering. The proposed algorithm considers the combination of multiple learning algorithms for better predictive performance. The obtained results clearly indicate that using only simple features like n-gram, Ensemble can boost efficiency of algorithm at significant level.

Keywords: Online review · E-commerce portal · Spam review
Ensemble · Machine learning

1 Introduction

With the rapid development of Internet, the impact of online reviews increases continuously. E-commerce Portals are getting increasingly popular to share customer views. Online reviews are important for a variety of reasons. With the help of online reviews, merchants can enhance their business by getting feedback of their products and customers can check reviews before placing an order for a product. However, besides this, online reviews have a negative impact too.

E-commerce websites have been a common target of spammers for writing fraud reviews with a view to affect the reputation of any product. There is a plethora of online sites like Amazon, Flipkart, Limeroad etc. which provide online reviews of customers on specific products. Although these reviews are vital source of information, customers can write anything about the products. But this freedom to write anything about a product can degrade the quality of reviews. Review spam refers to fake or any

P. Perner (Ed.): MLDM 2018, LNAI 10935, pp. 198–209, 2018.
https://doi.org/10.1007/978-3-319-96133-0_15

irrelevant review which are written to promote or defame a product. Sometimes, fake reviews may mislead customers, as spammers do this either to promote a particular product or demote their competitor's product. This leads to loss of consumers' trust on online reviews. Many researchers have made valuable contributions in this field.

Jindal and Liu [9] have classified the reviews in three groups. They are explained as following.

- **Untruthful reviews**- These reviews are generally meant for misleading the customers either by exaggerating about the product with a view to promote or demote a product.
- **Reviews on brands-** These reviews focus on a particular brand and not on the various features of the product. Since these are not related to products, they are considered as spam reviews.
- **Non-reviews-** These reviews involve mainly advertisements or any irrelevant reviews containing no opinions like questions, answers, links or any aimless texts (link, random text).

Jindal and Liu [9] proposed three basic techniques for detection of spam reviews.

- **Review Centric Approach-** This approach basically concentrates on the content of reviews written by reviewers. This method tries to find out relationship between reviews by considering each of them. The features, which are used in this method, are length of review, number of feedbacks, similarity between reviews, percentage of numerals, capitals etc.
- **Reviewer Centric Approach-** This method focuses on the behaviour of reviewers. It considers information about users and all reviews written by them. Based on the behaviours, reviewers have been classified as spammers and non-spammers. Features used in this method, are reviewer id, deviation from average rating and so on.
- **Product Centric Approach-** This method mainly emphasizes on the information related to any product. Features used in this method are price of product, sales rank of product etc.

Determining whether a given review is genuine or not is quite difficult from the human point of view. It is difficult for humans to distinguish between genuine reviews and fake reviews manually. So, various methods have been devised for detecting spam reviews. Most of the existing methods of detecting spam reviews have used supervised learning algorithm. This method makes use of labeled data sets which contain features for identifying whether the given review is spam or not. In the paper, ensemble learning techniques are used which combine multiple classifiers to create a robust classifier. Moreover, ensemble techniques have been shown to increase classifier performance.

Rest of the paper is organized as follows: Sect. 2 summarizes previous works related to the review spam detection techniques; Sect. 3 describes our proposed method; Sect. 4 explains experimental results and analysis; finally, Sect. 5 includes conclusion and scope of future work.

2 Related Works

Analysis of online opinions has become a popular research topic these days. Since the main motive of the proposed work is to detect spam in reviews. Numerous researchers have previously worked in this domain. Existing works have made many progresses in spam review detection. Jindal and Liu [9] covered supervised learning technique that has been applied for the detection of online spam reviews in their paper. They have extracted 5.8 million reviews, 2.14 million reviewers and 6.7 million products from Amazon website and divided them into three categories: untruthful reviews, reviews on brand and non-reviews. Features using review text, reviewer and product have been used to differentiate between spam and non-spam opinions. Logistic Regression using statistical package R have been found to be effective in detecting of type 2 and type 3 spam reviews. The result of Area under curve (AOC) shows 78% accuracy. Heredia et al. [8] proposed an ensemble technique combining multiple learners for detecting spam reviews. Three methods have been used: Bagging, Boosting and Random Forest. Base learners that have been used are Multinomial Naïve Bayes (MNB), C4.5, Logistic Regression (LR) and Support Vector Machine (SVM). Li et al. [13] worked on a dataset consisting of reviews on three domains including hotels, doctors and restaurants. They used LIWC, POS and unigram features. By SVM they yielded an accuracy of 65%. String to Word Vector in [7] has been used for creation of bag of words feature set. AUC has been used as a performance metric which plots graph between true and false rate of model. The result shows that Ensemble technique has increased the performance of SVM, LR and C4.5. Boosting technique with MNB has been found to have highest AUC curve and lowest standard deviation. It shows that SVM, LR and C4.5 have performed better by using boosting technique. MNB approach has been found to be faster and less costly.

Ott et al. [16] created a dataset consisting of 800 reviews with positive sentiment towards hotels. Naive Bayes and SVM classifiers have been used. They have used three approaches: Genre Identification, Psycholinguistic deception detection and Text Categorization. They concluded that both the contextual parameters (e.g. BIGRAM) and motivational parameters should be considered in deceptive opinion spam detection. Using LIWC and bigram features, 89.8% accuracy was reported. This is quite an interesting result but as proposed by Mukherjee et al. [14], the previous method is not suited for real world dataset. It yielded an accuracy of 67.8% on real world yelp dataset. Shojaee et al. [21] worked on same hotel dataset and using stylometry features, they yielded an accuracy of 84%. Srivastava et al. [23] proposed an algorithm for automatic product review analysis. They presented a Two-Parse algorithm with a training dataset of approximately 7,000 keywords. Authors proposed a Weighted k-Nearest Neighbor (Weighted k-NN) Classifier which was better than the classical k-Nearest Neighbor classifiers in terms of efficiency.

Peng and Zhong [18] proposed sentiment analysis techniques for spam review detection. They extracted 5000 reviews from Resellerrating.com. Firstly, a sentiment lexicon combined with general sentiment lexicon and sentiment lexicon special for product has been built. Then they proposed a method to compute the sentiment score from the natural language text by a shallow dependency parser. A set of discriminative rules have been presented through instinctive observation. The discriminative rules have been combined with the time series method to find out suspicious stores.

Many studies have been done to detect spammers through analysis of their behaviors. Qian and Liu [19] have used supervised learning method to identify multiple user ids of same reviewer. Mukherjee et al. [15] proposed a novel method to observe spammer behaviors. They have used real world amazon review dataset for their experiments. Due to difficulty of having trained dataset for supervised technique, they have proposed an unsupervised model called Author Spamcity Model (ASM). Feng et al. [4] have introduced a strategy to create a pseudo gold standard dataset that is labeled automatically based on different types of distributional footprints. In their study, they assumed that there is a set of distributions of review rating scores for a given domain. With this assumption, they concluded that spammers will definitely distort this distribution of review scores. Bajaj et al. [2] analyses e-commerce portals reviews to focuses on personal characteristics of the reviewer rather than reviews. For detecting fake reviews, they used four indicators: Gmail authentication, IP address, location and spam dictionary. In the proposed framework, the user can only post a review once from a particular email id and can use the same IP address twice. The user can use same location thrice to post reviews. Authors used 100 reviews for testing purpose. The proposed framework gives 65% accurate results. But this proposed framework is unable to find spammers in few cases such as spammers change their location using softwares. Sometimes, it may not detect the correct geo-coordinates. And the spammers route the device to other IP is another case in which this proposed method fails. Gaurav and Kumar [5] developed a Consumer Satisfaction Rating System (CSRS) which tells about specific features of the product being reviewed based on sentiment analysis of consumer reviews on the basis of product's feature. Prabhat et al. [11] proposed a mechanism to deal with social economic attacks on the e-commerce business. The mechanism worked against the spammers who tries to defame the product to affect their opponents. Prabhat et al. [12] have proposed an algorithm to mitigate the number of fake orders that can hamper the profit of the seller.

3 Proposed Work

This section includes the proposed method for detecting spam reviews that either they are honest or not. For this purpose, *n*-gram (unigram + bigram) feature is used. It is a contiguous sequence of n items from a given text. Unigram takes the most frequent words from a review text. In bigrams, two consecutive words are extracted from a text. Larger sizes are sometimes referred by the value of *n*.

Next, the analysis was done in two phases named as First Phase and Second Phase. In First Phase, a primary analysis was done in which the 10-fold cross validation was used to train and test the data. The classification algorithms used in First Phase were Naive Bayes, Support Vector Machine (SVM), and Random Forest. Cross validation was used as it uses full dataset for training and testing. In 10-fold cross validation, the dataset is split into ten folds. Each time, nine are used for training purpose and remaining one is used for testing at a time. In Second Phase, two ensemble techniques were introduced, Simple Majority Voting ensemble (Voting) and Stacked ensemble to provide a better classification model.

3.1 First Phase

Naïve Bayes

Naive Bayes classifiers are a family of simple probabilistic classifiers based on applying Bayes' theorem with strong independence assumptions between the features. It simply assumes that the presence of a particular feature in a class is unrelated to the presence of any other feature. It is a fast, highly scalable algorithm that can be used for binary and multiclass classification. A systemic problem with Naive Bayes is that features are assumed to be independent. As a result, even when words are dependent, each word contributes evidence individually [20]. Bayes theorem provides a way of calculating posterior probability P(c | x) from P(c), P(x) and P(x | c). It is assumed that all features are independent of each other.

$$\mathrm{P(c\,|\,x)} = \frac{\mathrm{P(x\ |\,c) * P(c)}}{P(x)}$$

Where

P(c | x) is the posterior probability of class (c, target) given predictor (x, attributes).
P(c) is the prior probability of class.
P(x | c) is the likelihood which is the probability of predictor given class.
P(x) is the prior probability of predictor.

Support Vector Machine

Support vector machines (SVMs) are supervised learning models with associated learning algorithms that analyze data used for classification and regression analysis. In this algorithm, each data item as a point in n-dimensional space is plotted (where n is number of features) with the value of each feature being the value of a particular coordinate. Then, classification is performed by finding the hyper-plane that differentiates the two classes. It finds the hyperplane that gives the largest minimum distance with the training examples. It is effective in cases where number of dimensions is greater than the number of samples. In the case of large data set, it fails to perform well because the required training time is higher. SVMs have empirically been shown to give good generalization performance on a wide variety of problems such as hand-written character recognition, face detection, pedestrian detection, and text categorization [6]. Among all the possible representations of hyperplane, the following equation is chosen.

$$|\,\beta^0 + \beta^T x\,| = 1$$

Where x is the training examples closest to the hyperplane
β^0 is the bias
β^T is weight vector.

Random Forest

Random forests or random decision forests are an ensemble learning method for both classification and regression. It operates by constructing a multitude of decision trees at training time and outputting the class that is the mode of the classes (classification) of the individual trees. These are a combination of tree predictors such that each tree depends on the values of a random vector sampled independently and with the same distribution for all trees in the forest [1]. It has methods for balancing error in class population unbalanced datasets.

3.2 Second Phase

In this phase, results which are produced by evaluating all the models in phase one are used. Multiple learning algorithms are used to obtain better predictive performance than that could be obtained from any of the constituent learning algorithms alone. Based on this, two models are proposed.

Voting Ensemble

Simple Majority Voting Ensemble has been used in this study to combine output of first phase to achieve a better combined result. Bagging is a voting scheme in which n models, usually of same type, are constructed. For an unknown instance, each model's predictions are recorded. Then a class is assigned which is having the maximum vote among all the predictions from the models. It trains each model in the ensemble using a randomly drawn subset of the training set [10] (Fig. 1).

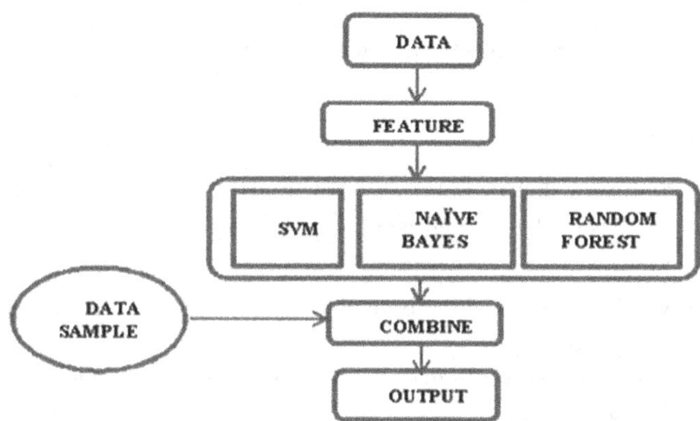

Fig. 1. Block diagram of voting ensemble

Stacking Ensemble

Stacking which is more frequently known as stacking generalization trains, an algorithm in its learning phase and combines the predictions of several further learning algorithms. The first step is to train all other algorithm with available data, which is further trained by a combiner algorithm for a final prediction while utilizing all predictions of other algorithms as additional inputs.

There are two step functionalities of stacking. For the first step, every first level classifier is taking part in the j- fold cross validation training which return a vector form < (y'0… y'm), yj > where y'm is the predicted output of the mth classifier and yj is the expected output for the classifier. The above model depicts level-0 data model. For the second step the input of first step is provided for the meta learning algorithm which optimizes the classification of combined model by adjusting the errors in the model. Now the obtained model is called as the level-1 data model. The process is then repeated for k fold cross validation to produce the final stacked generalization model [3] (Fig. 2).

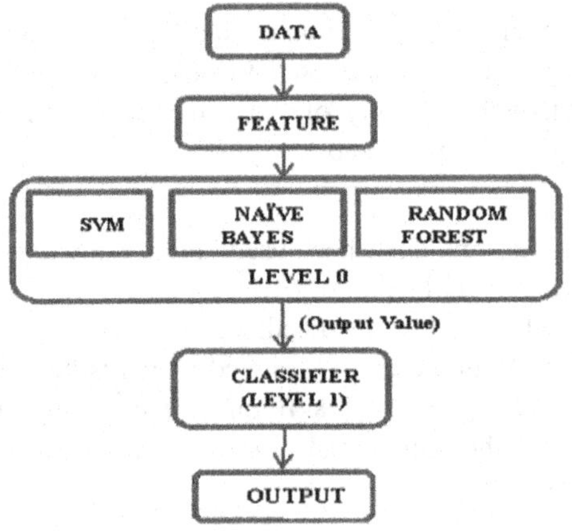

Fig. 2. Block diagram of stacking ensemble

3.3 Performance Metric

In this study, confusion matrix [22] has been used to evaluate the performance of classifiers on a set of test data. It is basically a two-dimensional matrix which results in four values:

- True Positive (TP) - True positive basically depicts that both the predicted and the actual label are spam.
- True Negative (TN) - True negative shows that both the predicted and the actual label are non-spam.
- False Positive (FP) - False positive concludes out to be a spam for predicted label while a non-spam for the actual label.
- False negative (FN) – False negative shows non-spam behaviour for predicted label while spam behaviour for actual label.

Based on the confusion matrix, precision, recall and accuracy are measured.

- Precision (P) - It is defined as the ratio of relevant instances among the retrieved instances. It is also known as positive predictive value.

$$P = TP/(TP + FP)$$

- Recall (R) - It is defined as the ratio of relevant instances that have been retrieved over total relevant instances. It is also known as sensitivity.

$$R = TP/(TP + FN)$$

- Accuracy (A) - It is defined as the degree of correctness of a quantity or expression.

$$A = (TP + TN)/(TP + TN + FP + FN)$$

4 Results and Analysis

This section includes the results and analysis of the proposed methodology for detecting spam review. This paper uses the dataset found in Ott et al. [17] for analysis purpose. This dataset consists of 1600 reviews for a set of 20 popular hotels in Chicago. It was named as "Gold Standard Dataset" which consists of 800 fake reviews using Amazon Mechanical Turk (AMT) and 800 genuine reviews from popular online review communities. In Subsects. 4.1 and 4.2, the confusion matrices and results are shown for the First and Second Phase respectively. These confusion matrices and results are generated using the mentioned classifiers which are SVM, Naïve Bayes and Random Forest.

4.1 First Phase

In the first phase, training and testing of features is done using 10-fold cross validation. The confusion matrix and results are generated using the classification models (SVM, Naïve Bayes and Random Forest).

Confusion Matrix Using SVM

Table 1 shows confusion matrix using SVM algorithm on the dataset mentioned above. From this table it can be inferred that out of 800 non-spam reviews 648 reviews and out of 800 spam reviews 680 reviews have been identified correctly. Whereas 152 non-spam reviews and 120 spam reviews have been miss-classified as spam and non-spam respectively.

Table 1. SVM

Actual	Predicted	
	Non-spam	Spam
Non-spam	648	152
Spam	120	680

Confusion Matrix Using Naïve Bayes

Table 2 shows the confusion matrix using Naïve Bayes algorithm on the dataset mentioned above. From this table it can be inferred that out of 800 non-spam reviews 687 reviews and out of 800 spam reviews 707 reviews have been identified correctly. Whereas 113 non-spam reviews and 93 spam reviews have been miss-classified as spam and non-spam respectively.

Table 2. Naïve Bayes

Actual	Predicted	
	Non-spam	Spam
Non-spam	687	113
Spam	93	707

Confusion Matrix using Random Forest

Table 3 shows confusion matrix using Random forest model of dataset mentioned above. Through this table it can be inferred as out of 800 non-spam reviews 664 non-spam reviews and out of 800 spam reviews 694 spam reviews have been successfully identified. Whereas 136 non-spam reviews and 106 spam reviews have been miss-classified as spam and non-spam respectively.

Table 3. Random forest

Actual	Predicted	
	Non-spam	Spam
Non-spam	664	136
Spam	106	694

Results of First Phase

From the above table, it is clear that Naïve Bayes performs best among all the three algorithms as it gives the most accurate results. The accuracy is 87.12%. The precision and recall of spam and non-spam reviews is also high using the Naïve Bayes algorithm (Table 4).

Table 4. Results of first phase

	SVM	Naïve Bayes	Random forest
Precision (Spam)	0.81	0.86	0.83
Precision (Non-spam)	0.84	0.88	0.86
Recall (Spam)	0.85	0.88	0.86
Recall (Non-spam)	0.81	0.85	0.83
Accuracy (%)	83.0	87.12	84.87

4.2 Second Phase

In this phase the output of first phase is used as the input and are analyzed using Voting and Stacking Ensemble technique. The confusion matrices and results are produced using these Ensembles.

Confusion Matrix using Voting Ensemble
Table 5 shows confusion matrix of the result in second phase of algorithm using, Voting Ensemble on the dataset. From this table it can be inferred that out of 800 non-spam reviews 686 reviews and out of 800 spam reviews 713 reviews have been correctly identified. From this table it can be inferred that out of 800 non-spam reviews 686 reviews and out of 800 spam reviews 713 reviews have been correctly identified. Whereas 114 non-spam and 87 spam reviews have been miss-classified as spam and non-spam respectively.

Table 5. Voting Ensemble

Actual	Predicted	
	Non-spam	Spam
Non-spam	686	114
Spam	87	713

Confusion Matrix using Stacking Ensemble
Table 6 shows confusion matrix of the result in second phase of algorithm using Stacking Ensemble on the mentioned dataset. From this table it can be inferred that out of 800 non-spam reviews 718 reviews and out of 800 spam reviews 685 reviews have been successfully identified. Whereas 82 non-spam reviews and 115 spam reviews have been miss-classified as spam and non-spam respectively.

Table 6. Stacking Ensemble

Actual	Predicted	
	Non-spam	Spam
Non-spam	718	82
Spam	115	685

Result (Voting Ensemble) Table 7 shows the voting ensemble result where it is clear that the proposed method correctly classifies spam and non-spam up to 86% and 88% of time correctly. Again from 100% spam and non-spam, it correctly identifies 89% and 85% of time. Using this technique 87.43% accuracy is achieved.

Result (Stacking Ensemble)
Table 8 shows that using Stacking Ensemble in second phase, the proposed algorithm correctly detected spam and non-spam up to 89% and 86% of time respectively. Again, spam and non-spam, it correctly identifies 85% and 89% of time. This technique gives 87.68% accurate results.

Table 7. Voting Ensemble

	Majority voting
Precision (Spam)	0.86
Precision (Non-spam)	0.88
Recall (Spam)	0.89
Recall (Non-spam)	0.85
Accuracy (%)	87.43

Table 8. Stacking Ensemble

	Random forest
Precision (Spam)	0.89
Precision (Non-spam)	0.86
Recall (Spam)	0.85
Recall (Non-spam)	0.89
Accuracy (%)	87.68

5 Conclusion and Future Work

The work presented in this paper proposes a spam review detection algorithm, which detects that a review is genuine or fake with respect to a product. This, indeed, is helpful for consumers who are intending to make decisions regarding purchase of a product based on reviews. In the proposed work, simple n-gram (unigram + bigram) feature has been used. In the first phase of analysis, three classification algorithms SVM, Naïve Bayes, and Random Forest are used for classification of reviews as spam or non-spam. Naïve Bayes gave the most accurate results among all three classification algorithms by achieving 87.12% of accuracy. In the second phase, two ensemble techniques have been used in which the Stacking Ensemble technique performed better with 87.68% of accuracy. These findings further suggest the importance of considering ensemble techniques. Research efforts in future may be based on improving the accuracy of finding spam reviews. The proposed method emphasizes only on detecting the fake reviews. So, a mechanism can be proposed for reducing the fake reviews in future.

References

1. Breiman, L.: Random forests. Mach. Learn. **45**(1), 5–32 (2001)
2. Bajaj, S., Garg, N., Singh, S.K.: A novel user-based spam review detection. Procedia Comput. Sci. **122**, 1009–1015 (2017)
3. Džeroski, S., Ženko, B.: Is combining classifiers with stacking better than selecting the best one? Mach. Learn. **54**(3), 255–273 (2004)
4. Feng, S., Xing, L., Gogar, A., Choi, Y.: Distributional footprints of deceptive product reviews. ICWSM **12**, 98–105 (2012)

5. Gaurav, K., Kumar, P.: Consumer satisfaction rating system using sentiment analysis. In: Kar, A.K., et al. (eds.) I3E 2017. LNCS, vol. 10595, pp. 400–411. Springer, Cham (2017). https://doi.org/10.1007/978-3-319-68557-1_35
6. Gunn, S.R.: Support vector machines for classification and regression. ISIS Tech. Rep. **14** (1), 5–16 (1998)
7. Hall, M., Frank, E., Holmes, G., Pfahringer, B., Reutemann, P., Witten, I.H.: The WEKA data mining software: an update. ACM SIGKDD Explor. Newsl. **11**(1), 10–18 (2009)
8. Heredia, B., Khoshgoftaar, T.M., Prusa, J., Crawford, M.: An investigation of ensemble techniques for detection of spam reviews. In: 15th IEEE International Conference on Machine Learning and Applications (ICMLA), pp. 127–133. IEEE, December 2016
9. Jindal, N., Liu, B.: Opinion spam and analysis. In: Proceedings of the 2008 International Conference on Web Search and DataMining. ACM (2008)
10. Kim, H.C., Pang, S., Je, H.M., Kim, D., Bang, S.Y.: Constructing support vector machine ensemble. Pattern Recognit. **36**(12), 2757–2767 (2003)
11. Kumar, P., Dasari, Y., Nath, S., Sinha, A.: Controlling and mitigating targeted socio-economic attacks. In: Dwivedi, Y.K., et al. (eds.) I3E 2016. LNCS, vol. 9844, pp. 471–476. Springer, Cham (2016). https://doi.org/10.1007/978-3-319-45234-0_42
12. Kumar, P., Dasari, Y., Jain, A., Sinha, A.: Fake order mitigation: a profile based mechanism. In: Kar, A.K., et al. (eds.) I3E 2017. LNCS, vol. 10595, pp. 276–288. Springer, Cham (2017). https://doi.org/10.1007/978-3-319-68557-1_25
13. Li, J., Ott, M., Cardie, C., Hovy, E.H.: Towards a general rule for identifying deceptive opinion spam. In: ACL, vol. 1, pp. 1566–1576, June 2014
14. Mukherjee, A., Venkataraman, V., Liu, B., Glance, N.: What yelp fake review filter might be doing?. In: Seventh International AAAI Conference on Weblogs and Social Media, June 2013
15. Mukherjee, A., Kumar, A., Liu, B., Wang, J., Hsu, M., Castellanos, M., Ghosh, R.: Spotting opinion spammers using behavioral footprints. In: Proceedings of the 19th ACM SIGKDD International Conference on Knowledge Discovery and Data Mining, pp. 632–640. ACM, August 2013
16. Ott, M., Choi, Y., Cardie, C., Hancock, J.T.: Finding deceptive opinion spam by any stretch of the imagination. In: Proceedings of the 49th Annual Meeting of the Association for Computational Linguistics: Human Language Technologies, vol. 1, pp. 309–319. Association for Computational Linguistics, June 2011
17. Ott, M., Cardi, C., Hancock, J.T.: Negative deceptive opinion spam. In: HLT- NAACL (2013)
18. Peng, Q., Zhong, M.: Detecting spam review through sentiment analysis. JSW **9**(8), 2065–2072 (2014)
19. Qian, T., Liu, B.: Identifying multiple userids of the same author. In: EMNLP, pp. 1124–1135, October 2013
20. Rennie, J.D., Shih, L., Teevan, J., Karger, D.R.: Tackling the poor assumptions of naive bayes text classifiers. In: Proceedings of the 20th International Conference on Machine Learning (ICML-2003), pp. 616–623 (2003)
21. Shojaee, S., Murad, M.A.A., Azman, A.B., Sharef, N.M., Nadali, S.: Detecting deceptive reviews using lexical and syntactic features. In: 2013 13th International Conference on Intelligent Systems Design and Applications (ISDA), pp. 53–58. IEEE, December 2013
22. Sokolova, M., Lapalme, G.: A systematic analysis of performance measures for classification tasks. Inf. Process. Manag. **45**(4), 427–437 (2009)
23. Srivastava, A., Singh, M.P., Kumar, P.: Supervised semantic analysis of product reviews using weighted k-NN classifier. In: 2014 11th International Conference on Information Technology: New Generations (ITNG), pp. 502–507. IEEE, April 2014

Automatic Keyphrase Extraction Using Recurrent Neural Networks

Johannes Villmow(✉), Marco Wrzalik, and Dirk Krechel

RheinMain University of Applied Sciences, Wiesbaden, Germany
{johannes.villmow,marco.wrzalik,dirk.krechel}@hs-rm.de

Abstract. Automatic Keyphrase Extraction describes the process of extracting keywords or keyphrases from the body of a document. To our knowledge until now all algorithms rely on a set of manually crafted statistical features to model word importance. In this paper we propose an end-to-end neural keyphrase extraction algorithm using a siamese LSTM network, eliminating the need for manual feature engineering. We train and evaluate our model on the *Inspec* [6] dataset for keyphrase extraction and achieve comparable results to state-of-the-art algorithms.

Keywords: Keyphrase extraction · Keyword extraction
Neural network · LSTM · Siamese network
Natural language processing

1 Introduction

Automatic keyphrase extraction is defined as the automatic selection of important, topical phrases from within the body of a document [15]. The keyphrases can then be used for summarizing or indexing the document. Most approaches initially extract a set of candidate phrases out of a document. These candidates are then ranked by an algorithm and the top ranked candidates will be returned as keyphrases for this document. Until now all algorithms use manually constructed statistical features to assign a score to each keyphrase.

One can divide the approaches to keyphrase extraction into supervised and unsupervised algorithms. Supervised approaches treat the problem as binary classification or ranking of the candidate keyphrases. This is done by training a machine learning algorithm, that is able to rank a candidate using a manually constructed feature set as input. To our knowledge all supervised approaches use an ensemble of structural, syntactical or external resource-based features of the candidate as input to the algorithm. Witten et al. proposed 1999 the KEA [17] keyphrase extraction algorithm. It consists of a naive bayes classifier, that uses the following feature set: *tf*idf*, *word distance* of a candidate to the start of a document, and *supervised keyphraseness*, e.g. the times a phrase appears as a keyphrase in the training set. Hulth proposed in 2003 a bagging model [6] that additionally uses Part-Of-Speech tags (POS-tags) of a candidate as a feature. In 2009 Medelyan et al. released the commonly used Maui [10] algorithm, which is

P. Perner (Ed.): MLDM 2018, LNAI 10935, pp. 210–217, 2018.
https://doi.org/10.1007/978-3-319-96133-0_16

a bagged decision tree model with a bigger feature set, adding six features to the feature set of KEA. The authors include external knowledge from Wikipedia as input features.

At the same time various unsupervised approaches have been proposed. In 2004 Mihalcea et al. proposed TextRank [11], a graph-based ranking approach that ranks candidates by their associations in the graph. Liu et al. proposed in 2009 [8] an unsupervised clustering-based approach (KeyCluster), that clusters similar words using Wikipedia and co-occurrence based statistics [4]. Afterwards they select all phrases from the document as keyphrases, that contain one or more words from the cluster centers and have a certain POS-tag structure. In 2010 Rose et al. [13] proposed RAKE, in which candidates are selected using a keyphrase-adjacency stoplist and further on ranked using a graph of co-occurrences. Wang et al. propose WordAttractionRank [16], an extension to TextRank incorporating background knowledge from word embeddings, which have been a factor of success for deep learning models in natural language processing.

2 Dataset

There are multiple corpora for keyword extraction, that vary in domain and length of the documents[1]. We train and evaluate our model on the *Inspec* dataset published by Hulth in 2003 [6], as it is with 2000 documents the biggest publicly available dataset for keyword extraction. Furthermore it has been used for evaluation of most of the proposed algorithms.

The dataset consists of 2000 abstracts from journal papers of the *Inspec* database, from the disciplines Computer Science and Information Technology. For each of these abstracts there are two sets of assigned keywords: A set of controlled keyphrases that are restricted to the *Inspec* thesaurus and a set of uncontrolled keyphrases which can be any suitable terms. Following the work of Hulth, we only used the set of uncontrolled keyphrases as 76.2% of them appear in the abstracts (compared to 18.1% for the controlled keyphrases) [6]. The dataset is split in three parts: A training set consisting of 1000 documents, a validation set consisting of 500 documents and a test set for final evaluation consisting again of 500 documents.

The document length varies from very short documents with 30 tokens to long documents with 540 tokens, with a median around 180 in the training set. The amount of keyphrases varies strongly from 0 up to 37 keyphrases per document. We only focus on the assigned keyphrases, that appear in the text of the document and thus discard the ones that don't.

3 Neural Keyphrase Extraction

To extract keyphrases from a document we initially construct a set of candidate keyphrases for this document. We can then rank each candidate as a keyphrase

[1] Hasan and Ng give a tabular overview over available keyword extraction datasets [4].

in context of the document. This enables us to treat the keyphrase extraction as a binary regression problem. Additionally when doing this, we increase the amount of training samples. Positive candidates we define as the ones that are labeled as keyphrases, whereas negative candidates are no keyphrases. To have a well balanced dataset while training our, we construct training samples from a document by randomly choosing as many negative as positive candidates. When testing, we rank every candidate with our neural network model and return the ones with the highest scores.

3.1 Candidate Generation

In our approach, candidates to a document are one or more consecutive words, known as n-grams, inside sentence boundaries that comply with the following two heuristics:

1. Candidates consist of at most n words. For our experiments we chose $n = 4$.
2. Candidates can not begin or end with a stopword. We use the Fox stopword list [2].

This is a slightly modified version of the candidate generation method of KEA. KEA is more restrictive as they also state that candidates cannot be proper names. The method we used, has initially been used in Maui [9]. Table 1 shows that using $n = 4$ is a good trade-off between the amount of extracted candidates and amount of keyphrases inside the set of candidates. It also reveals, that 9.9% of the keyphrases are either longer than 4 words, or begin/end with a stopword and so can't be found by our algorithm.

Table 1. Impact of candidate generation parameter n to the amount of generated candidates and percentage of keyphrases in that set of candidates.

Candidate length (n)	Extracted candidates (#)	Keyphrase recall (%)
2	39,413	67.6
3	56,334	85.7
4	**74,986**	**90.1**
5	93,325	91.5
6	111,313	92.0

3.2 Network

We use a siamese neural network architecture with an attention mechanism, which has been successfully applied to compare two tweets regarding their humor [1]. The attention mechanism has been introduced by Yang et al. to classify documents [18]. The architecture of our classifier is shown in Fig. 1.

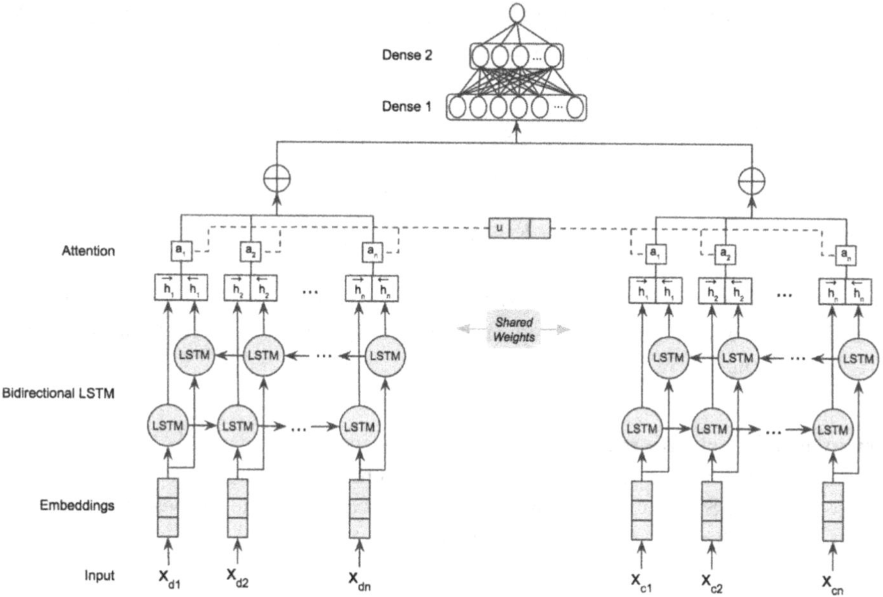

Fig. 1. Architecture of our neural keyphrase classifier (based on [1]).

A sample fed into our classifier consists of two inputs: The sequence of words of a document $X_d = (x_{d1}, x_{d2}, \ldots, x_{dt})$ and the sequence of words of the to be classified candidate $X_c = (x_{c1}, x_{c2}, \ldots, x_{ct})$. Here t is a parameter of our network, that describes the maximum length of words in each sequence. Both X_d and X_c will be padded beforehand to maximum length t with a padding token. In this siamese network both inputs will be read independently by the same recurrent neural network (RNN). An attention mechanism then detects important words and produces a fixed sized output for each sequence. Those two outputs are then compared by a two-layer MLP and the probability that the candidate is a keyphrase for the document is returned.

Word Encoding. An embedding layer projects all words from both sequences X_d and X_c into d-dimensional vector space. We use GloVe embeddings [12] with $d = 100$, that are pretrained on a Wikipedia dump. Words that are not part of the embedding vocabulary will be replaced by an unknown token. The embedding matrix W_e is fixed and will be not optimized during training. Formally, every one-hot encoded word x_i for $i \in [1 \mathinner{..} t]$ is projected into $\mathbb{R}^d$ by

$$v_i = W_e x_i \tag{1}$$

Sequence Encoding. The embedded words are fed into an RNN. We use bidirectional long short-term memory (BLSTM) units as RNN cells. A BLSTM consists of two LSTM units [5] with a hidden state dimensionality L: One LSTM unit reads the embedded input sequence from left to right, the other from right to left.

$$\overrightarrow{h_i} = \text{LSTM}(\overrightarrow{h_{i-1}}, v_i) \tag{2}$$

$$\overleftarrow{h_i} = \text{LSTM}(\overleftarrow{h_{i-1}}, v_{t-i+1}) \tag{3}$$

$$h_i = \overrightarrow{h_i} \otimes \overleftarrow{h_i} \tag{4}$$

Therefore, at each time step i, two hidden states $\overrightarrow{h_i}, \overleftarrow{h_i} \in \mathbb{R}^L$ emerge from the LSTM units, which concatenated (noted by $\otimes$) form the actual hidden state $h_i \in \mathbb{R}^{2L}$ of the BLSTM.

Attention Mechanism. Following the practice of Yang et al. [18], we point out informative words using an attention mechanism, that weights the hidden states of the BLSTM with a context or query vector u. The process of assigning an attention weight $a_i \in [0, 1]$ to every hidden state h_i is as following:

$$s_i = tanh(W_1 h_i + b_1) \tag{5}$$

$$a_i = softmax(s_i^T u) = \frac{exp(s_i^T u)}{\sum_i exp(s_i^T u)} \tag{6}$$

$$r = \sum_i a_i h_i \tag{7}$$

Initially we feed h_i through an one-layer feed-forward neural network to obtain another hidden representation s_i. The importance of a word is measured by its similarity to the context vector u. We can compute the word attention weights by taking the softmax of words similarity score. The final representation of the input sequence is then the weighted sum of the hidden states.

Output Layer. After the two input sequences X_d and X_c are transformed into two fixed-sized vectors r_c and r_d, those vectors will be concatenated fed into a two-layer fully-connected neural network to rank the candidate.

$$z_1 = tanh(W_2(r_d \otimes r_c) + b_2) \tag{8}$$

$$z_2 = tanh(W_3 z_1 + b_3) \tag{9}$$

$$y = sigmoid(W_4 z_2 + b_4) \tag{10}$$

We used a hidden size of 25 neurons for layer z_1 and z_2, the output layer consists of one neuron with sigmoid activation to output a probability.

Settings and Generalization. To minimize overfitting we use dropout [14]. Dropout randomly deactivates a configurable percentage of units of that layer by setting it to zero. In our experiments we use a dropout probability of

0.3. We randomly remove input words with dropout in the embedding layer [3], units of the recurrent layer and neurons of the two fully-connected output layers. We use binary cross-entropy as loss function. The network is trained by optimizing the parameters of the LSTM and the other parameters $W_1, b_1, u, W_2, b_2, W_3, b_3, W_4, b_4$ using the Adam optimizer [7].

4 Evaluation

We evaluated our model against the test set of the *Inspec* dataset consisting of 500 abstracts. Before comparing the returned keyphrases against the ground truth, we stem both of them. We measure precision, recall and F-measure as evaluation statistics. Since our model simply returns a sorted list of all ranked candidates, the final prediction still has to be determined. We experimented with both the selection of the N most confident keyphrases and a threshold T selecting those candidates whose confidence is higher than that threshold. We optimize N and T on the validation set. Figure 2 shows that in 40% the highest scored keyphrase is labeled as such. While increasing N, the precision decreases while recall increases and we find a f-measure maximum at $N = 13$. The threshold evaluation shows the same effect and reveals two peaks in f-measure at around $T = 0.73$ and $T = 0.77$. We use the later for evaluation on test set, as we favor precision over recall. For comparison the reported results of Hulth, TextRank, RAKE, WordAttractionRank and KeyCluster are included in Table 2.

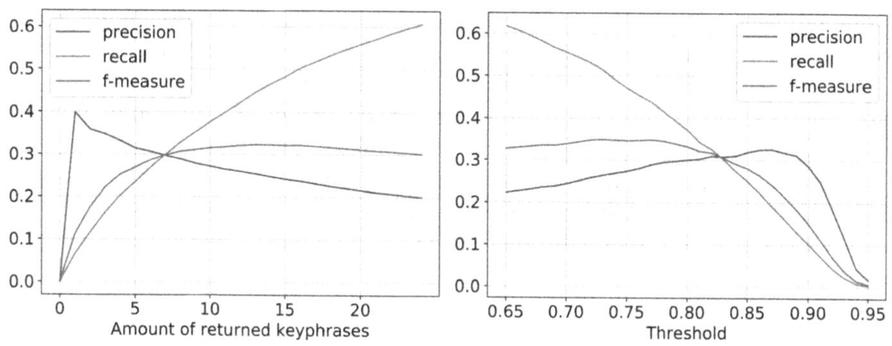

Fig. 2. Effect of parameters N and T to precision, recall and f-measure evaluated on the validation split.

Table 2. Comparison of the performance of our neural model against the reported evaluation results of established algorithms [6,10,13].

Evaluation					
Algorithm	Dataset	Method	Precision	Recall	F-measure
Hulth	*Inspec*	*n*-gram w. tag	25.2	51.7	33.9
TextRank	*Inspec*	Undir., Co-occ.win. = 2	31.2	43.1	36.2
RAKE	*Inspec*	KA stoplist (df>10)	33.7	41.5	37.2
WordAttractionRank	*Inspec*	F = 1, N = 10	37.1	50.3	42.6
KeyCluster	*Inspec*	Spectral clustering	35.0	66.0	45.7
Neural	*Inspec*	threshold: 0.77	31.0	47.4	37.5
		top: 13	28.7	47.5	35.8

5 Conclusion

Our method may not achieve state-of-the-art results, but is to our knowledge the first model that is trained end-to-end without manual feature engineering. All required features for ranking a candidate are learned while training our neural network. We see our approach as an initial proposal of a neural network architecture that is able to learn the complex task of ranking a candidate regarding it's suitability as a keyphrase. We leave it to other researchers to try other network architectures, such as convolutional or recursive neural networks. The results of our model may be improved by pretraining the word embeddings on a big domain specific dataset, like research papers. Furthermore one could not only supply word embeddings, but also use or learn character embeddings to improve information about words, that are outside of the vocabulary. Our model focuses on extracting keyphrases from very short documents. Scaling that model to bigger documents requires different techniques and research. We would like to thank Anette Hulth for open sourcing her dataset, but at the same time would like to point out the need for an even bigger dataset, to train more sophisticated models.

References

1. Baziotis, C., Pelekis, N., Doulkeridis, C.: Datastories at semeval-2017 task 6: Siamese lstm with attention for humorous text comparison. In: Proceedings of the 11th International Workshop on Semantic Evaluation (SemEval-2017), pp. 381–386. Association for Computational Linguistics, Vancouver (2017)
2. Fox, C.: A stop list for general text. SIGIR Forum **24**(1–2), 19–21 (1989). https://doi.org/10.1145/378881.378888
3. Gal, Y., Ghahramani, Z.: A theoretically grounded application of dropout in recurrent neural networks. In: Advances in Neural Information Processing Systems, pp. 1019–1027 (2016)
4. Hasan, K.S., Ng, V.: Automatic keyphrase extraction: a survey of the state of the art (2014)

5. Hochreiter, S., Schmidhuber, J.: Long short-term memory. Neural Comput. **9**(8), 1735–1780 (1997). https://doi.org/10.1162/neco.1997.9.8.1735
6. Hulth, A.: Improved automatic keyword extraction given more linguistic knowledge. In: Proceedings of the 2003 Conference on Empirical Methods in Natural Language Processing, pp. 216–223. Association for Computational Linguistics (2003)
7. Kingma, D.P., Ba, J.: Adam: a method for stochastic optimization. CoRR abs/1412.6980 (2014). http://arxiv.org/abs/1412.6980
8. Liu, Z., Li, P., Zheng, Y., Sun, M.: Clustering to find exemplar terms for keyphrase extraction. In: Proceedings of the 2009 Conference on Empirical Methods in Natural Language Processing, vol. 1, pp. 257–266. Association for Computational Linguistics (2009)
9. Medelyan, O.: Human-competitive automatic topic indexing. Ph.D. thesis, The University of Waikato (2009)
10. Medelyan, O., Frank, E., Witten, I.H.: Human-competitive tagging using automatic keyphrase extraction. In: Proceedings of the 2009 Conference on Empirical Methods in Natural Language Processing, vol. 3, pp. 1318–1327. Association for Computational Linguistics (2009)
11. Mihalcea, R., Tarau, P.: Textrank: bringing order into text. In: Proceedings 9th Conference on Empirical Methods in Natural Language Processing (EMNLP 2004) (2004)
12. Pennington, J., Socher, R., Manning, C.D.: GloVe: global vectors for word representation. https://nlp.stanford.edu/projects/glove/
13. Rose, S., Engel, D., Cramer, N., Cowley, W.: Automatic keyword extraction from individual documents. Text Mining: Applications and Theory, pp. 1–20 (2010)
14. Srivastava, N., Hinton, G., Krizhevsky, A., Sutskever, I., Salakhutdinov, R.: Dropout: a simple way to prevent neural networks from overfitting. J. Mach. Learn. Res. **15**, 1929–1958 (2014)
15. Turney, P.D.: Learning algorithms for keyphrase extraction. CoRR cs.LG/0212020 (2002). http://arxiv.org/abs/cs.LG/0212020
16. Wang, J., Liu, W., McDonald, C.: Corpus-independent generic keyphrase extraction using word embedding vectors (2015)
17. Witten, I.H., Paynter, G.W., Frank, E., Gutwin, C., Nevill-Manning, C.G.: KEA: practical automatic keyphrase extraction. In: Proceedings of the Fourth ACM Conference on Digital Libraries, pp. 254–255. ACM (1999)
18. Yang, Z., Yang, D., Dyer, C., He, X., Smola, A.J., Hovy, E.H.: Hierarchical attention networks for document classification (2016)

The Wild Bootstrap Resampling in Regression Imputation Algorithm with a Gaussian Mixture Model

Aisyah Mat Jasin, Daniel Neagu(✉), and Attila Csenki

University of Bradford, Bradford BD7 1DP, UK
{A.MatJasin, D.Neagu, A.Csenki}@bradford.ac.uk

Abstract. Unsupervised learning of finite Gaussian mixture model (FGMM) is used to learn the distribution of population data. This paper proposes the use of the wild bootstrapping to create the variability of the imputed data in single missing data imputation. We compare the performance and accuracy of the proposed method in single imputation and multiple imputation from the R-package Amelia II using RMSE, R-squared, MAE and MAPE. The proposed method shows better performance when compared with the multiple imputation (MI) which is indeed known as the golden method of missing data imputation techniques.

Keywords: Missing data imputation · Gaussian mixture model
Bootstrap

1 Introduction

Missing data can occur in data records for various reasons, such as: data entry errors, system failures, or respondents who avoid answering questions within a survey. Various methods have been proposed to deal with the missing data problem. The standard technique is discarding observations or variables that contain missing values. The deletion method is inappropriate when the missing proportions are high, resulting in inefficient parameter estimates, and estimated results tend to be underestimated. To deal with these issues, imputation methods can be used to substitute missing values with plausible values. For example, the single mean imputation consists of replacing the missing values with the mean, median or mode value. However, this simple approach produces biased analysis results. The multiple imputation method introduced in [1] is a complex approach where missing data are filled-in by drawing multiple sets of complete data that contain different plausible values. This method is complicated and computationally expensive [2], especially for large data sets because execution processes are implemented through three phases in several iterations. The improved version of the single imputation technique such as conditional mean imputation, which incorporates the statistical and machine learning methods with multivariate Gaussian mixture models (GMM) [3] have gained interest in many years [4].

The conditional mean imputation (also known as ordinary least square, OLS) or regression imputation can preserve the data distribution, according to Di Zio [5].

P. Perner (Ed.): MLDM 2018, LNAI 10935, pp. 218–230, 2018.
https://doi.org/10.1007/978-3-319-96133-0_17

The conventional OLS $\hat{y}_i = \beta_0 + \sum_{j=1}^{J} \beta_j x_j + \varepsilon_i$ implementation requires the use of random error ε_i which can be obtained in two ways [6]: (1) draw a random error with underlying assumption that it is independent and identically distributed, that follows a Gaussian distribution with zero mean and finite variance; (2) draw a random error with replacement from the empirical distribution of the estimated residuals $\varepsilon_i = y_i - \hat{y}_i$ [7]. Problems can occur in the random error and residual ε_i in method (1) that will create the sparsity problem whereas the random ε_i generation will be either too large or too small although the normality distribution assumption is met. The sparsity of data in method (2) will be inconsistent if the data distribution has different clusterings and each cluster consists of a different density. The sparsity of data creates some problems such as increases in the variance between the imputed and original data.

The conditional mean imputation proposed in [5] does not consider adding the residuals. Although this method may preserve the data distribution, it will underestimate the variability, introduce the bias on imputed data and the result of imputed data will be highly inaccurate. The additional steps are required to improve data sparsity in the random error ε_i generated in the OLS to obtain a better predicted missing value.

The main objective of this study is to investigate the random error and employ the wild bootstrap [8, 9] on the missing data prediction using regression imputation on the Gaussian mixture model. The wild bootstrap is used to improve the variance in heteroscedasticity issue when the data variance is not homoscedastic [8, 9]. Further details about the wild bootstrap approach are discussed in the next section that introduces the modelling framework.

In this paper, we employ the wild bootstrap to the single imputation technique in missing value prediction, since the GMM framework is flexible to learn multimodal data distribution. We combine the GMM model with the proposed missing data prediction method. We also employ the wild bootstrap to investigate the effect of the sparsity of imputed data in a different mixture data distribution case. Thus, we would like to show that the performance of single imputation may perform well, and as good as the implementation of MI. We assume that the data is missing data at random (MAR).

This paper is organized as follows: in Sect. 2, we present the Gaussian mixture model framework and the proposed regression imputation with wild bootstrap technique. In Sect. 3 we discuss the experimental evaluation and experimental results. Section 4 concludes the paper and identifies further directions for research and study.

2 Modelling Framework

GMM is a powerful probabilistic model used in predicting specifically in data clustering [5]. This model is flexible to learn from different data distributions by fitting the probability density function (PDF) to represent different clusters [3]. The well-known strategy for finding the Maximum Likelihood (ML) parameter estimation uses the Expectation-Maximization (EM) algorithm [10]. GMM applications to missing data problems have been studied extensively for example in [4, 5, 11].

2.1 Definitions

Suppose the data set **X** having N units of independent and identically distributed (i.i.d) data points with p-column vectors can be written as follows:

Figure 1 illustrates a data set that contains missing values (highlighted with *NA* in the relevant cells). Let $\mathbf{X} = \{X_1, X_2, \ldots, X_p\}$ be the random variable of the $N \times p$ data matrix. In the imputation process, Rao and Shao [12] suggested to create a set of respondents $\mathbf{X}^O$ and a set of non-respondents $\mathbf{X}^M$ separately. The variable $\mathbf{X}^O$ denotes the $n_1 \times p$ matrix where n_1 is the size of observed data while $\mathbf{X}^M$ denote the $n_0 \times p$ matrix where $n_0 = N - n_1$ is the number of missing values that occur in $\mathbf{x}_l$. Let $\mathbf{x}_l$ of size $n_1 \times 1$ vector contain observed data and n_0 be the size of missing values in $\mathbf{x}_l$.

Observation (index)	X_1	X_2	..	X_l	..	X_p
1	x_{11}^O	x_{12}^O	..	x_{1l}^O	..	x_{1p}^O
..	:	:	..	..	..	:
n_1	$x_{n_1 1}^O$	$x_{n_1 2}^O$	..	$x_{n_1 l}^O$	..	$x_{n_1 p}^O$
n_1+1	$x_{(n_1+1)1}^M$	$x_{(n_1+1)2}^M$	..	*NA*	..	$x_{(n_1+1)p}^M$
:	:	:	..		..	:
N	x_{N1}^M	x_{N2}^M	..	*NA*	..	x_{Np}^M

Fig. 1. A sample data set with missing values

2.2 Multivariate Gaussian Mixture Model

The Maximum Likelihood (ML) is an approach to estimate the parameters of the distribution from multivariate GMM using the Expectation-Maximization EM algorithm [10]. The data in GMM are distributed by different k Gaussian components and estimated as follows:

$$f(\mathbf{x}; \boldsymbol{\Phi}) = \sum_{k=1}^{K} \pi_k f(\mathbf{x} \mid \theta_k) \quad (1)$$

where $f(\mathbf{x} \mid \boldsymbol{\theta}_k)$ is the density of p-variate Gaussian distribution with the k component. The vector $\boldsymbol{\Phi}$ contains the full set of parameters in the mixture model $\boldsymbol{\Phi} = (\pi_1, \ldots, \pi_K; \boldsymbol{\theta}_1, \ldots, \boldsymbol{\theta}_K)$, where $\boldsymbol{\theta}_k$ is the vector of unknown parameters of mean vector $\boldsymbol{\mu}_k$ and covariance matrix $\boldsymbol{\Sigma}_k$.

The mixing coefficients (or weights) π_k for the kth component must satisfy the conditions $0 < \pi_k < 1$, and $\sum_{k=1}^{K} \pi_k = 1$. The GMM is a dynamic model where it is not required to specify any column vector to be an input or output particularly.

2.3 The General EM Algorithm

The EM algorithm is a statistical tool to find the maximum likelihood estimates of the set parameters such as mean, variances, covariances and regression coefficients of a model. The optimisation algorithm introduced by Dempster et al. [10] starts with an initial estimate of $\mathbf{\Phi}$ and iteratively executes the process until it satisfies the convergence criteria. The iterative process has two steps known as the E-step and the M-step. The E-step computes the probability membership τ_{ik} for all data points x_i of mixture component k. The M-step will update the value of the parameter $\mathbf{\Phi}$ with respect to the k Gaussian component. Let denote q as an iteration counter, the expected values of the posterior distribution are computed by:

$$\hat{\tau}_{ik}^{(q)} = \frac{\hat{\pi}_k f(\mathbf{x}_i^o \mid \hat{\boldsymbol{\mu}}_k, \hat{\boldsymbol{\Sigma}}_k)}{\sum_{j=1}^{K} \hat{\pi}_j f(\mathbf{x}_i^o \mid \hat{\boldsymbol{\mu}}_j, \hat{\boldsymbol{\Sigma}}_j)} \tag{2}$$

In the M-step, we use the expected values in the posterior distribution (2) to re-estimate the means, covariances and mixing coefficients. The new set of parameters $\mathbf{\Phi}^{(q+1)}$ are updated as follows:

$$\hat{\pi}_k^{(q+1)} = \frac{N_k}{N} \text{ for } k = 1, \ldots, K, \tag{3}$$

$$\hat{\mu}_k^{(q+1)} = \frac{1}{N_k} \sum_{i=1}^{N} \tau_{ik} \hat{\mathbf{x}}_{ik} \tag{4}$$

$$\hat{\boldsymbol{\Sigma}}_k^{(q+1)} = \frac{1}{N_k} \sum_{i=1}^{N} \tau_{ik} [(\hat{\mathbf{x}}_{ik} - \boldsymbol{\mu}_k)(\hat{\mathbf{x}}_{ik} - \boldsymbol{\mu}_k)^T + \hat{\boldsymbol{\Sigma}}_{ik}^{MM}] \tag{5}$$

The algorithm then iterates the E-step and M-step until convergence is achieved.

2.4 The Least Square Method

The conditional mean imputation is also known as regression imputation [13]. The imputed values are regressed from independent variables $\mathbf{X}_p$. Let consider the following linear regression model:

$$x_{il} = \beta_0 + \beta_1 x_i + \varepsilon_i, \quad i = 1, 2, \ldots, n \tag{6}$$

where the response variable x_{il} is predicted from regression coefficients β_0 and β_1 with random error $\varepsilon_i \sim N(0, \sigma^2)$ i.i.d. and uncorrelated. The matrix development of Eq. (6) is presented as follows:

$$\mathbf{x}_l = \begin{bmatrix} x_{1l} \\ \vdots \\ x_{Nl} \end{bmatrix}, \mathbf{X} = \begin{bmatrix} 1 & x_{11} & \cdots & x_{1p} \\ \vdots & \vdots & \cdots & \vdots \\ 1 & x_{N1} & \cdots & x_{Np} \end{bmatrix}, \boldsymbol{\beta} = \begin{bmatrix} \beta_1 \\ \vdots \\ \beta_p \end{bmatrix}, \boldsymbol{\varepsilon} = \begin{bmatrix} \varepsilon_1 \\ \vdots \\ \varepsilon_N \end{bmatrix}$$

In general, $\mathbf{x}_l$ is an $N \times 1$ vector of the dependent variable contains missing values, $\mathbf{X}$ is a $N \times p$ matrix of observed variables, $\boldsymbol{\beta}$ is a $p \times 1$ vector of the regression coefficients and $\boldsymbol{\varepsilon}$ is a $N \times 1$ vector of random errors. The general least square estimator of $\boldsymbol{\beta}$ based on observed values is:

$$\hat{\boldsymbol{\beta}}^O = \left(\mathbf{X}^T\mathbf{X}\right)^{-1}\mathbf{X}^T\mathbf{X}_l \tag{7}$$

In the presence of missing data, the imputed values are obtained by the conditional mean imputation technique which corresponds to imputed values generated from a set of regression equation calculated in (7) as discussed in [13, 14]. There are two ways to generate the random error component ε_i. The random error component ε_i can be generated either with $\varepsilon_i \sim N(0, \sigma^2)$ or residual.

2.5 Fundamentals of the Bootstrap Method

The bootstrap non-parametric resampling technique was proposed by Efron [15] for estimating a standard error, confidence interval in various types of distributions. This method was extended in [16, 17] to generate the random error ε_i in the regression model. Let $\mathbf{X} = \{\mathbf{x}_1, \mathbf{x}_2, \ldots, \mathbf{x}_{n_1}\}$ is a random sample from p-variate normal distribution K where n_1 refers to the size of observed data $\mathbf{X}^O$ as shown in Fig. 1. Let $\mathbf{X}^{(b_k)}$ denote the bootstrap resampled data generated by sampling with replacement from the original dataset $\mathbf{X}_k$ where b indicates the counter $b = 1, \ldots, B$ of drawing samples of bootstrap and k refers to the current Gaussian component. In this study, the resampling and parameter estimation are implemented on the observed data $\mathbf{X}_k^O$ where the superscript O refers to observed data.

2.6 The Wild Bootstrap

Wu [8] introduced the wild bootstrap to deal with the heteroscedasticity issue. Later, a better approximation of the wild bootstrap was proposed by Liu [9]. The wild bootstrap is based on the modification of the bootstrap residual approach of the least square estimation. Wu [8] improved the resampling residual with replacement in bootstrap by drawing a value of t_i^* that follow a standard normal distribution with zero mean and unit variance:

$$x_{il}^b = x_i^T\hat{\beta} + t_i^*\frac{\hat{\varepsilon}_i}{\sqrt{1 - w_i}} \tag{8}$$

where $w_i = x_i^T(\mathbf{X}^T\mathbf{X})x_i$. However, the error variance $t_i^*\hat{\varepsilon}_i$ are inconsistent. Therefore, authors in [18] proposed to compute t_i^* by drawing a sample a_i with replacement:

$$t_i^* = a_i = \frac{\hat{\varepsilon}_i - \bar{\hat{\varepsilon}}_i}{\sqrt{n_{1k}^{-1} \sum_{i=1}^{n_{1k}} (\hat{\varepsilon}_i - \bar{\hat{\varepsilon}})^2}} \tag{9}$$

where $\bar{\hat{\varepsilon}} = n_{1k}^{-1} \sum_{i=1}^{n_{1k}} \hat{\varepsilon}_i$.

The second wild bootstrap technique employed in this study is the Liu's bootstrap [9]. Liu [9] proposed t_i^* in Wu [8] by resampling a set of central residual with zero mean and unit variance that has third central moments equal to one. Liu proposed two procedures to draw random numbers t_i^*. However, we consider the second procedure as it is appropriate for normal distribution. Liu's bootstrap is conducted by drawing random numbers:

$$t_i = D_1 D_2 - E(D_1)E(D_2) \tag{10}$$

where D_1 and D_2 are random i.i.d that follows normal distribution with means $0.5 * (\sqrt{17/6} + \sqrt{1/6})$ and $0.5 * (\sqrt{17/6} - \sqrt{1/6})$ respectively, and variance 0.5.

2.7 The Non-parametric Wild Bootstrap Applied in Missing Data Imputation

The bootstrap procedure based on the resample approach in the GMM is described in the following steps:

1. Initiate the set of parameters $\mathbf{\Phi}$ with K-means algorithm.
2. Compute the residual for each Gaussian component:
   a. Fit Gaussian mixture model using the parameter values from the step 1.
   b. Compute the residual: $\hat{\varepsilon}_k = \mathbf{X}_{lk}\hat{\boldsymbol{\beta}}_k$ where k is the Gaussian component $k = 1, .., K$.
3. For b = 1, .., B
   a. Draw a vector $\hat{\varepsilon}_k$ of n_{1k} i.i.d sample with a simple random sampling with replacement. The vector $\hat{\varepsilon}_k$ is generated from step 2b with respect to the option of the Wu's [8] or Liu's [9] bootstrap procedure as discussed in the Sect. 2.6.
   b. Fit Gaussian mixture model using the parameter values from the step 1.
   c. In the E-step,
      i. Compute the posterior probabilities vector τ_{ik} in Eq. (2) on the observed data.
   d. In the M-step,
      i. Impute the missing values of size n_{0k} using a linear regression model (6) based on OLS estimator $\hat{\boldsymbol{\beta}}^{(O_k)}$ in (12):

$$x_{il} = \hat{\beta}_0^{(O_k)} + \hat{\beta}_1^{(O_k)} x_i + t_i^* \hat{\varepsilon}_i / \sqrt{1 - w_i}$$

      where the residual t_i^* taken from the step 3a.
      ii. Update the new parameter $\mathbf{\Phi}$ for each component in GMM as shown in (3), (4) and (5).

3 Experiments and Discussion of Results

In this section, the numerical results are presented on real and simulated datasets.

3.1 The Non-parametric Wild Bootstrap Applied in Missing Data Imputation

Dataset: We applied various evaluation criteria on one real dataset and one artificial dataset with two variables and two Gaussian classes. The first case study is the Old Faithful Geyser dataset [19]. This dataset contains 272 records on the waiting time between geyser eruptions (waiting) and the duration of eruptions (eruptions) in Yellowstone National Park, USA.

For the artificial case study, the values are randomly sampled with 1000 observations of two Gaussian classes with different position mean values and positive-negative correlation. Data are drawn with normal distribution using the following parameters:

$$\pi_1 = 0.5,\ \pi_2 = 0.5$$

$$\mu_1 = (4,2)',\ \mu_2 = (-2,6)'$$

$$\Sigma_1 = \begin{pmatrix} 1 & -0.7 \\ -0.7 & 1 \end{pmatrix},\ \Sigma_1 = \begin{pmatrix} 3 & 0.9 \\ 0.9 & 3 \end{pmatrix}$$

Software: the proposed method in these experiments were conducted using Matlab version 2017a. The proposed method is compared with multiple imputation available in the R-package Amelia II. The comparisons are conducted based on the artificial missing data generated with different missing data percentages (MDP): 5%, 10%, 15% and 20%.

Imputation implementation: the missing data are imputed based on the regression imputation. Prior to the imputation process, the K-means algorithm is used to determine initial parameter values of mixing proportion π_k, mean μ_k and covariance matrix Σ_k in GMM. The stopping criteria is based on a selected threshold where the different iterations were less than 10^{-6}.

Evaluation criteria: these experiments are designed to measure the performance and prediction accuracy between predicted and actual values. RMSE computes the deviation between predicted and actual values that employed by most missing data imputation studies. The greater the deviation means the greater variance between them. Therefore, the lower value shows better performance:

$$RMSE = \sqrt{\frac{\sum_{i=1}^{N} (\hat{y}_i - y_i)^2}{N}} \tag{11}$$

MAPE was used to measure the average relative error of the imputation accuracy:

$$MAPE = \frac{100}{N} \times \sum_{i=1}^{N} \left| \frac{y_i - \hat{y}_i}{y_i} \right| \tag{12}$$

MAE was used to measure the average error of each different in imputation:

$$MAE = \frac{1}{N} \times \sum_{i=1}^{N} |y_i - \hat{y}_i| \tag{13}$$

R-squared values were used to describe the variance in goodness-of-fit for the regression models between observed data and the expected values of the dependent variable. The range of R-squared is between 0 and 1:

$$R^2 = \frac{\sum_{i=1}^{n} (y_i - \hat{y}_i)^2}{\sum_{i=1}^{n} (y_i - \bar{y})^2} \tag{14}$$

3.2 Experimental Results

In this study, we compare the imputation accuracy using MAPE and MAE whilst measuring the performance using RMSE and R-Squared of three methods: single regression imputation combined with Wu's and Liu's wild bootstrap and MI. The better results are highlighted in bold font.

Table 1 summarizes the performance and prediction accuracy of the three methods on the Old Faithful Geyser dataset while Table 2 shows the result estimation on the random data generation. The result of the proposed methods in RMSE shows better performance and significantly different between the MI with the proposed Wu's and Liu's method in all MDP proportions. This is shown in the 5% MDP, Wu and Liu method yielded 7.8225 and 7.8879 respectively while MI gained 9.8719. It is also found in 10%, 15% and 20% MDP where the Wu's and Liu's method have outperformed the MI where the result of Wu's shows 7.0955, 6.6819 and 6.7349 while Liu shows 7.8746, 7.0150 and 7.2354 in RMSE. In contrast, the MI obtained 8.4187, 8.7004 and 8.9103 higher than Wu's and Liu's method in 10%, 15% and 20% MDP respectively.

The R-squared values are used to quantify the overall model performance of variance in response variable explained by the independent variables. The larger the R-squared means the more variability is explained by the linear regression model. The result of R-squared presented in Table 1 showed that the proposed method gives the best performance with 0.6338% for 5% of MDP proportion followed by 0.6836, 0.7683 and 0.7127 for Wu, while Liu's obtained 0.6894, 0.6869 and 0.7050 for the 10%, 15% and 20% of MDP respectively on the Faithful data set. The R-squared obtained by the proposed method in the random generation data in the Table 2 showed less than 0.6% for all MDP percentages. In contrast, the MI in Amelia gives a lower variance than the proposed method in all MDP proportions with R-squared ranging from 0.03 to 0.2.

The imputation accuracy is measured based on the average relative error between predicted missing data and the original data using mean absolute percentage error (MAPE) and mean absolute error (MAE).

Table 1. The MAPE, MAE, R-square and RMSE estimates on the Old Faithful Geyser dataset

		MAPE	MAE	R square	RMSE
5%	Amelia	0.6220	6.5928	0.4960	9.8719
	Wu	**0.2758**	**2.6453**	**0.6338**	**7.8225**
	Liu	**0.1713**	**1.7281**	**0.4212**	**7.8879**
10%	Amelia	0.0827	1.5636	0.6450	8.4187
	Wu	**0.0379**	**0.6959**	**0.6836**	**7.0955**
	Liu	**0.0190**	**0.3594**	**0.6894**	**7.8746**
15%	Amelia	0.0346	1.0379	0.5184	8.7004
	Wu	**0.0012**	**0.0342**	**0.7683**	**6.6819**
	Liu	**0.0167**	**0.5018**	**0.6869**	**7.0150**
20%	Amelia	0.0432	1.6841	0.4959	8.9103
	Wu	**0.0054**	**0.2144**	**0.7127**	**6.7349**
	Liu	**0.0104**	**0.3994**	**0.7050**	**7.2354**

Table 2. The MAPE, MAE, R-square and RMSE estimates on the randomly generated data

		MAPE	MAE	R square	RMSE
5%	Amelia	1.5798	0.7748	0.2439	2.0230
	Wu	**0.0469**	**0.1055**	**0.2593**	**1.8642**
	Liu	**0.0721**	**0.1604**	**0.3754**	**1.8066**
10%	Amelia	0.1259	0.1477	0.2206	2.2926
	Wu	0.1344	0.5811	0.3977	2.1557
	Liu	**0.0089**	**0.0377**	**0.5803**	**1.6889**
15%	Amelia	0.1674	0.3105	0.0272	2.2817
	Wu	**0.0272**	**0.1812**	**0.1316**	**2.3463**
	Liu	**0.0203**	**0.1282**	**0.5197**	**1.7214**
20%	Amelia	0.0501	0.1159	0.1206	2.7461
	Wu	**0.0435**	0.3528	**0.1954**	**2.1858**
	Liu	**0.0127**	**0.1070**	**0.4586**	**1.8340**

The result of MAE in the Table 1 showed that the Wu's and Liu's methods are consistently outperformed the MI method on the Old Faithful Geyser dataset. In contrast, in the Table 2, the Liu's method offered consistent and better accuracy than MI method. Meanwhile the Wu's method showed inconsistent improvement in the measure of average error magnitude to MI method on the random data generation.

As can be observed from the MAPE values obtained in Table 1, the proposed method of Wu's and Liu's performed better imputation on the Old Faithful Geyser data set.

Meanwhile, by observing the MAPE values gained in the Table 2, Liu' method showed consistent to defeat the MI method compared to Wu's method.

Plots of the results shown in Fig. 2 compare the outcome between multiple imputation technique in r-package Amelia II and the proposed methods.

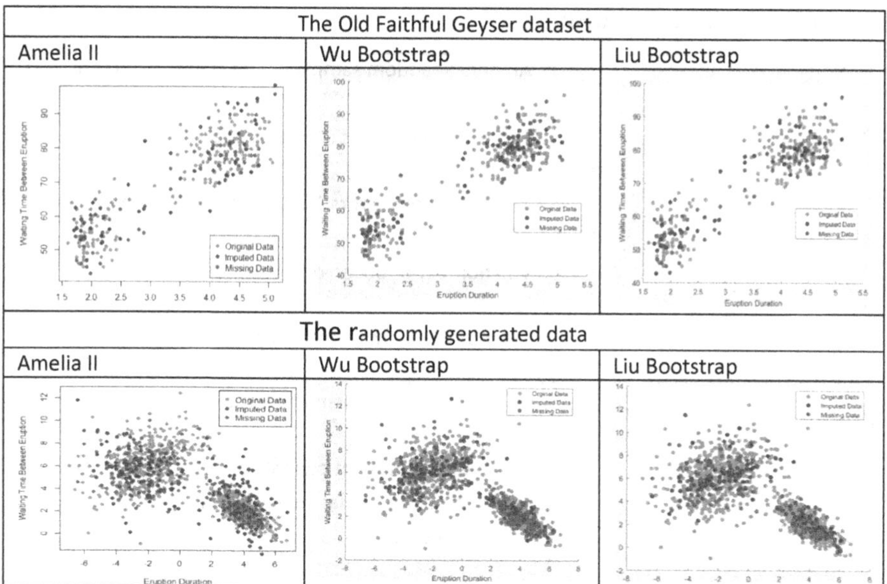

Fig. 2. The scatter plot of two datasets using R Amelia II and the proposed methods

4 Conclusions

In this paper, we proposed a method for single imputation that incorporates wild bootstrap in order to create the variability of imputed data as for example Multiple Imputation (MI) does. The MI is indeed known to be the preferred method in handling missing data problems over the years compared to the single imputation methods.

The imputation process in MI involves several steps while single imputation has simpler implementation compared to MI. The missing data in MI are imputed for M times with different plausible values and combine appropriately in the analysis stage. The sparsity of imputed data is a matter of concern because it will reflect the variance and measurement error between predicted and original data. Thus, the main purpose of this comparison is to show that the performance of single imputation in the Gaussian mixture model may perform well and as good as the implementation of MI.

The performance of this method is measured by the RMSE, R-squared, MAE, and MAPE. Based on the results, we summarize that the single missing data imputation combined with the wild bootstrap is preferrable over the MI technique for the data containing several Gaussian distributions. Furthermore, the imputation process on the Gaussian mixture model could be relevant to preserve the originality of data distribution.

Since this study is implemented on bivariate data with two Gaussian components, in the future work we will focus on multivariate data with multiple Gaussian components.

Appendix A: The Notation List

$\mathbf{X}$	An entire random sample of size N and p-column.
$\mathbf{X}^O$	The observed values of random vector $\mathbf{X}$
$\mathbf{X}^M$	The p-feature vectors $\mathbf{X}$ contains missing values occur in $\mathbf{X}_l$
n_1	The number of observed data in $\mathbf{X}^O$
n_0	The number of missing data in $\mathbf{X}^M$
K	Total number Gaussian of components
θ_k	Parameter theta that consists of parameter mean vector $\boldsymbol{\mu}_k$ and covariance matrix $\boldsymbol{\Sigma}_k$
$\boldsymbol{\mu}_k$	The mean vector
$\boldsymbol{\Sigma}_k$	The covariance matrix
π	Mixing proportion of the current Gaussian component
$0 \le \pi_k \le 1,$	The probability of mixing coefficient must be between 0 and 1.
$\sum_{k=1}^{K} \pi_k = 1$	The sum of mixing coefficient of each component must be equal to one
$\Phi = (\pi_1, \ldots, \pi_K; \theta_1, \ldots, \theta_K)$	The vector Φ containing the set of parameters π_K and θ_K
$f(\mathbf{x}; \Phi)$	The mixture density containing all the parameters of mixture model
$f(\mathbf{x} \mid \theta_k)$	The mixture of density function of vector $\mathbf{X}$ conditioned on parameter estimation theta.
$f(\mathbf{x}; \Phi) = \sum_{k=1}^{K} \pi_k f(\mathbf{x} \mid \theta_k)$	The probability density function governed by the set of parameters mixing coefficient and theta with K-component mixture density
τ_{ik}	The posterior probability or responsibility for each data point that belongs to the kth component
β_0	Beta 0 is represented as the intercept of regression coefficient
β_1	Beta 1 is represented as the slope of regression coefficient
$x_{il} = \beta_0 + \beta_1 x_i + \varepsilon_i$	The ordinary leas square model (OLS)
$\hat{\boldsymbol{\beta}}^O = (\mathbf{X}^T\mathbf{X})^{-1}\mathbf{X}^T\mathbf{X}_l$	The least square estimator of $\hat{\boldsymbol{\beta}}^O$

$\varepsilon_i \sim N(0,\sigma^2)$	The random error component follows he normal distribution with mean 0 and variance
$\mathbf{X}^{(b_k)}$	The sample data after bootstrapping
$\mathbf{X}_k^O$	The observed data based on the k current Gaussian component
$b = 1,\ldots,B$	b is a counter value for bootstrap iterative process until B times
t_i^*	The non-parametric bootstrap resampled residual $\hat{\varepsilon}_i = x_{il} - \hat{x}_{il}$
$w_i = x_i^T(\mathbf{X}^T\mathbf{X})x_i$	The leverage is the ith diagonal element of Hat Matrix
$a_i = \frac{\hat{\varepsilon}_i - \bar{\hat{\varepsilon}}_i}{\sqrt{n_{1k}^{-1}\sum_{i=1}^{n_{1k}}(\hat{\varepsilon}_i - \bar{\hat{\varepsilon}})^2}}$	The non-parametric bootstrap resampled residual proposed to improve the random draw of Wu's algorithm
D_1 and D_2	The i.i.d that follows normal distribution
$\mathbf{X}_{n1_k}^O$	The observed data of n_1 size and k Gaussian component
$\hat{\varepsilon}_i = x_{il} - \hat{x}_{il}$	The estimated residual fitted by OLS model
q	A counter in EM algorithm iteration

References

1. Rubin, D.B.: Multiple imputations in sample surveys - a phenomenological Bayesian approach to nonresponse. In: Proceedings of the Survey Research Methods Section of the American Statistical Association, pp. 20–34 (1978)
2. Honaker, J., King, G., Blackwell, M.: Amelia II: a program for missing data. J. Stat. Softw. **45**(7), 47 (2006)
3. McLachlan, G., Peel, D.: Finite Mixture Models. Wiley, New York (2000)
4. Ghahramani, Z., Jordan, M.I.: Supervised learning from incomplete data via an EM approach. In: Advances in Neural Information Processing Systems, vol. 6, pp. 120–127 (1994)
5. Di Zio, M., Ugo, G., Luzi, O.: Imputation through finite Gaussian mixture models. Comput. Stat. Data Anal. **51**(11), 5305–5316 (2007)
6. Paik, M.: Fractional Imputation. Unpublished Doctoral thesis, Iowa State University, USA (2009)
7. Srivastava, M.S., Dolatabadi, M.: Multiple imputation and other resampling schemes for imputing missing observations. J. Multivar. Anal. **100**(9), 1919–1937 (2009)
8. Wu, C.F.J.: Jackknife, bootstrap and other resampling methods in regression analysis. Ann. Stat. **14**(4), 1261–1295 (1986)

9. Liu, R.Y.: Bootstrap procedures under some non-i.i.d. models. Ann. Stat. **16**(4), 1696–1708 (1988)
10. Dempster, A.P., Laird, N.M., Rubin, D.B.: Maximum likelihood from incomplete data via the EM algorithm. J. R. Stat. Soc. Ser. B **39**(1), 1–38 (1977)
11. Eirola, E., Lendasse, A., Vandewalle, V., Biernacki, C.: Mixture of Gaussians for distance estimation with missing data. Neurocomputing **131**, 32–42 (2014)
12. Rao, J.N.K., Shao, J.: Jackknife variance estimation with survey data under hot deck imputation. Biom. Trust **79**(4), 811–822 (1992)
13. Enders, C.K.: Applied Missing Data Analysis. The Guilford Press, New York (2010)
14. Harvey, C.R.: The specification of conditional expectations. J. Empir. Financ. **8**(5), 573–637 (2001)
15. Efron, B.: Bootstrap methods: another look at the jackknife. Ann. Stat. **7**(1), 1–26 (1979)
16. Hardle, W., Mammen, E.: Comparing Nonparametric versus Parametric Regression Fits. Ann. Stat. **21**, 1926–1947 (1993)
17. Zadkarami, M.: Bootsrapping: a nonparametric approach to identify the effect of sparsity of data in the binary regression models. J. Appl. Sci. **8**(17), 2991–2997 (2008)
18. Cribari-Neto, F., Zarkos, S.G.: Bootstrap methods for heteroskedastic regression models: evidence on estimation and testing. Econom. Rev. **18**(2), 211–228 (1999)
19. Azzalini, A., Bowman, A.W.: A look at some data on the old faithful geyser. J. R. Stat. Soc. **39**(3), 357–365 (1990)

Association Rule Mining in Fuzzy Political Donor Communities

Scott Wahl(✉) and John Sheppard

Gianforte School of Computing, Montana State University,
Bozeman, MT 59717, USA

Abstract. Social networks can be found in many domains. While community analysis can help users understand relationships within the network, it can be difficult to analyze the results without considerable effort. However, using the communities to partition underlying data, it is possible to use association rule mining to more easily facilitate meaningful analysis. In this paper we use a real-world dataset drawn from political campaign contributions. The network of donations is treated as a social network and fuzzy hierarchical community detection is applied to the data. The resulting communities are then analyzed with association rule mining to find distinguishing features within the resulting communities. The results show the mined rules help identify notable features for the communities and aid in understanding both shared and differing community characteristics.

1 Introduction

While complex social networks arise from many areas, these networks can be difficult to analyze and understand when taken as a whole. A useful tool for gaining insight into these networks is partitioning the data by dense subnetworks, commonly referred to as community detection. Ideally, communities should have properties in common with one another. There has been a considerable amount of work performed in developing methods for community detection in social networks. Early research focused on splitting the nodes of networks into distinct and separate communities [1–4]. Spectral clustering has been shown to be popular due to its performance on more complex networks [5,6]. A notable limitation with crisp communities, however, is that they do not capture the ability for individuals to belong to multiple communities. A person can have bonds with different groups of friends, as an example. Additionally, hierarchical structure is often found in networks where communities in the network form larger groups at different levels. More recent approaches attempt to handle this by allowing fuzzy clusters as well as creating a hierarchical structure for the communities [7–12].

Many real world sources for networks exist that exhibit community structure. Prior research has yielded such structure in genetics [13], neuroscience [14], and Internet communities [15] as a small sample. Another area where social networks

P. Perner (Ed.): MLDM 2018, LNAI 10935, pp. 231–245, 2018.
https://doi.org/10.1007/978-3-319-96133-0_18

occur is political science. Some past research in this area covers social interaction and its effect on political participation [16–18]. Other work focuses on elitism and the behavior of corporations in politics [19]. Specifically related to campaign finance, the geography of donations has been shown to be useful for predicting donations [20].

The research reported here focuses more directly on campaign finance networks. Billions of dollars are now poured into political campaigns, and there are many cases where finance limits are being repealed. Using data drawn from the National Institute on Money in State Politics, it is possible to create a transactional database describing the donation of money to political actors. These transactions include additional information about the donor and recipient. As examples, the recipient information includes party, incumbency, status of the election, district, as well as other information. Donor information varies by state according to what is required to be filed, but includes address, employer, occupation, and industry codes. Then these transactions can be used to establish relationships between nodes to generate a social network.

Past work has shown the effectiveness of finding hierarchical fuzzy communities within these networks in relation to ideological estimates [21]. Even still, a better analysis of the patterns within each of the communities is desired. To that end, we apply association rule mining to the information contained withing the transactions. The rule mining is conditioned on the fuzzy community assignment of the entities within those transactions. Motivating this approach is that donations can be made for a variety of reasons. Some of this is captured in the descriptive information regarding both the donor and the recipient. One individual may focus donations on incumbents within a set of districts, for example. By treating each of the components of an individual transaction as elements in a market basket, the rules found highlight associations in the pattern of donations. This allows us to obtain rules characterizing these communities. Given the hierarchy of communities, more importantly it is possible to analyze the differences in rules arising between communities, especially in those that share a parent in the hierarchy.

2 Related Work

Political fundraising is a multi-billion dollar a year activity, and the question of how such money may impact legislators and legislation is a very important one. Researchers in the area of political science are conflicted on how much, if any, impact contributions have on the legislative process. While explicit *quid pro quo* does not appear likely, other forms of influence have been studied in the literature. Kalla and Broockman show some of this influence in a randomized field study [22]. In their experiment, it was asked whether or not campaign contributions to a legislator improved the chance of gaining access to that legislator, and if donating to different legislators helped gain access to legislators to which they did not donate. Using a political organization representing constituents who had donated previously, the group attempted to arrange meetings with legislators and those constituents. Whether or not they revealed if the attendees had donated was decided randomly, allowing a better analysis on how donations impact access. The results showed a five-fold increase in the rate of successfully arranging a meeting if the legislator was told the attendee had donated previously. Also notable was that meetings for those who donated were more likely to be with a senior staffer.

Other areas of research in political science have used game theory to determine how campaign contribution legislation can impact how lobbyists should be willing to spend money [23]. Another notable line of research noted that the majority of campaign contributions do not come from areas that represent the United States as a whole well when considering ideology or political positions [24]. Together, these works highlight the importance of finding trends and subgroups within campaign finance who may wield more influence than most of the populace, and thus skew the political process.

A possible tool for helping find these trends is social network analysis. Prior work in finding fuzzy hierarchical communities showed that the resulting communities are correlated with previous ideological estimates. Further, in cases where the communities do not reflect ideology as well as in other areas, the resulting communities still contain useful information on patterns of donations. Part of the issue with these communities, however, is their "understandability." Analyzing the resulting communities, especially in how they are different from each other, can be difficult and sometimes only the obvious is discovered. Rule mining is proposed in this paper as a possible method for assisting in that area.

An early rule mining algorithm uses the concept that a subset of a frequent itemset must also be large by definition [25]. The Apriori algorithm performs multiple passes, starting with finding the large 1-itemsets. The next passes generate new candidate itemsets based on the superset of previous itemsets. Then any non-frequent items are removed and the algorithm continues. A similar process can be used to create rules from the data based on these itemsets. One of the augmentations to association rule mining was to integrate classification with the rule mining [26]. As noted in their work, the framework they developed was intended to help solve an "understandability problem" where rules produced by classification are difficult to understand. The algorithm Classification Based on Associations, or CBA, contains a rule generator and a classifier builder. First, a set of frequent *ruleitems* is found from within a transactional database. One example listed is $\langle \{(V_1, 1), (V_2, 1)\}, (class, 1)\rangle$ where V_1 and V_2 are attributes. From these set of frequent rules, multiple passes are performed over the data to generate candidate rules. This list is refined to create the final rules for use in a classifier, the results of which gave an improvement on C4.5.

3 Background

As mentioned above, a variety of techniques have been proposed for detecting communities in social networks. These communities, or partitions, are often validated against modularity. The general idea behind this measure is to compare the fraction of links that connect any nodes in a community, C_i to any other community, C_j. This ratio of edges is compared against a null model. This null model is a graph where each individual node maintains the same degree, but each edge is reassigned randomly. For a community partitioning to be considered a good partition, the fraction of links within a community should be higher than the fraction of links leaving the community. However, it should be noted that simply putting all nodes into a single community would satisfy this constraint, so more must be done.

For this, first define **A** as the adjacency matrix for a network where $a_{i,j} = 1$ if there is an edge between nodes i and j. Assume there exists a community partitioning C represents the set of communities where C_i contains the nodes belonging to that community. Let $k = |C|$ and define **E** as a $k \times k$ matrix that contains data regarding the communities. Specifically, for any two communities $c_i \in C$ and $c_j \in C$, element e_{ij} represents the fraction of edges that connect nodes in c_i to c_j. This leaves the diagonal as the ratio of edges that connect nodes within the community to other nodes in that same community. From this representation, determining the value of a partitioning of the graph into communities relies on the trace of **E**, namely $\mathrm{Tr}\left(\mathbf{E}\right) = \sum_i \mathbf{e}_{ii}$. Furthermore, we define the value $d_i = \sum_j e_{ij}$, which gives the ratio of edges within the graph that connect to all the vertices within C_i. Then the modularity is then given by

$$Q = \sum_i \left(e_{ii} - d_i^2\right) = \mathrm{Tr}\left(\mathbf{E}\right) - \left|\left|\mathbf{E}^2\right|\right|. \tag{1}$$

This can alternatively be written using the adjacency matrix for the social network **A** directly as

$$Q = \frac{1}{2m} \sum_{i,j \in V} \left[a_{ij} - \frac{\deg_i \cdot \deg_j}{2m}\right] \delta_{C_i,C_j} \tag{2}$$

where m is the number of edges in **A**, $\deg_i$ is the degree of node i, and δ_{C_i,C_j} is 1 when i and j are in the same community and 0 otherwise. For crisp communities, these values help in determining the quality of the found community assignments. Modifications must be made to this equations to handle fuzzy partitioning and is discussed along with the algorithm.

4 Algorithm

Within this paper, the network used is based on political donations amongst candidates, committees, and donors. Each edge in the network represents a transaction between two entities. Since candidates and committees can donate and transfer money between themselves, the network is not simply bipartite, but instead a general graph. Our approach for finding fuzzy clusters is based on spectral clustering work of Ng et al. [5] and Zhang et al. [27]. The first step is to set the number of communities and then apply spectral clustering to the social network in a hierarchical fashion to extract the communities.

Algorithm 1. Finding number of communities

1: **function** NumCommunities(**A**)
2: $\Lambda =$ set of eigenvalues $(\mathbf{A}) : \lambda_i >= 1$
3: $\Lambda' =$ sort (Λ)
4: **for all** $\lambda_i \in \Lambda' : i < |\Lambda'|$ **do**
5: $\delta_i = \lambda_{i+1} - \lambda_i$
6: $t = \text{aad}(\Delta) \times 1.482$
7: $ind =$ first index $(\Delta_i \in \Delta : \Delta_i \geq t)$
8: $k = |\Delta| - ind + 1$

4.1 Determining Number of Communities

From prior work, it was shown that the eigenvalues can be used to estimate the number of clusters/communities [21]. Other work shows that a network with k communities will have k large eigenvalues [28–30]. The largest eigenvalue for a network is approximately the average degree of the nodes in **A** [31]. For a network with edges assigned randomly with probability p between n nodes, this can be estimated as $\lambda_{\max} \approx n \times p$. Then assuming k communities in the random network, two nodes are connected with some probability p if they belong to the same community. Otherwise, the two nodes are connected with probability q where $q < p$. Then there exist eigenvalues corresponding to $s\,(p - q)$ where s is the size of the community.

These principles form the basis for estimating the number of communities among the contributors and candidates. To utilize these principles, the gap in eigenvalues is used to determine an appropriate number of communities. This is accomplished by looking for outliers in the gaps between eigenvalues. For our approach, average absolute deviation was selected over other outlier detection methods as it was more consistent over different datasets in preliminary work. The procedure for finding the number of communities is shown in Algorithm 1.

For this procedure, first find the eigenvalues Λ of the adjacency matrix **A** limited to those eigenvalues greater than 1 as described on line 2. These values are sorted to get Λ'. The difference between successive values in Λ' form the set of eigen-gaps Δ. Since we are interested in outliers in the eigen-gap, we use t as a threshold, which is found by calculating the average absolute deviation on line 6 where $aad\,(\Delta) = \frac{1}{|\Delta|}\sum_{i=1}^{|\Delta|} |\delta_i - average\,(\Delta)|$. The constant from line 6 is from the default multiplier in the R implementation, and attempts to vary this value did not improve the results. The location of the first eigen-gap over this threshold is then used to determine the number of communities k.

4.2 Fuzzy Spectral Clustering

After determining the cutoff for the number of clusters using the eigen-gap analysis described above, fuzzy spectral clustering (FSC) is performed on network **A** with n nodes as shown in Algorithm 2. The parameters of this function are the adjacency matrix **A** and the number of communities desired, k. Lines 2 and

Algorithm 2. Fuzzy Spectral Clustering

1: **function** FSC($\mathbf{A}, k$)
2: $\quad \mathbf{D} \leftarrow \{d_i = \sum_1^n a_i j\}$
3: $\quad \mathbf{L} = \mathbf{D}^{-1/2}\mathbf{A}\mathbf{D}^{-1/2}$
4: $\quad \mathbf{V} = eigenvectors(\mathbf{L}, k)$
5: $\quad$ **for all** $\mathbf{v}_i \in \mathbf{V}$ **do**
6: $\qquad$ **for all** $v_j \in \mathbf{v}_i$ **do**
7: $\qquad\quad X_j, i = v_j$
8: $\quad$ **for all** row $\mathbf{x}_i \in \mathbf{X}$ **do**
9: $\qquad \mathbf{x}_i = \mathbf{x}_i / \|\mathbf{x}_i\|$
10: $\quad \mathbf{U} = \mathrm{FCM}\,(\mathbf{X})$

Algorithm 3. Hierarchical Generation

function HFSC($\mathbf{A}, k$)
$\quad$ **for** $i = 2$ to k **do**
$\qquad C_i = \mathrm{FSC}\,(\mathbf{A}, i)$
$\quad$ **for all** $C_{i,m} \in C_i : i > 1$ **do**
$\qquad$ **for all** $C_{i-1,n} \in C_{i-1}$ **do**
$\qquad\quad P_{i,m,n} = sim\,(C_{i,m}, C_{i-1,n})$

3 calculate the Laplacian **L** of adjacency matrix **A**. Line 4 determines the k largest eigenvectors of the Laplacian **L**. The eigenvectors **V** are used as columns for matrix **X**, and the loop starting on line 5 explicitly defines how each $\mathbf{v}_i$ becomes a column in **X**. Each row is then normalized in the loop starting on line 8. With each row forming a data point, fuzzy c-means clustering (FCM) is used over the normalized rows of **X** to obtain **U**. This matrix **U** has dimensions $n \times k$, where $u_{i,j}$ is the community assignment value of node i to community j.

To obtain hierarchical structure, the process is repeated with a varying number of communities corresponding to the number of clusters in each hierarchical level, shown in pseudo-code in Algorithm 3. The communities are calculated for each level using FSC to create set of communities on a hierarchical level. Each level is connected to its parent level by calculating the fuzzy Jaccard similarity of the communities given by

$$sim\,(C_1, C_2) = \sum_{i \in C_1 \cup C_2} \frac{\min\,(C_{1,i}, C_{2,i})}{\max\,(C_{1,i}, C_{2,i})}. \tag{3}$$

In practice, the possible parent of a child that has the highest similarity is selected as the parent of the community. However, due to this construction there can be cases where a child would be better characterized as having more than one parent and all similarity values are kept.

Then in order to validate the communities discovered with fuzzy spectral clustering, a generalization of modularity is used that was created independently by Nepusz et al. and Shen et al.

$$Q = \frac{1}{2m} \sum_{i,j \in V} \left[a_{ij} - \frac{\deg_i \cdot \deg_j}{2m} \right] s_{ij} \tag{4}$$

where $s_{ij} = \sum_{c \in C} \alpha_{ic} \alpha_{jc}$ and α_{ic} is the fuzzy community assignment of i to community c. The change of the indicator function from Eq. 2 to s_{ij} allows use of fuzzy community scores instead of strictly crisp communities.

4.3 Finding Association Rules Across Communities

The final analysis of the results relies on association rule mining. As mentioned earlier, the political network is determined by a set of political donations. These donations are a set of transactions between a donor and a recipient. For the association rule mining, each of these transactions are tagged with the relevant metadata fields to create lists of features. Table 1 has example rows from the data.

Table 1. Example transactions for political donations

District	Status	Party	Office	Incumn.	Donor Type	Industry	Zip
Assembly district 042	Lost	D	House	C	Individual	Uncoded	92220
Assembly district 031	Won	D	House	I	Non-individual	Health	95814
Senate district 029	Won	D	Senate	O	Non-individual	Party	94518

Using the fuzzy community values from the clustering procedure, the transactions within the full dataset are separated based on the fuzzy community assignments of the donors within a community. We allow membership for node i in a community j if $u_{i,j} \geq 0.3$. This creates overlapping communities and overlapping partitions of the underlying transactions. Using the Apriori Association Rule Mining package within R, association rules are found within the data based on the membership values of the donors. This procedure is performed for each community at each level of the hierarchy. Other algorithms were tested on the data as well, though the rules found did not differ appreciably from those given by Apriori.

Since we are in particular interested in discriminatory rules between communities, the focus of the analysis are on communities who share a parent. Any pair of communities could be analyzed in the same manner, however. Consider two sibling communities, $C_{i,m}$ and $C_{i,n}$. First, association rules are discovered for the transactions belonging to each of those communities, generating rule sets $R_{i,m}$ and $R_{i,n}$. Next, the two rule sets are compared against each other to generate categories of rules. First, the rules in common can be found by taking the intersection of the two sets. Such rules help identify overall trends in the data, but are not as useful as discriminatory information Table 2.

More interesting is the set of conflicting rules between the sets. This conflict set can be determined from the rule sets by using the intersection of the

Table 2. Fields used for rule association

Field	Description	Examples
Donor Type	An individual or non-individual	Individual, non-individual, other
Industry	Category for donor industry	Health, labor, agriculture, etc.
ZipCode	ZIP Code reported by the donor	94131, 94028, etc.
District	The area a candidate represents	Assembly 027, Senate 029, etc.
Office Type	Legislative body for the office	House, senate, gubernatorial, etc.
Party	Party affiliation of the candidate	Democratic, republican, etc.
Status	Candidate won or lost their election	Won, lost, withdrew, etc.
Incumbency	Incumbency status of candidate	Open, incumbent, challenger

antecedents as $\mathcal{A} = \text{ant}(R_{i,m}) \cap \text{ant}(R_{i,n})$. For each of the antecedents in common, we determine if there is a conflict in the consequents between the rules in $R_{i,m}$ and $R_{i,n}$. The resulting conflict sets provide information on discriminative donation patterns within that community. Additional information can be gleaned from the rules where the consequent is the same, but the antecedent is different.

4.4 Data Preparation

The data used here is compiled from the National Institute on Money in State Politics. This dataset contains information on donors and candidates and other political actors for all fifty U.S. states as well as U.S. federal elections. This data is a set of transactions describing money given to political campaigns that includes additional information such as party of recipient, industry categorization of donors, etc. For the analysis below, a variety of fields were selected for mining association rules for both donors and recipients. While there are more fields within the data, many of them are generalizations or specializations of other fields, and including them would create rules defining those relationships instead of finding more interesting relationships.

To create the network, donors and recipients are used as nodes. Donations between any two entities are considered relationships and result in edges being added to the network. The sets of donors and recipients are not mutually exclusive. Candidates can donate to other candidates. In addition, various party and candidate committees can receive donations and will also pass money between themselves. Thus, the resulting network is not bipartite. While it may be possible to separate a single entity into multiple nodes to represent donations and receiving, allowing the network to remain non-bipartite allows for other types of associations to be added in the future.

For this research, data for each state is analyzed separately. We do not comingle the state-level data sets since much of the analysis is focused on state elections. In addition, communities found in combined state data tend to cluster around each state instead of across map boundaries. Following prior work in political science, the network of relationships is subject to filtering [32]. The primary filter is that any entity who gave or received money only once is removed from consideration. This eliminates any nodes connected to the network by a single edge. As a second filter, any isolated subnetwork which is not connected to the largest set of connected nodes is also removed since any such group becomes its own community by default.

5 Results

For state elections, California sees the most money donated to candidates and committees within that state[1]. For this reason, we focus on that state for the analysis presented here. For just candidates and committees involved in 2016 elections, California had nearly one billion dollars in contributions. Out of that money, non-individuals outspent individuals at a rate of almost five to one. Using the 2016 data for initial analysis, the network of relationships for California in that year total 5372 nodes (of which 232 are recipients) and 32,309 edges.

As a baseline, association rule mining based on Apriori is performed over the transactions without partitioning the data by community. Many of the rules discovered here are not especially insightful. Most are things already well known, such as incumbents have a much higher chance of winning their elections, winning candidates raise more money, non-individuals spend more widely than individuals, etc. In all, 120 different rules were discovered for the entire set of transactions. A rendering of the groups of rules discovered is shown in Fig. 1 where the rules are grouped by common antecedents. In these figures, the size of the circle is relative to the support of the rules and darker shades indicate higher lift. Additionally, due to the length of the labels, Incumbency and Donor Type are abbreviated as ICO and DType respectively.

It is possible to get more interesting results after splitting the data by community. After performing fuzzy spectral clustering over the California data, the resulting communities are fairly evenly split. Using the donor membership values, the transactional data is split into two overlapping datasets. Using 0.3 as a threshold for membership, there are 3154 entities (141 candidates) in $C_{2,1}$ and 3382 entities (180) candidates in $C_{2,2}$. Rule groups found are shown in Figs. 2 and 3.

The results for mining rules over two communities provides additional information. For $C_{2,1}$, it becomes immediately clear that the members of this community donate to Democrats since the rule $\{\} \Rightarrow Party = Democratic$ has 0.91 support. Another rule discovered within this community is that of $\{Incumbency = I\} \Rightarrow Status = Won$ at a support of 0.35 and confidence 0.94. As expected based on prior knowledge is that the incumbent is more likely to win their election.

What is more interesting is when these rules are compared with those for $C_{2,2}$. One expectation is that for this evenly split data where one community

[1] Based on 2016 data from https://www.followthemoney.org.

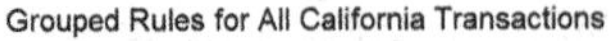

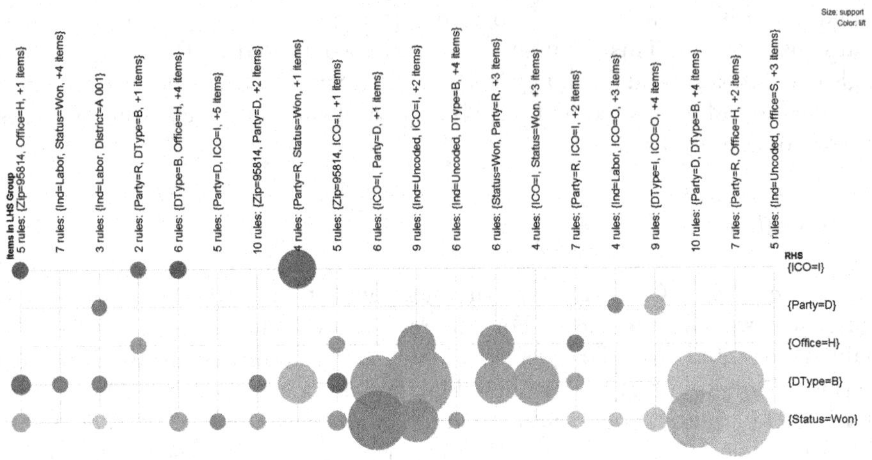

Fig. 1. Rules visualization for all California 2016 transactions

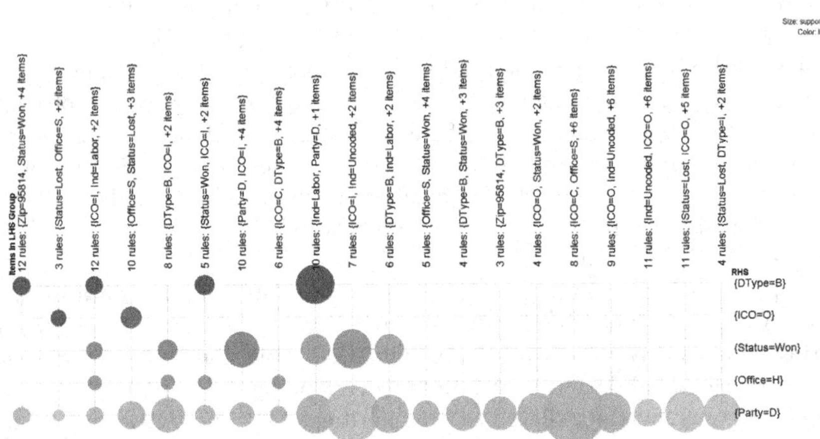

Fig. 2. Rules visualization for California 2016 community $C_{2,1}$

reflects Democrats is that the other community should be comprised mostly of Republicans. However, this is not the case. Instead, this group consists of mostly candidates who won their election, regardless of party. Many of these donations in C_{2_2} come from non-individuals, as shown by the rules

$$\{\text{Party} = \text{Democrat}\} \Rightarrow \text{Donor Type} = \text{Non-Individual} \quad (5)$$

$$\{\text{Party} = \text{Republican}\} \Rightarrow \text{Donor Type} = \text{Non-Individual} \quad (6)$$

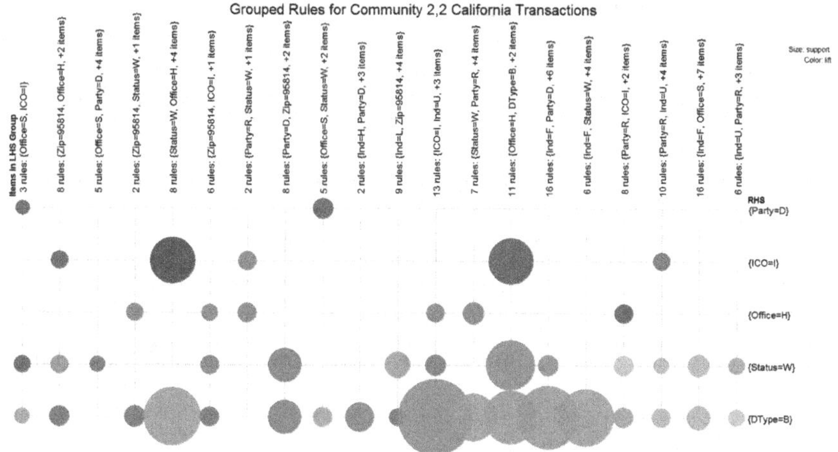

Fig. 3. Rules visualization for California 2016 community $C_{2,2}$

While both of these rules were extracted, the support for the former is 0.63 and the latter is 0.24. Community $C_{2,1}$ has the only rule referencing individuals of

$$\{\text{Donor Type} = \text{Individual}\} \Rightarrow \text{Party} = \text{Democratic} \tag{7}$$

with 0.39 support and 0.87 confidence. This helps show the difference between the two communities where $C_{2,1}$ contained more individuals donating to Democrats whereas $C_{2,2}$ is made of non-individuals donating to winning candidates independent of party, highlighting the differing implicit strategies of the types of donors.

Moving down the hierarchy provides even more information. Communities $C_{3,2}$ and $C_{3,3}$ both are composed of many Democrat candidates. In order to compare these communities, we calculate the intersection of the rules that were discovered in each community individually. The intersection of the rules are shown in Table 3. In particular, we look at the rules with the highest lift amongst the two communities. The top rules help show that both of these communities contain useful patterns and rules showing how non-individuals, especially in the labor industry, gave to winning candidates. Unsurprisingly, in both data sets, incumbents tended to win. The important information here is that it confirms this is an integral part of the parent community.

Also important are the ways in which the communities $C_{3,2}$ and $C_{3,3}$ differ. Tables 4 and 5 contain the highest lift rules that were not shared by the two communities. The high lift rules in $C_{3,2}$ immediately highlight behaviors related to donations to elections where there is no incumbent. Notably, these rules also refer mostly to individuals. There is no similar rule in the other community. This helps highlight that, although both communities contain primarily Democratic candidates, there is considerable difference in the donating habits between individuals and non-individuals. Confirming this are the high

Table 3. Common rules for $C_{3,2}$ and $C_{3,3}$

Antecedent	Consequent	$C_{3,2}$ Lift	$C_{3,3}$ Lift
Status = Won Industry = Labor	Donor Type = Non-Individual	2.039	1.421
Status = Won Party = D Industry = Labor	Donor Type = Non-Individual	2.039	1.421
Office Type = H Industry = Labor	Donor Type = Non-Individual	2.036	1.421
Party = D Office Type = H Industry = Labor	Donor Type = Non-Individual	2.036	1.421
Industry = Labor	Donor Type = Non-Individual	2.025	1.421
Party = D Industry = Labor	Donor Type = Non-Individual	2.025	1.421
Party = D Incumbency = I	Status = Won	1.574	1.226
Party = D Office Type = H Incumbency = I	Status = Won	1.557	1.221
Incumbency = I	Status = Won	1.520	1.188
Incumbency = I Donor Type = Non-Individual	Status = Won	1.515	1.204

lift rules for $C_{3,3}$. These rules focus on non-individuals and highlight that they are frequently associated with winning candidates. This is in particular true for two different types of donors: those from the labor industry, and those from ZIP Code 95814. ZIP code 95814 corresponds the state capital building and area. The rules indicate nearly all donations from that area were from non-individuals (0.997 confidence) and that this money went to winning candidates (0.853 confidence). Much of this came from party committees, in support of candidates likely to win. While this is mostly incumbents as indicated by the rule $\{Incumbency = I\} \Rightarrow Status = Won$ with 0.53 support, the rule $\{Incumbency = O\} \Rightarrow Party = Democrat$ at support 0.33 indicates that money was funneled to those going for open seats as well, but it seems they were less likely to win. All of this helps to demonstrate the ability for rule finding to improve analysis of the communities.

Table 4. Rules only in $C_{3,2}$

Antecedent	Consequent	Lift
Status = Lost, Party = D, Office Type = S	Incumbency = O	1.736
Status = Lost, Office Type = S Donor Type = Individual	Incumbency = O	1.719
Status = Lost, Office Type = S	Incumbency = O	1.709
Party = D, Office Type = S, Industry = Uncoded Donor Type = Individual	Incumbency = O	1.617
Office Type = S, Industry = Uncoded Donor Type = Individual	Incumbency = O	1.607
Party = D, Office Type = S Donor Type = Individual	Incumbency = O	1.596
Party = D, Incumbency = C	Office Type = H	1.585
Office Type = S, Donor Type = Individual	Incumbency = O	1.585
Status = Lost, Party = D, Industry = Uncoded Donor Type = Individual	Incumbency = O	1.575
Party = D, Office Type = S, Industry = Uncoded	Incumbency = O	1.573

Table 5. Rules only in $C_{3,3}$

Antecedent	Consequent	Lift
Status = Won, Party = D Office Type = H, Industry = Labor	Incumbency = I	1.459
Status = Won, Party = D, Office Type = H Donor Type = Non-Individual, Industry = Labor	Incumbency = I	1.459
Status = Won, Office Type = H, Industry = Labor	Incumbency = I	1.451
Status = Won, Office Type = H, Industry = Labor Donor Type = Non-Individual	Incumbency = I	1.451
Industry = Labor, ZipCode = 95814	Donor Type = Non-Individual	1.423
Party = D, Industry = Labor, ZipCode = 95814	Donor Type = Non-Individual	1.423
Status = Won, Office Type = H, Industry = Labor	Donor Type = Non-Individual	1.420
Status = Won, Party = D Office Type = H, Industry = Labor	Donor Type = Non-Individual	1.420
Incumbency = I, ZipCode = 95814	Donor Type = Non-Individual	1.420
Status = Won, ZipCode = 95814	Donor Type = Non-Individual	1.420

6 Conclusion

Using fuzzy spectral hierarchical clustering, we created overlapping clusters of entities in political contribution networks. Using the additional data provided with the transactions forming the links between individuals in the network, association rule mining found additional discriminatory information for the communities. The resulting rules aid in providing insight into the donation patterns of groups within the data beyond what was readily apparent from the rules found using the dataset in its entirety. While this work relied on Apriori for rule mining as a proof of concept, future work can use more sophisticated algorithms for determining rules as well. Additionally, future work will explore calculating communities using a heterogeneous clustering techniques that includes the additional transaction data such as party of the recipient, district, and others. By accounting for those values directly, it should be possible to better group individuals into communities based on donation patterns.

References

1. Blondel, V., Guillaume, J., Lambiotte, R., Mech, E.: Fast unfolding of communities in large networks. J. Stat. Mech.: Theor. Exp. **10**, P10008 (2008)
2. Newman, M.E.J.: Modularity and community structure in networks. Proc. Natl. Acad. Sci. **103**, 8577–8582 (2006)
3. Newman, M.E.J., Girvan, M.: Finding and evaluating community structure in networks. Phys. Rev. E. **69**, 026113 (2004)
4. Pons, P., Latapy, M.: Computing communities in large networks using random walks. J. Graph Algorithms App. **10**, 284–293 (2004)
5. Ng, A.Y., Jordan, M.I., Weiss, Y.: On spectral clustering: analysis and an algorithm. In: Advances in Neural Information Processing Systems, pp. 849–856. MIT Press (2001)
6. Pothen, A., Simon, H., Liou, K.: Partitioning sparse matrices with eigenvectors of graphs. SIAM J. Matrix Anal. Appl. **11**, 430–452 (1990)
7. Bandyopadhyay, S.: Automatic determination of the number of fuzzy clusters using simulated annealing with variable representation. In: Hacid, M.-S., Murray, N.V., Raś, Z.W., Tsumoto, S. (eds.) ISMIS 2005. LNCS (LNAI), vol. 3488, pp. 594–602. Springer, Heidelberg (2005). https://doi.org/10.1007/11425274_61
8. Devillez, A., Billaudel, P., Lecolier, G.V.: A fuzzy hybrid hierarchical clustering method with a new criterion able to find the optimal partition. Fuzzy Sets Syst. **128**(3), 323–338 (2002)
9. Liu, J.: Fuzzy modularity and fuzzy community structure in networks. Eur. Phys. J. B **77**(4), 547–557 (2010)
10. Palla, G., Derényi, I., Farkas, I., Vicsek, T.: Uncovering the overlapping community structure of complex networks in nature and society. Nature **435**(7043), 814–8 (2005)
11. Torra, V.: Fuzzy c-means for fuzzy hierarchical clustering. In: Proceedings of the 14th IEEE International Conference on Fuzzy Systems (FUZZ 2005), pp. 646–651, May 2005
12. Xie, J., Szymanski, B., Liu, X.: SLPA: uncovering overlapping communities in social networks via a speaker-listener interaction dynamic process. In: 2011 IEEE 11th International Conference on Data Mining Workshops (ICDMW), pp. 344–349, December 2011
13. Zhang, B., Horvath, S.: A general framework for weighted gene co-expression network analysis. Stat. Appl. Genet. Mol. Biol. **4**(1) (2005). Article 17

14. Power, J.D., Cohen, A.L., Nelson, S.M., Wig, G.S., Barnes, K.A., Church, J.A., Vogel, A.C., Laumann, T.O., Miezin, F.M., Schlaggar, B.L., Petersen, S.E.: Functional network organization of the human brain. Neuron **72**(4), 665–678 (2011)
15. Flake, G.W., Lawrence, S., Giles, C.L., Coetzee, F.M.: Self-organization and identification of web communities. IEEE Comput. **35**, 66–71 (2002)
16. Aldrich, J.H., Gibson, R.K., Cantijoch, M., Konitzer, T.: Getting out the vote in the social media era: are digital tools changing the extent, nature and impact of party contacting in elections? Party Polit. **22**(2), 165–178 (2015)
17. La Due Lake, R., Huckfeldt, R.: Social capital, social networks, and political participation. Polit. Psychol. **19**(3), 567–584 (1998)
18. Quintelier, E., Stolle, D., Harell, A.: Politics in peer groups: exploring the causal relationship between network diversity and political participation. Polit. Res. Q. **65**(4), 868–881 (2012)
19. Mizruchi, M.S.: Similarity of political behavior among large American corporations. Am. J. Sociol. **95**(2), 401–424 (1989)
20. Gimpel, J.G., Lee, F.E., Kaminski, J.: The political geography of campaign contributions in American politics. J. Polit. **68**(3), 626–639 (2006)
21. Wahl, S., Sheppard, J.: Hierarchical fuzzy spectral clustering in social networks using spectral characterization. In: 28th International Florida Artificial Intelligence Research Society Conference (2015)
22. Kalla, J.L., Broockman, D.E.: Campaign contributions facilitate access to congressional officials: a randomized field experiment. Am. J. Polit. Sci. **60**(3), 545–558 (2016)
23. Fox, J., Rothenberg, L.: Influence without bribes: a noncontracting model of campaign giving and policymaking. Polit. Anal. **19**(3), 325–341 (2011)
24. Akey, P.: Valuing changes in political networks: evidence from campaign contributions to close congressional elections. Rev. Financ. Stud. **28**(11), 3188–3223 (2015)
25. Agrawal, R., Srikant, R.: Fast algorithms for mining association rules in large databases. In: Proceedings of the 20th International Conference on Very Large Data Bases. VLDB 1994, pp. 487–499. Morgan Kaufmann Publishers Inc., San Francisco (1994)
26. Liu, B., Hsu, W., Ma, Y.: Integrating classification and association rule mining. In: Proceedings of the Fourth International Conference on Knowledge Discovery and Data Mining. KDD 1998, pp. 80–86. AAAI Press (1998)
27. Zhang, S., Wang, R.S., Zhang, X.S.: Identification of overlapping community structure in complex networks using fuzzy c-means clustering. Phys. A: Stat. Mech. Appl. **374**(1), 483–490 (2007)
28. Chauhan, S., Girvan, M., Ott, E.: Spectral properties of networks with community structure. Phys. Rev. E **80**, 056114 (2009)
29. Sarkar, S., Dong, A.: Community detection in graphs using singular value decomposition. Phys. Rev. E **83**, 046114 (2011)
30. Sarkar, S., Henderson, J.A., Robinson, P.A.: Spectral characterization of hierarchical network modularity and limits of modularity detection. PLoS ONE **8**(1), e54383 (2013)
31. Farkas, I.J., Derényi, I., Barabási, A.L., Vicsek, T.: Spectra of "real-world" graphs: beyond the semicircle law. Phys. Rev. E **64**, 026704 (2001)
32. Bonica, A.: Mapping the ideological marketplace. Am. J. Polit. Sci. **58**(2), 367–386 (2014)

Educational Data Mining: An Application of Regressors in Predicting School Dropout

Rafaella Leandra Souza do Nascimento(✉), Ricardo Batista das Neves Junior, Manoel Alves de Almeida Neto, and Roberta Andrade de Araújo Fagundes

University of Pernambuco, Recife, Pernambuco, Brazil
{rlsn,rbnj,maan}@ecomp.poli.br, roberta.fagundes@upe.br

Abstract. School dropout is one of the great challenges for the educational system. Educational data mining seeks to study and contribute with results that aim to hidden problems and find possible solutions. Considering its importance, this work aims to use two nonparametric techniques, Quantile Regression and Support Vector Regression, to predict the results of school dropout in the Brazilian scenario. The development of the work followed the phases of CRISP-DM. The evaluation metric of the models is the mean of the absolute error. The results show more significant results for Support Vector Regression.

Keywords: Education Data Mining · SVR · NPQR · Prediction School dropout

1 Introduction

School dropout represents a major problem that needs to be studied and wrestled. There is a governmental effort to reduce this rate, which interferes in the country's educational development indexes, as well as directly impacts the personal and professional scope of Brazilian youth. The phenomenon of school dropout is seen as one of the greatest problems in any level of education [1], and knowing the reasons behind it, institutions can create mechanisms to reduce it.

The National Institute for Educational Studies and Research Anísio Teixeira (INEP) is responsible for collecting and disseminating information about education in Brazil, at all stages of education, through assessments and indicators [2]. According to INEP [3], new researches show that 12.9% and 12.7% of the students enrolled in the first and second years of high school, respectively, evaded school as shown in the School Census between the years 2014 and 2015. The 9th year of elementary education has the third highest dropout rate 7.7%, followed by the 3rd year of high school with 6.8%. Considering all high school grades, dropout sums to 11.2% of total students at this stage of education.

It is notable that combating evasion is still one of the great challenges for education, becoming a very relevant issue, which has been expanded and addressed

P. Perner (Ed.): MLDM 2018, LNAI 10935, pp. 246–257, 2018.
https://doi.org/10.1007/978-3-319-96133-0_19

in many studies. Some of the major research approaches to school evasion are prediction, clustering, relationship mining, discovery with models, and data processing for decision support.

One of the factors that helps the development of this work is the availability of the data related to education, increasing the applicability of Education Data Mining (EDM). By applying EDM, it is possible to effectively and accurately understand students, the role of the context in which learning takes place and other factors that influence the teaching-learning process [4]. This is of extreme interest to a wide variety of people, including educators, students, institutions, government, parents and the general public [5].

EDM focusing on finding answers to specific education questions related to learning processes, development of instructional materials, monitoring and predicting. From obtaining important information and behavior patterns it is possible to support certain pedagogical practices [4]. EDM can be interpreted as a process where the goal is not only to transform data into knowledge, but also to filter knowledge to help make decisions about how to modify the educational environment [6]. It is a field that exploits statistical, machine-learning, and data mining algorithms over the different types of educational data [7]. There are several lines of research in the area of education and many of them derived from the data mining area, such as predictive, grouping or association tasks [4].

Recently, EDM research papers have focused on predictive models to maximize student retention [8], enrollment prediction models based on admission data [9], student performance forecast [10] and school dropout [11]. An accurate predictive model can be used to gain insight into success and risk factors in relation to the educational environment.

In the field of prediction, techniques can be applied to discover structures or associations in data set and make predictions. Among them, regression models are emphasized, which are mathematical model and have as one of the objectives to predict the value of the dependent variable (Y) from the information coming from a set of independent variables (X) [12]. Therefore, this type of technique can estimate the educational benefits and problems.

In this way, this paper aims to predict school dropout by means of educational databases provided by INEP. The techniques used in the experiments were support vector regression (SVR) and nonparametric quantile regression (NPQR). The study followed the phases of the Cross-Industry Standard for Data Mining (CRISP-DM) [13] model.

This paper is divided as it follows: Sect. 2 presents works related to the theme of this article; Sect. 3 the methodology, where the information about the databases and the techniques used in the development of this work are presented; Sect. 4 shows the results of the experiments performed; and, finally, Sect. 5 composes the conclusion of this work developed after the analysis of the results obtained at the end of the experiments.

2 Related Works

EDM has grown in recent years and the information provided by that can serve as a subsidy to improve education practices, as well as being an important tool to enable education qualification. More and more studies address the educational issue, especially school dropout.

In relation to the techniques of data mining applied to education, Machado et al. [14] made a bibliography review focused on to identify the papers which approach the problem of school dropout using data mining techniques. That study allowed to identify the main methods used in this subject, which are decision trees, neural networks, logistic regression and clustering algorithms.

In Martinho et al. [15], a Neural Network was used to present the prediction of the group of the students at risk of dropping out in higher education classroom courses. In Meedech et al. [11], it also addresses evasion and applies decision trees and rules induction models to discover student data knowledge. Such as in the research by Quadri and Kalyankar [16] that applies the decision tree technique to choose the best forecast and analysis.

Regarding distance education, Cambruzzi et al. [17] address the dropout rates observed in these courses which are very significant. The research presents a system of Learning Analysis developed to deal with the problem of abandonment in distance education courses in university education. Data visualization and text mining are used.

Márquez-Vera et al. [18] proposed the application of data mining techniques to predict school failure and school dropout. From actual data of Mexican students, they conducted experiments using induction rules and decision trees. Veitch [19] justifies the use of decision trees to predict school dropout because this method is designed to sift a set of predictor variables and successively divides a set of data into subgroups to improve the prediction (classification) of a target (dependent) variable. In addition to decision trees, the research by da Cunha et al. [20] implements grouping algorithms in the school dropout scenario.

In the work of Rodrigues et al. [21], the objective was to investigate the feasibility of using the linear regression model to obtain inferences in the initial stages of on-line courses, as a way of supporting decision making by teachers and managers. The results obtained demonstrated that it is possible to use the linear regression technique to obtain inferences with good accuracy rates.

However, few papers are still used in its technical experiments different techniques from those that were approached by Machado et al. [14] in his bibliography review. This is the case of quantile regression [22]. Quantile regression differs from the other regression types since it allows the use of several curves (or quantiles) to obtain a more complete view of the relationship between the studied variables. For nonparametric quantile regression model, the Gaussian kernel can be applied by adjusting the parameter bandwidth that controls the degree of smoothness of the estimated function.

Another technique for prediction is the Support Vector Machine (SVM) and it can also be used for regression, retaining all the key functions of the algorithm in which are, generating the maximum number of support vectors in order to maximize the separation of data classes and to maximize the margins of these support vectors. The extension of the SVM technique for regression is the support vector regression (SVR). SVR algorithms perform well in nonlinear problems, such as time series [23]. The advantage of using SVR over other techniques because it focuses on finding the global optimal value. In addition, its model is easier to understand comparing to others well known techniques, such as artificial neural network.

In order to approach these presented works, the accomplishment of this research brings relevant contribution since the data mining for the educational environment is an extremely important subject and that needs to be studied deeply. The topic of evasion is much addressed in the works, so it is necessary to apply other techniques and verify the performance of the results. Thus, using the techniques NPQR and SVR the predictive analyzes of school dropout will be constructed.

3 Methodology

One of the most popular methodologies to increase the success of data mining processes is CRISP-DM [13]. The methodology defines a non-rigid sequence of six phases which allows the construction and implementation of a mining model to be used in a real environment, helping business decisions [24]. Therefore, the development of this paper follows the phases of CRISP-DM that is shown in Fig. 1 and described in the following subsections.

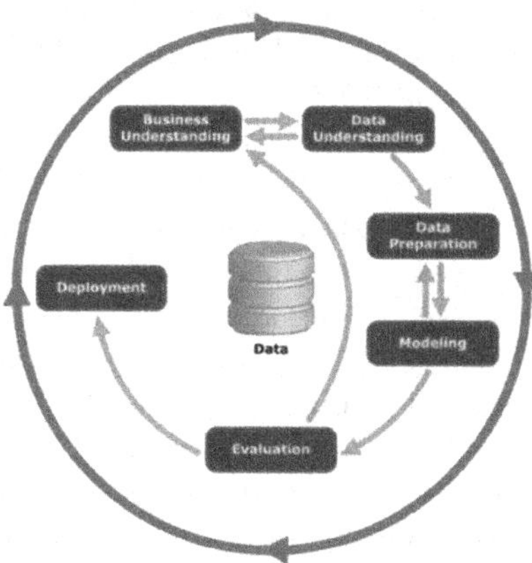

Fig. 1. Steps of CRISP-DM [13].

3.1 Business Understanding

The initial phase focuses on understanding project goals and requirements and converts this knowledge into a data mining problem definition and a preliminary plan designed to achieve the objectives.

A bibliographical research was carried out, checking all material already elaborated and related to educational scenarios, educational data mining and regression models. After that, the main variables involving the research subject were listed, which consists of predicting school dropout.

3.2 Data Understanding

The data understanding phase begins with initial data collection and proceeds with activities that allow to familiarize with the data, identify data quality problems, first discover insights about the data, and detect interesting subsets to form hypotheses about hidden information.

Databases Description. The databases used in this study were obtained from the open web portal of the INEP, referred to the microdata of the School Census and School Performance Rate, both from the year 2016.

The School Census is a survey of educational information of national scope carried out annually and made available by INEP. This research aims to carry out a survey on the basic education schools in the country. It covers beyond the stages of regular education, special education, youth and adult education and professional education. The information collected is classified into four major dimensions, schools, students, teachers and classes.

In this paper, it is used the schools dimension, where data are collected on infrastructure such as basic sanitation, electricity, science and computer labs, accessibility, as well as on stages and schooling modalities offered, location and administrative dependency. The original basis of the 2016 School Census of the schools dimension has a total of 279,359 instances and 166 variables.

INEP also provides educational indicators which attribute statistical value to the quality of teaching, attending not only to students' performance but also to the economic and social context in which schools are inserted. They consider information such as access, permanence and student learning. The educational indicator used in this study refers to the database Performance Rate. The database has information on student approval, disapproval and dropout.

These data can be obtained in different levels such as national, regional or school. Thus, the school level database is used since it is necessary to relate the values of this base to each school present in the base of the School Census. The original basis of this indicator has a total of 139,823 instances and 63 variables.

3.3 Data Preparation

At this point, tasks include record and attribute selection as well as transformation and data cleaning for modeling tools. In this phase, the following activities were carried out:

- Identification of scenarios in the data.
- Selection of variables.

- Checking for missing or blank values.
- Data cleaning.
- Data normalization.

Through each school's identifier, the school dropout rate present in the School Income database have been turned into a column of the School Census database. Then, the aspects of school infrastructure were correlated with the dropout rate. The data considered for study are of high school in the scope of the state of Pernambuco that contains 3,215 instances.

We used the Random Forest technique focusing on to obtain the degree of importance of each explanatory variable (X) in relation to the response variable (Y). Then, the 9 most important variables were selected for the model. At the end of this data preparation phase, the following data mapping was obtained, shown in the Table 1. After that, the correlation matrix of the selected variables was created, as shown in Fig. 2.

Table 1. Variables selected for study.

Ref	Variable	Description
NSU	NU_SALAS_UTILIZADAS	Total rooms used
NC	NU_COMPUTADOR	Total computers
NSE	NU_SALAS_EXISTENTES	Total rooms that the school has
NCAD	NU_COMP_ADMINISTRATIVO	Total administrative computers
NCA	NU_COMP_ALUNO	Total computers for students
IBC	IN_BANHEIRO_CHUVEIRO	Bathroom presence indicator
TLD	TP_LOCALIZACAO_DIFERENCIADA	Type of school location
NF	NU_FUNCIONARIOS	Total employees
IAF	IN_AGUA_FONTE_RIO	Indicator of presence of water (river)
EM	EVASAO_MEDIO	School dropout values

By analyzing the Table 1, the variables related to the existence of computers in schools were identified and selected. The presence of computers in schools is a very important differential. This may indicate a better school structure as well as a concern of the school to develop technological teaching and improve digital access to information by students. Variables such as the existence of rooms, restrooms and the number of employees may also relate to the administrative capacity of the school, and consequently its structure. Such as the correlation matrix (Fig. 2) was analyzed, there is a high degree of correlation between these variables (NSU, NC, NSE, NCA, NCAD and NF references, with coefficients from 0.34 to 0.92), except for the presence of the variable of bathrooms in schools (IBC reference).

The location of the schools is also an important factor. Students who study far or in schools that are difficult to access may be likely to be evaded. The last aspect to be analyzed is the issue of water supplied to school. The selected

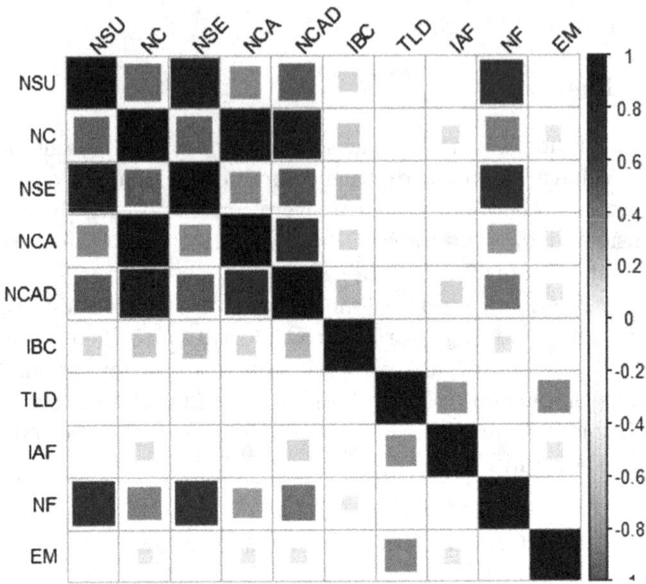

Fig. 2. Matrix of correlation of the selected variables.

variable is related to the water supply through the river. This may be an indicator of schools that are not in areas near water stations, or further away from the urban environment. The correlation matrix shows a higher correlation between these two variables (TLD and IAF references, with coefficient 0.32) compared to the others variables. The correlation matrix still informs that the variable with the highest correlation with school dropout (MS reference) is the location of the school (TLD), with coefficient 0.35.

3.4 Modeling

In this step, several modeling techniques were selected and applied, and their parameters were calibrated to optimal values. It represents the development of the models for the problem, based on the data that were already suitable to be used. The techniques used in this work consisted of NPQR and SVR.

Nonparametric Quantile Regression: To the use of the NPQR, the Gaussian Kernel was performed, and considered the quantile 0.5. This quantile was used because it presents better results in analyzes performed and it is considering the median of the data.

Support Vector Regression: The Grid Search approach was used to adjust SVR parameters. It was chosen: Radial Base Function Kernel; ϵ 1e-13, 1, 10; γ 1e0, 1, 1e-3, 1e-14; and C 1, 10, 100, 1000. These parameters are the most discussed in the literature.

3.5 Evaluation

At this step of the project, the developed model is evaluated and the steps taken to create it are reviewed to be sure that it adequately achieves the defined objectives.

One of the most used performance indexes in the prediction techniques is the calculation based on the prediction error. The performance index used in this work is the absolute mean error (MAE) that is a useful measure widely used in model evaluations [25].

3.6 Deployment

All the knowledge obtained through the work of mining became subsidies for the development of strategies that solve the proposed problem. In this work, strategies will be listed for the scenario studied after all stages of CRISP-DM.

4 Results of Experiments

In linear techniques, the predictors are combined in a linear fashion to model the effect on the response. This linearity may be insufficient to capture the data structure. However, there are techniques that allow a more flexible regression modeling that combines the predictors in a nonparametric way. Thus, nonparametric regression techniques, NPQR and SVR were modeled seeking a response that best fits the data.

The results obtained through the experiments were organized in a table and graph box-plot, in relation to the Median Absolute Error (MAE). The configurations of the techniques were described in the previous sections and performed 30 independent simulations for each experiment to form a significant sample. Therefore, the mean of the executions of the techniques was extracted as it shows on Table 2.

It can be verified that the SVR obtained smaller error comparing to NPQR. In analysis of the box-plot graph shown in the Fig. 3, the results of the techniques were compared side by side. One can note that SVR obtained smaller median in comparison to NPQR and the level of variability of the data contained in the NPQR sample is greater. This result shows that, indeed, SVR was able to find the best combination of its parameters in all executions for predicting the school output.

Both techniques used in this work use a Kernel function to perform the estimation of the model. However, the results achieved by each technique are significantly different. The SVR is able to achieve better results because it searches for the optimal value. Another analysis is that the result of NPQR depends directly on how well estimated the bandwidth value is. However, estimating this value is not an easy task. This may cause lower NPQR performance.

Through these analyzes, the SVR is doing a better work than the NPQR model at predicting school dropout. The Fig. 4 shows the visual plot for the

Table 2. Results NPQR X SVR

Technique	Mean error	SD error
SVR	0.015665	0.0
NPQR	0.0223757	0.00324569

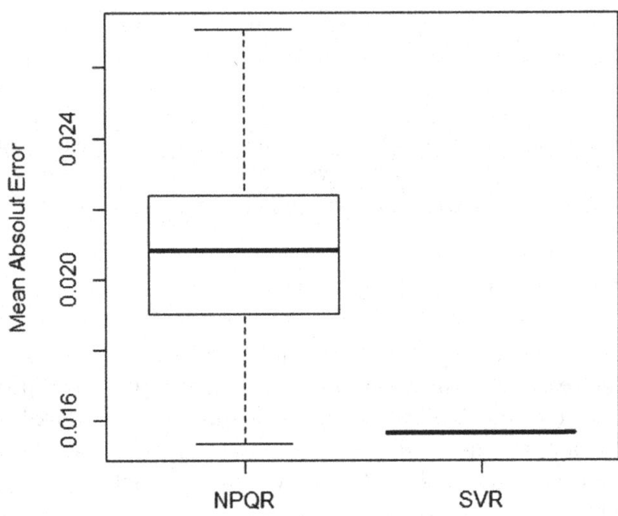

Fig. 3. Boxplot NPQR x SVR.

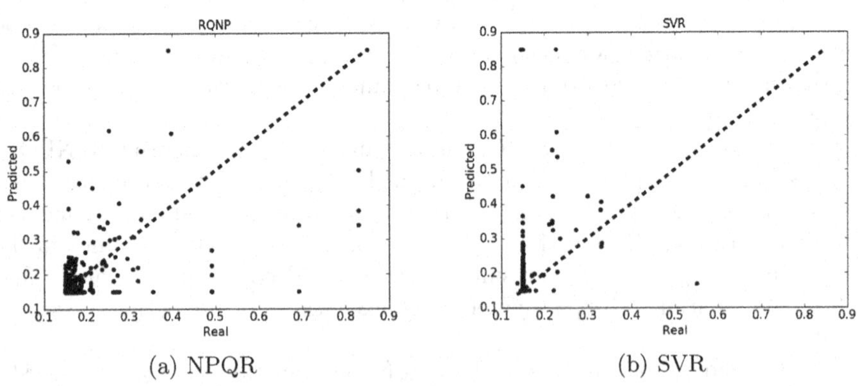

(a) NPQR (b) SVR

Fig. 4. Plot Predicted x Real of techniques.

performance of the techniques on the test set with predicted and real values. By visually inspecting the plot we can see that the predictions made by the SVR model are, in general, more concentrated around the line with less outliers than those made by the NPQR model. In this way, we can conclude that SVR made a more accurate prediction.

With this result, it can be observed that nonparametric regression techniques can be applied in education. The power of adjustments and flexibility to the data justifies the applicability of this type of modeling when the parametric techniques are insufficient. The educational area has several challenges. Using techniques associated with EDM to better estimate the variables related to these scenarios brings a great gain to the literature and to those interested in the area, such as students, educators and government, for example. Despite its advantages, in the literary this type of modeling is still little approached in educational problems.

5 Conclusion

EDM enables the prior identification of aspects that may need improvements and better investments, thus improving teaching, learning and mitigating problems such as school dropout. In this scenario, this study seeks to contribute to the area of Education in terms of understanding and explaining the characteristics of schools that may be related to school dropout. Public databases provided by INEP were used in this paper.

Through the Random Forest technique the variables that are most related to the problem of school dropout in high school were identified, according to the database provided. The existence of computers in schools, the number of employees, rooms and the location of schools were highlighted as the most important for the model of prediction of evasion and were selected for the study.

In addition, it consisted of investigating new different prediction techniques from those most used in the literature for EDM. Thus, experiments with the SVR and NPQR techniques were used. The NPQR was used to allow the use of several curves (quantiles) to obtain a more complete view of the relationship between the studied variables. The advantage of using SVR is that it searches for the optimal value and the model is easier to understand. The development of this work followed the phases of CRISP-DM.

The results of the experiments indicated that SVR obtained better results in comparison to NPQR. This means that SVR was able to minimize the prediction error for the school dropout problem.

The nonparametric techniques used bring a more flexible regression modeling, looking for a response that best matches the data. These techniques bring a gain to the educational area towards a better response when parametric techniques are insufficient. With that being said, predicting variables related to teaching-learning can make research more accurate. These results made it possible to investigate aspects in which the public policies for the scenario studied can be better applied.

In this way, for future work, it is sought to refine forecasting techniques, as well as explore other educational aspects, expand the scope of study to other levels of education and to other scenarios such as regional or national.

References

1. Lobo, M.B.C.M.: Panorama da evasão no ensino superior brasileiro: aspectos gerais das causas e soluções. Associação Brasileira de Mantenedoras de Ensino Superior. Cadernos, no. 25 (2012)
2. INEP: Instituto Nacional de Estudos e Pesquisas Educacionais Anísio Teixeira. http://portal.inep.gov.br. Accessed 22 Dec 2017
3. INEP: Instituto Nacional de Estudos e Pesquisas Educacionais Anísio Teixeira: INEP divulga dados inéditos sobre fluxo escolar na educação básica. http://portal.inep.gov.br/. Accessed 07 Jan 2018
4. Baker, R., Isotani, S., Carvalho, A.: Mineração de dados educacionais: oportunidades para o Brasil. Braz. J. Comput. Educ. **19**(02), 03 (2011)
5. Guy, M.: The open education working group: bringing people, projects and data together. In: Mouromtsev, D., d'Aquin, M. (eds.) Open Data for Education. LNCS, vol. 9500, pp. 166–187. Springer, Cham (2016). https://doi.org/10.1007/978-3-319-30493-9_9
6. Romero, C., Ventura, S.: Data mining in education. Wiley Interdisc. Rev.: Data Min. Knowl. Discov. **3**(1), 12–27 (2013)
7. Sharma, M., Mavani, M.: Accuracy comparison of predictive algorithms of data mining: application in education sector. In: Unnikrishnan, S., Surve, S., Bhoir, D. (eds.) ICAC3 2011. CCIS, vol. 125, pp. 189–194. Springer, Heidelberg (2011). https://doi.org/10.1007/978-3-642-18440-6_23
8. Yu, C.H., DiGangi, S., Jannasch-Pennell, A., Kaprolet, C.: A data mining approach for identifying predictors of student retention from sophomore to junior year. J. Data Sci. **8**(2), 307–325 (2010)
9. Yadav, S.K., Pal, S.: Data mining application in enrollment management: a case study. Int. J. Comput. Appl. **41**(5), 1–6 (2012)
10. Ramaswami, M., Bhaskaran, R.: A CHAID based performance prediction model in educational data mining. arXiv preprint arXiv:1002.1144 (2010)
11. Meedech, P., Iam-On, N., Boongoen, T.: Prediction of student dropout using personal profile and data mining approach. Intelligent and Evolutionary Systems. PALO, vol. 5, pp. 143–155. Springer, Cham (2016). https://doi.org/10.1007/978-3-319-27000-5_12
12. Montgomery, D.C., Peck, E.A., Vining, G.G.: Introduction to Linear Regression Analysis, vol. 821. Wiley, Hoboken (2012)
13. Chapman, P., Clinton, J., Kerber, R., Khabaza, T., Reinartz, T., Shearer, C., Wirth, R.: CRISP-DM 1.0 step-by-step data mining guide. CRISP-DM Consortium (2000)
14. Machado, R.D., Benitez, E., Corleta, J., Augusto, G.: Estudo bibliométrico em mineração de dados e evasão escolar. In: Apresentado na XI Congresso Nacional de Excelência em Gestão, Rio de Janeiro, RJ (2015)
15. Martinho, V.R.D.C., Nunes, C., Minussi, C.R.: An intelligent system for prediction of school dropout risk group in higher education classroom based on artificial neural networks. In: 2013 IEEE 25th International Conference on Tools with Artificial Intelligence (ICTAI), pp. 159–166. IEEE (2013)
16. Quadri, M.M., Kalyankar, N.: Drop out feature of student data for academic performance using decision tree techniques. Global J. Comput. Sci. Technol. **10**(2), 1–4 (2010)
17. Cambruzzi, W.L., Rigo, S.J., Barbosa, J.L.: Dropout prediction and reduction in distance education courses with the learning analytics multitrail approach. J. UCS **21**(1), 23–47 (2015)

18. Márquez-Vera, C., Morales, C.R., Soto, S.V.: Predicting school failure and dropout by using data mining techniques. IEEE Rev. Iberoam. Tecnol. Aprendiz. **8**(1), 7–14 (2013)
19. Veitch, W.R.: Identifying characteristics of high school dropouts: data mining with a decision tree model (2004). (Online Submission)
20. da Cunha, J.A., Moura, E., Analide, C.: Data mining in academic databases to detect behaviors of students related to school dropout and disapproval. New Advances in Information Systems and Technologies. AISC, vol. 445, pp. 189–198. Springer, Cham (2016). https://doi.org/10.1007/978-3-319-31307-8_19
21. Rodrigues, R.L., de Medeiros, F.P., Gomes, A.S.: Modelo de regressão linear aplicado à previsão de desempenho de estudantes em ambiente de aprendizagem. In: Brazilian Symposium on Computers in Education (Simpósio Brasileiro de Informática na Educação-SBIE), vol. 24, p. 607 (2013)
22. Koenker, R., Bassett Jr., G.: Regression quantiles. Econom.: J. Econom. Soc. **46**(1), 33–50 (1978)
23. Chen, Y., Xu, P., Chu, Y., Li, W., Wu, Y., Ni, L., Bao, Y., Wang, K.: Short-term electrical load forecasting using the support vector regression (SVR) model to calculate the demand response baseline for office buildings. Appl. Energy **195**, 659–670 (2017)
24. Moro, S., Laureano, R., Cortez, P.: Using data mining for bank direct marketing: an application of the CRISP-DM methodology. In: Proceedings of European Simulation and Modelling Conference-ESM 2011, pp. 117–121. Eurosis (2011)
25. Chai, T., Draxler, R.R.: Root mean square error (RMSE) or mean absolute error (MAE)?-arguments against avoiding RMSE in the literature. Geosci. Model Dev. **7**(3), 1247–1250 (2014)

Reinforcement Learning for Computer Vision and Robot Navigation

A. V. Bernstein(✉), E. V. Burnaev, and O. N. Kachan

Skolkovo Institute of Science and Technology, Moscow, Russia
{A.Bernstein,E.Burnaev,Oleg.Kachan}@skoltech.ru

Abstract. Nowadays, machine learning has become one of the basic technologies used in solving various computer vision tasks such as feature detection, image segmentation, object recognition and tracking. In many applications, various complex systems such as robots are equipped with visual sensors from which they learn the state of a surrounding environment by solving corresponding computer vision tasks. Solutions of these tasks are used for making decisions about possible future actions. Reinforcement learning is one of the modern machine learning technologies in which learning is carried out through interaction with the environment. In recent years, reinforcement learning has been used both for solving robotic computer vision problems such as object detection, visual tracking and action recognition as well as robot navigation. The paper describes shortly the reinforcement learning technology and its use for computer vision and robot navigation problems.

Keywords: Reinforcement learning · Computer vision
Robot navigation

1 Introduction

The general goal of Computer vision (CV) is high-level understanding of the content of images (recognizing objects, scenes, or events; classification of detected objects, discovering geometric configuration and relations between objects). In other words, machine vision models the world from images and videos received from visual sensing system by recognizing an environment and estimating its required features.

In real applications, this high-level understanding of the content of images, received from certain visual sensing system, is only the first key step in solving various specific tasks such as mobile robot navigation in uncertain environments, road detection in autonomous driving systems, etc. In other words, specific decision-making problems are solved on the basis of information about surrounding environment extracted from captured images. Therefore, the efficiency of used image understanding technologies should be estimated from the efficiency of decisions made on basis of the resulted "surrounding environment understanding"; such decisions, in turn, are determined by results of previous

P. Perner (Ed.): MLDM 2018, LNAI 10935, pp. 258–272, 2018.
https://doi.org/10.1007/978-3-319-96133-0_20

actions of a system. For example, navigation decisions for a mobile robot [1] are made based on estimated robot localization, which in turn is a result of specific CV task, namely, passive vision-based robot position estimation [2,3].

Nowadays, one of the main drivers of high-level understanding of surrounding environment from images is the application of machine learning methods to CV tasks (image registration, segmentation, classification; 3D reconstruction; object detection and tracking). Vision tasks are formulated as learning problems, and machine learning is an essential and ubiquitous tool for automatic extraction of patterns or regularities from images. Thus, machine learning is now a core part of machine vision. Reinforcement learning (RL) is one of modern machine learning technologies in which learning is carried out through interaction with the environment and allows taking into account results of decisions and further actions based on solutions of corresponding CV tasks.

In the paper, we consider the use of RL technologies in solving various CV tasks concerning processing and analysis of visual information by an intelligent agent. The paper is organized as follows. Section 2 provides a short description of the RL technology. Section 3 contains examples of applying this technology to typical CV tasks and robot navigation.

2 Reinforcement Learning Technology

RL has gradually become one of the most active research areas in machine learning, artificial intelligence, and neural network research, and has developed strong mathematical foundations and impressive applications. The RL technology is based on a simple idea to create a learning system that wants something, that adapts its behavior in order to maximize a special signal from its environment [4,5]. RL is close to such important direction in statistics as adaptive design of experiments [6,7].

The main element is a learning of an active decision-making autonomous agent (or, simply, the agent), which learns its behavior through trial-and-error interactions with a dynamic environment (described as a dynamic process) to achieve a goal despite uncertainty about the environment. At successive time steps (t), the agent makes an observation of the environment state (s_t), selects an action (a_t) and applies it back to the environment, modifying the state (s_{t+1}) at next moment ($t + 1$). The goal of the agent is to find adequate actions for controlling this process.

Formally, RL can be modeled as a Markov decision process, consisting of:

- a set of environment states $S = \{s\}$, plus a distribution of starting states $p(s_0)$ at starting moment $t = 0$;
- a set of actions $A = \{a\}$;
- a transition dynamics (probability function) $T(s_{t+1}|s_t, a_t)$ that map a pair, consisting of a state s_t and an action a_t at time t onto a distribution of states s_{t+1} at time $(t + 1)$.

Beyond this model, one can identify three main subelements of the RL system: a policy, a reward function, and a value function. A policy is defined as a map $\pi : S \times A \rightarrow [0, 1]$ that determines the probability $\pi(a|s) = \mathbb{P}(a_t = a|s_t = s)$ of taking action $a_t = a$ at moment (t) in state $s_t = s$.

A reward function is a map $r : S \times A \rightarrow \mathbb{R}$ that determines an immediate/instantaneous reward (return) $r_t = r(s_t, a_t)$ that is received (earned) at moment t, in going from state s_t to state s_{t+1} under action a_t. Denote by $p(s_t, a_t, s_{t+1})$ the probability of this transition with taking into account a randomness of the action a_t (with probability $\pi(a_t|s_t)$) and the transition $s_t \rightarrow s_{t+1}$ (with probability $T(s_{t+1}|s_t, a_t)$). For simplicity, we consider only deterministic (non-randomized) policies when the action a is a nonrandom function $a = \pi(s)$ of the state s; thus, $r_t = r(s_t, \pi(s_t))$.

Introduce a discount factor $\gamma \in (0, 1)$, which forces recent rewards to be more important than remote ones (i.e., lower values place more emphasis on immediate rewards). Under chosen policy π and starting state s, a value function

$$V_\pi(s) = \lim_{M \rightarrow \infty} \mathbb{E}\left[\sum_{t=0}^{M} \gamma^t \times r(s_t, \pi(s_t))\big|s_0 = s\right] \tag{1}$$

is defined as expected discounted reward and intuitively denotes the total discounted reward earned along an infinitely long trajectory starting at the state s, if policy π is pursued throughout the trajectory. Thus, $V_\pi(s) = \mathbb{E}[R|s, \pi]$, where $R = \sum_{t=0}^{\infty} \gamma^t \times r_t$ is a discounted reward.

The problem is to find a stationary policy π^* of actions $\{a_t = \pi^*(s_t)\}$ called optimal policy which maximizes the function $V_\pi(s)$ for all starting states: $V_{\pi^*}(s) = \max_\pi V_\pi(s) \equiv V^*(s)$. Under appropriate assumptions, it was proven [8] that optimal value function $V^*(s)$ is unique, although there can be more than a single optimal policy π^*. Dynamic Programming [9] is the theory behind evaluation of an optimal stationary policy π^* for the problem above and encompasses a large collection of techniques for the evaluation.

This optimal value function can be defined as the solution to the simultaneous equations

$$V^*(s) = \max_{a \in A}\left\{r(s, a) + \gamma \times \sum_{s' \in S} T(s'|s, a) \times V^*(s')\right\}, \tag{2}$$

and, given the optimal value function (2), we can specify the optimal policy as

$$\pi^*(s) = \arg\max_{a \in A}\left\{r(s, a) + \gamma \times \sum_{s' \in S} T(s'|s, a) \times V^*(s')\right\}. \tag{3}$$

Therefore, one way to find the optimal policy is to find the optimal value function (2) that can be determined by a simple iterative algorithm called value iteration, which can be shown to converge to the correct V^* values [9]. Under known model consisting of transition probability function $T(s'|s, a)$ and reward function $r(s, a)$, these techniques use various iteration schemes and Monte-Carlo techniques for obtaining the optimal policy.

The RL technology is primarily concerned with how to obtain the optimal policy when such model is not known in advance. The agent must interact with the environment directly to obtain information which by means of an appropriate algorithm can be processed to produce an optimal policy.

2.1 Q-Learning

The most popular technique in RL is Q-learning [10] which is a model-free method used to find an optimal action-selection policy for any given (finite) Markov decision process. It works by learning an action-valued function

$$Q^*(s,a) = r(s,a) + \gamma \times \sum_{s' \in S} T(s'|s,a) \times \max_{a'} Q(s',a'), \tag{4}$$

written recursively, that ultimately gives the expected utility of taking a given action in a given state and following the optimal policy thereafter. Then $V^*(s) = \max_{a \in A} Q^*(s,a)$ and $\pi^*(s) = \arg\max_{a \in A} Q^*(s,a)$.

Q-learning consists in approximating of the action-valued function $Q^*(s,a)$ (4) and uses an experience tuple (s, a, r, s') summarizing a single transition in the environment, in which s is the agent's state before the transition, a is its choice of action, r is the instantaneous reward it receives, and s' is its resulting state. There are various current policies how to choose the action. Initially, since the agent knows nothing about what it should do, it acts randomly. When it has some current version $Q(s,a)$ of the action-valued function, it can use a greedy policy, choosing the action which maximizes this function. Another policy called ε-greedy is to choose the greedy policy with probability $(1 - \varepsilon)$ and random action with probability ε.

Let $Q_{old}(s,a)$ be an old value of the action-valued function which is updated from the received experience tuple (s, a, r, s') resulting in a new value

$$Q_{new}(s,a) = Q_{old}(s,a) + \alpha \times \left[r(s,a) + \gamma \times \max_{a'} Q_{old}(s',a') - Q_{old}(s,a) \right], \tag{5}$$

here α is a learning rate, decreasing slowly. After some trial-and-error interactions, the agent begins to learn its task and performs better and better. If each action is executed in each state an infinite number of times on an infinite run and α is decaying with appropriate speed, then Q values converge with probability 1 to $Q^*(s,a)$ (4) [9,11].

2.2 Policy Gradient Methods

Until now we have reviewed indirect ways to estimate policies via estimating values of states and actions in this states. One can consider learning a parametrized policy to directly draw state-action pair from without referring to value function.

We consider methods for learning the policy parameters based on the gradient of some performance measure $J(\theta)$ with respect to the policy parameters, i.e. our goal is to maximize performance:

$$\theta_{t+1} = \theta_t + \alpha \widehat{\nabla_{\theta_t} J} \tag{6}$$

where $\widehat{\nabla_{\theta_t} J}$ is a stochastic estimate whose expectation approximates the gradient of the performance measure with respect to its argument θ_t. All methods that follow this general schema are called policy gradient methods, whether or not they also learn an approximate value function [4].

In REINFORCE [12] policy gradient algorithm each increment is proportional to the product of return G_t and a vector equal to the gradient of the probability of taking the action actually taken divided by the probability of taking that action, i.e.

$$\theta_{t+1} = \theta_t + \alpha G_t \frac{\nabla_\theta \pi(A_t \mid S_t, \theta_t)}{\pi(A_t \mid S_t, \theta_t)}. \tag{7}$$

REINFORCE is from a class of Monte Carlo reinforcement learning methods in a sense that it uses a full episode to evaluate the policy, thus it can not be applied in a step-by-step manner.

2.3 Actor-Critic Methods

RL algorithms can be summarized as methods that learn value function (critic), policy function (actor), both of which have their own strengths and weaknesses. The benefit of policy gradient methods is the ability to naturally handle continuous action space, but they suffer from slow learning and high variance of estimates.

Actor-critic methods learn both policy $\pi(a \mid s)$ and value function $V(s)$. The actor is able to update with lower variance gradients, using information of estimated expected returns, provided by the critic. This leads to lower variance, thus faster learning and better convergence properties.

2.4 Approximation in Reinforcement Learning

Although RL had some successes in the past, previous approaches lacked scalability and were inherently limited to fairly low-dimensional problems. Deep learning [13] is a modern tool which, relying on the powerful function approximation and representation learning properties of neural networks [14,15], has accelerated progress in RL defining the field of Deep Reinforcement Learning (DRL) [16]. Deep learning enables RL to scale to decision-making problems that were previously intractable, i.e., settings with high-dimensional state and action spaces.

DRL algorithms have already been applied to a wide range of problems, such as robotics, where control policies for robots can now be learned directly from camera inputs in the real world, succeeding controllers that used to be hand engineered or learned from low-dimensional features of robot's states.

3 Reinforcement Learning in Computer Vision Tasks

Computer vision tasks such as image processing and image understanding involve solving decision-making problems, and recent research has shown that RL can be applied to such problems. We give a short overview of successful applications of RL to solving various computer vision tasks, closely related to robot operating in indoor or outdoor environments.

3.1 Object Detection and Region Proposal

Object detection task has a goal to identify the spatial location of a bounding box of an object within a given image. While this task has been solved by heuristic algorithms [17,18] in the past, currently object detection is addressed by region proposal networks [19]. This model combines proposals from a vast number of class-independent regions, further to be classified with another model.

One of the first known attempts to solve this problem with RL was proposed in [20]. Candidate proposals are selected by the learned localization policy, with intuitive actions to transform proposal bounding box in a sequential way to find the best focus on the object. An agent starts an episode with a bounding box covering whole or most of the image and then narrows the bounding box to perfectly match it against the object of interest.

The set of actions consists of eight transformations to move a box in a horizontal and vertical directions, change its scale, aspect ratio and the trigger action which is performed to indicate that the object is correctly localized within the current bounding box. The state is represented by the information of the currently visible region obtained from a pre-trained CNN and past actions taken place. The reward function is proportional to the improvement after an agent performs a particular action and measured as Intersection-over-Union (IoU[1]) between prediction $\hat{y}$ and target object y. The reward is positive if IoU improved from state s_t to state s_{t+1} and negative otherwise. Also, the reward implicitly considers the number of steps as a cost due to the discount factor term in the objective of DQN [18] algorithm used to train the model.

3.2 Visual Tracking

Visual tracking is a hard task in computer vision due to object appearance, occlusion, and computational demands. Recently, tracking by detection approach which employs convolutional neural networks lead to great improvements in tracking accuracy, but only because of the improved detection subtask. Tracking decisions such as where to look next, how to manage the cases when the tracked object is lost are mostly heuristics and are not trained in an end-to-end manner. Another requirement for visual tracking is continuous and accurate predictions in both spatial and temporal domain over a long period of time, which is not properly addressed by existing algorithms.

[1] $IoU(\hat{y}, y) = \frac{|\hat{y} \cap y|}{|\hat{y} \cup y|}$.

In [21] authors formulate single-object tracking as a sequential decision problem, predicting the object location for each frame. Deep RL Tracker model, the first known attempt to solve the visual tracking problem with reinforcement learning, is designed to address tracking performance on the long run and trained end-to-end with backpropagation and REINFORCE [12] algorithms for the visual encoder and policy network components respectively.

Training a reinforcement learning algorithm requires a reward function, which is presented by two instances, each of which is used on different stages of the learning procedure. In the early stage reward function $r_t = -avg(|\hat{y} - y|) - max(|\hat{y} - y|)$ is used, where avg and max are pixel-wise average and maximum between prediction and ground truth. In the last stage $r_t = IoU(\hat{y}, y)$ is used to ensure IoU metric maximization.

Active tracking, where an agent controls a camera to follow the object of interest is also have been addressed with RL. Usually for such type of tracking a camera is mounted on a mobile robot or robotic arm. In [22] tracking for such task is trained end-to-end with A3C [23] reinforcement learning algorithm. An agent's observations are given by raw visual input and it is able to take actions whether to move left, forward, right or abstain from action. Virtual environment simulator UnrealCV [24] is used to generate visual states and receive agent actions.

Network architecture is presented by a combination of convolutional and recurrent neural networks to process spatial and temporary domains respectively. Visual feature vectors obtained with a CNN are flattened and thus serve as inputs to a recurrent policy network. The reward function, following the intuition that an agent should closely follow object of interest is chosen in such a way that the maximum reward is achieved when the object is located perfectly in front of an agent within given distance and exhibits no rotation.

3.3 Action Detection

Action detection algorithm should provide temporary bounds of an occurrence of a certain action in a video. Existing algorithms address this by applying exhaustive sliding-window search using frame-level classifiers on different time scales. This kind of brute-force search leads to inefficiency in both computation and accuracy.

One can reformulate this problem using reinforcement learning framework, following the intuition that the action detection process is an iterative refinement process: after observing an occurring action an algorithm can skip ahead and back to determine time frames within which the action is happening. Refinement based on such approach is shown [25,26] to be more effective that traditional approaches.

The model [25] architecture consists of two networks. The observation network, using video frame and normalized temporal location of the observation as input data outputs a feature vector, encoding where and when given action was seen. The recurrent network by processing observation network output produces three outputs at each time step: a detection window vector (consisting of the start and end frames and model certainty about them), binary indicator signaling whether to emit detection window as a prediction, and temporal location, indicating the frame to observe next.

While the detection window is trained using standard backpropagation, prediction indicator and next observation location are non-differentiable and cannot be trained in such fashion. Instead, REINFORCE [12] algorithm is used, enabling learning of non-differentiable model parameters.

As the goal is to learn policies for named parameters that lead to action detection with both high recall and high precision, authors construct the reward function to maximize true positive and minimize false positive for the observation location and prediction indicator, while the total reward is provided on the final time step.

As an alternative [26] refines a temporary window through continuously adjusting the span of the current window and not predicting the bounds directly and provide dense reward on each step measuring an improvement of action localization accuracy by Intersection-over-Union (IoU) between the currently predicted temporal window and the ground truth.

3.4 Robot Navigation

Classically robot navigation problem is solved in a modular way, by dividing it into independent subproblems of mapping, localization, and planning. An agent has a model, often a metric map of an environment or progressively recover visual geometry of a scene, recognizing obstacles during interaction with the environment and then use sensor observations to localize itself within a map and use path planning algorithms to navigate to the desired destination. Having the map allows navigation in environments with complex dead-end obstacles, but maintaining a map from sensory inputs can be problematic, due to localization drift, scene dynamics, and sensor noise. On the other side a map is registered in a purely geometrical way, therefore an agent is not able to learn from visual patterns and generalize to new environments.

Deep neural networks demonstrated some success in navigational tasks, directly predicting actions from raw pixel observations of an agent [27,28], capable of performing localization [28], navigation by translating natural language command sequences to sequences of action [29], etc. An obvious drawback of this class of models is their demand for supervision, while collecting training data for navigational tasks is expensive, tedious and requires a lot of manual labor.

Reinforcement Learning for Robot Navigation. Learning to navigate can be formulated as reinforcement learning problem. So an agent will learn in self-supervised manner by trial-and-error, guided only by reward function.

An agent operates in an environment E, indoor or outdoor space, often discretized to a regular grid or a graph. The interaction is modeled by Markov decision process, consisting of a set of states S, a set of navigation actions A, a set of actions $A(s) \subseteq A$ for each state $s \in S$ that can be performed in that state. Generally, an agent does not have access to a state s directly, instead it receives observations $o_{t,n}(s)$ at time t from n sensors with some certainty, given by conditional probability $p_n(o \mid s)$. A reward function R incentivize an agent to reach the desired destination, while it can be additionally engineered to ensure certain agent behavior, like obstacle avoidance, curiosity for exploration, maximizing speed, etc. A policy is approximated with a deep neural network, using observations $o_{t,n}(s)$ as inputs.

A simplest class of navigation is obstacle avoidance, where agent's goal is to move within a scene while avoiding collisions with obstacles. In [30] authors propose a reinforcement learning approach to this problem as opposed to conventional mapping and planning algorithms. The architecture is comprised of two neural networks (Fig. 1a). At first visual observations are inputted to a fully convolutional neural network [31], predicting depth from RGB observations. Second, network, trained with DoubleDQN [32] algorithm solves obstacle avoidance task in simulated environments.

Actions are defined in discretized form to control linear and angular velocities. Reward function $r = v \cdot \cos(\omega) \cdot \delta t$, where v and ω are local linear and angular velocities respectively and δt is the time of each training loop, is engineered to incentivize the robot to move as fast as possible and to penalize if rotating on the spot. In case of a collision the agent is imposed an additional penalty of -10 and the episode is terminated. An episode is terminated with no additional penalty after 500 steps, hence robot is required to move within a scene as long as possible.

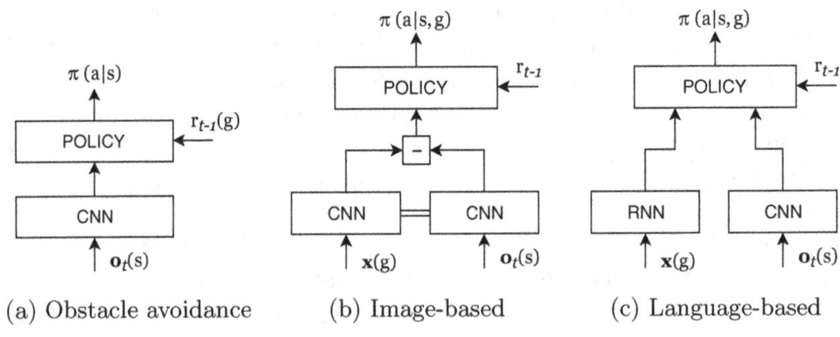

(a) Obstacle avoidance (b) Image-based (c) Language-based

Fig. 1. Examples of navigation architectures

The described setup allows an agent to successfully navigate within a scene, while avoiding obstacles, but not to reach a certain destination. To achieve that reward function may be engineered to reward an agent if it decreases the distance to a destination or to a set of landmarks [33] on each step. While a lot of reinforcement learning problems, including 2D and 3D mazes, are solved in a such a way, this setup is inflexible. In this case, the goal gets hardcoded in the weights θ of deep neural network approximating parametrized policy $\pi_\theta(a \mid s)$. Therefore goal change requires re-training of the model.

To avoid this one can additionally condition a policy on g: $\pi_\theta(a \mid s, g)$ where g is the goal or a target to navigate a robot, hence reformulating navigation problem being *target-driven.* A target can be set with explicit location coordinates, distances to landmarks [33], or given an image of destination [34], image of object to locate within a scene, or an instruction in natural language [29,35], thus learning a mapping from a 2D image or a sequence in natural language to a policy.

In [34] authors train an agent to navigate to new targets without re-training, specifying the navigation destination as inputs to the model, instead of hard-coding the target in model reward function. Both observations at each time step $\mathbf{o}_t(s)$ and the target image $\mathbf{x}(g)$ are given by the agent's RGB camera serve as inputs to convolutional siamese network [36]. It consists of two identical twin networks with shared parameters that transform inputs to the same embedding space and then fuse them using absolute difference. The obtained representation which can be viewed as a notion of similarity of an observation and a goal is then inputted to the scene-specific policy network (Fig. 1b). A reward is constructed giving a goal-reaching reward of 10 and small penalty (-0.01) on each step is imposed to encourage shorter navigation trajectories.

In language-based visual navigation agent given natural language instruction $\mathbf{x}(g) = \{x_1, \ldots, x_n\}$ is expected to navigate to a goal position g. The agent's action set A consists of linear and angular movement actions and a special action *stop* ending an episode. Language instruction $\mathbf{x}(g)$ and visual observations $\mathbf{o}_t(s)$ at each time step are encoded with recurrent and convolutional neural networks respectively and inputted to a policy network, approximating policy $\pi_\theta(a \mid s, g)$ (Fig. 1c). A reward is given in terms of the final navigation error, which is the distance to the target's position [35].

Challenges. Solving visual navigation problem with reinforcement learning are effective in environments of ever-increasing complexity [37] – simple mazes, 3D games [38], photorealistic simulated environments [24,39], city maps [33] and real environments [30]. Nevertheless, reinforcement learning for robot navigation faces a number of challenges.

Reinforcement learning has *low sample efficiency*, learning by trial-and-error with *high-dimensional* state and action spaces guided only with reward provided by the environment requires millions of episodes. Often the rewards are *sparse*, in extreme case being an indication whether an agent has succeeded or failed to reach the desired goal. Moreover training a real robot to navigate in the physical world pose additional challenges, being inefficient in terms of costs, time, risk of self-damage, property damage or even be a harm to humans. Usually, real environments are *partially observable*, thus an agent does not have a perfect knowledge of the state, receiving observations from sensors from which state can be only estimated. A robot must operate in the environments on *large horizons*, augmenting perception with reasoning.

Reinforcement learning benefits from recent success in deep learning. Nowadays, representation learning of high-dimensional visual inputs and policy approximation are largely addressed with deep neural networks. Generally, to deal with low sample efficiency, learning is moved to simulations and pre-trained models further are fine-tuned on real environments. To improve statistical efficiency various techniques to carefully re-use experience pool are developed, with hindsight experience replay [40] allowing an agent to learn from failures being a great example.

Reward engineering may shorten effective problem horizon by introducing immediate rewards. In many cases sparse binary rewards can be reformulated to more dense ones, as an example instead of a binary reward for reaching a destination one can reward an agent on each step if the agent decreases the distance to it. Another approach is to augment loss with *auxiliary tasks*, as predicting visual modality from raw RGB data, loop closures [37], intersections or line crossings [41], providing more dense reward signals.

An agent, able to reason how to tackle long-term dependencies should possess memory at different timescales. Augmenting policy networks with *external memory* shown increased agent' abilities to navigate, extrapolating knowledge to unseen environments.

We will describe simulators, knowledge transfer and memory-augmented architectures in more detail.

Simulators and Transfer Learning. Until recently visual data generated with simulators was not realistic-looking [42], so a virtual-to-real model transfer is highly desirable. The latest generation of simulators [24,39,41,43] aims to narrow the gap between real and virtual world and offers realistic-looking visual data in different modalities. Besides RGB data most simulators do provide depth, normals, object and semantic segmentation. In this environments, agents can be equipped with different numbers of cameras, able to simulate different setups, including monocular and stereo ones. Simulators can leverage existing open-source 3D synthetic [44] or real [45] datasets or using proprietary 3D scenes [24,41,43], be either multi-purpose or dedicated to specific scenarios like autonomous driving [41] or UAV navigation [43].

Transfer learning involves a transfer of knowledge of a model trained on a *source domain* $\mathcal{A}$ to a *transfer domain* $\mathcal{B}$. In case of reinforcement learning for visual robot navigation this includes transferring knowledge from a domain of simulated visual observations generated by a simulator to a domain of real-world visual sensor inputs. Another task is transferring knowledge between old and new domains of actions, i.e. leveraging prior knowledge to perform different navigational tasks and learning to navigate in unseen environments.

There are a number of transfer learning strategies exist. *Domain randomization* [46,47], while being conceptually simple, comes with insight that if a model was trained on a large number of simulated environments, and have seen enough variations in data, with different textures, lighting conditions, occlusions and scene dynamics, even it is not realistic, it would be able to generalize to perform well on real data [30,47]. For navigation tasks this involves randomization of actions allowing an agent to over-explore the state-action space, thus further being applied to the desired task the agent is capable to use knowledge from a larger experience pool.

In *domain adaptation*, a model pre-trained on synthetic data, either RGB input, depth or semantic segmentation, is then fine-tuned on real data, allowing to learn on a smaller amount of real data. This approach is a mainstream, being a successful strategy in a variety of tasks [48]. As an alternative [49,50] proposed to sidestep it by converting non-realistic virtual image input from a simulator into realistic one using image-to-image translation model, which takes synthetic image generated by a simulator and outputs realistic image, using semantic segmentation as interim representation, thus allowing to simplify or even abandon further fine-tuning.

External Memory. A self-navigating robot should possess intelligent behavior, equipped both with perception and reasoning. It should be capable of handling long-term spatio-temporal dependencies, which is not the case for feedforward architecture of a policy network which is inherently reactive. Recurrent neural networks architectures such that LSTM [51] improve agent capabilities to reason, but have a set of disadvantages. Memory is tied up in the activations of the latent state of the networks, hence to increase the amount of memory one needs to increase the size of the network and therefore its' computational cost. On the other side, memory tends to be fragile, due to residing in the space of the activations, that are being constantly updated [52].

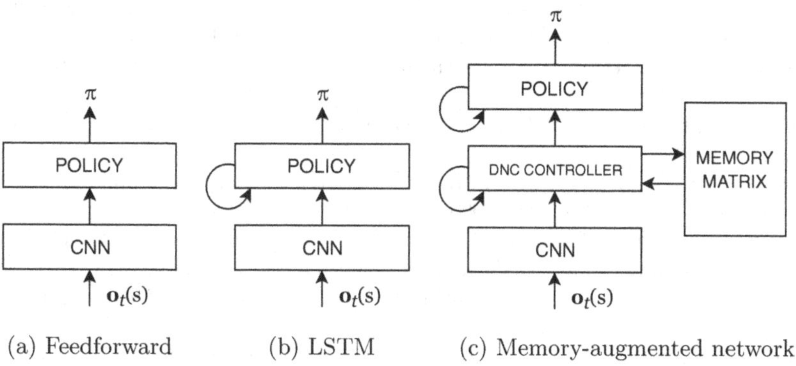

(a) Feedforward (b) LSTM (c) Memory-augmented network

Fig. 2. Policy networks

Memory-augmented neural networks [53,54] with Differentiable Neural Computer [52] being the most prominent one notably improve handling long-term dependencies. DNC architecture (Fig. 2c) allows features to be stored in the external memory matrix, by imposing learnable read and write operations. Information retrieval can be performed by content similarity or in a sequential manner. A policy network augmented with external memory have shown to solve navigation problems in environments with dead-end obstacles [55], which pose difficulties for simpler LSTM policy networks (Fig. 2b).

Other than general external memory, a large body of work is devoted to specific neural memory architectures for RL-based visual navigation, powering agents with abilities to explicitly memorize map [56,57] of the environment, and further localize [58] and plan [59], thus reiterating classical ideas within the reinforcement learning framework.

4 Conclusions

We briefly reviewed the RL technology. We provided a short overview of modern applications of reinforcement learning to computer vision and robot navigation tasks.

Acknowledgement. The work was supported by the Skoltech NGP Program No. 1-NGP-1567 "Simulation and Transfer Learning for Deep 3D Geometric Data Analysis" (a Skoltech-MIT joint project).

References

1. Bonin-Font, F., Ortiz, A., Oliver, G.: Visual navigation for mobile robots: a survey. J. Intell. Robot. Syst. **53**(3), 263–296 (2008)
2. Kuleshov, A., Bernstein, A., Burnaev, E.: Mobile robot localization via machine learning. In: Perner, P. (ed.) MLDM 2017. LNCS (LNAI), vol. 10358, pp. 276–290. Springer, Cham (2017). https://doi.org/10.1007/978-3-319-62416-7_20
3. Kuleshov, A., Bernstein, A., Burnaev, E., Yanovich, Yu.: Machine learning in appearance-based robot self-localization. In: 16th IEEE International Conference on Machine Learning and Applications (ICMLA). IEEE Conference Publications (2017)
4. Sutton, R., Barto, A.: Reinforcement Learning: An Introduction. MIT Press, Cambridge (1998)
5. Kaelbling, L.P., Littman, M.L., Moore, A.P.: Reinforcement learning: a survey. J. Artif. Intell. Res. **4**, 237–285 (1996)
6. Burnaev, E., Panov, M.: Adaptive design of experiments based on Gaussian processes. In: Gammerman, A., Vovk, V., Papadopoulos, H. (eds.) SLDS 2015. LNCS (LNAI), vol. 9047, pp. 116–125. Springer, Cham (2015). https://doi.org/10.1007/978-3-319-17091-6_7
7. Burnaev, E., Panin, I., Sudret, B.: Efficient design of experiments for sensitivity analysis based on polynomial chaos expansions. Ann. Math. Artif. Intell. **81**, 187–207 (2017)
8. Puterman, M.L.: Markovian Decision Processes - Discrete Stochastic Dynamic Programming. Wiley, New York (1994)
9. Bellman, R.: Dynamic Programming. Princeton University Press, Princeton (1957)
10. Watkins, C.J.C.H., Dayan, P.: Q-learning. Mach. Learn. **8**(3), 279–292 (1992)
11. Jaakola, T., Jordan, M., Singh, S.: On the convergence of stochastic iterative dynamic programming algorithms. Neural Comput. **6**(6), 1185–1201 (1994)
12. Williams, R.J.: Simple statistical gradient-following algorithms for connectionist reinforcement learning. Mach. Learn. **8**(3–4), 229–256 (1992)
13. LeCun, Y., Bengio, Y., Hinton, G.: Deep learning. Nature **521**(7553), 436–444 (2015)

14. Burnaev, E.V., Erofeev, P.D.: The influence of parameter initialization on the training time and accuracy of a nonlinear regression model. J. Commun. Technol. Electron. **61**(6), 646–660 (2016)
15. Burnaev, E.V., Prikhod'ko, P.V.: On a method for constructing ensembles of regression models. Autom. Remote Control **74**(10), 1630–1644 (2013)
16. Li, Y.: Deep reinforcement learning: an overview, pp. 1–70 (2017). [cs.LG]
17. Uijlings, J.R., Van De Sande, K.E., et al.: Selective search for object recognition. Int. J. Comput. Vis. **104**(2), 154–171 (2013)
18. Viola, P., Jones, M.: Rapid object detection using a boosted cascade of simple features. In: Proceedings of the 2001 IEEE Computer Society Conference on Computer Vision and Pattern Recognition, CVPR 2001, vol. 1, p. I. IEEE (2001)
19. Ren, S., He, K., Girshick, R., Sun, J.: Faster R-CNN: towards real-time object detection with region proposal networks. In: Advances in Neural Information Processing Systems, pp. 91–99 (2015)
20. Caicedo, J.C., Lazebnik, S.: Active object localization with deep reinforcement learning. In: 2015 IEEE International Conference on Computer Vision (ICCV), pp. 2488–2496. IEEE (2015)
21. Zhang, D., Maei, H., et al.: Deep reinforcement learning for visual object tracking in videos. Preprint arXiv (2017)
22. Luo, W., Sun, P., Mu, Y., Liu, W.: End-to-end active object tracking via reinforcement learning. Preprint (2017)
23. Mnih, V., Badia, A.P., et al.: Asynchronous methods for deep reinforcement learning. In: International Conference on Machine Learning, pp. 1928–1937 (2016)
24. Qiu, W., Zhong, F., et al.: UnrealCV: virtual worlds for computer vision. In: ACM Multimedia Open Source Software Competition (2017)
25. Huang, J., Li, N., et al.: A Self-Adaptive Proposal Model for Temporal Action Detection based on Reinforcement Learning (2017)
26. Yeung, S., Russakovsky, O., et al.: End-to-end learning of action detection from frame glimpses in videos, vol. 10, no. 1109, pp. 2678–2687 (2016)
27. Giusti, A., Guzzi, J., et al.: A machine learning approach to visual perception of forest trails for mobile robots. IEEE Robot. Autom. Lett. **1**, 661–667 (2016)
28. Maqueda, A.I., Loquercio, A., et al.: Event-based Vision meets Deep Learning on Steering Prediction for Self-driving Cars, April 2018. ArXiv e-prints
29. Anderson, P., Wu, Q., et al.: Vision-and-language navigation: Interpreting visually-grounded navigation instructions in real environments. CoRR abs/1711.07280 (2017)
30. Xie, L., Wang, S., et al.: Towards monocular vision based obstacle avoidance through deep reinforcement learning. CoRR abs/1706.09829 (2017)
31. Shelhamer, E., Long, J., Darrell, T.: Fully convolutional networks for semantic segmentation. In: 2015 IEEE Conference on Computer Vision and Pattern Recognition (CVPR), pp. 3431–3440 (2015)
32. van Hasselt, H., Guez, A., Silver, D.: Deep reinforcement learning with double Q-learning. In: AAAI (2016)
33. Mirowski, P., Grimes, M.K., et al.: Learning to navigate in cities without a map. arXiv preprint arXiv:1804.00168 (2018)
34. Zhu, Y., Mottaghi, R., et al.: Target-driven visual navigation in indoor scenes using deep reinforcement learning. In: 2017 IEEE International Conference on Robotics and Automation (ICRA), pp. 3357–3364 (2017)
35. Wang, X., Xiong, W., et al.: Look Before You Leap: Bridging Model-Free and Model-Based Reinforcement Learning for Planned-Ahead Vision-and-Language Navigation. ArXiv e-prints, March 2018

36. Koch, G., Zemel, R., Salakhutdinov, R.: Siamese neural networks for one-shot image recognition. In: ICML Deep Learning Workshop, vol. 2 (2015)
37. Mirowski, P.W., Pascanu, R., et al.: Learning to navigate in complex environments. CoRR abs/1611.03673 (2016)
38. Kempka, M., Wydmuch, M., et al.: ViZDoom: A Doom-based AI Research Platform for Visual Reinforcement Learning. ArXiv e-prints, May 2016
39. Savva, M., Chang, A.X., et al.: Minos: Multimodal indoor simulator for navigation in complex environments. arXiv preprint arXiv:1712.03931 (2017)
40. Andrychowicz, M., Wolski, F., et al.: Hindsight experience replay. In: Advances in Neural Information Processing Systems, pp. 5048–5058 (2017)
41. Dosovitskiy, A., Ros, G., et al.: Carla: An open urban driving simulator. arXiv preprint arXiv:1711.03938 (2017)
42. Koenig, N.P., Howard, A.: Design and use paradigms for gazebo, an open-source multi-robot simulator. In: 2004 IEEE/RSJ International Conference on Intelligent Robots and Systems (IROS), vol. 3, pp. 2149–2154 (2004)
43. Shah, S., Dey, D., Lovett, C., Kapoor, A.: AirSim: high-fidelity visual and physical simulation for autonomous vehicles. In: Hutter, M., Siegwart, R. (eds.) Field and Service Robotics. SPAR, vol. 5, pp. 621–635. Springer, Cham (2018). https://doi.org/10.1007/978-3-319-67361-5_40
44. Song, S., Yu, F., et al.: Semantic scene completion from a single depth image. In: IEEE Conference on Computer Vision and Pattern Recognition (2017)
45. Chang, A., Dai, A., et al.: Matterport3d: Learning from RGB-D data in indoor environments. arXiv preprint arXiv:1709.06158 (2017)
46. Tobin, J., Fong, R., et al.: Domain randomization for transferring deep neural networks from simulation to the real world. In: 2017 IEEE/RSJ International Conference on Intelligent Robots and Systems (IROS), pp. 23–30 (2017)
47. Sadeghi, F., Levine, S.: CAD2RL: Real single-image flight without a single real image. CoRR abs/1611.04201 (2017)
48. Yosinski, J., Clune, J., et al.: How transferable are features in deep neural networks? In: NIPS (2014)
49. You, Y., Pan, X., Wang, Z., Lu, C.: Virtual to real reinforcement learning for autonomous driving. CoRR abs/1704.03952 (2017)
50. Zhang, J., Tai, L., et al.: Vr goggles for robots: Real-to-sim domain adaptation for visual control. CoRR abs/1802.00265 (2018)
51. Hochreiter, S., Schmidhuber, J.: Long short-term memory. Neural Comput. **9**(8), 1735–1780 (1997)
52. Graves, A., Wayne, G., et al.: Hybrid computing using a neural network with dynamic external memory. Nature **538**(7626), 471–476 (2016)
53. Sukhbaatar, S., Szlam, A., et al.: End-to-end memory networks. In: NIPS (2015)
54. Graves, A., Wayne, G., Danihelka, I.: Neural turing machines. CoRR abs/1410.5401 (2014)
55. Khan, A., Zhang, C., et al.: Memory augmented control networks. In: International Conference on Learning Representations (2018)
56. Parisotto, E., Salakhutdinov, R.: Neural map: Structured memory for deep reinforcement learning. CoRR abs/1702.08360 (2017)
57. Savinov, N., Dosovitskiy, A., Koltun, V.: Semi-parametric topological memory for navigation. arXiv preprint arXiv:1803.00653 (2018)
58. Chaplot, D.S., Parisotto, E., Salakhutdinov, R.: Active neural localization. CoRR abs/1801.08214 (2018)
59. Karkus, P., Hsu, D.F.C., Lee, W.S.: QMDP-net: deep learning for planning under partial observability. In: NIPS (2017)

A Fast Two-Level Approximate Euclidean Minimum Spanning Tree Algorithm for High-Dimensional Data

Xia Li Wang[1(✉)], Xiaochun Wang[2], and Xiaqiong Li[2]

[1] School of Information Engineering, Changan University, Xi'an 710061, China
xlwang@chd.edu.cn

[2] School of Software Engineering,
Xi'an Jiaotong University, Xi'an 710049, China
xiaocchunwang@mail.xjtu.edu.cn,
xiaqiongli@stu.xjtu.edu.cn

Abstract. Euclidean minimum spanning tree algorithms run typically with quadratic computational complexity, which is not practical for large scale high dimensional datasets. In this paper, we propose a new two-level approximate Euclidean minimum spanning tree algorithm for high dimensional data. In the first level, we perform outlier detection for a given data set to identify a small amount of boundary points and run standard Prim's algorithm on the reduced dataset. In the second level, we conduct a k-nearest neighbors search to complete an approximate Euclidean Minimum Spanning Tree construction process. Experimental results on sample data sets demonstrate the efficiency of the proposed method while keeping high approximate precision.

Keywords: Euclidean minimum spanning tree · Minimum spanning tree
Approximate minimum spanning tree · Nearest neighbor search

1 Introduction

Finding a minimum spanning tree (MST) for a given connected graph is a fundamental problem with diverse application domains and many efficient MST algorithms have been developed. In today's MST tasks, usually, a set of N d-dimensional data points is given and the problem is commonly solved in the Euclidean setting, giving rise to the so-called Euclidean minimum spanning tree (EMST) problem. In this case, there are $V = N$ vertices and $E = N(N-1)/2$ edges in the complete graph, and standard EMST algorithms, such as Kruskal's [1] and Prim's [2], have a time complexity roughly equal to $O(dN^2)$. For large-scale high-dimensional datasets, standard EMST algorithms will lose its time performance. Fortunately, in many practical applications, an exact EMST can be generally replaced by an approximate one without degrading the quality of the final application.

Being a compact data representation of a given data set, EMST has been extensively used in image segmentation [3, 4], cluster analysis [5–7], classification [8], and manifold learning [9]. In particular, we are interested in EMST construction for cases

P. Perner (Ed.): MLDM 2018, LNAI 10935, pp. 273–287, 2018.
https://doi.org/10.1007/978-3-319-96133-0_21

where data consist of well separated clusters and the number of clusters is significantly smaller than the number of data points [3, 10, 11]. In these situations, in-cluster edges have weights that are significantly smaller than those longest edges corresponding to cluster breakers and can be found quickly by fast k-nearest neighbors search algorithms. However, the location of the longest edges in such EMSTs can not be determined quickly and is often the bottle neck for fast EMST algorithms.

In this paper, we propose a new clustering-inspired EMST algorithm that is both computationally efficient and competent with the state-of-the-art EMST algorithms. Basically, our novel data-dependent clustering-inspired EMST algorithm tries to identify the relatively small number of longest edges in an EMST by finding and retaining the relatively small number of boundary points and outliers before the complete EMST is constructed. Being composed of two-level structures, the working principle behind our novel data-dependent EMST algorithm is illustrated in Fig. 1.

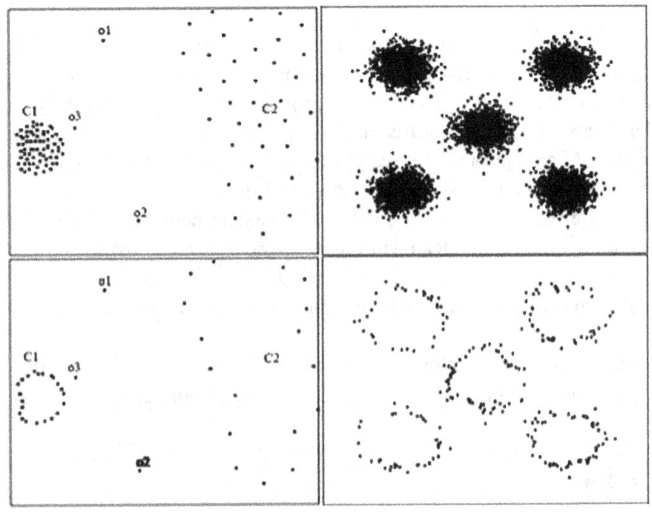

Fig. 1. An illustration of the simple idea.

During the first level, a relatively small group of the boundary points (including outliers) are identified using angle based outlier factors (ABOF) and retained, upon which the Prim's EMST algorithm is applied to locate the small number of longest edges. During the second level, we perform an index-based fast search algorithm to find k-nearest neighbors (kNN) for each data point, based on which cluster-wise EMSTs are formed. To be as general as possible, our proposed algorithm has no specific requirements on the dimensionality of the data sets and the format of the distance measure, although Euclidean distance is used as the edge weight in our experiments. Experiments conducted on sample datasets show that our proposed algorithm can obviously improve the run time performance of EMST while keeping high precision.

The rest of this paper is organized as follows. Section 2 gives a review of some related work. Section 3 presents the proposed data-dependent EMST method. Section 4 shows the experimental comparisons. In Sect. 5, we give the conclusion and future work.

2 Related Work

Work related to the method presented in this paper falls into two main categories: EMST algorithms and density-based outlier detection methods.

2.1 MST Algorithms

For a given connected and weighted graph G = (E, V), Bor°uvka's algorithm begins with each vertex of a graph being a tree, and for each consecutive iteration, it selects the shortest edge from a tree to another tree and combines them. This process continues until all the trees are combined into one tree [12]. Proposed independently by Jarn´ık [13], Prim [2] and Dijkstra [14] in 1930, 1957 and 1959, respectively, the famous Prim's algorithm first arbitrarily selects a vertex as a tree, and then repeatedly adds the shortest edge that connects a new vertex to the tree, until all the vertices are included. Proposed in 1956, Kruskal's algorithm starts with sorting all the edges by their weights in a non-decreasing order, treats each vertex as a tree, and iteratively combines the trees by adding edges in the sorted order excluding those leading to a cycle until all the trees are combined into one tree [1]. The time complexity of these classic MST algorithms is O(ElogV).

To construct an MST in the Euclidean setting, standard Prim's algorithm requires a quadratic running time. To be more efficient, in 1978, Bentley and Friedman [15] proposed to use a kd-tree in Prim's algorithm to enhance the search for the next edge to add to the tree, which can reach an O(NlogN) running time for most data distributions. In 1985, Preparata and Shamos [16] gave a lower bound for the EMST problem of (NlogN), which has been the tightest known lower bound. In 1993, Callahan and Kosaraju's proposed Well-Separated Pair Decomposition (WSPD) [17] which forms the basis of most recent EMST algorithms. The WSPD partitions data points into a set of pairs of tree nodes such that the nodes in any pair are farther apart than the diameter of either node. It can be shown that the WSPD has O(N) pairs of nodes, and that the MST is a subset of the edges formed between the closest pair of points in each pair of nodes. In 2000, Narasimhan and Zachariasen applied WSPD to compute neighbors of components for Boruvka's algorithm to find edges of the MST [18]. However, the constant in the O(N) size of the WSPD grows exponentially with the data dimension and is often very large in practice. In 2010, March et al. presented a new dual-tree algorithm for efficiently computing the EMST [19], which is superficially similar to the method in [18] except that the WSPD is replaced by the new dual-tree data structure and referred to in the following as FEMST algorithm. They used adaptive algorithm analysis to prove the tightest (and possibly optimal) runtime bound for the EMST problem to-date. Experiments conducted demonstrated the scalability of their method on astronomical data sets.

In addition to exact EMST problem, approximate EMST algorithms have been also proposed. In 1988, Vaidya [20] employed a group of grids to partition a data set into cubical boxes of identical size. For each box, a representative point was determined. Within a cubical box, points were connected to the representative. Any two representatives of two cubical boxes were connected if corresponding edge length was between two specific thresholds. In 1993, Callahan and Kosaraju [17] proposed to utilize WSPD of a data set to extract a sparse graph from the complete graph and then apply an exact MST algorithm to it. More recently, in 2009, Wang et al. [21] employed a divide-and-conquer scheme to construct an approximate EMST, which is superficially similar to WSPD method. Their goal was to detect longest edges in an EMST at an early stage for clustering. At the same year, Lai et al. proposed a two-stage Hilbert curve based approximate EMST algorithm for clustering [22]. In 2014, Wang et al. proposed a fast EMST algorithm [23], which is superficially similar to the method in [15] except that the kd-tree is replaced by the iDistance indexing structure for fast kNN search in high-dimensional datasets. The authors argue that the algorithm has an expected O(NlogN) running time, but do not prove this rigorously. In 2015, Zhong et, al. proposed a fast two-stage Euclidean minimum spanning tree (FEMST) algorithm which employs a divide-and-conquer scheme to produce an approximate EMST with theoretical time complexity of $O(N^{1.5})$ [24]. In the first stage, K-means is employed to partition a dataset into $N^{1/2}$ clusters. Then an exact EMST algorithm is applied to each cluster and the produced $N^{1/2}$ EMSTs are connected to form an approximate EMST. In the second stage, the clusters produced in the first stage form $N^{1/2} - 1$ neighboring pairs, and the dataset is repartitioned so that the neighboring boundaries of a neighboring pair are put into a cluster. With these $N^{1/2} - 1$ clusters, another approximate EMST is constructed. Finally, the two approximate EMSTs are combined into a graph and a more accurate EMST is generated from it.

2.2 Outlier and Boundary Point Detection

Being an important branch of data clustering techniques, density-based clustering, such as DBSCAN [25], produces clusters that consist of a set of data objects spread in the data space over a contiguous region of high density of objects, and that are separated from each other by contiguous regions of low density of objects. Data objects located in low-density regions are typically considered as noise (i.e., outliers) and border points. For our purpose, we need to detect outliers and border points which are objects in the sparse areas and are required to separate clusters.

The existence of outliers can make data modeling more difficult and imprecise. Many efforts have been devoted to detecting them. It has been generally agreed that distance based outlier definitions can detect more globally-orientated outliers and regard being an outlier as a binary property while density based outlier definitions can detect more locally distributed outliers. A classic example of global vs. local outliers is given in the left plot of Fig. 1, where o1 and o2 are global outliers while o3 is a local one. To deal with this situation, Breunig et al. pioneered the density-based outlier detection research by assigning to each object a degree of being an outlier, called the local outlier factor (LOF) [26]. Rather than manipulating distances of a data object to its k-nearest neighbors, density-based approaches consider the outlying status of an object by using the ratio between the local density around the object and the local density around its neighboring objects. For an object, LOF captures the degree to which it is an outlier by looking at the densities of its neighbors. The higher the difference between the density in the local neighborhood of an object and those around its k-nearest neighbors, the higher its LOF value, and the more distinctly the object is considered to be an outlier. It has been proved that a LOF value of approximately 1 indicates that the corresponding object is located within a region of homogeneous density (i.e., the object is deep inside a cluster). Following the notion of local outlier factor, several extensions and refinements to the basic LOF model have been proposed. To name a few, a connectivity-based outlier factor (COF) [27], local outlier integral (LOCI) [28], SLOM [29], INFLO [30], a resolution-based outlier factor (ROF) [31], a robust local density estimation based outlier factor [32], local distance-based outlier factor (LDOF) [33], and RBDA [34] were proposed in 2002, 2003, 2004, 2006, 2006, 2007, 2009, and 2011, respectively. Most of the outlier factors presented so far are based on *k*-nearest neighbors. However, in high-dimensional data space, these methods are bound to deteriorate due to the notorious "curse of dimensionality" where concepts like proximity, distance, or nearest neighbour become less meaningful with increasing dimensionality. The issue exists whether the notion of these outliers is still meaningful for high-dimensional data.

In current research, we are particularly interested in an angle-based outlier detection (ABOD) method designed especially for high dimensional data. It was proposed by Kriegel [35] and plays an important role in identifying outliers in high-dimensional spaces. The angle-based outlier factor using the variance of angles is formulated as follows:

Definition 1. Given a point set S of size N and a point p in S. For a random pair of different points a and b, let Θapb denote the angle between the difference vectors a-p and b-p. The angle-based outlier factor ABOF(p) is the variance of Θapb:

$$ABOF(p) = Var[\Theta_{apb}] = Var\left[\frac{\langle \overrightarrow{pa}, \overrightarrow{pb} \rangle}{||\overrightarrow{pa}|| \bullet ||\overrightarrow{pb}||}\right] \quad (1)$$

It is obvious that the ABOF measure is entirely free of parameters and therefore is suitable for unsupervised outlier detection methods. The naive ABOD algorithm computes the ABOF for each point of the data set and returns the top points having the smallest ABOF's as outliers. Although outliers and boundary points are different by definition, they are generally located around the margin of the data set with high density, such as in a cluster. Thus, this paper focuses on simultaneously identifying outliers and boundary points of clusters based on local geometrical measures. An illustration of three different kinds of points (i.e., inner points, border points and outliers) and the variances of angles for different kinds of points are illustrated in Fig. 2. However, the time complexity of the naive ABOD algorithm is in $O(dN^3)$ and it will be very difficult to mine outliers in very large data sets. To improve, a novel random projection-based technique proposed in 2012 is able to estimate the angle-based outlier factors for all data points in time near linear in the size of the data [36].

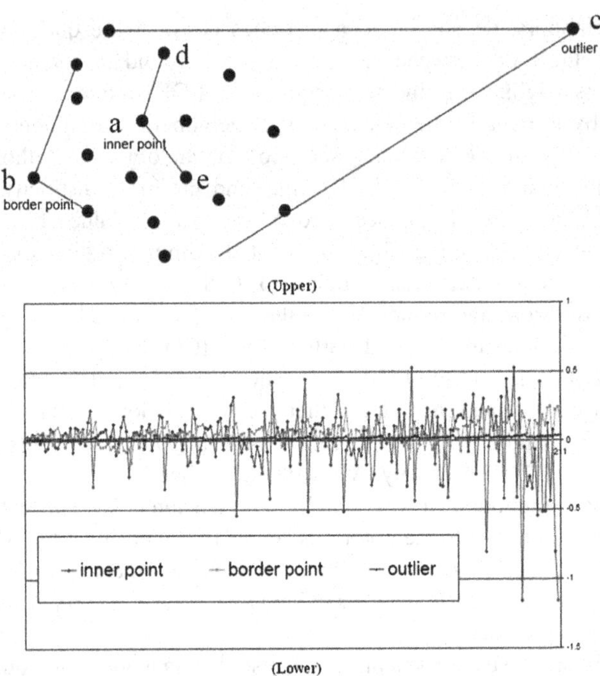

Fig. 2. (Upper) An illustration of three different kinds of points. (Lower) The variances of angles for different kinds of points.

In this paper, we utilize the ABOD method for simultaneously identifying outliers and boundary points of clusters. The proposed method is based on the idea that boundary points and outliers are characterized by a smaller angle variance.

3 The Proposed Fast Approximate EMST Algorithm

3.1 A Simple Idea

EMST-based applications often require the construction of an EMST in the first place. To speed up, in this section, we describe a new hybrid scheme that integrates kNN searching, the detection of a small number of longest edges, and EMST construction into one process, to enhance EMST computation for high-dimensional datasets. Basically, the design of this more efficient scheme is motivated by the following two observations. Firstly, for some application problems, such as clustering, an EMST consists of a major amount of in-cluster edges which have weights that are significantly smaller than those edges corresponding to cluster breakers and can be found quickly by fast kNN search algorithms. This is because, according to Boruvka's algorithm, in the Euclidean setting, at least half of data points' edge to its nearest neighbor must be in an EMST. Secondly, for clustering-based applications, EMST algorithms can be more efficient if a small number of boundary points (including outliers) can be correctly identified and quickly retained for the efficient localization of a few longest edges which correspond to the cluster breakers.

Data points in a given dataset for density-based clustering purpose can be categorized into three different types, that is, core points (i.e., inner points), boundary (or border) points and outliers. Most of the times, the number of boundary points and outliers is significantly less than the number of inner points. If these small amount of boundary points (including outliers) can be separated from inner points quickly, it will be more efficient on the whole to perform standard Prim's algorithm on this much size-reduced data set to fully identify the small number of longest edges. Then the rest edges in the EMST can be found quickly through k-nearest neighbors search. The assumption underlying the proposed method is that boundary points and outliers are located around the margin of densely distributed data such as a cluster and are generally characterized by smaller angle variances.

Different from distance-based and density-based outlier definitions which categorize a data point to be an outlier by capturing the degree a data point is away from its nearest neighbors in terms of their distances and local densities, respectively. However, comparing distances become less meaningful with increasing high-dimensional space, causing the performance of methods to deteriorate. To get away from a k-nearest neighbor setting, it has been observed that inliers (i.e., the object is deep inside a cluster) are well surrounded by other data objects in all directions and have a large variance of angles to other data objects while outliers are surrounded by other data objects only in a few certain directions and, correspondingly, have a small variance of angles to other data objects. Therefore, for high dimensional data, it has been proposed to use the measure of the angle variances to solve the outlier detection problem to some extent. Larger ABOF (angle based outlier factor) values indicate that the corresponding objects are located within a cluster while small ABOF values indicate that the objects are close to the cluster boundary or more distinctly considered to be an outlier.

Our clustering-inspired EMST algorithm augments Prim's algorithm with angle variance-based boundary points' detection and retaining. It consists of two levels. The first level starts by computing a ABOF score for each data object and then identifying a relatively small number of ABOF-based boundary points (including outliers), which are subsequently retained for the location of a few longest edges corresponding to cluster separations. For our purpose, we would be more interested in those data points whose ABOF scores are smaller than a threshold. The results of the first level are a small number of identified boundary data points (i.e., the data points whose ABOF score is smaller than the threshold), upon which an EMST segment is constructed. Then at the second level, a fast k-nearest neighbors based EMST algorithm is applied to the size-reduced data points (i.e., the inner points) to locate the large number of intra-cluster edges. If data clusters are connected, that is, all the clusters overlap, the first level is not necessary.

3.2 Determination of the Threshold Value

Let the ABOF scores of all points in dataset be computed, based on which the average of the scores, mean, and the corresponding standard deviation, std, are calculated. The threshold for finding potential boundary points and outliers is defined as,

$$threshold = mean - f \times std \tag{2}$$

where f can be 0, 1, 1.5, 2, 2.5, 3 and more.

3.3 Construction of a Simple iDistance

The problem with the proposed approach is its high computation cost. To significantly reduce the running time, a simplest case of iDistance is used in our algorithm to find k-nearest neighbors for the construction of cluster-wise EMST. Proposed first in 2001 [37] and more comprehensively presented in 2005 [38], iDistance is an efficient B+ -tree based indexing method, and designed to process k-nearest neighbors queries more efficiently in high-dimensional spaces. It has been used in many applications and proven to be especially good for skewed data distributions. To build an iDistance index, there are two steps: (1) a number of reference points in the data space should be first chosen; (2) the distance between a data point and its closest reference point is then calculated, which plus a scaling value is called the point's iDistance. By this means, points in a multi-dimensional space are mapped to one-dimensional values, and their nearest neighbors can be obtained earlier with a very high probability than a sequential search.

According to iDistance, a number of reference points have to be decided and the corresponding search structure has to be constructed. With no a priori assumptions on the distribution model underlying the data, for simplicity, we consider only one reference point here. This reference point is not freely chosen. To find it, we first take a random data point, and calculate its distance with the rest of the dataset. The data point with the largest distance is selected as the reference point. By this strategy, a larger (if not the largest) distance span of the data distribution can be obtained.

To construct the search facility, the distances between the reference point and the rest of dataset are calculated as the index. Next, the data points are sorted according to their distances to the reference point in a non-decreasing order and their original sequential position in the dataset is remembered. To be used as a search structure, given a data point, the search starts from its position in the sorted distances and proceeds bi-directionally. Only those data points which are within a searching radius are considered. Data points outside of the radius can be excluded in the calculation. The radius can be updated if a smaller one is encountered during the searching process. Thus, the use of the search structure is expected to reduce the O(N) time full search by a faster O (n < N) time partial search. It is easy to see the search structure can be constructed by any one dimensional sorting algorithm.

To run the search algorithm, an initial search radius for each data point has to be determined, and the smaller the better. Given the root node and its k-nearest neighbors, the search radius for each of its k-nearest neighbors can be computed from these k + 1 data points, and can provide an upper bound for the distances of each data item to its k-nearest neighbors. This initialization strategy can be applied to each node subsequently encountered.

To remember k-nearest neighbors of data points currently used for the calculation of clusterwise EMST, in the implementation, for each data point, a priority queue of size k + 1 is used to record the distances of the data point to its k-nearest neighbors, and an index array of size k, is used to record the positions of k nearest neighbors in the sequential order in which they are read in.

3.4 Our Clustering-Inspired MST Algorithm

Standard EMST-based clustering algorithms usually take a quadratic running time to ensure the properties of MSTs to be satisfied. From our point of view, this is neither efficient nor necessary. This observation is based on the following facts.

First of all, a vertex can usually have a limited number of nearest neighbors in an EMST. At any point of EMST construction, data points can be classified into two different kinds of tree nodes, those already in the tree and those not in it yet. The nodes in the tree have a limited number of edges to the nodes not in it. For example, at the beginning, the number of data points added to the tree is relatively small (compared to the whole database) and the next edge to add to the tree must be the nearest neighbor of one of the data points in the tree from the data points not in it. With the help of some index structure, the nearest neighbor of each node in the tree from the nodes not in it can be calculated efficiently.

Secondly, after certain clusters significantly far away from other data points are identified, that is, EMST segments, the data points in them can retire from the rest of calculation and memory used to store their kNNs can be released. If the index structure can be reconstructed with the reduced (i.e., the remaining) data set quickly, the number of data points involved in the nearest neighbor search may be reduced to save some distance computations.

Based on these findings, our clustering-inspired approximate EMST algorithm can be summarized in the following,

(1) Calculate the ABOF scores for all the data points in time near linear in the size of the dataset using the method presented in [36] and compute the threshold according to Eq. (2). Next, a Boolean array is used to mark those data points done if their scores are smaller than the threshold. To locate the few longest edges corresponding to cluster separations, a small number of identified boundary data points (i.e., the data points whose ABOF score is smaller than the threshold) are retained, and then the standard Prim's EMST algorithm is applied to the size-reduced data points (i.e., the boundary points and outliers).
(2) To establish the search structure, a random data point not marked done yet (using the Boolean array) is taken, and its distances to the rest data points which are not marked done either (using the Boolean array) are calculated. The data point with the largest distance is selected as the reference point and the tree root node, and then marked done (using the Boolean array). The distances between the reference point and the rest of dataset not marked done yet (in the Boolean array) are calculated as the index. Next, the data points are sorted according to their distances to the reference point in a non-decreasing order and their original sequential positions in the dataset are remembered in the sorted order.
(3) Initialize the search radii of kNNs of the tree nodes, find all the needed kNNs and mark them done (using the Boolean array).

(4) Add the edge corresponding to the nearest neighbor of the tree node and the data point at the other end of the edge to the tree, if some data points are not marked done yet, search through the distance array for the edge with the smallest distance to the tree, remember the vertex at the other end of this edge, select this vertex as the tree node candidate, and goto (3), otherwise, if all the nearest neighbors of the data points currently in tree are added to the tree, we can declare the end of a detected cluster and the start of another cluster, and if some data points are not marked done in the Boolean array, goto (2), otherwise, goto (5).

(5) Finally combine the EMST segments using Prim's algorithm and return an approximate EMST.

We use two arrays to record the final approximate EMST, namely, the distance array and the parent array, to remember the edge weight (i.e., distance) and the corresponding parent vertex index (i.e., the data point at the other end of the edge in an EMST) for each data point, respectively. Physically, the resource consumed by our algorithm includes the space for the whole data set to reside in memory, some temporary space to store the k-nearest neighbors of some data items' for calculating the short edges in the EMST, the space for distance and parent arrays, and the space for the indexing structure. Unlike traditional iDistance, the selection of reference points in our approach is dynamically adapted to the data being analyzed. This data dependency hopes to produce improved performance over traditional iDistance transforms.

3.5 Time Complexity Analysis

From the description in the previous subsections, our final clustering-inspired approximate EMST algorithm includes a running time near linear in the size of the data to estimate the angle-based outlier factors for all data points, an $O(M^2)$ $(M \ll N)$ Prim's algorithm where M is the number of data points whose ABOFs are smaller than the threshold at the end of the first stage, and an O(NlogN) for calculating short edges in cluster-wise EMST in the second stage. We expect our proposed efficient approximate EMST algorithm to run with a $O(N + LQ\log Q + M^2)$ time complexity, where L denotes the number of clusters and Q denotes the number of data points in the largest cluster in the second stage, and an O(QlogQ) is the time complexity of the index-based search for calculating intra-cluster EMST edges for each data point in the largest cluster since an approximate Q number of data points is involved in the dynamically updated search structure. The worst case time complexity of our proposed final approximate EMST algorithm could degenerate to $O(N^2)$, particularly when a data set contains only a few large clusters.

4 A Performance Study

In this section, two sets of experiments are conducted to evaluate the proposed approximate EMST algorithm. In the first set of experiments, we evaluate the performance of our proposed EMST algorithm on three 2-dimensional synthetic data sets to investigate the influences of the ABOF scores on the detection of boundary points and outliers. The performance is also compared with those of two other outlier definitions, INFLO and RBDA, to show the possible impact of different outlier definitions on the detection of boundary points. With some sensible knowledge of the impact of ABOF scores on the detection of boundary and outlying points, in the second set of experiments, we evaluate our EMST algorithm on two real high-dimensional data sets and compare the overall performance with the state-of-the-art approximate EMST algorithms to check the technical robustness of this study. We implemented all the algorithms in C++. All the experiments were performed on a computer with Intel Core 2 Duo Processor E6550 2.33 GHz CPU and 2 GB RAM. The operating system running on this computer is Windows XP. We use the timer utilities defined in the C standard library to report the CPU time. All the data sets used in this set of experiments are briefly summarized in Table 1.

Table 1. The sets of data

Data set	Data size (N)	Dimensionality	# of clusters
Data1	6,142	2	6
Data2	7,634	2	9
Data3	6,566	2	8
ISOLET	7,797	617	/
MNIST	10 000	784	/

4.1 Performance on Small 2-Dimension Data

To find an EMST of a data set using our algorithm, we need to specify the threshold value of ABOF scores for the detection of boundary points (including outliers). Therefore, we first focus our study on the behavior of the ABOF scores on the performance of our algorithm. The performance of the proposed method is evaluated on three CHAMELEON datasets obtained from [39] and consisting of clusters of different size, shape and orientation. The geometric shapes of three datasets are displayed in the plots on the first row of Fig. 3 together with the corresponding EMSTs.

Shown in rest three rows of Fig. 3 are plots of values of ABOF, INFLO and RBDA for each point when the number of nearest neighbors, k, is set to be 30, respectively. The red dots denote those data points whose outlier score is smaller (for our case) and larger (for the other two cases) than the corresponding thresholds. The blue ones denote the relatively normal data points in terms of the corresponding thresholds.

It can be seen from the figures that overall ABOF values can identify boundary points (including outliers) more accurately than two other methods. To summarize, using ABOF can very easily detect the boundary points (including outliers). For the datasets in this experiment, we observe that an estimated lower bound number of data points to retain to identify the longest edges amounts to a fifth of the original data size.

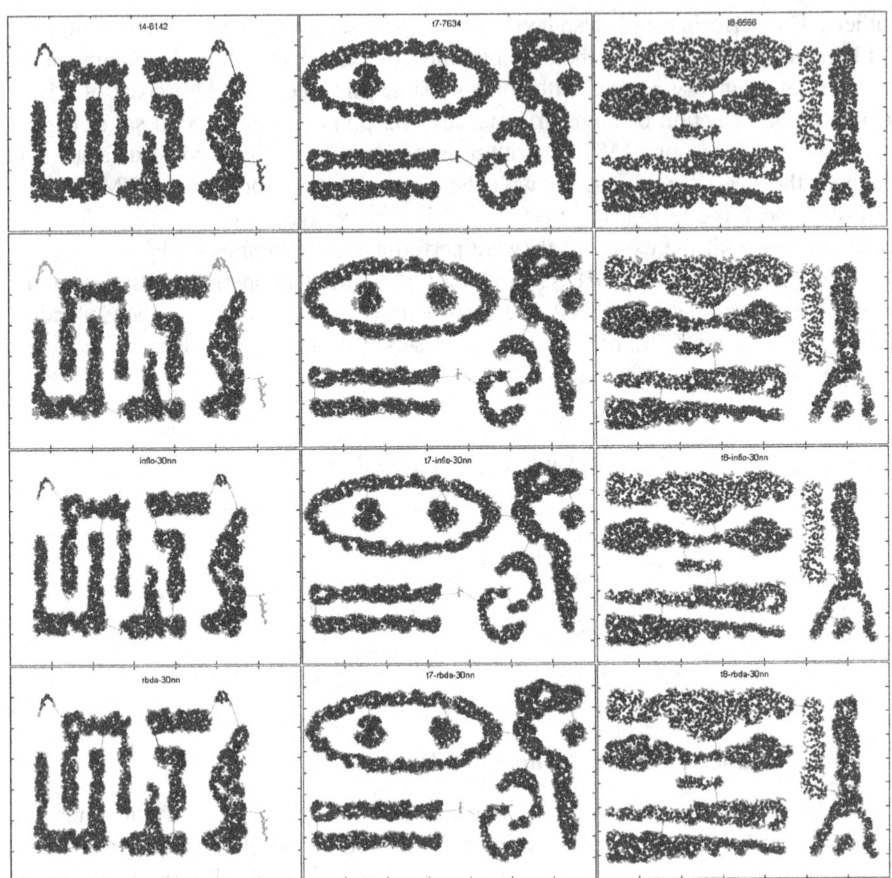

Fig. 3. (First row) three original datasets, (Second row) the thresholded ABOD values, (Third row) the thresholded INFLO values, (Fourth row) the thresholded RBDA values. (Color figure online)

4.2 Performance on High-Dimensional Real Datasets

In this subsection, we evaluate the efficiency of our algorithm versus three other EMST algorithms on two real high-dimensional data sets, ISOLET and MNIST, which are from the UCI KDD Archive [40]. The ISOLET dataset consists of data extracted from the recorded spoken name of each letter of the alphabet and contains 7797 instances with 617 attributes. The MNIST dataset consists of data extracted from the handwritten digits and contains 10000 instances with 784 attributes.

The running time performances of our algorithm, FEMST algorithm, Prim's algorithm (i.e., MST(BF)), FAEMST algorithm are summarized in Table 2. From the table, we can see that our algorithm outperforms all other three algorithms. It is faster than Prim's algorithm by a factor of 5 and than FAEMST algorithm by a factor of 3 for ISOLET dataset. For MNIST dataset, it is faster than Prim's algorithm by a factor of 2, but has a similar performance with FAEMST. The FEMST method loses its power for these two high dimensional datasets and takes a longer time even than the Prim's algorithm.

Table 2. Running time performance in seconds

Data name	FEMST	MST (BF)	FAEMST	Ours
ISOLET	519	417	231	84
MNIST	1 135	286	164	138

5 Conclusions

EMST algorithms are of growing importance in many application domains. A central problem in such algorithms for modern large high-dimensional real datasets is the classic quadratic time complexity on the construction of an EMST. In this paper, we present a more efficient method that can quickly identify the longest edges in an EMST so as to save some computations. We conducted experiments to evaluate our algorithm against Prim's algorithm and two other state-of-the-art EMST algorithms on two UCI data sets. The experimental results show that our proposed EMST algorithm is very effective and stable. In the future, we will further study the rich properties of the existing EMST algorithms and adapt our proposed EMST algorithm to larger data sets, particularly when the whole dataset can not fit into the main memory.

Acknowledgment. The authors would like to thank the Chinese National Science Foundation for its valuable support of this work under award 61473220 and all the anonymous reviewers for their valuable comments.

References

1. Kruskal, J.B.: On the shortest spanning subtree of a graph and the traveling salesman problem. Proc. Am. Math. Soc. **7**, 48–50 (1956)
2. Prim, R.C.: Shortest connection networks and some generalizations. Bell Syst. Tech. J. **36**, 567–574 (1957)
3. An, L., Xiang, Q.S., Chavez, S.: A fast implementation of the method for phase unwrapping. IEEE Trans. Med. Imaging **19**(8), 805–808 (2000)
4. Xu, Y., Uberbacher, E.C.: 2D image segmentation using minimum spanning trees. Image Vis. Comput. **15**, 47–57 (1997)
5. Zahn, C.T.: Graph-theoretical methods for detecting and describing gestalt clusters. IEEE Trans. Comput. **C20**, 68–86 (1971)
6. Xu, Y., Olman, V., Xu, D.: Clustering gene expression data using a graph-theoretic approach: an application of minimum spanning trees. Bioinformatics **18**(4), 536–545 (2002)
7. Zhong, C., Miao, D., Wang, R.: A graph-theoretical clustering method based on two rounds of minimum spanning trees. Pattern Recognit. **43**(3), 752–766 (2010)
8. Juszczak, P., Tax, D.M.J., Pe¸kalska, E., Duin, R.P.W.: Minimum spanning tree based one-class classifier. Neurocomputing **72**, 1859–1869 (2009)

9. Yang, C.L.: Building k edge-disjoint spanning trees of minimum total length for isometric data embedding. IEEE Trans. Pattern Anal. Mach. Intell. **27**(10), 1680–1683 (2005)
10. Gower, J.C., Ross, G.J.S.: Minimum spanning trees and single linkage cluster analysis. Appl. Stat. **18**(1), 54–64 (1969)
11. Balcan, M., Blum, A., Vempala, S.: A discriminative framework for clustering via similarity functions. In: Proceedings of ACM Symposium on Theory of Computing, pp. 671–680 (2008)
12. Bor°uvka, O.: O jist´em probl´emu minim´aln´ım (About a Certain Minimal Problem). Pr´ ace moravsk´e pˇr´ırodovˇedeck´e spoleˇcnosti v Brnˇe. III, pp. 37–58 (1926). (in Czech with German summary)
13. Jarn´ık, V.: O jist´em probl´emu minim´aln´ım (About a Certain Minimal Problem). Pr´ace moravsk´e pˇr´ırodovˇedeck´e spoleˇcnosti v Brnˇe VI, pp. 57–63 (1930). (in Czech)
14. Dijkstra, E.W.: A note on two problems in connexion with graphs. Numer. Math. **1**(1), 269–271 (1959)
15. Bentley, J., Friedman, J.: Fast algorithms for constructing minimal spanning trees in coordinate spaces. IEEE Trans. Comput. **27**, 97–105 (1978)
16. Preparata, F.P., Shamos, M.I.: Computational Geometry. Springer, New York (1985). https://doi.org/10.1007/978-1-4612-1098-6
17. Callahan, P., Kosaraju, S.: Faster algorithms for some geometric graph problems in higher dimensions. In: Proceedings of 4th Annual ACM-SIAM Symposium on Discrete Algorithms, pp. 291–300 (1993)
18. Narasimhan, G., Zachariasen, M., Zhu, J.: Experiments with computing geometric minimum spanning trees. In: Proceedings of ALENEX 2000, pp. 183–196 (2000)
19. March, W.B., Ram, P., Gray, A.G.: Fast euclidean minimum spanning tree: algorithm, analysis, and applications. In: Proceedings of 16th ACM SIGKDD International Conference on Knowledge Discovery and Data Mining (KDD), Washington, pp. 603–612 (2010)
20. Vaidya, P.M.: Minimum spanning trees in k-dimensional space. SIAM J. Comput. **17**(3), 572–582 (1988)
21. Wang, X., Wang, X., Wilkes, D.M.: A divide-and-conquer approach for minimum spanning tree-based clustering. IEEE Trans. Knowl. Data Eng. **21**(7), 945–958 (2009)
22. Lai, C., Rafa, T., Nelson, D.E.: Approximate minimum spanning tree clustering in high-dimensional space. Intell. Data Anal. **13**, 575–597 (2009)
23. Wang, X., Wang, X.L., Zhu, J.: A new fast minimum spanning tree based clustering technique. In: Proceedings of the 2014 IEEE International Workshop on Scalable Data Analytics, 14–17 December, Shenzhen, China (2014)
24. Zhong, C., Malinen, M., Miao, D., Fränti, P.: A fast minimum spanning tree algorithm based on K-means. Inf. Sci. **295**(C), 1–17 (2015)
25. Ester, M., Kriegel, H.-P., Sander, J., Xu, X., Simoudis, E., Han, J., Fayyad, U.M. (eds.): A density-based algorithm for discovering clusters in large spatial databases with noise. In: Proceedings of the Second International Conference on Knowledge Discovery and Data Mining (KDD-96), pp. 226–231. AAAI Press (1996)
26. Breunig, M.M., Kriegel, H.P., Ng, R.T., Sander, J.: LOF: identifying density-based local outliers. In: Proceedings of the 2000 ACM SIGMOD International Conference on Management of Data, pp. 93–104 (2000)
27. Tang, J., Chen, Z., Fu, A.W.-C., Cheung, David W.: Enhancing effectiveness of outlier detections for low density patterns. In: Chen, M.-S., Yu, P.S., Liu, B. (eds.) PAKDD 2002. LNCS (LNAI), vol. 2336, pp. 535–548. Springer, Heidelberg (2002). https://doi.org/10.1007/3-540-47887-6_53

28. Papadimitriou, S., Kitagawa, H., Gibbons, P.B., Faloutsos, C.: LOCI: fast outlier detection using the local correlation integral. In: Proceedings of the IEEE 19th International Conference on Data Engineering, Bangalore, India, pp. 315–328 (2003)
29. Sun, P., Chawla, S.: On local spatial outliers. In: Proceedings of the 4th International Conference on Data Mining (ICDM), Brighton, UK, pp. 209–216 (2004)
30. Jin, W., Tung, A.K.H., Han, J., Wang, W.: Ranking outliers using symmetric neighborhood relationship. In: Ng, W.-K., Kitsuregawa, M., Li, J., Chang, K. (eds.) PAKDD 2006. LNCS (LNAI), vol. 3918, pp. 577–593. Springer, Heidelberg (2006). https://doi.org/10.1007/11731139_68
31. Fan, H., Zaïane, O.R., Foss, A., Wu, J.: A nonparametric outlier detection for effectively discovering top-n outliers from engineering data. In: Ng, W.-K., Kitsuregawa, M., Li, J., Chang, K. (eds.) PAKDD 2006. LNCS (LNAI), vol. 3918, pp. 557–566. Springer, Heidelberg (2006). https://doi.org/10.1007/11731139_66
32. Latecki, L.J., Lazarevic, A., Pokrajac, D.: Outlier detection with kernel density functions. In: Perner, P. (ed.) MLDM 2007. LNCS (LNAI), vol. 4571, pp. 61–75. Springer, Heidelberg (2007). https://doi.org/10.1007/978-3-540-73499-4_6
33. Zhang, K., Hutter, M., Jin, H.: A new local distance-based outlier detection approach for scattered real-world data. In: Theeramunkong, T., Kijsirikul, B., Cercone, N., Ho, T.-B. (eds.) PAKDD 2009. LNCS (LNAI), vol. 5476, pp. 813–822. Springer, Heidelberg (2009). https://doi.org/10.1007/978-3-642-01307-2_84
34. Huang, H., Mehrotra, K., Mohan, C.K.: Rank-based outlier detection. J. Stat. Comput. Simul. 1–14 (2011)
35. Kriegel, H.-P., Schubert, M., Zimek, A.: Angle-based outlier detection in high-dimensional data. In: Proceedings KDD 2008, pp. 444–452 (2008)
36. Pham, N., Pagh, R.: A near-linear time approximation algorithm for angle-based outlier detection in high-dimensional data. In: Proceedings of the 18th ACM SIGKDD Conference on Knowledge Discovery and Data Mining, pp. 877–885 (2012)
37. Yu, C., Chin Ooi, B., Tan, K.L., Jagadish, H.V.: Indexing the distance: an efficient method to KNN processing. In: Proceedings of the 27th International Conference on Very Large Data Bases, Roma, Italy, pp. 421–430 (2001)
38. Jagadish, H.V., Chin Ooi, B., Tan, K.L., Yu, C., Zhang, R.: iDistance: an adaptive B+-tree based indexing method for nearest neighbor search. ACM Trans. Data Base Syst. (ACM TODS) **30**(2), 364–397 (2005)
39. Karypis, G., Han, E.H., Kumar, V.: CHAMELEON: a hierarchical clustering algorithm using dynamic modeling. IEEE Comput. **32**(8), 68–75 (1999)
40. http://archive.ics.uci.edu/ml/datasets.html

Rule Induction Partitioning Estimator

A New Deterministic Algorithm for Data Dependent Partitioning Estimate

Vincent Margot(✉), Jean-Patrick Baudry, Frederic Guilloux, and Olivier Wintenberger

Laboratoire de Probabilités, Statistique et Modélisation, Sorbonne Universités, Campus Pierre et Marie Curie, Tour 16-26, 1er étage, 4, Place Jussieu, 75005 Paris, France
vincent.margot@upmc.fr

Abstract. RIPE is a novel deterministic and easily *understandable* prediction algorithm developed for continuous and discrete ordered data. It infers a model, from a sample, to predict and to explain a real variable Y given an input variable $X \in \mathcal{X}$ (features). The algorithm extracts a sparse set of hyperrectangles $\mathbf{r} \subset \mathcal{X}$, which can be thought of as rules of the form *If-Then*. This set is then turned into a partition of the features space $\mathcal{X}$ of which each cell is explained as a list of rules with satisfied their *If* conditions.

The process of RIPE is illustrated on simulated datasets and its efficiency compared with that of other usual algorithms.

Keywords: Machine learning · Data mining · Interpretable models
Rule induction · Data-dependent partitioning · Regression models

1 Introduction

To find an easy way to describe a complex model with a high accuracy is an important objective for machine learning. Many research fields such as medicine, marketing, or finance need algorithms able to give a reason for each prediction made. Until now, a common solution to achieve this goal has been to use induction rule to describe cells of a partition of the features space $\mathcal{X}$. A rule is an *If-Then* statement which is understood by everyone and easily interpreted by experts (medical doctors, asset managers, etc.). We focus on rules with a *If* condition defined as a hyperrectangle of $\mathcal{X}$. Sets of such rules have always been seen as decision trees, which means that there is a one-to-one correspondence between a rule and a generated partition cell. Therefore, algorithms for mining induction rules have usually been developed to solve the *optimal decision tree* problem [8]. Most of them use a greedy splitting technique [3,5,6,11] whereas others use an approach based on Bayesian analysis [4,9,12].

P. Perner (Ed.): MLDM 2018, LNAI 10935, pp. 288–301, 2018.
https://doi.org/10.1007/978-3-319-96133-0_22

RIPE (Rule Induction Partitioning Estimator) has been developed to be a *deterministic* (identical output for an identical input) and easily *understandable* (simple to explain and to interpret) predictive algorithm. In that purpose, it has also been based on rule induction. But, on the contrary to other algorithms, rules selected by RIPE are not necessarily disjoint and are independently identified. So, this set of selected rules does not form a partition and it cannot be represented as a decision tree. This set is then turned into a partition. Cells of this partition are described by a set of activated rules which means that their *If* conditions are satisfied. So, a same rule can explain different cells of the partition. Thus, RIPE is able to generate a fine partition whose cells are easily described, which would usually require deeper decision tree and less and less *understandable* rules. Moreover, this way of partitioning permits to have cells which are not a hyperrectangles.

The simplest estimator is the constant one which predicts the empirical expectation of the target variable. From it, RIPE searches rules which are *significantly* different. To identify these, RIPE works recursively, searching more and more complex rules, from the most generic to the most specific ones. When it is not able to identify new rules, it extracts a set of rules by an empirical risk minimization. To ensure a covering of $\mathcal{X}$, a *no rule satisfied* statement is added to the set. It is defined on the subset of $\mathcal{X}$ not covered by the union of the hyperrectangles of the extracted rules. At the end, RIPE generates a partition spanned by these selected rules and builds an estimator. But the calculation of a partition from a set of hyperrectangles is very complex. To solve this issue, RIPE uses what we called the *partitioning trick* which is a new algorithmic way to bypass this problem.

1.1 Framework

Let $(\mathbf{X}, Y) \in \mathcal{X} \times \mathbb{R}$, where $\mathcal{X} = \mathcal{X}_1 \times \cdots \times \mathcal{X}_d$, be a couple of random variables with unknown distribution P.

Definition 1.
1. *Any measurable function* $g : \mathcal{X} \to \mathbb{R}$ *is called a* predictor *and we denote by* $\mathbb{G}$ *the set of all the predictors.*
2. *The accuracy of* g *as a predictor of* Y *from* $\mathbf{X}$ *is measured by the quadratic* risk, *defined by*

$$\mathcal{L}(g) = \mathbb{E}_{(\mathbf{X},Y)\sim\mathbb{P}}\left[(g(\mathbf{X}) - Y)^2\right]. \tag{1}$$

From the properties of the conditional expectation, the optimal predictor is the regression function (see [1,7] for more details):

$$g^* := \mathbb{E}\left[Y|\mathbf{X}\right] = \underset{g\in\mathbb{G}}{\arg\min}\, \mathcal{L}(g) \;\text{ a.s.} \tag{2}$$

Definition 2. *Let* $D_n = ((\mathbf{X}_1, Y_1), \ldots, (\mathbf{X}_n, Y_n))$ *be a sample of independent and identically distributed copies of* $(\mathbf{X}, Y)$.

Definition 3. *The empirical risk of a predictor g on D_n is defined by:*

$$\mathcal{L}_n(g) = \frac{1}{n}\sum_{i=1}^{n}(g(\mathbf{X}_i) - Y_i)^2 \tag{3}$$

Equation (2) provides a link between prediction and estimation of the regression function. So, the purpose is to produce an estimator of g^* based on a partition of $\mathcal{X}$ that provides a good predictor of Y. However, the partition must be simple enough to be *understandable.*

1.2 Rule Induction Partitioning Estimator

The RIPE algorithm is based on rules. Rules considered in this paper are defined as follows:

Definition 4. *A rule is an* If-Then *statement such that its* If *condition is a hyperrectangle $\mathbf{r} = \prod_{k=1}^{d} I_k$, where each I_k is an interval of $\mathcal{X}_k$.*

Definition 5. *For any set $E \subset \mathcal{X}$, the empirical conditional expectation of Y given $X \in E$ is*

$$\mu(E, D_n) := \frac{\sum_{i=1}^{n} y_i \mathbf{1}_{\mathbf{x}_i \in E}}{\sum_{i=1}^{n} \mathbf{1}_{\mathbf{x}_i \in E}},$$

where, by convention, $\frac{0}{0} = 0$.

The natural estimator of g^* on any $\mathbf{r} \subset \mathcal{X}$ is the empirical conditional expectation of Y given $X \in \mathbf{r}$. A rule is completely defined by its condition $\mathbf{r}$. So, by an abuse of notation we do not distinguish between a rule and its condition.

A set of rules $\mathcal{S}_n$ is selected based on the sample D_n. Then $\mathcal{S}_n$ is turned into a partition of $\mathcal{X}$ denoted by $\mathcal{K}(\mathcal{S}_n)$ (see Sect. 2.1). To make sure to define a covering of the features space the *no rule satisfied* statement is added to the set of hyperrectangles.

Definition 6. *The* no rule satisfied *statement for a set or rules $\mathcal{S}_n$, is an* If-Then *statement such that its* If *condition is the subset of $\mathcal{X}$ not covered by the union of the hyperrectangles of $\mathcal{S}_n$.*

One can notice that it is not a rule according to the Definition 4 because it is not necessarily defined on a hyperrectangle.

Finally, an estimator $\hat{g}^{\mathcal{S}_n}$ of the regression function g^* is defined:

$$\hat{g}^{\mathcal{S}_n} : (\mathbf{x}, D_n) \in \mathcal{X} \times (\mathcal{X} \times \mathbb{R})^n \mapsto \mu(K_n(\mathbf{x}), D_n), \tag{4}$$

with $K_n(\mathbf{x})$ the cell of $\mathcal{K}(\mathcal{S}_n)$ which contains $\mathbf{x}$.

The partition itself is *understandable.* Indeed, the prediction $\hat{g}^{\mathcal{S}_n}(\mathbf{x}, D_n)$ of Y is of the form "*If rules ... are satisfied, then Y is predicted by ...*" The cells of $\mathcal{K}(\mathcal{S}_n)$ are explained by sets of satisfied rules, and the values $\mu(K_n(\mathbf{x}), D_n)$ are the predicted values for Y.

2 Fundamental Concepts of RIPE

RIPE is based on two concepts, the *partitioning trick* and the *suitable rule.*

2.1 Partitioning Trick

The construction of a partition from a set of R hyperrectangles is time consuming and it is an exponential complexity operation and this construction occurs several times in the algorithm. To reduce the time and complexity we have developed the *partitioning trick.*

First, we remark that to calculate $\mu(K_n(\mathbf{x}), D_n)$, it is not necessary to build the partition, it is sufficient to identify the cell which contains $\mathbf{x}$. Figure 1 is an illustration of this process. To do that, we first identify rules activated by $\mathbf{x}$, i.e that $\mathbf{x}$ is in their hyperrectangles (Fig. 1, to the upper left). And we calculate the hyperrectangle defined by their intersection (Fig. 1, to the lower left). Then, we calculate the union of hyperrectangles of rules which are not activated (Fig. 1, to the upper right). To finish, we calculate the cell by difference of the intersection and the union (Fig. 1, to the lower right). The generated subset is the cell of the partition $\mathcal{K}(\mathcal{S}_n)$ containing $\mathbf{x}$.

Proposition 1. *Let $\mathcal{S}_n$ be a set of R rules selected from a sample D_n. Then, the complexity to calculate $\mu(K_n(\mathbf{x}), D_n)$ for a new observation $\mathbf{x} \in \mathcal{X}$ is $O(nR)$.*

Proof. It is sufficient to notice that $\mu(K_n(\mathbf{x}), D_n)$ can be express as follows:

$$\mu(K_n(\mathbf{x}), D_n) = \frac{\sum_{j=1}^{n} y_j k(\mathbf{x}, \mathbf{x}_j, \mathcal{S}_n)}{\sum_{j=1}^{n} k(\mathbf{x}, \mathbf{x}_j, \mathcal{S}_n)}, \tag{5}$$

with $k(\mathbf{x}, \mathbf{x}_j, \mathcal{S}_n) = \prod_{i=1}^{R} \left(\mathbf{1}_{\mathbf{x} \in \mathbf{r}_i} \mathbf{1}_{\mathbf{x}_j \in \mathbf{r}_i} + \mathbf{1}_{\mathbf{x} \notin \mathbf{r}_i} \mathbf{1}_{\mathbf{x}_j \notin \mathbf{r}_i} \right)$.

In (5), the complexity $O(nR)$ appears immediately. ∎

2.2 Independent Suitable Rules

Each dimension of $\mathcal{X}$ is discretized into m_n classes such that

$$\frac{(m_n)^d}{n} \to 0, \qquad n \to \infty. \tag{6}$$

To do so empirical quantiles of each variable are considered (when it has more than m_n different values). Thus, each class of each variable covers about $100/m_n$ percent of the sample. This discretization is the reason why RIPE deals with continuous and ordered discrete variables only.

It is a theoretical condition. However, it indicates that m_n must be inversely related to d: The higher the dimension of the problem, the smaller the number of modalities. It is a way to avoid *overfitting.*

We first define two crucial numbers:

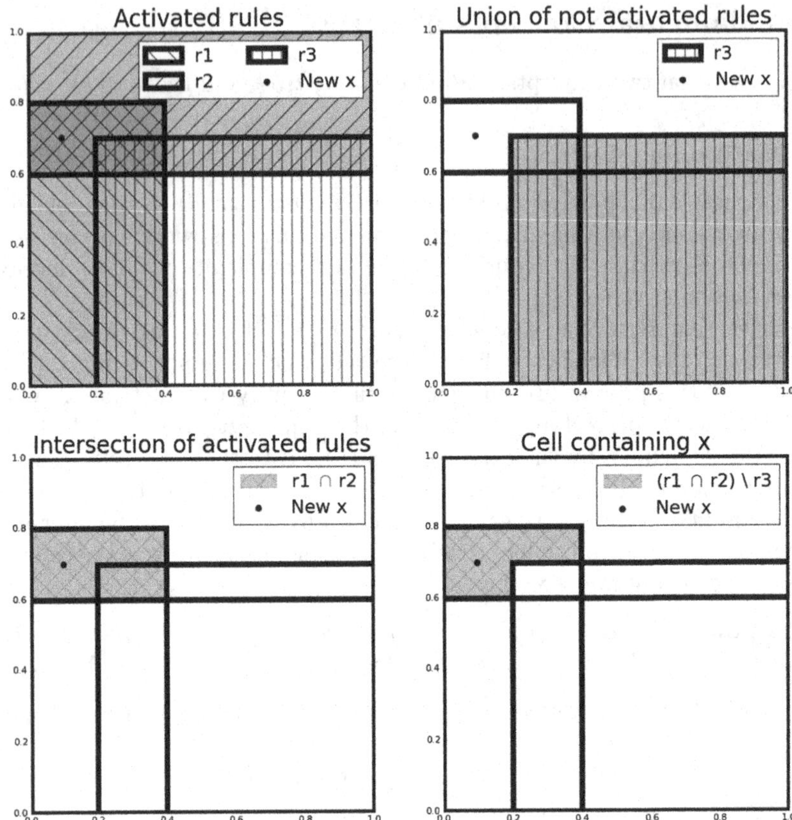

Fig. 1. Different steps of the *partitioning trick* for a set of three hyperrectangles $\{\mathbf{r}_1, \mathbf{r}_2, \mathbf{r}_3\}$ of $[0, 1]^2$ and a new observation $\mathbf{x} = (0.1, 0.7)$. It is important to notice that the cell containing $\mathbf{x}$, $(\mathbf{r}_1 \cap \mathbf{r}_2) \setminus \mathbf{r}_3$, is not a hyperrectangle so it does not define a rule.

Definition 7. *Let* $\mathbf{r} = \prod_{k=1}^{d} I_k$ *be a hyperrectangle.*

1. The number of activations *of* $\mathbf{r}$ *in the sample* D_n *is*

$$n(\mathbf{r}, D_n) := \sum_{j=1}^{n} \mathbf{1}_{\mathbf{x}_j \in \mathbf{r}}. \tag{7}$$

2. The complexity *of* $\mathbf{r}$ *is*

$$cp(\mathbf{r}) = d - \#\{1 \leq k \leq d; I_k = \mathcal{X}_k\}, \tag{8}$$

We are now able to define a *suitable rule.*

Definition 8. *A rule* $\mathbf{r}$ *is a* suitable rule *for a sample* D_n *if and only if it satisfies the two following conditions:*

1. ***Coverage condition.***

$$\frac{n(\mathbf{r}, D_n)}{n} \leq \frac{1}{\ln(m_n)}, \tag{9}$$

2. ***Significance condition.***

$$|\mu(\mathbf{r}, D_n) - \mu(\mathcal{X}, D_n)| \geq z(\mathbf{r}, D_n, \alpha), \tag{10}$$

for a chosen $\alpha \in [0, 1]$, *and a chosen function* z.

The coverage condition (9) ensures that the coverage ratio $n(\mathbf{r}, D_n)/n$ of a rule tends toward 0 for $n \to \infty$. It is a necessary condition to prove the consistency of the estimator which it is the purpose of a companion paper.

The threshold in the significance condition (10) is set such that the probability of falsely rejecting the null hypothesis $\mu(\mathbf{r}, D_n) = \mu(\mathcal{X}, D_n)$ is less than α. As mentioned before RIPE searches rules that are different from the simplest estimator. This is why we compare $\mu(\mathbf{r}, D_n)$ and $\mu(\mathcal{X}, D_n)$. The parameter α permits to control the number of suitable rules. The higher α, the higher the number of suitable rules. We present in Appendix, two functions z used in practice.

RIPE generates rules of complexity $c \geq 2$ by a *suitable intersection* of rules of complexity 1 and rule of complexity $c - 1$.

Definition 9. *Two rules* $\mathbf{r}_i$ *and* $\mathbf{r}_j$ *define a* suitable intersection *if and only if they satisfy the two following conditions:*

1. ***Intersection condition:***

$$\begin{aligned} &\mathbf{r}_i \cap \mathbf{r}_j \neq \emptyset, \\ &n(\mathbf{r}_i \cap \mathbf{r}_j, D_n) \neq n(\mathbf{r}_i, D_n), \\ &n(\mathbf{r}_i \cap \mathbf{r}_j, D_n) \neq n(\mathbf{r}_j, D_n) \end{aligned} \tag{11}$$

2. ***Complexity condition:***

$$cp(\mathbf{r}_i \cap \mathbf{r}_j) = cp(\mathbf{r}_i) + cp(\mathbf{r}_j). \tag{12}$$

The intersection condition (11) avoids adding a useless condition for a rule. In other words, to define a *suitable intersection* $\mathbf{r}_i$ and $\mathbf{r}_j$ must not be satisfied for the same observations of D_n. And the complexity condition (12) means that $\mathbf{r}_i$ and $\mathbf{r}_j$ have no marginal index k; $\mathbf{1}_k \subsetneq \mathcal{X}_k$, in common of $\mathcal{X}$.

3 RIPE Algorithm

We now describe the methodology of RIPE for designing and selecting rules. The *Python* code is available at https://github.com/VMargot/RIPE.

The main algorithm is described as Algorithm 1. The methodology is divided into two parts. The first part aims at finding all suitable rule and the second one aims at selecting a small subset of suitable rules that estimate accurately the objective g^*.

The parameters of the algorithm are:

- m_n, the sharpness of the discretization, which must fulfill (6);
- $\alpha \in [0, 1]$, which specifies the false rejecting rate of the test;
- z, the significance function of the test;
- and $M \in \mathbb{N}$, the number of rules of complexity 1 and $c - 1$ used to define the rules of complexity c.

Algorithm 1. Main

Global parameters : m_n, α, z and M;
Input:
- $(\mathbf{X}, \mathbf{y}) \in \mathbb{R}^{n(d+1)}$: data

Output:

- $\mathcal{S}$: the set of selected rules

Set $h_n = 1/\ln(m_n)$ the maximal coverage ratio of a rule;
$\tilde{\mathbf{X}} \leftarrow Discretize(\mathbf{X}, m_n)$ discretization in m_n modalities;
$\mathcal{R} \leftarrow Calc\_cp1((\tilde{\mathbf{X}}, \mathbf{y}), h_n)$;
for $c = 2, \ldots, d$ **do**
    $\mathcal{R}' \leftarrow Calc\_cpc((\tilde{\mathbf{X}}, \mathbf{y}), \mathcal{R}, c, h_n)$;
    if $len(\mathcal{R}') = 0$ **then**
        Break;
    else
        $\mathcal{R} \leftarrow append(\mathcal{R}, \mathcal{R}')$;
end
$\mathcal{R} \leftarrow Sort\_by\_risk(\mathcal{R}, (\tilde{\mathbf{X}}, \mathbf{y}))$;
$\mathcal{S} \leftarrow Select(\mathcal{R}, (\tilde{\mathbf{X}}, \mathbf{y}))$;
Return $\mathcal{S}$;

3.1 Designing Suitable Rules

The design of suitable rules is made recursively on their complexity. It stops at a complexity c if no rule is suitable or if the maximal complexity $c = d$ is achieved.

Complexity 1: The first step is to find suitable rules of complexity 1. This part is described as Algorithm 2. First notice that the complexity of evaluating all rules of complexity 1 is $O(ndm_n^2)$.

Rules of complexity 1 are the base of RIPE search heuristic. So all rules are considered and just suitable are kept, i.e rules that satisfied the coverage condition (9) and the significance condition (10). Since rules are considered regardless of each others, the search can be parallelized.

At the end of this step, the set of suitable rules is sorted by their empirical risk (3), $\mathcal{L}_n(\hat{g}^{\{\mathbf{r}\}})$, with $\hat{g}^{\{\mathbf{r}\}}$ the predictor based on exactly one rule $\mathbf{r}$.

Algorithm 2. Calc_cp1

Global parameters : m_n, α and z;
Input:
- $(\mathbf{X}, \mathbf{y}) \in \mathbb{R}^{n(d+1)}$: data
- h_n: parameter

Output:
- $\mathcal{R}$: the set of all suitable rules of complexity 1;

```
R ← ∅;
for i = 1, ..., d do
    x_i ← X[i], the i^th feature ;
    for b_min = 0, ..., m_n do
        for b_max = b_min, ..., m_n do
            Set r = ∏_{k=1}^{d} I_k with { I_k = [0, m_n], k ≠ i ; I_i = [b_min, b_max] } ;
            if is_suitable(r, (X, y), h_n, z, α) then
                R ← append(R, r);
        end
    end
end
Return R
```

Complexity c**:** Among the suitable rules of complexity 1 and $c-1$ sorted by their empirical risk (3), RIPE selects the M first rules of each complexity (1 and $c-1$). Then it generates rules of complexity c by pairwise *suitable intersection* according to the Definition 9. It is easy to see that the complexity of evaluating all rules of complexity c obtained from their intersections is then $O(nM^2)$.

The parameter M is to control the computing time and it is fixed by the statistician. This part is described as Algorithm 3.

3.2 Selection of Suitable Rules

After designing suitable rules, RIPE selects an optimal set of rules. Let $\mathcal{R}_n$ be the set of all suitable rules generated by RIPE. The optimal subset $\mathcal{S}_n^* \subset \mathcal{R}_n$ is defined by

$$\mathcal{S}_n^* := \operatorname*{arg\,min}_{\mathcal{S} \subset \mathcal{R}_n} \mathcal{L}_n(\hat{g}^{\{\mathcal{S}\}}) \tag{13}$$

is the empirical risk (3) of the predictor based on $\mathcal{S}$.

Each computation of the empirical risk (13) requires the partition from the set $\mathcal{S}$ of rules, as described in Sect. 2.1. The complexity to solve (13) naively, comparing all the possible sets of rules, is exponential in the number of suitable rules.

To work around this problem, RIPE uses Algorithm 4, a greedy recursive version of the naive algorithm: it does not explore all the subsets of $\mathcal{R}_n$. Instead,

Algorithm 3. Calc_cpc

Global parameters : α, z and M;

Input:

- $(\mathbf{X}, \mathbf{y}) \in \mathbb{R}^{n(d+1)}$: data
- $\mathcal{R}$: set of rules of complexity up to $c - 1$
- c: complexity
- h_n: parameter

Output:

- $\mathcal{R}_c$: set of suitable rules of complexity c

```
R_c ← ∅;
R ← Sort_by_risk(R, (X, y));
R_1 ← the M first rules of complexity 1 in R;
R_{c-1} ← the M first rules of complexity c − 1 in R;
if R_1 ≠ ∅ and R_{c-1} ≠ ∅ then
    for r_1 in R_1 do
        for r_2 in R_{c-1} do
            if is_suitable_intersection(r_1, r_2) then
                Set r = r_1 ∩ r_2;
                if is_suitable(r, (X, y), h_n, z, α) then
                    R_c ← append(R_c, r);
        end
    end
Return R_c
```

it starts with a single rule, the one with minimal risk, and iteratively keeps/leaves the rules by comparing the risk of a few combinations of these rules. More precisely, suppose that

- $\mathbf{r}_1, \ldots, \mathbf{r}_N$ are the suitable rules, sorted by increasing empirical risk;
- $\mathbf{r}_1, \ldots, \mathbf{r}_k$, $k < N$, have already been tested;
- j of them, say $\mathcal{S} \subset \{\mathbf{r}_1, \ldots, \mathbf{r}_k\}$ have been kept, the $k - j$ other being left.

Then $\mathbf{r}_{k+1}$ is tested in the following way :

- Compute the risk of $\mathcal{S}$, $\mathcal{S} \cup \{\mathbf{r}_{k+1}\}$ and of all $\mathcal{S} \cup \{\mathbf{r}_{k+1}\} \backslash \{\mathbf{r}\}$ for $\mathbf{r} \in \mathcal{S}$;
- Keep the rules corresponding to the minimal risk;
- Possibly leave *once for all*, the rule in $\{\mathbf{r}_1, \ldots, \mathbf{r}_{k+1}\}$ which is not kept at this stage.

Thus, instead of testing the 2^N subsets of rules, we make N steps and at the kth step we test at most $k + 2$ (and usually much less) subsets, which leads to a theoretical overall maximum of $O(N^2)$ tested subsets. The heuristic of this strategy is that rules with low risk are more likely to be part of low risk subsets of rules; and the minimal risk is searched in subsets of increasing size.

Algorithm 4. Select

Input:
- $\mathcal{R}$: set of rules sorted by increasing risk
- $(\mathbf{X}, \mathbf{y}) \in \mathbb{R}^{n(d+1)}$: data

Output:
- $\mathcal{S}$: subset of selected rules approaching the argmin (13) over all subsets of $\mathcal{R}$;

```
Set S = {R(1)};
for i = 2,...,len(R) do
    Set 𝔖 = { S ; S∪{R(i)} };
    for j=1,...,len(S) do
        𝔖 ← append( 𝔖 ; S∪{R(i)}\{S(j)} )
    end
    𝔖 ← Sort_by_risk(𝔖, (X, y));
    S ← 𝔖(1)
end
Return S;
```

4 Experiments

The experiments have been done with *Python*. To assure reproducibility the *random seed* has been set at 42. The codes of these experiments are available in **GitHub** with the package RIPE.

4.1 Artificial Data

The purpose here it is to understand the process of RIPE, and how it can explain a phenomenon. We generate a dataset of $n = 5000$ observations with $d = 10$ features. The target variable Y depends on two features X_1 and X_2 whose are identically distributed on $[-1, 1]$. In order to simulate features assimilated to white noise, the others variables follow a centered- reduced normal distribution $\mathcal{N}(0, 1)$. The model is the following

$$y_i = F^*(\mathbf{x}_i) + \epsilon_i \tag{14}$$

with $\epsilon_i \sim \mathcal{N}(0, 1)$ and

$$F^*(\mathbf{x}_i) = -2 \times \mathbf{1}_{\{x_{1,i}^2 + x_{2,i}^2 > 0.8\}} + 2 \times \mathbf{1}_{\{x_{1,i}^2 + x_{2,i}^2 < 0.5\}} \tag{15}$$

The dataset is randomly split into training set D_m^1 and test set D_l^2 such as D_m^1 represents 60% of the dataset. RIPE uses significance test based on (17) with a threshold $\alpha = 0.05$ and $m_n = 5$.

On Fig. 2, we have on the left, the true model (15) according to X_1 and X_2 with the realization of Y not used during the learning. On the right, the model inferred by RIPE.

On Table 1 we represent the set of selected rules. In this case rules form a covering of the features space. So it is not necessary to add the *no rule satisfied* statement (see Definition 6).

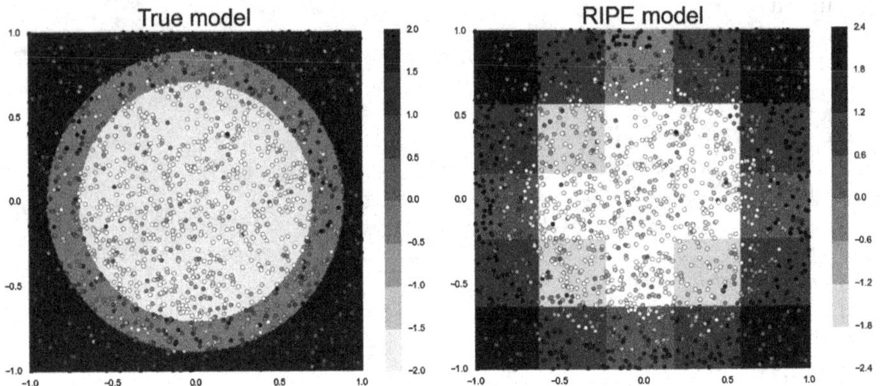

Fig. 2. The true model vs the model inferred by RIPE.

4.2 High Dimension Simulation

In this simulation, we use the function *make_regression*, from the **Python** package *sklearn* [10], to generate a random linear regression model with n observations and d variables. Among these variables, p are informative and the rest are gaussian centered noise.

In this example we take $n = 500$, $d = 1000$ and $p = 5$ to simulate a noisy high dimensional problem. The data are randomly split into a training set and a test set, with a ratio of 60% \40%, respectively.

Table 1. Summary of selected rules with conditions interval express as modalities

Rule	Conditions	Coverage	Prediction	Z	MSE
R 0(2)-	$X0 \in [1.0, 3.0]$ & $X1 \in [1.0, 3.0]$	0.36	−0.91	0.09	2.14
R 1(2)-	$X0 \in [1.0, 3.0]$ & $X1 \in [2.0, 4.0]$	0.36	−0.57	0.09	3.31
R 2(2)-	$X0 \in [2.0, 4.0]$ & $X1 \in [1.0, 3.0]$	0.35	−0.54	0.09	3.40
R 3(1)-	$X0 \in [2.0, 3.0]$	0.40	−0.46	0.08	3.48
R 4(1)-	$X0 \in [1.0, 2.0]$	0.40	−0.43	0.08	3.55
R 5(1)-	$X1 \in [1.0, 2.0]$	0.40	−0.43	0.08	3.55
R 6(1)+	$X0 = 4.0$	0.20	0.63	0.11	3.65
R 7(1)+	$X1 = 4.0$	0.20	0.56	0.12	3.74
R 8(1)+	$X1 \in [0.0, 1.0]$	0.40	0.19	0.08	3.95
R 9(1)+	$X0 \in [0.0, 1.0]$	0.40	0.15	0.08	3.99

We use two others algorithms in this case: Decision Tree (DT) [3] without pruning and Random Forests (RF) [2], all from the package of python **sklearn** [10]. In order to evaluate the performance of our model, the normalized mean square error ($NMSE$) is computed.

Results are summarized in Table 3. Difference between the $NMSE$ of the training and the $NMSE$ of test indicates that Decision Tree and Random Forests overfit in this context. Conversely, RIPE infers a model which is more general (see Table 3). Indeed, RIPE is able to describe the model with only 14 rules (see Table 2) which have conditions on only seven variables from 1000. Among these selected variables only two are very important X_{976} and X_{298} (see Table 4).

In this case, RIPE discretizes each variable in 5 modalities from 0 to 4. Table 2 presents the selected rules with their conditions. The rule $R14$ is the *no rule satisfied* statement (see Definition 6).

Table 2. Summary of selected rules with conditions interval express as modalities

Rule	Conditions	Coverage	Prediction	Z	MSE
R 1(2)-	$X_{976} \in [0.0, 2.0]$ & $X_{298} \in [0.0, 2.0]$	0.35	-0.83	0.28	7808.24
R 2(1)-	$X_{976} = 0.0$	0.20	-1.07	0.46	8907.38
R 3(1)+	$X_{976} = 4.0$	0.20	0.88	0.41	10081.34
R 4(2)-	$X_{976} \in [0.0, 1.0]$ & $X_{336} \in [0.0, 1.0]$	0.19	-0.89	0.40	10245.30
R 5(2)+	$X_{298} \in [2.0, 4.0]$ & $X_{976} \in [2.0, 3.0]$	0.24	0.65	0.27	10781.00
R 6(1)+	$X_{298} = 4.0$	0.20	0.73	0.43	10813.83
R 7(1)-	$X_{298} = 0.0$	0.20	-0.73	0.42	10822.09
R 8(2)-	$X_{298} \in [0.0, 1.0]$ & $X_{336} \in [0.0, 1.0]$	0.20	-0.66	0.38	11109.50
R 9(2)+	$X_{976} \in [2.0, 4.0]$ & $X_{564} = 4.0$	0.14	0.77	0.45	11253.10
R 10(2)+	$X_{976} \in [2.0, 4.0]$ & $X_{163} = 4.0$	0.13	0.75	0.44	11419.93
R 11(2)-	$X_{976} \in [0.0, 1.0]$ & $X_{945} = 2.0$	0.10	-0.87	0.51	11427.16
R 12(2)-	$X_{976} \in [0.0, 1.0]$ & $X_{733} = 1.0$	0.10	-0.84	0.58	11524.60
R 13(1)+	$X_{976} = 3.0$	0.20	0.55	0.31	11548.05
R 14	No rule activated	0.02	-0.35	0.45	12440.40

Table 3. Performance results of RIPE compared to two supervised learning algorithms: The Decision Tree (DT) and the Random Forests (RF).

Algorithm	Parameters	$NMSE$ training	$NMSE$ test	Nb of rules	Complexity max
DT	/	0.0	0.46	350	14
RF	m_tree = 200 m_try = $d/3$	0.04	0.39	128.25[a]	21
RIPE	M=300 z: see (17) α=0.05	0.13	0.30	14	2

[a]It is the mean of the number of rules of each tree

Table 4. Count of variable occurencies in rules selected by RIPE.

Variable	X_{976}	X_{298}	X_{336}	X_{163}	X_{945}	X_{565}	X_{733}
Count	10	5	2	1	1	1	1

5 Conclusion and Future Work

In this paper we present a novel *understandable* predictive algorithm, named RIPE. Considering the regression function is the best predictor RIPE has been developed to be a simple and accurate estimator of the regression function. The algorithm identified a set of *suitable rules*, not necessary disjoint, of the form *If-Then* such as their *If* conditions are hyperrectangles of the features space $\mathcal{X}$. Then, the estimator is built on the partition generated by the *partitioning trick*. Its computational complexity is linear in the data dimension $O(dn)$.

RIPE is different from existing methods which are based on a space-partitioning tree. It is able to generate a fine partition from a set of *suitable rules*, reasonably quickly such that their cells are explained as a list of *suitable rules*. Whereas there is a one-to-one correspondence between a rule and a cell of a partition provided by a decision tree. So to have a finer partition decision trees must be deeper and rules become less and less *understandable*. Furthermore, on the contrary to decision trees, the partition generated by RIPE can have cells which are not hyperrectangles.

A paper on the universal consistency of RIPE under some technical conditions is in preparation.

Appendix: Examples of Significance Function

Here, we present two functions z used in practice.

$$z(\mathbf{r}, D_n, \alpha) = \frac{(M - m)\sqrt{\ln(2/\alpha)}}{\sqrt{2n(\mathbf{r}, D_n)}}, \tag{16}$$

where $M = \max_{i \in \{1,\dots,n\}} y_i$ and $m = \min_{i \in \{1,\dots,n\}} y_i$.

$$z(\mathbf{r}, D_n, \alpha) = \frac{1}{6n(\mathbf{r}, D_n)} \left(M \ln\left(\frac{2}{\alpha}\right) + \sqrt{M^2 \ln\left(\frac{2}{\alpha}\right)^2 + 72v \ln\left(\frac{2}{\alpha}\right)} \right), \tag{17}$$

where $M = \max_{i \in \{1,\dots,n\}} y_i$ and $v = \sum_{i=1}^{n} y_i^2$.

References

1. Arlot, S., Celisse, A.: A survey of cross-validation procedures for model selection. Stat. Surv. **4**, 40–79 (2010)
2. Breiman, L.: Random forests. Mach. Learn. **45**(1), 5–32 (2001)
3. Breiman, L., Friedman, J., Olshen, R., Stone, C.: Classification and Regression Trees. CRC Press, Boca Raton (1984)
4. Chipman, H., George, E., McCulloch, R.: Bayesian cart model search. J. Am. Stat. Assoc. **93**(443), 935–948 (1998)
5. Dembczyński, K., Kotłowski, W., Słowiński, R.: Solving regression by learning an ensemble of decision rules. In: Rutkowski, L., Tadeusiewicz, R., Zadeh, L.A., Zurada, J.M. (eds.) ICAISC 2008. LNCS (LNAI), vol. 5097, pp. 533–544. Springer, Heidelberg (2008). https://doi.org/10.1007/978-3-540-69731-2_52
6. Friedman, J., Popescu, B.: Predective learning via rule ensembles. Ann. Appl. Stat. **2**, 916–954 (2008)
7. Györfi, L., Kohler, M., Krzyzak, A., Walk, H.: A Distribution-Free Theory of Non-parametric Regression. Springer, Heidelberg (2006)
8. Hyafil, L., Rivest, R.L.: Constructing optimal binary decision trees is NP-complete. Inf. Process. Lett. **5**(1), 15–17 (1976)
9. Letham, B., Rudin, C., McCormick, T., Madigan, D.: Interpretable classifiers using rules and Bayesian analysis: building a better stroke prediction model. Ann. Appl. Stat. **9**(3), 1350–1371 (2015)
10. Pedregosa, F., Varoquaux, G., Gramfort, A., Michel, V., Thirion, B., Grisel, O., Blondel, M., Prettenhofer, P., Weiss, R., Dubourg, V., Vanderplas, J., Passos, A., Cournapeau, D., Brucher, M., Perrot, M., Duchesnay, E.: Scikit-learn: machine learning in Python. J. Mach. Learn. Res. **12**, 2825–2830 (2011)
11. Quinlan, J.R.: C4.5: Programs for Machine Learning. Morgan Kaufmann Publishers Inc., San Francisco (1993)
12. Yang, H., Rudin, C., Seltzer, M.: Scalable Bayesian rule lists. In: Proceedings of the 34th International Conference of Machine Learning (ICML 2017) (2017)

A CNN Based Transfer Learning Model for Automatic Activity Recognition from Accelerometer Sensors

Belkacem Chikhaoui[1(✉)], Frank Gouineau[2], and Martin Sotir[2]

[1] Department of Science and Technology, TELUQ University, Montreal, Canada
belkacem.chikhaoui@teluq.ca

[2] Computer Research Institute of Montreal, Montreal, Canada
{Frank.Gouineau,Martin.Sotir}@crim.ca

Abstract. Accelerometers are become ubiquitous and available in several devices such as smartphones, smartwaches, fitness trackers, and wearable devices. Accelerometers are increasingly used to monitor human activities of daily living in different contexts such as monitoring activities of persons with cognitive deficits in smart homes, and monitoring physical and fitness activities. Activity recognition is the most important core component in monitoring applications. Activity recognition algorithms require substantial amount of labeled data to produce satisfactory results under diverse circumstances. Several methods have been proposed for activity recognition from accelerometer data. However, very little work has been done on identifying connections and relationships between existing labeled datasets to perform transfer learning for new datasets. In this paper, we investigate deep learning based transfer learning algorithm based on convolutional neural networks (CNNs) that takes advantage of learned representations of activities of daily living from one dataset to recognize these activities in different other datasets characterized by different features including sensor modality, sampling rate, activity duration and environment. We experimentally validated our proposed algorithm on several existing datasets and demonstrated its performance and suitability for activity recognition.

Keywords: Transfer learning · Deep learning · Accelerometer data
Activity recognition · CNN

1 Introduction

Human activity recognition is a challenging and well-researched problem [1]. With the emergence of wearable devices and accelerometers, activity recognition is applied in different domains including assisted living, healthcare, sport, human-computer interaction, smart cities, and security [2,3]. Most researchers use machine learning algorithms for activity recognition, which require substantial amount of labeled data to produce satisfactory results under diverse circumstances, and significant efforts are required to apply the learned models to

P. Perner (Ed.): MLDM 2018, LNAI 10935, pp. 302–315, 2018.
https://doi.org/10.1007/978-3-319-96133-0_23

different environments [3]. Besides, collecting real labeled data in new environments is costly, time consuming and error prone process.

Transfer learning is a challenging research problem that consists in designing systems which can leverage experience from previous tasks into improved performance in a new task which has not been encountered before [1]. Transfer learning has several potential benefits which can be summarized as follows: (1) learning new tasks requires less time, (2) less information is required from experts, (3) less labeled data is required for activity recognition, and (4) more situations in different environments can be handled effectively. These benefits have led researchers to deeply investigate and apply transfer learning techniques with varying degrees of success [1,4]. However, most past research on transfer learning assumed that the distribution of the data in the new environment is the same as that used in the learning process (old environment), which may not hold and can be difficult to satisfy in many real-world applications [3].

With the emergence and successful deployment of deep learning techniques for activity recognition, more research is directed from traditional machine learning techniques to deep learning techniques in different applications such as activity recognition [5], speech recognition [6], health monitoring [7], home automation [8], and automatic security surveillance [9]. Deep learning techniques such as Convolutional Neural Networks (CNNs) have the potential to automatically extract features by stacking several convolutional operators to create a hierarchy of progressively more abstract features [10]. CNNs are well suited for classification problems and very often outperforming existing approaches [11]. For example, in computer vision, features learned by CNNs in the first layers represent low level representations of the input data, and can be potentially applicable to different datasets and domains [12]. The rational of using CNNs is that features learned by CNNs represent invariant factors underlying the input data, which have great potential to be generalized through transfer learning. Therefore, kernels in the lower layers of the CNN architecture are rather generic such as edge detectors in image recognition, while kernels in upper layers are more specialized and specific that represent higher level features [10,12]. Besides, CNN is suitable for inferring long term repetitive activities given its capability to learn deep features contained in recursive patterns [13,14].

In this paper, we are interested in investigating whether the representations learned by CNNs are transferable in accelerometer activity recognition. Activity recognition from accelerometer sensors poses several challenges such as (1) sensor modality, (2) sensor placement, (3) sensor sampling rate, (4) environment, and (5) users types i.e. young, older adults, male, and female. To overcome these challenges, and in order to characterize the feasibility and benefits of transfer learning using CNNs, we deeply analyze kernel transfer between users, applications, environments, and accelerometer modalities, sampling rate and placements. The contributions of this paper are as follows:

- Investigate the specialization of features transfer learning in CNNs layer-by-layer.

- Identify the best number of layers and CNN parameters to be retained for successful transfer learning for activity recognition.
- Evaluate the benefits of transfer learning for activity recognition from accelerometer data under different scenarios.
- Conduct extensive experiments over a variety of publicly available datasets to validate the transfer learning using CNNs.

The rest of the paper is organized as follows. First, we give an overview of related work in Sect. 2. Section 3 describes the CNN model for automatic feature extraction and analyzes the specialization of features for transfer learning. The results of our experiments on real datasets with different scenarios are presented in Sect. 4. Finally, Sect. 5 presents our conclusions and highlights future work directions.

2 Related Work

Transfer learning is an emerging topic in different domains and applications such as computer vision [15], speech recognition [16] and more specifically in human activity recognition [10]. Researchers applied transfer learning to tackle different challenges categorized as follows: (1) sensor modalities, (2) environments, (3) availability of labeled datasets, and (4) knowledge representation [1]. Several models have been proposed on transferring knowledge representation such as data instances [17], feature representations [18,19], or model parameters [10,20]. Despite the scarcity of transfer learning models for activity recognition from accelerometers, research has demonstrated its feasibility using conventional machine learning techniques [21,22]. Recent progress in transfer learning has been analyzed in [4,23] and a survey on transfer learning for activity recognition using conventional methods can be found in [1]. Conventional transfer learning techniques for activity recognition present several limitations such as (1) feature space in the different domains must be the same, (2) user intervention for parameter setting, (3) the number of sensors in source and target domain must be the same, and (4) the distribution of the sensor data in the target domain must be the same as the source domain. These limitations pushed researchers to investigate other directions such as deep learning techniques for transfer learning.

Very little work has been done on transfer learning for activity recognition from accelerometer data using deep learning techniques. For example, Morales and Roggen [10] studied feature transfer using CNNs across mobile activity recognition domains. The authors found than kernels in the first layers tend to be more generic in the same application domain. However, the authors did not investigate different sensor modalities such as different sampling rates. To the best of our knowledge, the work of [10] is the only the work that studies CNNs for transfer learning for activity recognition from wearable devices. In consequence, our work represents an exhaustive investigation of CNNs for transfer learning for activity recognition from accelerometer data in different scenarios. Generic representations using sparse coding have been used for transfer learning [24]. Sparse coding allows to model sensor signals from a set of basic vectors capturing latent structures in the data [10,24]. sparse coding makes it possible to understand the variability of input data by visualizing and analyzing the signals representations. It has been demonstrated that sparse coding is suitable to overcome the problem of different sensor modalities [24].

This paper represents the first exhaustive investigation of transfer learning with CNNs for activity recognition from accelerometer sensors. Our goal is to understand, through several scenarios related to challenging problems, the characteristics and conditions of CNNS features to ensure successful transfer learning. Furthermore, we will experimentally show how the number of hidden layers and epochs could affect the transfer learning performance.

3 Transfer Learning with CNNs

This section presents how transfer learning is performed using CNNs. We define first transfer learning, and then we present the CNN architecture for transfer learning.

According to [4], transfer learning can be formally defined as follows: given a source domain D_s and learning task T_s, a target domain D_t and learning task T_t, transfer learning aims to help improve the learning of the target predictive function $f_t(.)$ in D_t using the knowledge in D_s and T_s where $D_s \neq D_t$, or $T_s \neq T_t$. When source and target domains are the same i.e., $D_s = D_t$, and their learning tasks are the same i.e., $T_s = T_t$, the learning problem becomes a traditional machine learning problem [4].

The CNN architecture comprises usually convolutional, pooling, and fully-connected layers to perform classification or regression tasks as shown in Fig. 1. According to [13], CNN has two advantages over other models: local dependency and scale invariance. Local dependency means the nearby signals in activity recognition are likely to be correlated, while scale invariance refers to the scale-invariant for different frequencies. Two important aspects to be considered when applying CNNs to activity recognition from sensors: (1) input adaptation, and (2) pooling. Input adaptation consists in adapting the input data to create virtual images given that sensors produce time series readings such as acceleration signal, which is temporal multidimensional 1D readings. To do so, we treat each dimension as a channel, and perform a 1D convolution on them. After convolution and pooling, the outputs of each channel are flattened as shown in Fig. 1 to prepare the input for the dense layers. The advantage of this approach is that we treat the 1D sensor reading as a 1D image, which is simple and easy to implement. The convolution-pooling is very common in CNNs, and most approaches performed max or average pooling functions after convolution [25,26]. In this paper, we performed max pooling after convolution as shown in Fig. 1. Before the softmax classification layer, we make use of dense (fully-connected) layers in which every input is connected to every output by a weight.

To perform transfer learning, we train a CNN model using a source dataset. Then, we transfer the learned features (kernels) at each layer to train a target dataset by freezing their weights, i.e. transferring kernels without changing their weights and without backpropagating errors during the training. The number of kernels to be transferred is determined empirically based on the recognition accuracy in the target dataset. We performed experiments with a number of hidden layers ranging from one (1) to eight (8) layers. Note that the purpose of transferring learned features is to avoid training a new model on the target dataset. Therefore, the task will consist in fine-tuning the trained model using the target dataset.

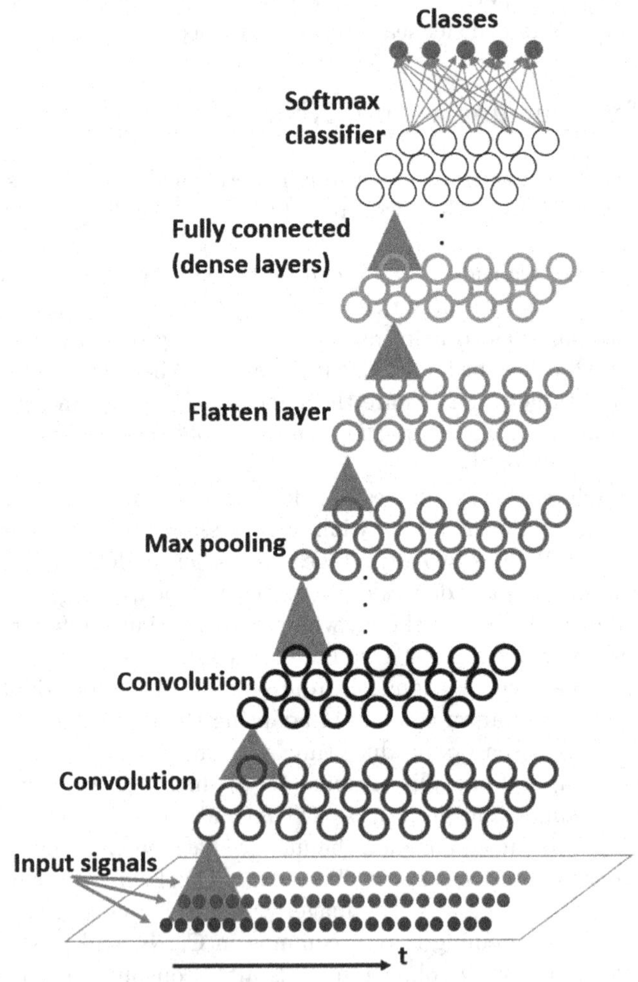

Fig. 1. CNN architecture with convolution layers, max pooling layer, flatten layer, dense layers and softmax classifier layer.

4 Experimental Setup

We evaluate our transfer learning model using three publicly available datasets as described in the Datsets section. We evaluate the performance of our transfer learning model using the F-score measure. We used four (4) seconds non overlapping segments as input to our model for all datasets independently of the sampling rate. We always train the source model using 100 epochs. Target models are fine-tuned with 50 epochs, which is determined empirically as shown in the following sections. All experiments were run on a GPU GeForce GTX 1080 Ti with 3584 cores, 1582 MHz boost clock and 11 GB GDDR5X RAM. The CNN models are developed under the Tensorflow library. Datasets and experimental results are presented in the following sections.

4.1 Datasets

To experimentally evaluate the transfer learning with CNNs, we selected three publicly available wearable sensor datasets with different characteristics as shown in Table 1. These datasets contain multimodal activities such as walking, sitting, standing, ascending and descending stairs, jumping, jogging, and biking, with different timescales and multiple sensor modalities, placements and sampling rates. These datasets make a real testbed for the transfer learning activity recognition scenarios. We used six features: three accelerometer features and three gyroscope features. Table 1 summarizes the three datasets with their characteristics.

Table 1. Datasets used in our experiments.

	Datasets		
Characteristics	Heterogeneity HAR [27]	RealWorld (HAR) [28]	MobiAct [29]
Sensor modality	2 embedded sensors 8 smartphones 4 smart watches	Smart phone + smart watch	One smartphone
Sensor placement	Waist (smartphone) Arm (smart watch)	Head, chest, upper arm, waist, forearm, thigh, and shin	Trousers' pocket
Sampling rate	200 Hz, 150 Hz, 100 Hz	50 Hz	20 Hz
Number of activities	6	8	9
Number of subjects	9	15	57
Environment	Indoor and outdoor	Indoor and outdoor	Indoor
Males	-	8	42
Females	-	7	15
Age	25–30	16–62	20–47

These datasets gather the overall challenges we discussed before, which will help experimentally evaluate different scenarios of transfer learning using CNNs. To find the optimal number of epochs for training our models, we performed transfer learning experiments with different number of epochs ranging from 5 to

100 epochs to find the optimal number of epochs for our experiments. Table 2 shows transfer learning results between the three datasets by varying the number of epochs.

Table 2. Transfer learning results between all datasets with different number of epochs.

Number of epochs	MobiAct → RealWorld	MobiAct → Heterogeneity
	F-Measure	F-Measure
5	0.745	0.778
10	0.755	0.803
15	0.776	0.815
20	0.792	0.836
30	0.792	0.856
50	0.803	0.862
100	0.752	0.842

As shown in Table 2, with 50 epochs we get the best transfer learning results. Despite the fact that with 20 epochs the results do not change significantly, more specifically for the RealWorld dataset, we take 50 as the optimal number of epochs in our experiments to fine tune our transferred models on target datasets.

4.2 Transfer Learning Across Users

In this section we evaluate the CNN transfer learning among users in different datasets. The goal is to determine if the CNN learned features are able to be generalized across users with different ages and genders. This is a challenging problem in activity recognition as usually young people perform activities with daily living faster than older people for example. To do so, we do a learning on the heterogeneity dataset (D_s) that comprises mainly young users where ages range from 25 to 30, and then we perform a transfer learning using the RealWorld dataset (D_t) as target as this dataset comprises different user ages ranging from 16 to 62 years. We then switch both datasets to evaluate transfer learning from RealWorld dataset to heterogeneity dataset. We perform experiments by varying the number of hidden layers transferred from 1 to 8 layers. The results obtained as shown in Table 3. We used only the forearm sensor placement for these experiments. Experiments with other sensor placements are presented in Sect. 4.3.

As shown in Table 3, transfer learning from RealWorld dataset to heterogeneity dataset shows good recognition results compared to the transfer learning from heterogeneity dataset to RealWorld dataset. This can be explained by the fact that RealWorld dataset contains a variety of users including young and older adults, whereas the heterogeneity dataset is composed mainly of young users.

Table 3. Transfer learning results between Heterogeneity and RealWorld datasets with different number of layers.

Number of layers	Heterogeneity → RealWorld F-Measure	RealWorld → Heterogeneity F-Measure
1	0.8	0.917
2	0.562	0.9
3	0.8	0.873
4	0.83	0.867
5	0.645	0.893
6	0.654	0.866
7	0.66	0.681
8	0.678	0.521

Therefore, we have more variability in RealWold dataset compared to the heterogeneity dataset. We observe also that the transfer learning results using the first four to five layers are better compared to those obtained with more that five layers. Consequently, the first four to five layers represent the generic features extracted by the CNN model that can be transferred to other datasets. The remaining layers are more dataset specific layers and not suitable to be transferred.

4.3 Transfer Learning with Different Sensor Placements

Wearable sensors can be placed at different locations of the human body as shown in Fig. 2.

For example, as shown in Table 1, the RealWorld dataset is collected by placing sensors at seven (7) different locations such as head, chest, upper arm, waist, forearm, thigh, and shin, whereas in the MobiAct dataset, the smartphone was placed in the trouser's pocket. Therefore, sensor placement plays an important role in data collection and significantly influences the activity recognition. In our work, we train a CNN model on the MobiAct dataset (D_s) and perform a transfer learning on the RealWorld dataset (D_t) for each sensor placement among the seven locations. We inverse the source and target datasets to evaluate transfer learning in both directions. Tables 4 and 5 show the transfer learning results obtained for each sensor placement. For space limitation, we choose only four sensor placements such as Waist, Head, Chest and Forearm.

As shown in Table 4, the transfer learning results from MobiAct to RealWorld are better than those obtained from RealWorld to MobiAct. This can be explained by the fact that the waist placement captures more variations in the activities compared to the forearm placement, which mainly captures variations in hand movements. Therefore, transfer learning from a more generic placement to a more specific placement is possible since the first placement captures generic

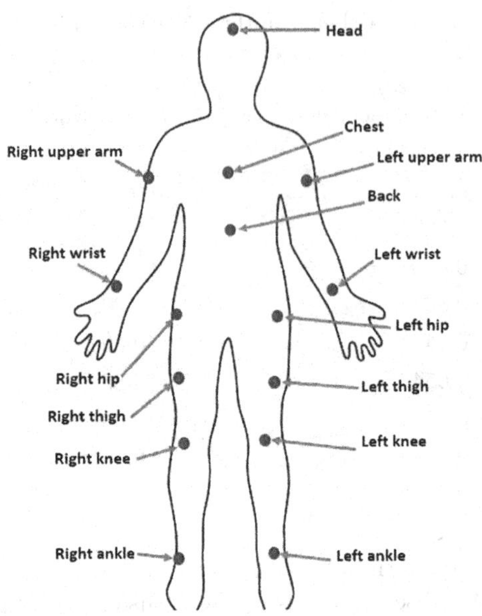

Fig. 2. Sensor placements on the human body for data collection.

Table 4. Transfer learning results with different sensor placements (Waist ↔ Forearm).

Number of layers	MobiAct → RealWorld Sensor placement: Waist → Forearm F-Measure	RealWorld → MobiAct Sensor placement: Forearm → Waist F-Measure
1	0.842	0.626
2	0.796	0.658
3	0.811	0.665
4	0.812	0.685
5	0.829	0.632
6	0.806	0.664
7	0.75	0.602
8	0.744	0.585

features of activities. The same observation applies for the head sensor placement in RealWorld dataset, as shown in Table 5. We performed also experiments for the two other sensor placements: Waist and Chest by varying the number of layers. Note that here we performed transfer learning only in one direction from MobiAct dataset to RealWorld dataset. The transfer learning results are presented in Table 6.

As we can see from the results obtained, transfer learning between different sensor placements is possible since all the results obtained showed an F-score greater that 0.5. The results are very promising between Waist sensor placement

Table 5. Transfer learning results with different sensor placements (Waist ↔ Head).

Number of layers	MobiAct → RealWorld Sensor placement: Waist → Head F-Measure	RealWorld → MobiAct Sensor placement: Head → Waist F-Measure
1	0.842	0.626
2	0.796	0.658
3	0.811	0.665
4	0.812	0.685
5	0.829	0.632
6	0.806	0.664
7	0.75	0.602
8	0.744	0.585

and Head and Forearm sensor placements with an average F-score of about 0.8. Sensor placement could help transfer learning by placing sensors in places so that they can capture more variations in activities such as the Waist placement as shown in the results obtained.

The transfer learning results obtained between the Waist and Chest sensor placements are overall similar. This is because the Waist and Chest placements captures relatively similar acceleration patterns for the whole body, which are more generic and could be transferred to other datasets.

Table 6. Transfer learning results with Waist and Chest sensor placements (Waist ↔ Chest and Waist ↔ Waist).

Number of layers	MobiAct → RealWorld Sensor placement: Waist → Chest F-Measure	MobiAct → RealWorld Sensor placement: Chest → Waist F-Measure
1	0.53	0.559
2	0.496	0.568
3	0.615	0.573
4	0.545	0.596
5	0.546	0.543
6	0.601	0.567
7	0.577	0.55
8	0.528	0.555

4.4 Transfer Learning with Different Sampling Rates, Sensor Modalities and Environments

Sampling rate is intrinsically connected to the accelerometer device. While a low sampling rate saves considerable energy, as well as transmission bandwidth and storage capacity, it is also prone to omitting relevant signal details that are of interest for contemporary analysis tasks [30]. The sampling rate plays an important role in determining the sliding window size when analyzing signals. Moreover, the use of different sensor modalities such as sensor type (smartphone, smartwatch, device, etc.) creates another challenge on how to take into account these sensor modalities to perform transfer learning. In addition, each dataset is collected in a specific environment. Activities of daily living such as eating, sitting, standing, reading and sleeping are usually performed in a home environment. However, activities like walking, jogging, and biking are performed outside the home environment. Most of the publicly available datasets perform all these activities in a laboratory setting as it is the case for the MobiAct dataset. In this scenario, our aim is to evaluate the transfer learning for activity recognition across multiple settings such as different sampling rates, sensor modalities and environments. For example, we would like to validate whether activities simulated in a laboratory match real activities. In the datasets we used, the sampling rate is different for each dataset as shown in Table 1 with different sensor modalities, which makes the transfer learning more complicate. Since the sampling rate and sensor modalities are different in all datasets, we evaluate transfer learning between the MobiAct and Heterogeneity datasets as they have very different sampling rates. Therefore, we train a CNN model on MobiAct dataset (D_s) with 20 Hz sampling rate and perform a transfer learning using Heterogeneity dataset as target domain (D_t) with sampling rate between 100 Hz and 200 Hz. We perform also transfer learning in both directions. We show the results by taking into account different hidden layers. We trained the source dataset with 100 epochs. In our experiments, we show how the first layer of a CNN architecture could be used to overcome the problem of sampling rates in accelerometer devices. Table 7 shows the transfer learning results.

Table 7. Transfer learning results by taking into account different sampling rates, sensor modalities and environments.

Number of layers	MobiAct → Heterogeneity Sampling rate: 20 Hz → 100 Hz to 200 Hz F-Measure	Heterogeneity → MobiAct Sampling rate: 100 Hz to 200 Hz → 20 Hz F-Measure
1	0.941	0.681
2	0.922	0.706
3	0.899	0.632
4	0.911	0.662
5	0.914	0.633
6	0.906	0.597
7	0.905	0.557
8	0.931	0.48

The transfer learning results obtained with different sampling rates indicate that it is possible to transfer features even if the sampling rate, sensor modality and environment are different. Indeed, the first layers of a CNN model tries to capture high level features independently of the sampling rate, which is demonstrated by the good results obtained by transferring layers from 1 to 5. However, the results decrease with the increase of the number of layers, which tend to be more specific features. We can also see that the transfer learning results obtained from MobiAct dataset to heterogeneity dataset are more promising compared to the results obtained in the other direction. This can be explained by: (1) the number of training samples in the MobiAct dataset is high, which allows to extract meaningful generic features of activities compared to when we have small training sample size. (2) single activity event detection requires a lower sampling rate than multi-activity recognition. Therefore, lower sampling rates allow to extract fine-grained features for each single activity in addition to the generic features, which explain the good transfer learning results obtained from MobiAct dataset with 20 Hz sampling rate to heterogeneity dataset with 100 Hz to 200 Hz sampling rates. This is supported by the findings obtained in [31].

5 Conclusion

In this paper we investigated transfer learning with CNNs for activity recognition from accelerometer sensors. We studied different scenarios corresponding to different challenging transfer learning problems in literature such as sensor modalities, environments, users, and sensor placement. We experimentally showed that first layers learned by CNNs could be transferred across datasets as they represent generic features. We have also demonstrated empirically that CNNs are suitable models for transfer learning under different scenarios. We found that the model performance increases when transfer learning is performed from low sampling rates to high sampling rates. We also noticed a drastic degradation in performance using different sensor placements even when changing the number of layers. In addition, first layers in this case may be specific also compared to other scenarios.

In the future, it would be interesting to identify automatically the number of layers to be transferred across datasets. In addition, it would be interesting to test other deep learning techniques such as RNNs, LSTM, and GANs and compare their performances in transfer learning for activity recognition with the same scenarios.

References

1. Cook, D., Feuz, K.D., Krishnan, N.C.: Transfer learning for activity recognition: a survey. Knowl. Inf. Syst. **36**(3), 537–556 (2013)
2. Chikhaoui, B., Ye, B., Mihailidis, A.: Aggressive and agitated behavior recognition from accelerometer data using non-negative matrix factorization. J. Ambient Intell. Hum. Comput. 1–15 (2017)
3. Chen, W.-H., Cho, P.-C., Jiang, Y.-L.: Activity recognition using transfer learning. Sens. Mater. **29**(7), 897–904 (2017)

4. Pan, S.J., Yang, Q.: A survey on transfer learning. IEEE Trans. Knowl. Data Eng. **22**(10), 1345–1359 (2010)
5. Chikhaoui, B., Gouineau, F.: Towards automatic feature extraction for activity recognition from wearable sensors: a deep learning approach. In: 2017 IEEE International Conference on Data Mining Workshops, ICDM Workshops 2017, New Orleans, LA, USA, 18–21 November 2017, pp. 693–702 (2017)
6. Graves, A., Jaitly, N.: Towards end-to-end speech recognition with recurrent neural networks. In: Proceedings of the 31st International Conference on International Conference on Machine Learning - Volume 32, ICML 2014, pages II–1764–II–1772. JMLR.org (2014)
7. Zhao, R., Yan, R., Chen, Z., Mao, K., Wang, P., Gao, R.X.: Deep learning and its applications to machine health monitoring: a survey. CoRR, abs/1612.07640 (2016)
8. Gokul, V., Kannan, P., Kumar, S., Jacob, S.G.: Deep Q-learning for home automation. Int. J. Comput. Appl. **152**(6), 1–5 (2016)
9. Kim, J., Mo, Y.J., Lee, W., Nyang, D.: Dynamic security-level maximization for stabilized parallel deep learning architectures in surveillance applications. In: 2017 IEEE Symposium on Privacy-Aware Computing (PAC), pp. 192–193 (2017)
10. Morales, F.J.O., Roggen, D.: Deep convolutional feature transfer across mobile activity recognition domains, sensor modalities and locations. In: Proceedings of the 2016 ACM International Symposium on Wearable Computers, ISWC 2016, Heidelberg, Germany, 12–16 September 2016, pp. 92–99 (2016)
11. Morales, F.J.O., Roggen, D.: Deep convolutional and LSTM recurrent neural networks for multimodal wearable activity recognition. Sensors **16**(1), 115 (2016)
12. Yosinski, J., Clune, J., Bengio, Y., Lipson, H.: How transferable are features in deep neural networks? In: Advances in Neural Information Processing Systems 27: Annual Conference on Neural Information Processing Systems 2014, 8–13 December 2014, Montreal, Quebec, Canada, pp. 3320–3328 (2014)
13. Wang, J., Chen, Y., Hao, S., Peng, X., Hu, L.: Deep learning for sensor-based activity recognition: a survey. CoRR, abs/1707.03502 (2017)
14. Hammerla, N.Y., Halloran, S., Plötz, T.: Deep, convolutional, and recurrent models for human activity recognition using wearables. In: Proceedings of the Twenty-Fifth International Joint Conference on Artificial Intelligence, IJCAI 2016, pp. 1533–1540 (2016)
15. Sargano, A.B., Wang, X., Angelov, P., Habib, Z.: Human action recognition using transfer learning with deep representations. In: 2017 International Joint Conference on Neural Networks (IJCNN), pp. 463–469 (2017)
16. Kunze, J., Kirsch, L., Kurenkov, I., Krug, A., Johannsmeier, J., Stober, S.: Transfer learning for speech recognition on a budget. In: Proceedings of the 2nd Workshop on Representation Learning for NLP, Rep4NLP@ACL 2017, Vancouver, Canada, 3 August 2017, pp. 168–177 (2017)
17. Zheng, V.W., Hu, D.H., Yang, Q.: Cross-domain activity recognition. In: Proceedings of the 11th International Conference on Ubiquitous Computing, UbiComp 2009, pp. 61–70 (2009)
18. Ordóñez, F.J., Englebienne, G., de Toledo, P., van Kasteren, T., Sanchis, A., Kröse, B.: In-home activity recognition: Bayesian inference for hidden Markov models. IEEE Pervasive Comput. **13**(3), 67–75 (2014)
19. Rashidi, P., Cook, D.J.: Activity recognition based on home to home transfer learning. In: Proceedings of the 5th AAAI Conference on Plan, Activity, and Intent Recognition, AAAIWS 2010–05, pp. 45–52 (2010)

20. van Kasteren, T.L.M., Englebienne, G., Kröse, B.J.A.: Transferring knowledge of activity recognition across sensor networks. In: Floréen, P., Krüger, A., Spasojevic, M. (eds.) Pervasive 2010. LNCS, vol. 6030, pp. 283–300. Springer, Heidelberg (2010). https://doi.org/10.1007/978-3-642-12654-3_17
21. Kurz, M., Hölzl, G., Ferscha, A., Calatroni, A., Roggen, D., Tröster, G.: Real-time transfer and evaluation of activity recognition capabilities in an opportunistic system. In: Adaptive, pp. 73–78 (2011)
22. Calatroni, A., Roggen, D., Tröster, G.: Automatic transfer of activity recognition capabilities between body-worn motion sensors: training newcomers to recognize locomotion. In: Eighth International Conference on Networked Sensing Systems (INSS 2011) (2011)
23. Weiss, K., Khoshgoftaar, T.M., Wang, D.D.: A survey of transfer learning. J. Big Data **3**(1), 3–9 (2016)
24. Bhattacharya, S., Nurmi, P., Hammerla, N., Plötz, T.: Using unlabeled data in a sparse-coding framework for human activity recognition. Perv. Mob. Comput. **15**(C), 242–262 (2014)
25. Ha, S., Yun, J.M., Choi, S.: Multi-modal convolutional neural networks for activity recognition. In: 2015 IEEE International Conference on Systems, Man, and Cybernetics, pp. 3017–3022 (2015)
26. Pourbabaee, B., Roshtkhari, M.J., Khorasani, K.: Deep convolutional neural networks and learning ECG features for screening paroxysmal atrial fibrillation patients. IEEE Trans. Syst. Man Cybern.: Syst. **PP**(99), 1–10 (2017)
27. Stisen, A., Blunck, H., Bhattacharya, S., Prentow, T.S., Kjærgaard, M.B., Dey, A., Sonne, T., Jensen, M.M.: Smart devices are different: assessing and mitigatingmobile sensing heterogeneities for activity recognition. In: Proceedings of the 13th ACM Conference on Embedded Networked Sensor Systems, SenSys 2015, pp. 127–140 (2015)
28. Sztyler, T., Stuckenschmidt, H.: On-body localization of wearable devices: an investigation of position-aware activity recognition. In: 2016 IEEE International Conference on Pervasive Computing and Communications (PerCom), pp. 1–9 (2016)
29. Vavoulas, G., Chatzaki, C., Malliotakis, T., Pediaditis, M., Tsiknakis, M.: The mobiact dataset: recognition of activities of daily living using smartphones. In: Proceedings of the 2nd International Conference on Information and Communication Technologies for Ageing Well and e-Health, ICT4AgeingWell 2016, Rome, Italy, 21–22 April 2016, pp. 143–151 (2016)
30. Khan, A., Hammerla, N., Mellor, S., Plötz, T.: Optimising sampling rates for accelerometer-based human activity recognition. Pattern Recogn. Lett. **73**, 33–40 (2016)
31. Qi, X., Keally, M., Zhou, G., Li, Y., Ren, Z.: Adasense: adapting sampling rates for activity recognition in body sensor networks. In: 2013 IEEE 19th Real-Time and Embedded Technology and Applications Symposium (RTAS), pp. 163–172, April 2013

A Mixture of Personalized Experts for Human Affect Estimation

Michael Feffer(✉), Ognjen (Oggi) Rudovic, and Rosalind W. Picard

MIT Media Lab, Massachusetts Institute of Technology, Cambridge, MA 02139, USA
{mfeffer,orudovic,roz}@mit.edu

Abstract. We investigate the personalization of deep convolutional neural networks for facial expression analysis from still images. While prior work has focused on population-based ("one-size-fits-all") approaches, we formulate and construct personalized models via a mixture of experts and supervised domain adaptation approach, showing that it improves greatly upon non-personalized models. Our experiments demonstrate the ability of the model personalization to quickly and effectively adapt to limited amounts of target data. We also provide a novel training methodology and architecture for creating personalized machine learning models for more effective analysis of emotion state.

Keywords: Mixture of experts · Domain adaptation
Personalized machine learning · Residual networks

1 Introduction

In recent years, machine learning has become increasingly popular for performing analysis and generating predictions based on data in a variety of different areas, especially in healthcare and fields devoted to improving health and wellness [1,2]. In the realm of affective computing, it has been applied to tasks such as automated analysis of persons' engagement, personality, and affective states during human-computer [3] and human-robot interaction [4]. For instance, as robots become increasingly complex and integrated into daily life, it will be important for them to perceive and understand not only human biometrics but also human emotional metrics. Enhanced perception of human emotions could enable robots to avoid actions that would worsen a human's emotional state and perhaps even influence them to act in a way that could improve a human's well-being. Moreover, in the advent of powerful machine learning capabilities for mobile devices, it is possible nowadays to perform emotion analysis through smartphone cameras. Therefore, a smartphone application could be programmed to detect a user's emotions and recommend strategies for dealing with negative emotions or actively attempt to improve the user's mood. Lastly, emotion analysis could be used for emotion and engagement detection and coaching for individuals with autism, who have inherent difficulties in reading others' emotions and expressing their own in a way that can be easily understood by neurotypicals.

P. Perner (Ed.): MLDM 2018, LNAI 10935, pp. 316–330, 2018.
https://doi.org/10.1007/978-3-319-96133-0_24

Most machine learning approaches are successful because the models produced can generalize to unseen data and were trained on a plethora of existing data. However, most of the existing approaches ignore the fact that people express affect differently, even when they are part of the same culture. Therefore, learning a general predictor or classifier (the traditional "one-size-fits-all" approach) with data from one set of people typically underperforms when tested on people from a disjoint set but also on specific people within the training set. Moreover, it is difficult to obtain a large amount of training data for each target subject because providing labels for these data is costly in terms of time and resources [5]. Thus, improving machine learning approaches so that they can efficiently leverage small amounts of training data to adapt to each target subject is of large importance for increasing the model's effectiveness. To address these challenges, a number of works attempted model personalization [4]. The goal of model personalization is to leverage the individual-specific data in order to adapt a general classifier (the "one-size-fits-all" approach) learned from data of previously seen people (source subjects) to the profiles of specific individuals (target subjects). This has shown great improvements in a number of machine learning tasks related to human-data analysis (e.g., [1,2,6]).

In this paper, we have focused on a specific type of model personalization for estimation of facial affect (valence and arousal) using the notion of an ensemble of models and domain adaptation [7]. Specifically, we use the Mixture-of-Experts (MoEs) approach [8] to model the facial expression data (face images) of source subjects, for whom it is assumed that a large amount of image labels for the facial affect is readily available. We adopt MoEs where each expert represents one of the source subjects, which has greater modeling flexibility and improved ability to capture the large variation in facial expressions of different subjects compared to a single expert, which typically captures the average variation. While this approach performs better fitting of the source subjects, it is not guaranteed that this performance translates to previously unseen subjects (target), as confirmed in our experiments. To this end, we perform a supervised adaptation of the learned MoEs model using a varying portion of labeled data of all of the target subjects. We show: (i) that this approach achieves improved performance on the target subjects compared to a single expert model, and (ii) that it also outperforms the same model trained solely on the data of target subjects used to adapt the model. The latter is due to the ability of our approach to efficiently leverage the data of the source subjects. We demonstrate this in the context of deep neural networks that we tuned for "end-to-end" estimation of valence and arousal from still images of faces from the multimodal affect database REmote COLlaborative and Affective (RECOLA) database [9], used in the Audio/Visual Emotion Challenge and Workshop (AVEC) 2016 [10]. Note, however, that the focus of this work is not to outperform existing models in affect estimation but instead to examine how personalization and supervised domain adaptation can bolster current affect estimation models, an intersection of research areas that has yet to be explored.

In the sections that follow, we describe our approach to personalizing deep convolutional neural networks and a MoEs model. We first discuss related work both regarding the domain adaptation and MoEs. Then, we describe our mixture of personalized experts approach, followed by its experimental validation and derived conclusions.

2 Related Work

MoEs refers to a learning approach that involves training multiple "expert" subnetworks. These networks produce the same type of output for the same type of input, but they are trained on different subsets of data so that they are fine-tuned to specific contexts. At test time, all of them are given the same input, and an output for the overall network is created by combining the expert outputs in a certain way or somehow selecting the "best" output. Among the first proponents of this technique, [8] introduce a method of selecting an output by using a gating network that performs softmax activation to assign probabilities of randomly selecting the outputs of each of the expert networks for the current training example, after which an output is randomly chosen via a probability draw based on that distribution. Since then, it has been studied extensively for over twenty years, and in that time, expert models with other classifiers such as SVMs [11] and Gaussian Processes [12,13], have also been researched. With regard to deep neural networks, a myriad of different network architectures have been explored, ranging from hierarchical and ensemble experts [14] to networks with infinite numbers of experts [15] and nested mixtures of experts that allow for deep learning [16]. Although our work is built upon the same framework of MoEs, it differs from existing approaches as we personalize the experts to each subject. Furthermore, we combine the learning of MoEs with supervised domain adaptation (DA) [7] in order to efficiently adapt the model to previously unseen subjects. While there is a large body of work on DA, we do not intend here to improve upon existing DA methods. We rather use its notion when adapting the target MoE model that we propose for the model personalization to target subjects. For a detailed review of existing DA approaches, see [7]. Also, even though model personalization has been researched in several previous contexts (e.g., self-reported pain analysis [6] and robot therapy for autism [4]), none of these methods have been explored using a mixture of experts architecture in the context of model personalization. To this end, we have adopted this approach in our personalized framework. We have taken particular inspiration from [16] to utilize a mixture of experts approach that outputs a weighted combination of expert outputs based on a gating network learned form source subjects, as will be elaborated in the following sections.

3 Methodology

3.1 Notation

We consider the following setting: we are given a number of training subjects (source), denoted as $P^{(s)} = \{p_1^{(s)}, \ldots, p_{n_s}^{(s)}\}$, where $id^{(s)} = 1, \ldots, n_s$ represents

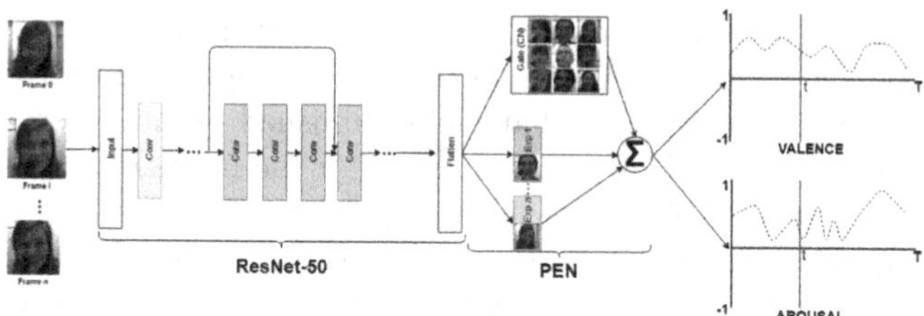

Fig. 1. The architecture of the proposed approach. The input is a subject's video and the outputs are his/her estimated valence and arousal levels. We first applied Faster R-CNN [17] to extract the face region from each raw image frame. The extracted faces were passed through a ResNet-50 [18], fine-tuned on source subjects' data. The obtained deep features were used as input to our personalized expert network (PEN) for automatic estimation of valence and arousal. This also contains a "gating network" (CN) that assigns different weights to each expert in the PEN during inference of new test images.

the subject *id*, and n_s is the number of the subjects. Data of these source subjects are used to learn a shared model optimized on these subjects. We are also given a number of test subjects (target) which were previously unseen by the shared model. These are denoted as: $P^{(t)} = \{p_1^{(t)}, \ldots, p_{n_t}^{(t)}\}$, and $id^{(t)} = 1, \ldots, n_t$. Then, starting from the shared model, the goal is to achieve the best possible performance on $P^{(t)}$. Furthermore, the data of each subject is stored as $p_i = \{X_i, Y_i\}$[1], where input features of the subject i are given by: $X_i = [x_{i1}; \ldots; x_{iN_i}] \in \mathcal{R}^{N_i \times D_x}$, where N_i is the number of available examples of the subject, and D_x is the input feature size. Note that these examples may be temporally correlated (in the case of video data) or be randomly sampled from independent observations of the subject. Likewise, the output labels (in our case, the levels of valence and arousal of the subject), are given as: $Y_i = [y_{i1}^v\, y_{i1}^a; \ldots; y_{iN_i}^v\, y_{iN_i}^a] \in \mathcal{R}^{N_i \times D_y}$, where $D_y = 2$. In what follows, we first describe how the shared model is learned from $P^{(s)}$, and used to estimate target affective states on $P^{(t)}$. Then, we propose an expert model, where each expert corresponds to one of the source subjects. The key to our approach is the model personalization step, where we propose a learning strategy designed to adapt the expert model to target subjects, using a varying portion of their (previously seen) data. Lastly, we provide details of the learning and inference in the personalized expert model.

3.2 Shared Model

We start by learning a shared model that is trained on data of all source subjects, without taking into account their *id*. This model is based on a deep architecture composed of the layers of a residual network (ResNet) [18], a pre-trained deep network commonly used to extract the most informative features for object classification [18]. We used the ResNet-50 architecture composed of multiple three-

[1] For notational simplicity, we drop here the dependence on the source/target subjects.

layer "bottleneck" building blocks (containing 50 layers in total) described in the original ResNet paper [18]. When beginning training, we initialized the layer weights corresponding to weights that yield published optimal performance on the ImageNet dataset [19]. However, we use all of the layers of the network but the last (i.e. the softmax layer) as it was optimized for classification of various object categories such as "laptop" and "orange". We instead replace the last layer with an data-uninformed fully-connected dense layer with linear activation, which we use to fine-tune the ResNet weights for the target task: the estimation of valence and arousal from face images (see Fig. 1). This architecture receives as input the face images of source subjects (x) and passes the most discriminative (deep) facial features (z) to the regression layer in the output through the following mapping: $x \rightarrow z \rightarrow \tilde{y}$, where $\tilde{y}$ are the estimated levels of valence and arousal. The training of the shared model is divided into two stages. First, the whole network (ResNet included) is optimized for estimation of y. Then, we freeze all of the (fine-tuned) layers of the ResNet ($W^{r-net} \in \mathcal{R}^{D_x \times D_z}$) and additionally fine-tune the last (regression) layer ($W^s \in \mathcal{R}^{D_z \times D_y}$). We experimented with different learning strategies and found that this one performed the best. This is because of the large number of parameters that need to be tuned simultaneously. Due to this, the network underfits the last layer, so we overcome this by additional tuning of the last layer. The resulting network, referred to as the shared network (SN), is used to initialize the expert network (EN), which is then adapted to the population of target subjects as described below.

3.3 Personalized Expert Network (PEN)

The learning of the expert network is accomplished using the Mixture-of-Experts (MoEs) approach, where an expert network (EN) is comprised of a set of layers called "experts" that are denoted as $e_1, \ldots, e_n$. Furthermore, an EN is also comprised of a "gating network" denoted as CN (which in our case is a person selector network). Its output is used to weight the contribution (relevance) of each expert during the inference stage. In our personalized model setting, during the model training on source subjects, each expert corresponds to one training subject (thus, n_s experts). This personalization yields a personalized expert network (PEN). Each expert is modeled using a feed-forward network with fully connected linear activations, as used for the SN, but each with their own parameters. Thus, the following mapping is learned: $x \rightarrow z \rightarrow \tilde{y}^e = [\tilde{y}_1^e \ \ldots \ \tilde{y}_{n^{(s)}}^e]$, where $\tilde{y}_i^e$ are the valence/arousal estimates by the i-th expert. Likewise, the CN aims to learn the mapping: $x \rightarrow z \rightarrow h \rightarrow \tilde{c} = [\tilde{c}_1 \ \ldots \ \tilde{c}_{n^{(s)}}]$, where $\tilde{c}_i$ is the (normalized) weight for the i-th expert during the model learning. More specifically, the CN is designed as a two-layer network. The first layer is a fully-connected feed-forward linear activation network. The linear activations are then passed through a softmax layer, providing the probability that the input sample comes from one of n_s source subjects and thus assuring that the outputs of the CN sum to one. More formally, given an input x to the ResNet, the output of the PEN is defined as:

$$\tilde{y} = \sum_{i=1}^{n_s} \tilde{c}_i \cdot \tilde{y}_i^e = \tilde{c} \otimes \tilde{y}^e, \tag{1}$$

where $\tilde{y}$ is the weighted combination of the individual experts. The output of the CN, given the activations (z) of the ResNet is obtained as:

$$\tilde{c} = \text{softmax}(h) \text{ and } h = \text{fcl}(z; W^s), \tag{2}$$

where z is passed through the fully connected layer (fcl), the output of which is fed into the softmax function to yield a categorical probability distribution over the source subjects. Note that during training of the PEN, the targets for the CN output are the subjects' ids encoded via the 1-hot encoding (e.g., $c = [0, 1, 0, \ldots, 0]$ for subject $i = 2$). Similarly, each expert $i = 1, \ldots, n^{(s)}$ produces estimates for target affective dimensions as:

$$\tilde{y}_i^e = \text{fcl}(z; W_i^e), \tag{3}$$

where z is multiplied by a trainable weight matrix W_i^e for expert i.

The network personalization is attained by using the prior knowledge about the source subjects: each expert is supposed to represent one of the subjects, and CN performs the selection of that expert during the model learning. Therefore, given the training data of $n^{(s)}$ source subjects, the overall loss is the sum of losses due to differences between y and $\tilde{y}$ (the weighted combination of expert outputs) as well as losses from the CN. However, in practice this loss does not enforce the sparsity on the selector's weights ($\tilde{c}$), which may result in the learned PEN expending too much modeling power of each expert on trying to fit the data of all source subjects. In turn, we may end up with an expert that is unable to specialize in individual characteristics of the subject, resulting in an average model that is suboptimal. To overcome this, we introduce the L-1 sparsity constraint on the output of the CN, but this cannot be done directly because the outputs of that layer always sum to one. Instead, we enforce the sparsity on the activations of the fcl of the SN. This leads to the following joint loss being optimized during the parameter learning:

$$\alpha = \alpha^y + \lambda_0 \alpha_c^s + \lambda_1 \alpha_r^s, \tag{4}$$

where α^y is the mean-squared error (MSE) loss between the PEN estimates and the ground-truth labels for valence and arousal (y). (λ_0, λ_1) are regularization parameters that are optimized on the validation data. They control the trade-off between the model performance and the penalty terms: α_c^s, which ensures that each expert focuses on its corresponding subject, and α_r^s further ensures this through the sparsity constraint. These individual losses are defined as:

$$\alpha^y = \frac{1}{N^{(s)}} \sum_{i=1}^{n^{(s)}} \sum_{j=1}^{N_i^{(s)}} (y_j^i - \tilde{y}_j^i)(y_j^i - \tilde{y}_j^i)^T, \tag{5}$$

and $N_i^{(s)}$ and $N^{(s)}$ are the number of training data per source subject and overall, respectively. The selector loss is given by:

$$\alpha_c^s = \frac{1}{N^{(s)}} \sum_{i=1}^{n^{(s)}} \sum_{j=1}^{N_i^{(s)}} H(c_j^i, \tilde{c}_j^i), \tag{6}$$

where $H(\cdot,\cdot)$ is the cross-entropy loss that is commonly used for discrete variables, as is the case here. Finally, the standard L-1 sparsity is enforced via:

$$\alpha_r^s = \frac{1}{N^{(s)}} \sum_{i=1}^{n^{(s)}} \sum_{j=1}^{N_i^{(s)}} |h_j^i|_{L_1}. \tag{7}$$

Note that this loss treats each activation/image frame independently, and therefore, no structure (prior information) about the source subjects is directly modeled. Nevertheless, this should still result in the learned activations being sparse, on average, for different face images of the subjects.

3.4 PEN: Supervised Adaptation to Target Population

Once the PEN parameters are optimized for the source subjects, our goal is to achieve the best performance on target subjects that the PEN has not seen before. In the traditional supervised machine learning approach, this would be evaluated using the network learned solely from the data of the source subjects. However, this typically leads to the learned model attaining a lower performance on target subjects than on the source subjects, as expected. Also, since the PEN is "tuned" to the latter, it may even more easily overfit those subjects, leading to the comparable to or worse performance than that of the SN trained on the target subjects. This is despite the modeling flexibility introduced by the local (person) experts, which allows the PEN to better fit the source subjects. Conversely, in the proposed personalized learning approach, we adopt the supervised domain adaptation approach [20], where the target subjects are treated as another domain assumed to be different from the one used to train the PEN. To investigate this, we pool the data of target subjects and use a varying portion of this data (with approximately equal amounts of data from each subject) to "tune" the PEN to those subjects. This is driven by two assumptions. First, we should be able to adapt the PEN to target subjects using a (significantly) smaller number of target data than originally used from the source. This is mainly because the network has been pre-trained on the latter and should be able to adapt to new subjects more easily. Second, while the same assumption may hold for the SN, the proposed PEN is more flexible in its modeling power (due to the multiple experts). This in turn should allow it to better adapt to the previously unseen subjects by focusing on their individual characteristics, thus, avoiding the limitations of the "one-size-fits-all" approach. Formally, this is attempted by fine-tuning the PEN to the population of target subjects via the following adaptation loss:

$$\alpha_{ad} = \alpha^{y^t} + \lambda_1 \alpha_r^{s,t}, \tag{8}$$

This supervised adaptation loss uses a small number of labeled data of target subjects (x^t, y^t) to adjust the PEN parameters. It is important to note that we do not use the ids of target subjects during the model adaptation, thus, the parameters of the CN are optimized in an unsupervised fashion by setting $\alpha_c^s = 0$. However, we still impose the sparsity constraint on the CN parameters. We also do not further optimize the ResNet parameters - these are rather used as the feature extractor during the adaptation stage. The benefits of this are two-fold. First, we preserve the privacy of the target subjects[2]. This is important in the context of many applications where the user does not want to reveal his/her identity, while still being able to receive estimates of target affective states. Second, instead of completely "overwriting" the previously learned PEN, the adapted model performs the actual adaptation of the model rather than learning it from scratch. This is motivated by the assumption that the PEN model has seen a large amount of labeled data from source subjects, thus encapsulating valuable knowledge about the affect expressions of different subjects. In this way, more robust adaptation of the PEN to new subjects is expected. Ideally, when $n^{(s)} > n^{(t)}$, PEN should be able to specialize one expert to each target subject, while compensating the lack of target subject data with the knowledge extracted from the source subjects.

3.5 Learning, Inference and Implementation Details

We first summarize the learning and inference in the proposed models, followed by the implementation details of our deep architecture. We start with the learning of the SN. The learning of the model parameters $\{W^{r-net}, W^s\}$ is performed in two steps: (i) The joint fine-tuning of the ResNet with the parameter optimization in the appended fcl, used for estimation of valence and arousal. (ii) The additional fine-tuning of the fcl parameters $\{W^s\}$, while freezing the ResNet parameters i.e., $\partial W^{r-net} = 0$. For (i) W^{r-net} was initialized to weights corresponding to optimal training on the ImageNet dataset, and W^s was initialized randomly. Our implementation appended the SN to the last flattening layer of the ResNet and trained all of these layers together. For (ii) we took our best model from (i) (based on validation set performance) and froze every layer in the architecture besides the SN. After more training, we saved this fully-adapted architecture and used it as the starting point for expert layers going forward.

To learn the PEN parameters, we first initialized each expert using the weights of the SN as $W^{e_i} \leftarrow W^s$, $i = 1, \ldots, n^s$. This ensures that individual experts do not overfit the corresponding source subjects, which could adversely affect their generalization to target subjects (we describe this below). The initial learning of the CN was done in isolation from the rest of the network. Only the

[2] For instance, only the ResNet features of target subjects need be provided as input to the adapted model, as original face images cannot be reconstructed from those features.

Algorithm 1. Personalized Experts Network (PEN)

Source Learning Input: Source persons data $P^{(s)} = \{p_1^{(s)}, \ldots, p_{n_s}^{(s)}\}$
step 1: Fine-tune ResNet weights (W^{r-net}) and optimize SN (W^s)
step 2: Freeze ResNet weights and fine-tune SN (W^s)
step 3: Initialize the experts ($W_i^e \leftarrow W^s$) and optimize PEN (W^s, W^c)
Target Adaptation Input: Target persons data $P_{ad}^{(t)} \leftarrow n\%$ of $P^{(t)} = \{p_1^{(t)}, \ldots, p_{n_t}^{(t)}\}$
step 1: Fine-tune PEN weights (W^s, W^c) using adaptation data $P_{ad}^{(t)}$

Inference Input: Unseen target persons data $P_{un}^{(t)} = P^{(t)} \cap P_{ad}^{(t)}$
Output: $(\tilde{y}^v, \tilde{y}^a) \leftarrow \text{PEN}(P_{un}^{(t)})$

outputs of the fine-tuned ResNet were used as input (z), and the $W^{c,0}$ was optimized by minimizing the loss $\alpha_c^s + \lambda_1 \alpha_r^s$ using the 1-hot encoding of the source subjects' ids as ground-truth labels (c). Then, the joint learning of the PEN was performed by minimizing the loss in Eq. (1). We noticed that the model with the individually tuned experts to data of each subject (thus, in isolation from the selector) generalized worse to target subjects than when only the joint learning of the selector and the experts was attempted. For this reason, we report our results only for the latter setting. Also, we noticed that doing the joint learning for a large number of epochs even led to overfitting of source subjects (on their left-out portion of the data). For this reason, we used the early stopping strategy, which prevented the model from overfitting after only five epochs. This model was subsequently used for further adaptation to target subjects by minimizing the PEN loss (Eq. (8)) on the adaptation data of target subjects – see Sect. 4. These learning and inference steps are summarized in Algorithm 1.

We implemented the PEN architecture using the Keras API [21] with the Tensorflow [22] back-end. For the soft-max and fcls, we used the existing implementations. The layer sizes were 2048 × 9 (for a total of 18441 parameters including the 9 offsets) for the CN and 2048 × 2 (for a total of 4098 parameters including the 2 offsets) for a given expert. The reweighting part of the weighted sum was performed via a custom Lambda layer that took the tensors output by the CN and the experts as input and scaled the outputs of each expert by the corresponding CN output. Afterwards, the scaled components were summed via an Addition layer and output as the overall network output. During training with source data, mean-squared error was used to train with this overall output, and categorical cross-entropy loss was used to train with the output of the CN. However, to implement the PEN loss, we created a custom loss function that performed L-1 regularization on the pre-softmax- activation CN output by taking the mean of the absolute value of the pre-activation tensor and summing it with the categorical cross-entropy loss. The parameter optimization was then performed using the standard back-propagation algorithm and Adadelta optimizer with the default parameters. The details of the employed validation settings are provided in the description of the experiments.

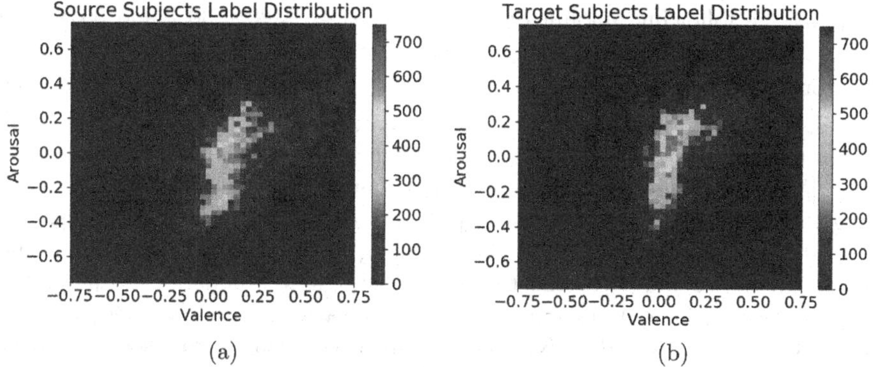

(a) (b)

Fig. 2. The joint distribution of the labels of affective dimensions: valence and arousal, in source and target subjects. By personalizing the PEN using adaptation data of target subjects, we reduce the difference between two distributions.

4 Experiments

To demonstrate the effects of the model personalization using the proposed PEN approach, we used image sequences from the multimodal affect database REmote COLlaborative and Affective (RECOLA) database [9], used in the Audio/Visual Emotion Challenge and Workshop (AVEC) 2016 [10] This database contains four modalities or sensor-signals: audio, video, electro-dermal activity (EDA), and electro-cardiogram (ECG). The data are synchronized with the video modality and coded by five human experts. Specifically, the gold standard labels (i.e., the aligned codings) for two affective dimensions - valence and arousal - are provided by the database creators. The time-continuous codings for each dimension are provided on a scale from -1 to $+1$. For our experiments, we used the publicly available data partitions from AVEC 2016, namely, training (9 subjects) and development (9 subjects) sets. We refer to these as source and target persons. The video of each person is 5 min long (25 fps), resulting in ~7k image frames per person after the face detection using Faster R-CNN [17]. We used the processed face images as input to the models, and the output was the estimated levels of valence and arousal per frame. The models' performance was evaluated in terms of Root-mean-square-error (RMSE) and concordance correlation coefficient (CCC), both of which were used as competition measures in AVEC and were computed on the pairs of model estimates and the gold-standard labels.

We performed the following experiments: (i) the SN and PEN models trained and tested on the source subjects ($P^{(s)}$), in order to evaluate the modeling power of the latter when fitting the data. We denote these models as s-SN and s-PEN. (ii) These models were then adapted using the adaptation data ($P_{ad}^{(t)}$), a varying portion of the data from target subjects, and evaluated on the non-overlapping data of the target subjects. The data of target subjects were split into the adaptation and test data at random. Specifically, we formed $P_{ad}^{(t)}$ by incrementally sampling the following amount of data: $n = 5\%, 10\%, 20\%, 30\%$ and 50% from each target subject and then combining the data across the subjects. For example, from 7k frames per target subject, $n = 5\%$ led to $P_{ad}^{(t)}$ of size $350 * 9 = 3150$ images, and for $n = 50\%$ the $P_{ad}^{(t)}$ size was ten times larger.

For testing, we always used the same (nonoverlapping) 50% of target subject data. The goal of these comparisons was to assess the models' behavior when using a different amount of target subject data to adapt the deep networks. The main premise here is that due to the modeling flexibility of the PEN, it would be able to better adapt to the target subjects than would SN. To show the benefits of using the data of source subjects to learn the (non-adapted) models, we also tested the SN and PEN model architectures trained from scratch on the 5 different adaptation sets $P_{ad}^{(t)}$ formed from the $n\%$ of target data as described above and evaluated on the test set, i.e. the left-out 50% of target subject's data. (iii). We refer to these settings as target SN (t-SN) and target PEN (t-PEN), respectively.

To form the base models, we first trained the s-SN together with the ResNet (see Algorithm 1). This was accomplished using 80% of the source data (evenly sampled from each subject) while the remaining 20% were left out for the model validation. We found that 10 epochs were sufficient to fine-tune the ResNet without overfitting it due to its large number of parameters and the limited number of data used to tune the network. Further optimization of the SN and PEN configurations was performed using 30 epochs for training the models, which was enough for the models' loss to converge. To select the regularization parameters when training s-PEN, we cross-validated (λ_0, λ_1) using the following values $\{10^{-4}, 10^{-3}, \ldots, 0, 1, 10, 100\}$, with 10^{-3} performing the best for both. During model adaptation and training of the target models, we used the same regularization parameters. Note that for these models, no subject id was provided during the adaptation/training, as we assumed that these are available only for the source subjects.

Table 1 compares our networks initially trained with source data, s-SN and s-PEN, with the t-SN and t-PEN, trained from scratch using only $n\%$ of target data, as mentioned above[3]. The results show that the initial training of the SN and PEN on the source data improved performance on target test data, compared to the t-SN and t-PEN. This evidences that the proposed models were able to efficiently leverage the data of the source subjects during the estimation. Moreover, the s-PEN approach eventually outperforms the SN architecture after as little as 10% of target (adaptation) data, due to its more flexible architecture that allowed it to easier adapt to target population data. Furthermore, we observe that with even as few as 5% of target subjects' data, both models' performance improves largely, with the s-PEN improving more advantage as more adaptation data become available. We assume that these (supervised) adaptation data are sufficient to constrain the feature/label space of the source models, rendering more efficient models for the target population. We also note the high

[3] Note, however, that the Resnet used to extract the features for these models was fine-tuned using the labeled source data.

Table 1. Performance on target test data in terms of CCC after adapting the networks with $n\%$ of (non-overlapping) target data.

$n[\%]$		0	5	10	20	30	50
Valence	s-SN	**0.72**	**0.80**	0.80	0.82	0.82	0.82
	s-PEN	0.71	**0.80**	**0.82**	**0.84**	**0.85**	**0.86**
	t-SN	N/A	0.71	0.77	0.8	0.81	0.82
	t-PEN	N/A	0.75	0.79	0.81	0.82	0.83
Arousal	s-SN	**0.66**	**0.79**	0.80	0.81	0.82	0.82
	s-PEN	0.65	**0.79**	**0.81**	**0.83**	**0.84**	**0.85**
	t-SN	N/A	0.73	0.77	0.79	0.81	0.81
	t-PEN	N/A	0.76	0.79	0.8	0.81	0.82

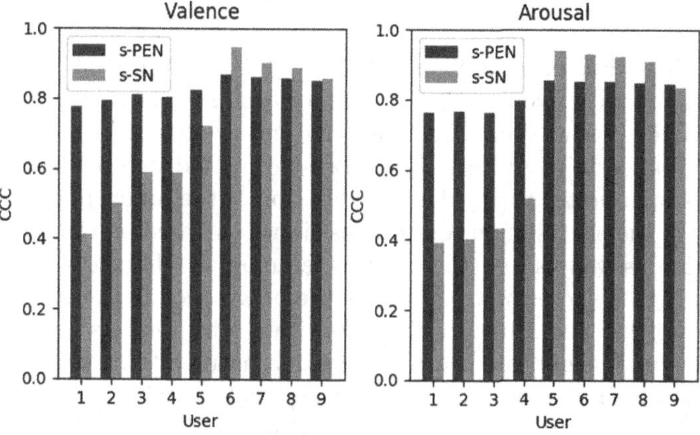

Fig. 3. Per-subject valence and arousal estimation performance on target test data of source-trained models adapted with limited target data. The s-PEN model has more consistent performance than the s-SN model over all of the target subjects.

performance of the t-SN and t-PEN, even with only 5% of the target data. We attribute this to the fact that they use the same ResNet feature extractor that was fine-tuned (via the s-SN) with the source labeled data. At the same time, we also note that s-SN and s-PEN have been trained on significantly more data. This, in turn, allowed the s-PEN to outperform the t-PEN by larger margin than is the case with their SN versions.

To analyze the effects of the model personalization to the target population, in Fig. 3 we show the CCC values of the s-SN and s-PEN models per target subject. We averaged the per-subject CCC values for both valence and arousal across all of the adaptation data sizes, and sorted the subjects based on the absolute difference in the models' performance for each subject. We observe that on 4/9 subjects, s-PEN largely outperforms s-SN in estimation of both valence and arousal levels. On the remaining subjects, s-SN outperforms t-PEN

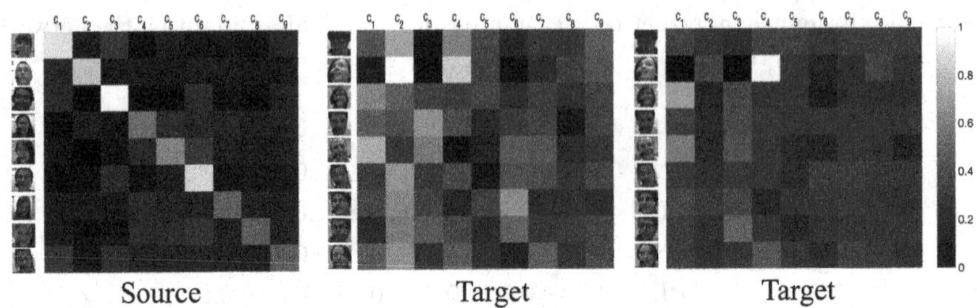

Fig. 4. Sparse combinations of experts via CN. Left: the selector learns the weighting of the outputs for the source subjects during source training. Center: the selector weights are effectively random yet sparse for the target subjects from source training. Right: after some fine-tuning on target, the selector begins to lose its sparse weighting despite regularization.

- however, these differences are less pronounced. For instance, the s-PEN model consistently produces average CCC values for each subject that are around 0.8. This experiment shows limitations of the "one-size-fits-all" machine learning architecture on certain subjects. By contrast, the modeling flexibility of the s-PEN allows it to better fit the data distribution of the target population.

We depict the effects of our custom loss function based on the L-1 (sparsity) regularization as well as how the CN was learning a combination of the source subjects in Fig. 4. These three plots show the progression of the average outputs of the CN weights per subject as our algorithm advances. As seen in the first image, the source weights are nearly an identity matrix after training only on the source subjects. This allowed the s-PEN to specialize each expert to one source subject, while also sharing the knowledge. However, when evaluated on target subjects, the selector produces different weights after unsupervised fine-tuning to those subjects (i.e. it is ignorant of the subject ids). Compared to the third image, these "subject" weights are still more sparse than when the training is done using data of target subjects. This is because the former uses the pre-trained selector network, resulting in the more sparse weights. By contrast, without leveraging this information, the s-PEN finds it more challenging to specialize in target subjects (as measured by the distribution of its CN weights ($c_1 \dots c_9$)) after the fine-tuning, despite the regularization of its weights (perhaps, due to the lack of subject ids). On the other hand, we found that by increasing the level of regularization adversely affects the model's performance, diminishing the role of the valence-arousal estimation loss.

Finally, in Table 2, we have included the s-PEN model's performance alongside the performance reported recently by [23], where a ResNet-50 with Gated Recurrent Unit (GRU) networks for sequence estimation is used for estimation of valence and arousal from videos of the same target subjects. We show that the proposed s-PEN exceeds their reported performance by large margin. However, these results may not directly be comparable because of possibly different evaluation settings.

Table 2. Comparison to End2You and AVEC 2016 Baseline [23].

Model	Valence	Arousal	Avg
s-PEN (0% of fine-tuning data)	0.71	0.65	0.68
s-PEN (5% of fine-tuning data)	0.80	0.79	0.80
End2You	0.58	0.41	0.50
AVEC 2016 Baseline	0.61	0.38	0.50

5 Conclusions

In summary, we propose a novel strategy for the personalization of deep convolutional neural networks for the purpose of valence and arousal estimation from face images. The key to our approach is the personalization of the mixture of experts architecture using a limited amount of data of target subjects. These personalized models have clear advantages over the compared single-expert ("one-size-fits-all") models in terms of how well are able to adapt to the target population when using limited amounts of labeled data from target subjects. Given the limitations involved in obtaining annotated valence and arousal data due to cost of expert labor and large variations in levels of expressiveness between people, model personalization can be key in working with limited data with many different domains. The audio-visual data we used come from sessions limited to 5 min, yielding ~7k image frames per subject, and we randomly sampled and split this data into non-overlapping training, adaptation, and test sets. However, ideally the system would have access to multiple sessions, allowing the proposed model to actively personalize as the interactions progress, and give us enough sessions so that we could draw samples that are further apart in time and less likely to be correlated. While minimizing correlation is not as critical in this work as in non-personalized situations, future work should explore how different methods of pseudo-random sampling of frames for constructing the adaptation and the hold-out test sets affect the results.

Acknowledgments. The work of O. Rudovic has been funded by the European Union H2020, Marie Curie Action - Individual Fellowship no. 701236 (EngageMe).

References

1. Peterson, K., Rudovic, O., Guerrero, R., Picard, R.W.: Personalized Gaussian processes for future prediction of Alzheimer's disease progression. In: NIPS Workshop on Machine Learning for Healthcare (2017)
2. Jaques, N., Rudovic, O., Taylor, S., Sano, A., Picard, R.: Predicting tomorrow's mood, health, and stress level using personalized multitask learning and domain adaptation. In: IJCAI Workshop (2017)
3. Zeng, Z., Pantic, M., Roisman, G.I., Huang, T.S.: A survey of affect recognition methods: audio, visual, and spontaneous expressions. IEEE TPAMI **31**, 39–58 (2009)

4. Rudovic, O., Lee, J., Dai, M., Schuller, B., Picard, R.: Personalized machine learning for robot perception of affect and engagement in autism therapy. arXiv preprint arXiv:1802.01186 (2018)
5. Mollahosseini, A., Hasani, B., Mahoor, M.H.: Affectnet: a database for facial expression, valence, and arousal computing in the wild. IEEE TAC (2017)
6. Martinez, D.L., Rudovic, O., Picard, R.: Personalized automatic estimation of self-reported pain intensity from facial expressions. In: IEEE CVPR'W (2017)
7. Csurka, G.: Domain adaptation for visual applications: a comprehensive survey. CoRR, abs/1702.05374 (2017). http://arxiv.org/abs/1702.05374
8. Jacobs, R.A., Jordan, M.I., Nowlan, S.J., Hinton, G.E.: Adaptive mixtures of local experts. Neural Comput. **3**(1), 79–87 (1991)
9. Ringeval, F., Sonderegger, A., Sauer, J., Lalanne, D.: Introducing the RECOLA multimodal corpus of remote collaborative and affective interactions. In: IEEE FG (Workshops) (2013)
10. Valstar, M., Gratch, J., Schuller, B., Ringeval, F., Lalanne, D., Torres Torres, M., Scherer, S., Stratou, G., Cowie, R., Pantic, M.: Avec 2016: depression, mood, and emotion recognition workshop and challenge. In: Proceedings of the 6th International Workshop on Audio/Visual Emotion Challenge. ACM (2016)
11. Collobert, R., Bengio, S., Bengio, Y.: A parallel mixture of SVMS for very large scale problems. In: NIPS (2002)
12. Shahbaba, B., Neal, R.: Nonlinear models using Dirichlet process mixtures. JMLR **10**, 1829–1850 (2009)
13. Theis, L., Bethge, M.: Generative image modeling using spatial LSTMS. In: NIPS (2015)
14. Yao, B., Walther, D., Beck, D., Fei-Fei, L.: Hierarchical mixture of classification experts uncovers interactions between brain regions. In: NIPS (2009)
15. Rasmussen, C.E., Ghahramani, Z.: Infinite mixtures of Gaussian process experts. In: NIPS (2002)
16. Shazeer, N., Mirhoseini, A., Maziarz, K., Davis, A., Le, Q., Hinton, G., Dean, J.: Outrageously large neural networks: the sparsely-gated mixture-of-experts layer. In: ICLR (2017)
17. Jiang, H., Learned-Miller, E.: Face detection with the faster R-CNN. In: IEEE FG (2017)
18. He, K., Zhang, X., Ren, S., Sun, J.: Deep residual learning for image recognition. In: IEEE CVPR (2016)
19. Krizhevsky, A., Sutskever, I., Hinton, G.E.: Imagenet classification with deep convolutional neural networks. In: NIPS, pp. 1097–1105 (2012)
20. Jiang, J.: A literature survey on domain adaptation of statistical classifiers. **3** (2008). http://sifaka.cs.uiuc.edu/jiang4/domainadaptation/survey
21. Chollet, F., et al.: Keras (2015)
22. Abadi, M., Barham, P., Chen, J., Chen, Z., Davis, A., Dean, J., Devin, M., Ghemawat, S., Irving, G., Isard, M., et al.: Tensorflow: a system for large-scale machine learning. In: OSDI (2016)
23. Tzirakis, P., Zafeiriou, S., Schuller, B.W.: End2you-the imperial toolkit for multimodal profiling by end-to-end learning. arXiv preprint arXiv:1802.01115 (2018)

Learning to Rank and Discover for E-Commerce Search

Anjan Goswami[1(✉)], Chengxiang Zhai[2], and Prasant Mohapatra[1]

[1] University of California, Davis, Davis, USA
{agoswami,pmohapatra}@ucdavis.edu
[2] University of Illinois Urbana-Champaign, Champaign, USA
cheng@uiuc.edu

Abstract. E-Commerce (E-Com) search is an emerging problem with multiple new challenges. One of the primary challenges constitutes optimizing it for relevance and revenue and simultaneously maintaining a discovery strategy. The problem requires designing novel strategies to systematically "discover" promising items from the inventory, that have not received sufficient exposure in search results while minimizing the loss of relevance and revenue because of that. To this end, we develop a formal framework for optimizing E-Com search and propose a novel epsilon-explore Learning to Rank (eLTR) paradigm that can be integrated with the traditional learning to rank (LTR) framework to explore new or less exposed items. The key idea is to decompose the ranking function into (1) a function of content-based features, (2) a function of behavioral features, and introduce a parameter epsilon to regulate their relative contributions. We further propose novel algorithms based on eLTR to improve the traditional LTR used in the current E-Com search engines by "forcing" exploration of a fixed number of items while limiting the relevance drop. We also show that eLTR can be considered to be monotonic sub-modular and thus we can design a greedy approximation algorithm with a theoretical guarantee. We conduct experiments with synthetic data and compare eLTR with a baseline random selection and an upper confidence bound (UCB) based exploration strategies. We show that eLTR is an efficient algorithm for such exploration. We expect that the formalization presented in this paper will lead to new research in the area of ranking problems for E-com marketplaces.

1 Introduction

One of the most critical components of an e-commerce (e-com) marketplace is its search functionality. The goal of an e-commerce search engine is to show the buyers a set of relevant and desirable products and facilitate the purchasing transactions that generate the revenue for the platform. Additionally, the e-com search also need to facilitate the discovery of the new or less exposed items to the buyers. This is in-fact critical for some categories such as apparel where new items are added periodically. However, a search ranking algorithm uses the behavioral signals such as sales, clicks, cart adds as features in its learning to

P. Perner (Ed.): MLDM 2018, LNAI 10935, pp. 331–346, 2018.
https://doi.org/10.1007/978-3-319-96133-0_25

rank algorithm. Therefore, the search engine may favor certain items that are purchased more by customers than other items in order to maximize the revenue, but the more an engine favors certain items, the higher those items would be ranked in the search. This creates a conflict between the revenue and discovery metrics since some less-favored items might never have a chance of being exposed to the users. It is also easy to see maximizing discovery can compromise relevance of search results since those "unseen products" may not be relevant to a user's interest. We thus see that an e-com search engine must deal with a much more challenging optimization problem dealing with optimizing relevance and revenue as well as providing a discovery mechanism for the buyers. We address this problem in this paper and have made three contributions: Firstly, we suggest a formal framework for optimizing E-Com search and define multiple objectives to form a theoretical foundation for developing effective E-Com search algorithms. Secondly, we propose a simple and practical framework for conducting regulated discovery in e-com search. We then provide an exploration algorithm (eLTR) with a theoretical guarantee that can be easily integrated with traditional learning to rank algorithms. We also discuss how existing multi-armed bandit algorithms such as upper confidence bound (UCB) can also be used to address this problem in e-com search. Thirdly, we suggest a possible evaluation methodology based on simulation with E-Com search log data and show the effectiveness of our proposed eLTR algorithm using our evaluation methodology with synthetically generated data. We also compare different exploration strategies and show the effectiveness of eLTR algorithm.

2 Related Work

There has been extensive research on learning to rank (LTR) algorithms particularly in the context of web search [11]. Most of the algorithms are designed to optimize a single metrics. Recently, Svore et al. [17] proposes a variant of LambdaMart [3] that can optimize multiple objectives particularly when two objectives are positively correlated. Authors conducted their experiments showing optimization of two different variants of normalized discounted cumulative gain (NDCG) [8] metrics that are based on judgments of human raters and based on click feedback respectively. In case of e-com search, aiming to maximize exploration can hurt the main business objectives and hence using this approach is not possible. The exploration algorithms are well researched in machine learning [2,7,16], particularly in the context of recommender systems [14,15], news content optimization problems [10]. However, in e-com search we also need to ensure maximization of revenue and the exploration needs to be well regulated to minimize the expected loss. Vermorel and Mohri [18] in their paper compared the effectiveness of several multi-armed bandit (MAB) algorithms including heuristics such as ϵ-explore, soft-max etc., and also approaches based on upper confidence bound (UCB) [1] which has nice theoretical regret guarantee. The authors suggested often simple heuristics can provide very good practical performance for exploration. In this paper, we use a sub-modular function [12,20] for exploration in order to have a nice theoretical guarantee for the exploration component. Our approach can be considered similar to the approaches used in learning adaptive sub-modular function [6]. However, we integrated this with a ranking function and propose a novel function and prove the monotonic sub-modularity of it.

3 Optimization of e-com Search

We consider the problem of optimizing an E-Com search engine over a fixed period of time $\{1, \cdots, T\}$. We assume that the search engine receives N queries, denoted by $Q = (q_1, q_2, \cdots, q_N)$ during this time. Let $\mathcal{Z} = \{\zeta_1, \cdots, \zeta_M\}$ be the set of M items during the same time. Let's denote the all the relevant items (recall set) for a query q_i by $R_i \subseteq \mathcal{Z}$. Consequently, we have $\mathcal{Z} = \bigcup_{i=1}^{N} R_i$. Now, we define a ranking policy by $\pi : 2^{(Q \times \mathcal{Z})} \rightarrow \Re$, where the input is the set of query item tuples where the items are from the recall set for the query and the output of the policy function is a subset of K items. These items are shown to the users and then an user browses the items one after another in the order they are shown. The user may click an item, add it to the shopping cart, and can also purchase. If a purchasing transaction happens then either the revenue generated or a sale can be considered to be a reward for the π. If the reward is designed to be using revenue it then needs to be real valued. If the reward is based on a sale, it can be a binary variable. It is also possible to construct the reward using clicks or cart-adds or a combination of all or some of these. An e-com search intends to maximize all these measures. However, the policy functions space is exponential and we require to formulate an optimization problem for e-com search. We don't generally have the knowledge when a purchasing transaction can happen. Hence, we introduce a binary random variable $\lambda_{ij} \in \{0, 1\}$ to indicate whether a purchasing transaction will happen with $\lambda_{ij} = 1$ meaning a purchasing event. Naturally, $p(\lambda_{ij} = 1|\zeta_j, q_i) + p(\lambda_{ij} = 0|\zeta_j, q_i) = 1$ for an item ζ_j shown for query q_i. The expected RPV for this query is then given by

$$RPV(q_t) = \sum_{\zeta_j \in \pi(q_i)} price(\zeta_j) \times N(\lambda_{ij} = 1|\zeta_j, q_i).$$

The total revenue defined on all the query results for the fixed period of time T when using policy π is thus

$$g_{RPV}(\pi) = \sum_{i=1}^{N} RPV(q_i)$$

Similarly, we can also define the relevance objective function as

$$g_{REL}(\pi) = \sum_{i=1}^{N} \rho(\pi(q_i))$$

where ρ can be any relevance measure such as nDCG, which is generally defined based on how well the ranked list $\pi(q)$ matches the ideal ranking produced based on human annotations of relevance of each item to the query. The aggregation function does not have to be a sum (over all the queries); it can also be, e.g., the minimum of relevance measure over all the queries, which would allow us to encode the preference for avoiding any query with a very low relevance measure score.

$$g_{REL}(\pi) = min_{i\in[1,p]}\rho(\pi(q_i))$$

We can now define the notion of discoverability of an e-commerce engine by considering a minimum number of impressions of items in a fixed period of time. The notion of discoverability is important because the use of machine learning algorithms in search engines tends to bias a search engine toward favoring the viewed items by users due to the use of features that are computed based on user behavior such as clicks, rates of "add to cart", etc. Since a learning algorithm would rank items that have already attracted many clicks on the top, it might overfit to promote the items viewed by users. As a result, some items might never have an opportunity to be shown to a user (i.e., "discovered" by a user), thus also losing the opportunity to potentially gain clicks. Such "undiscovered" products would then have to stay in the inventory for a long time incurring extra cost and hurting satisfaction of the product providers. To formalize the notion of discoverability, we say that the LTR function f is β-discoverable if all items are shown at least β times. Now, we can further define a β-discoverability rate as the percentage of items that are impressed at least β times in a fixed period of time. Let us now define again a binary variable γ_i for every item ζ_i and then assume that $\gamma_i = 1$ if the item got shown in the search results for β times and $\gamma_i = 0$ in case the item is not shown in the search results more than β times. We can express this as follows:

$$g_{\beta-discoverability} = \frac{\sum_{i=1}^{i=M} \gamma_i}{M}$$

Given these formal definitions, our overall optimization problem for the e-com search is to find an optimal ranking policy π that can simultaneously maximize all three objectives, i.e.,

$$\textbf{Maximize } g_{RPV}(\pi), g_{REL}(\pi), g_{\beta-dicoverability}.$$

The above is a multi-objective problem and maximizing simultaneously all of the above objective may not be possible and it may also not be a desirable business goal from the platform side. The optimal tradeoff between the different objectives would inevitably application dependent.

The challenging aspect of this multi-objective problem is that the objectives such as discovery requires exploration that can also hurt the relevance and revenue.

4 Strategies for Solving the Optimization Problem

Since there are multiple objectives to optimize, it is impossible to directly apply an existing Learning to Rank (LTR) methods to optimize all the objectives. However, there are multiple ways to extend an LTR method to solve the problem as we will discuss below.

4.1 Direct Extension of LTR

One classic strategy is to use a convex combination of multiple objectives to form one single objective function, which can then be used in a traditional LTR framework to find a ranking that would optimize the consolidated objective function. One advantage of this approach is that we can directly build on the existing LTR framework, though the new objective function would pose new challenges in designing effective and efficient optimization algorithms to actually compute optimal rankings. One disadvantage of this strategy is that we cannot easily control the tradeoff between different objectives (e.g., we sometimes may want to set a lower bound on one objective rather than to maximize it). Additionally, it does not have any exploration component and hence we can not ensure optimizing discovery with such algorithm.

4.2 Incremental Optimization

An alternative strategy is to take an existing LTR ranking function as a basis for a policy and seek to improve the ranking (e.g., by perturbation) so as to optimize multiple objectives as described above; such an incremental optimization strategy is more tractable as we will be searching for solutions to the optimization problem in the neighborhood. We can then construct such a perturbation by keeping a fixed number of x positions for exploration out of the K top results. Then, the rest of the $(K - x)$ items can be selected using a LTR function based on other criteria such as combination of revenue and relevance. This framework is so simple that it is very easy to realize in practice but it is possible to conduct exploration based on several strategies such that the regret in the form of loss of revenue and relevance can be minimized. In the next section of the paper, we discuss a few such strategies.

5 Exploration with LTR (eLTR)

Let us define the set from which the LTR function selects the items as $L_i \subset R_i$ for a given query q_i. We assume that all the items outside set L_i are not β-discoverable. Then, $L = \cup_{i=1}^{i=N} L_i$ is the set of all β-discoverable items. Hence, the set $E = R \setminus L$ can then be consisting of all the items that require exploration.

Now, we propose three strategies to incorporate discovery in an e-commerce search.

Random Selection Based Exploration from the Recall Set (RSE): This is a baseline strategy for continuous exploration with a LTR algorithm. In this, for every query q_i, we randomly select x items from the set $E \cap R_i$. Then, we put these x items on top of the other $(k - x)$ items that are selected using LTR from the set R_i. The regret here will be linear with the number of.

Upper Confidence Bound (UCB) Based Exploration from the Recall Set (UCBE): This is another simple strategy that uses a variant of UCB based algorithm for exploration instead of random sampling. Here, we maintain a MAB for each query. We consider each item in the set $E \cap R_i$ as an arm for the MAB corresponding to a query q_i. We maintain an UCB score for each of those items based on sales over impression for the query. If an item ζ_j is in the set $E \cap R_i$ and is shown b_j times in T iterations, and is sold a_j times in between, then the UCB score of the item ζ_j is $ucb_j = \frac{a_j}{b_j} + \sqrt{\frac{2\log_2 T+1}{b_j}}$. Note, this is for a specific query. We then select x items based on top UCB scores.

Explore LTR (eLTR In this, we define a function that we call explore LTR (eLTR) to select the x items. The rest of the items for top K can be chosen using the traditional LTR. Then, we can either keep the x items on top or we can rerank all K items based on eLTR.

The main motivation for the eLTR is the observation that there is inherent overfitting in the regular ranking function used in an e-com search engine that hinders exploration, i.e., hinders improvement of β-discoverability and STR. The overfitting is mainly caused by a subset of features derived from user interaction data. Such features are very important as they help inferring a user's preferences, and the overfitting is actually desirable for many repeated queries and for items that have sustained interests t users (since they are "test cases" that have occurred in the training data), but it gives old items a biased advantage over the new items, limiting the increase of β-discoverability and STR. Thus the main idea behind e-LTR is thus to separate such features and introduce a parameter to restrict their influences on the ranking function, indirectly promoting STR. Formally, we note that a ranking function can be written as follows:

$$y = f(\mathbf{X}) = g(f_1(\mathbf{X1}),\ f_2(\mathbf{X2}))$$

where $y \in \mathbb{R}$ denotes a ranking score and $\mathbf{X} \in \mathbb{R}^N$ is a N dimensional feature vector, $\mathbf{X1} \in \mathbb{R}^N 1$ and $\mathbf{X2} \in \mathbb{R}^N 2$ are two different groups of features such that $N1 + N2 = N$, $\mathbf{X1} \cup \mathbf{X2} = \mathbf{X}$. The two groups of features are meant to distinguish features that are unbiased (e.g., content matching features) from those that are inherently biased (e.g., clickthrough-based features). Here g is an aggregation function which is monotonic with respect to both arguments. It is easy to show that any linear model can be written as a monotonic aggregation function. It is not possible to use such representation for models such as additive trees. However, our previous techniques do not have such limitation since they are completely separated from the LTR. In this paper, we keep our discussion limited to linear models. We now define explore LTR (eLTR) function as follows:

$$y^e = f_e(\mathbf{X}) = g(f(\mathbf{X1}), \epsilon \times f(\mathbf{X2}))$$

where $y^e \in \mathbb{R}$ and $0 \leq \epsilon \leq 1$ is a variable in our algorithmic framework. Since, g is monotonic, $f_e(\mathbf{X}) <= f(X)$ when $\epsilon \leq 1$. Since feature set $X2$ is a biased feature set favoring old items, we can expect ranking based on f^e would be more in favor of new items in comparison with the original f, achieving the goal of emphasizing exploration of new items. Note that ϵ controls the amount of exploration: the smaller ϵ is, the more exploration (at the cost of exploitation). Since the maximum exploration is achieved when $\epsilon = 0$, in which case, ranking is entirely relying on f_1, the only loss in the original objective function is incurred by the removal of f_2. By controlling what features to be included in f_2, we can control the upper bound of the loss. In this sense, eLTR ensures a "safe" exploration strategy since f_1 is always active. Note, this function gradually can become very same as the LTR function when ϵ is close to 1. There can be various ways of constructing the ϵ. In this paper, we experimented with three different expressions for ϵ. These are as follows:

eLTR Basic Exploration (eLTRb): In this strategy, we keep $\epsilon = \frac{I}{T_{max}}$. Here, I is an iteration and T_{max} is a maximum number of iteration after which everything can be reset. This is a very simple strategy where the eLTR just increases the importances of the behavioral features gradually with every iteration.

eLTR UCB Weighted Exploration (eLTRu): In this strategy, we keep $\epsilon = \frac{ucb_j}{U_j}$. Here, U_j is a normalization factor and in our experiment it is chosen to be the maximum UCB score in the set $E \cap R_i$. This can be intuitively considered as the expected LTR score based on a sales estimation. It is motivated by adaptive sub-modular optimization in bandit setting [6] that has nice regret guarantee.

eLTR UCB Weighted Exploration and Reranking (eLTRur): This strategy first selects the top x items using eLTRu and it selects the remaining $(k-x)$ items using the classic LTR and then it reranks the k items using eLTRu.

6 Theoretical Analysis

In this section, we discuss the regret bounds of all the strategies. We express the regret in terms of total number of search session n in a fixed period of time T. Our first strategy RSE can be arbitrarily bad and can have a worst case regret proportional to $O(xn)$. However, it can have a fast discovery. The UCB is a better strategy compared to RSE. The regret of stochastic variant of UCB can be estimated as $O(\log(xn))$. The discovery in this algorithm will be not as good as the RSE and it can be worst if the MAB arms converge fast towards optimality. On the other hand, we can construct the eLTR function as monotonic sub-modular. Then, the regret for eLTR can be estimated as $O(1+1/e^{-\frac{|E|}{x}})$ times worst compared to the optimal. In our case, the optimal algorithm is LTR [13]. The eLTR algorithm is inspired from ϵ-greedy style MAB algorithm and hence can have better discovery compared to UCB. However, it is not clear if the regret is necessarily better than UCB based strategy. In Sect. 8 we conduct a simulation to understand how these algorithms compare with each other. Here, we now show that eLTR can be indeed monotonic sub-modular.

6.1 Monotonic Sub-modularity of eLTR

Let's call the ranking policy for selecting x items from the set E as

$$\pi^e : 2^{(Q \times \mathcal{E})} \rightarrow \Re,$$

where the cost function for our policy can be as follows:

$$c(E) = arg \max_{\zeta \in E} \sum_{i=1}^{i=x} y_i^e$$

We now show that this cost function is monotonic.

If we add a new item in set E, that will be added to the result of a query if the eLTR score for that query and that item is greater than the score of existing top x items. In that case, the cost of eLTR will increase. If the eLTR score for the query and the item is less than the existing top x items, the overall score from eLTR will be unchanged. Hence, the function is monotonically nondecreasing.

Now, we show that this function is sub-modular.

Let us assume that $A \subset B \subseteq E$. Let's also assume that there is an item $\zeta_g \notin (A \cup B)$. Consequently, we can have, $a = c(A \cup \{\zeta_g\}) - c(A)$ and $b = c(B \cup \{\zeta_g\}) - c(B)$.

There can then be three cases: case 1: $c(B) \geq c(A)$

In this case, there must be one or more high eLTR items in set B. Now, if we add the item ζ_g, it will either get added to the top x or not. If it is added to the top items in set B, that means it replaces at least the one item with the minimum eLTR score in top x items in set B. If there are no common items in the top x items for A and B, and since $c(B) \geq c(A)$, the new item has a higher eLTR value than any items in set A and will also replace an item in top x for A. Hence, $a = b$.

Now, if the item does not get added to top x items in B, that means the item does not have higher value compared to the top x items in B. Then, we have $b = 0$. Now, the item can be added in top items for A or not. If it is added in A then we will have $a > 0$ and if it is not added then we have $a = 0$.

case 2: $c(B) \leq c(A)$

This case will never happen since all the items in set A are also in set B and if there are top items in set A, all of those items will be in set B. Hence, unless there are items with higher eLTR compared to the top items in A, top items in B will never be different.

case 3: $c(B) = c(A)$

This is the simplest case. All top items are same and the new item will either get added in both or not since it need to replace one of the top items. Hence, $a = b$.

We have now shown that this cost function always have $a >= b$ and hence this function is sub-modular.

7 Evaluation Methodology

Due to the involvement of multiple objectives, the evaluation of E-Com search algorithms also presents new challenges. Here we discuss some ideas for evaluating the proposed e-LTR algorithm, which we hope will stimulate more work in this direction. The ideal approach for conducting such an evaluation would require simultaneously deploying all candidate methods to live user traffic, and computing various user engagement metrics such as click through rate,sales, revenue etc. However this strategy is difficult to implement in practice. Since user traffic received by each candidate method is different, we need to direct substantial amount of traffic to each method to make observations comparable and conclusions statistically significant. Deployment of a large number of experimental and likely sub-optimal ranking functions, especially when evaluating baselines, can result in significant business losses for e-Commerce search engines. Perhaps a good and feasible strategy is to design a simulation-based evaluation method using counterfactual techniques [9]. Here, we use historical search session data to replay the sessions for a fixed period of time. We then artificially make a set of items selected randomly as candidates for exploration where we do not have estimation of purchase probabilities. We keep these items in set E. We use the true purchase probabilities estimated from the data fro the items that have been shown sufficient number of time in our rank function but use zero values for the same probabilities for the items in set E.

On the surface, it appears that we may simply use the clicks or sales of the items to estimate the utility of each product. However, such a commonly used strategy would inherently favor already exposed items, and if an item has never been exposed, its utility would be zero, thus this strategy cannot be used for evaluating discoverability. To ensure discoverability for potentially *every* item in the collection, we can define the gold utility of a query product pair $u_{q,d}$ as a number randomly sampled between $[0, 1]$. Such a random sampling strategy would give every item a chance of being the underexposed target to be "discovered." Thus although the assigned utilities in this way may not reflect accurately real user preferences, the simulated utility can actually give more meaningful evaluation results than using click-throughs to simulate utility when comparing different exploration-exploitation methods where only the relative difference of these methods matters.

8 Experimental Results

In this section, we first construct a synthetic historical dataset with queries, items, and their prices. We also generate the true purchase probabilities and utility scores for the item and query pairs. Additionally, we use a specific rank function to simulate the behavior of a trained LTR model.

Then we conduct a simulation as described in Sect. 7 with various exploration strategies. During the simulation, we use the observed purchase probabilities estimated from the purchase feedback as the most important feature for the rank function but we use the true probabilities generated during the initial data generation phase to simulate the user behavior.

The main goal of this experimental study is to evaluate the behavior of the exploration strategies (a) with various different sets of number of queries and number of items, (b) with different values of β-discoverability at the beginning, (c) with different distributions of the utility scores representing different state of the inventory in an e-com company.

We evaluate our algorithms by running the simulation for T times. We compute RPV and β-discoverability at the end of T iterations. We also compute a purchase based mean reciprocal ranking [4] metric (MRR). This metric is computed by summing the reciprocal ranks of all the items that are purchased in various user visits for all queries. Moreover, we also discretize our gold utility score between 1 to 5 and generate a rating for each item. This also allows us to compute a mean NDCG score at k-th position for all the search sessions as a relevance metric.

We expect to see that the RPV and NDCG of the LTR function will be the best. However, the β-discoverability values will be better in ranking policies that use an exploration strategy. The new ranking strategies will incur a loss in RPV and NDCG and based on our theoretical analysis we expect the eLTR methods to have less loss compared to the RSE and UCB based approaches in those measures. We also expect to see a loss in MRR for all exploration methods. However, we mainly interested in observing how these algorithms perform in β-discoverability metric compared to LTR.

8.1 Synthetic Data Generation

We first generate a set of N queries and M items. We then assign the prices of the items by randomly drawing a number between a minimum and a maximum price from a multi-modal Gaussian distribution that can have up to 1 to 10 peaks for a query. We select the specific number of peaks for a query uniform randomly. We also assign a subset of the peak prices to a query to be considered as the set of it's preferred prices. This makes a situation where every query may have a few preferred price peak points where it may also have the sales or revenue peaks.

Now that we have the items and queries defined, we randomly generate a utility score, denoted by (u_{ij}) for every item ζ_j for a query q_i. In our set up, we use uniform random, Gaussian and a long-tailed distribution for selecting the utilities. These three different distributions represent three scenarios for a typical e-com company's inventory. Additionally, we generate a purchase probability between 0.02 to 0.40 for every item for every query. We generate these probabilities such that they correlate with the utility score. We generate these numbers in a way so that these are correlated with the utility scores with a statistically significant (p-value less than 0.10) Pearson correlation coefficient [19]. We also intend to correlate the purchase probability with the preferred peak prices for a query. Hence, we give an additive boost between 0 to 0.1 to the purchase probability in proportion to the absolute difference of the price of the item from the closest preferred mean price for that query. By generating the purchase probabilities in this way, we ensure that the actual purchase probabilities are related to the preferred prices for the queries, and also it is related to the utility scores of the items for a given query. Now, we define a β-discoverability rate $\beta = b$ and selects $b \times M$ items randomly from the set of all items. In our simulation, we assume that the estimated (observed) purchase probability for all the items in set E at the beginning can be zero. The rest of the items purchase probability is assumed to be estimated correctly at the beginning. Now, we create a simple rank function that is a weighted linear function of the utility score (u), the observed purchase probability (p_o), and the normalized absolute difference between the product price and the closest preferred mean price ($\hat{m}_p$ for the query such that $l = 0.60p_o + 0.20u + 0.20\hat{m}_p$. Here l denotes the score of the ranker. This ranker simulates a trained LTR function in e-com search where usually the sales can be considered the most valuable behavioral signal.

We now construct a user model. Here, a user browses through the search results one after another from the top and can purchase an item based on that item's purchase probability for that query. Note, in order to keep the simulation simple, we consider a user only purchases one item in one visit and leaves the site after that. We also can apply a discount to the probability of purchase logarithmically for each lower rank by multiplying $\frac{1}{\log_2(r+1)}$, where r is the ranking position of the item. This is to create an effect of the position bias [5].

8.2 Description of the Experimental Study

We conduct four sets of experiments with this simulated data.

In the first set of experiments, we use a small set of queries and a small set of products to understand the nature of the algorithms. The utility scores for all the products are generated from a uniform random distribution.

The Table 1 shows the RPV, NDCG@6, PMRR, and β-discoverability at the end of 10000 iterations of the simulation. The parameters of the simulation run are given in the caption of the corresponding table. We note that all the variants achieve high discoverability score with a relatively small loss in RPV, NDCG, and MRR. It is clear that eLTRur performs better than all other variants. It in-fact performs even better than the LTR algorithm in RPV metric along with doing well in discovery. The reason we see that all the variants can achieve high discoverability is that we have considered a very small set of items and we are running the experiments for many iterations. It is to be noted even with only 100 items in the inventory the LTR algorithm's discovery can only be about 37%. The LTR function thus seems to have not improved the discoverability much in this experiment. This conforms with our discussion on the weakness of LTR functions for optimizing for discoverability.

In the second set of experiments, we use a larger number of queries and products and we select a smaller starting value for β-discoverability. This simulates the scenario where an e-commerce business has very poor discovery and many items are not shown to the customers. We also run this simulation longer to

Table 1. Simulation of eLTR framework, with $|Q| = 10, |\mathcal{Z}| = 100, |L| = 50, \beta - d = 20\%, \beta = 50, K = 6, x = 3, T = 10000$.

Algorithms	RPV	NDCG	MRR	$\beta - d$
LTR	0.09	0.94	0.41	0.37
RSE	0.089	0.86	0.38	0.97
UCBE	0.09	0.87	0.39	0.96
eLTRb	0.09	0.88	0.39	0.97
eLTRu	0.09	0.88	0.39	0.97
eLTRur	0.092	0.88	0.40	0.98

Table 2. Simulation of eLTR framework, with $|Q| = 100, M = 5000, |L| = 200, \beta - d = 10\%, \beta = 50, K = 6, x = 3, T = 50000$.

Algorithms	RPV	NDCG	MRR	$\beta - d$
LTR	0.12	0.90	0.42	0.12
RSE	0.09	0.73	0.27	0.30
UCBE	0.10	0.73	0.27	0.66
eLTRb	0.11	0.91	0.32	0.68
eLTRu	0.11	0.92	0.32	0.68
eLTRur	0.11	0.92	0.32	0.68

understand the impact of the algorithms. In Table 2, we find the eLTR variants perform much better compared to the UCBE and RSE. We also observe that there is hardly any improvement of discoverability in LTR. Here, the eLTR variants perform better than the other approaches in terms of NDCG metric, and the NDCG drop is insignificant compared to LTR. In this case, the business needs to make a strategy if they want to maximize for revenue and keep the bias or by incurring some loss in revenue and relevance, they want to optimize discovery for growing their business.

In the third set of experiments, we use a Gaussian distribution with mean 0.5 and the variance 0.1 for generating the utility scores, but everything else is same as the previous experiment. We again see in Table 3 that eLTR variants perform well compared to UCBE and RSE and they also do better in terms of NDCG compared to LTR. Essentially, we see that if we have a normal utility distribution of the items, the NDCG for eLTR may not even drop since there are many unexposed "good" items. The Table 4 shows the convergence plots for the six competing algorithms for RPV, MRR, and the discovery.

In the fourth set of experiment, we use a power law distribution to generate the utility scores. This means that only a small set of items here can be considered valuable in this scenario (Table 4). The Table 5 shows the final metrics for this case and the Table 6 shows the convergence plots for RPV, NDCG, and discoverability for the six different algorithms. We notice that even with

Table 3. Simulation of eLTR framework, with $|Q| = 100, |\mathcal{Z}| = 5000, |L| = 200, \beta - d = 10\%, \beta = 50, K = 6, x = 3, T = 50000$. The distribution of utility scores is a Gaussian with mean 0.5 and the variance .1.

Algorithms	RPV	NDCG	MRR	$\beta - d$
LTR	0.10	0.92	0.44	0.10
RSE	0.08	0.86	0.28	0.29
UCBE	0.08	0.87	0.28	0.66
eLTRb	0.09	0.94	0.33	0.67
eLTRu	0.09	0.94	0.33	0.67
eLTRur	0.09	0.94	0.33	0.67

Table 4. Simulation of eLTR framework, with $|Q| = 100, |\mathcal{Z}| = 5000, |L| = 200, \beta - d = 10\%, \beta = 50, K = 6, x = 3, T = 50000$. Utility score distribution is Gaussian.

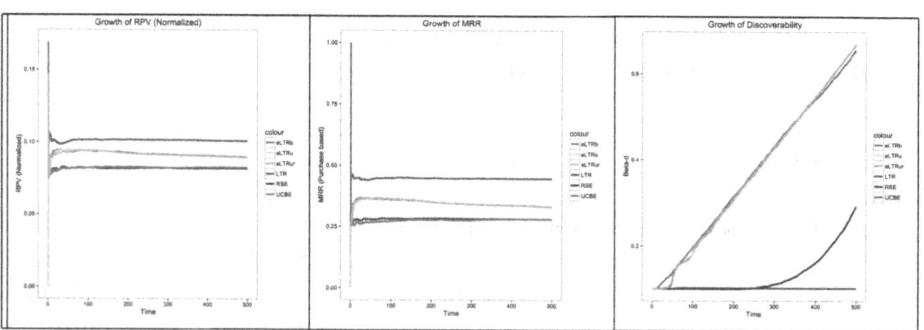

Table 5. Simulation of eLTR framework, with $|Q| = 100, |\mathcal{Z}| = 5000, |L| = 200, \beta - d = 10\%, \beta = 50, K = 6, x = 3, T = 50000$. The utility score distribution is a power law here.

Algorithms	RPV	NDCG	MRR	$\beta - d$
LTR	9.56	0.57	0.45	0.11
RSE	7.55	0.27	0.25	0.30
UCBE	7.55	0.27	0.26	0.66
eLTRb	8.2	0.33	0.31	0.67
eLTRu	8.3	0.33	0.31	0.67
eLTRur	8.4	0.33	0.32	0.67

this distribution of utility scores the eLTR variants have a smaller loss in RPV, NDCG, and in MRR. Note that in this distribution, the discoverability can be considered to be naturally not so useful since a large number of items are not that valuable. We expect in such situation, a nice discoverability algorithm can help to eliminate items that do not get sold after sufficient exposure and enable

Table 6. Simulation of eLTR framework, with $|Q| = 100, |\mathcal{Z}| = 5000, |L| = 200, \beta - d = 10\%, \beta = 50, K = 6, x = 3, T = 50000$. The distribution for the utility scores follow a power law.

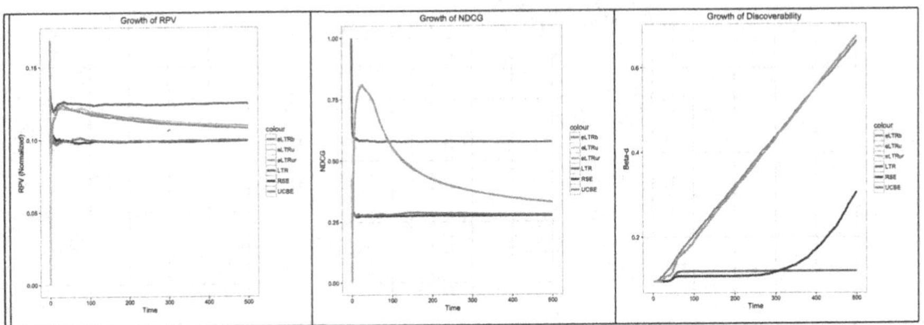

the e-com company to optimize its inventory. The Table 6 shows the convergence plots of all the algorithms in this scenario.

9 Conclusions

This paper represents a first step toward formalizing the emerging new E-Com search problem as an optimization problem with multiple objectives including the revenue per-visit (RPV), and discoverability besides relevance. We formally define these objectives and discuss multiple strategies for solving such an optimization problem by extending existing learning to rank algorithms. We also proposed a novel exploratory Learning to Rank (eLTR) method that can be integrated with the traditional LTR framework to explore new or less exposed items and discussed possible methods for evaluating eLTR. We show that selecting the items from a set of yet not discovered items using eLTR can be mapped to a monotonic sub-modular function and hence the greedy algorithm has nice approximation guarantees. We hope that our work will open up many new directions in research for optimizing e-com search. The obvious next step is to empirically validate the proposed eLTR strategy by using the proposed simulation strategy based on log data from an e-com search engine. The proposed theoretical framework also enables many interesting ways to further formalize the e-com search problem and develop new effective e-com search algorithms based on existing multi-armed bandit and sub-modular optimization theories. Finally, the proposed eLTR algorithm is just a small step toward solving the new problem of optimizing discoverability in e-com search; it is important to further develop more effective algorithms that can be applied with non-linear learning to rank algorithms.

References

1. Auer, P., Ortner, R.: UCB revisited: improved regret bounds for the stochastic multi-armed bandit problem. Period. Math. Hung. **61**(1–2), 55–65 (2010)
2. Bubeck, S., Cesa-Bianchi, N.: Regret analysis of stochastic and nonstochastic multi-armed bandit problems. Found. Trends Mach. Learn. **5**(1), 1–122 (2012)
3. Burges, C.J.: From ranknet to lambdarank to lambdamart: an overview. Learning **11**(23–581), 81 (2010)
4. Craswell, N.: Mean reciprocal rank. In: Liu, L., Özsu, M.T. (eds.) Encyclopedia of Database Systems, pp. 1703–1703. Springer, Boston (2009). https://doi.org/10.1007/978-0-387-39940-9
5. Craswell, N., Zoeter, O., Taylor, M., Ramsey, B.: An experimental comparison of click position-bias models. In: Proceedings of the 2008 International Conference on Web Search and Data Mining, pp. 87–94. WSDM 2008. ACM (2008)
6. Gabillon, V., Kveton, B., Wen, Z., Eriksson, B., Muthukrishnan, S.: Adaptive submodular maximization in bandit setting. In: Advances in Neural Information Processing Systems, pp. 2697–2705 (2013)
7. Gittins, J., Glazebrook, K., Weber, R.: Multi-armed Bandit Allocation Indices. Wiley, Hoboken (2011)
8. Järvelin, K., Kekäläinen, J.: Cumulated gain-based evaluation of IR techniques. ACM Trans. Inf. Syst. (TOIS) **20**(4), 422–446 (2002)
9. Li, L., Chen, S., Kleban, J., Gupta, A.: Counterfactual estimation and optimization of click metrics in search engines: a case study. In: Proceedings of the 24th International Conference on World Wide Web, pp. 929–934. ACM (2015)
10. Li, L., Chu, W., Langford, J., Schapire, R.E.: A contextual-bandit approach to personalized news article recommendation. In: Proceedings of the 19th International Conference on World Wide Web, pp. 661–670. ACM (2010)
11. Liu, T.Y.: Learning to rank for information retrieval. Found. Trends Inf. Retr. **3**(3), 225–331 (2009)
12. Lovász, L.: Submodular functions and convexity. In: Bachem, A., Korte, B., Grötschel, M. (eds.) Mathematical Programming The State of the Art, pp. 235–257. Springer, Heidelberg (1983). https://doi.org/10.1007/978-3-642-68874-4_10
13. Nemhauser, G.L., Wolsey, L.A., Fisher, M.L.: An analysis of approximations for maximizing submodular set functionsi. Math. Program. **14**(1), 265–294 (1978)
14. Park, S.T., Chu, W.: Pairwise preference regression for cold-start recommendation. In: Proceedings of the Third ACM Conference on Recommender Systems, pp. 21–28. ACM (2009)
15. Schein, A.I., Popescul, A., Ungar, L.H., Pennock, D.M.: Methods and metrics for cold-start recommendations. In: Proceedings of the 25th annual international ACM SIGIR Conference on Research and Development in Information Retrieval, pp. 253–260. ACM (2002)
16. Sutton, R.S., Barto, A.G.: Reinforcement Learning: An Introduction, vol. 1. MIT press, Cambridge (1998)
17. Svore, K.M., Volkovs, M.N., Burges, C.J.: Learning to rank with multiple objective functions. In: Proceedings of the 20th iNternational Conference on World Wide Web, pp. 367–376. ACM (2011)
18. Vermorel, J., Mohri, M.: Multi-armed Bandit algorithms and empirical evaluation. In: Gama, J., Camacho, R., Brazdil, P.B., Jorge, A.M., Torgo, L. (eds.) ECML 2005. LNCS (LNAI), vol. 3720, pp. 437–448. Springer, Heidelberg (2005). https://doi.org/10.1007/11564096_42

19. Wilcox, R.R.: Introduction to Robust Estimation and Hypothesis Testing. Academic Press, Cambridge (2011)
20. Yue, Y., Guestrin, C.: Linear submodular bandits and their application to diversified retrieval. In: Advances in Neural Information Processing Systems, pp. 2483–2491 (2011)

A Hybrid Neural Machine Translation Technique for Translating Low Resource Languages

Ebtesam H. Almansor[1,2](✉) and Ahmed Al-Ani[1]

[1] Faculty of Engineering and Information Technology, University of Technology Sydney, Sydney, Australia
EbtesamHussain.Almansor@student.uts.edu.au, Ahmed.Al-Ani@uts.edu.au
[2] Community College, Najran University, Najran, Saudi Arabia

Abstract. Neural machine translation (NMT) has produced very promising results on various high resource languages that have sizeable parallel datasets. However, low resource languages that lack sufficient parallel datasets face challenges in the automated translation filed. The main part of NMT is a recurrent neural network, which can work with sequential data at word and sentence levels, given that sequences are not too long. Due to the large number of word and sequence combinations, a parallel dataset is required, which unfortunately is not always available, particularly for low resource languages. Therefore, we adapted a character neural translation model that was based on a combined structure of recurrent neural network and convolutional neural network. This model was trained on the IWSLT 2016 Arabic—English and the IWSLT 2015 English—Vietnamese datasets. The model produced encouraging results particularly on the Arabic datasets, where Arabic is considered a rich morphological language.

Keywords: Low resource languages · Neural machine translation
Convolutional neural network · Recurrent neural network

1 Introduction

Neural machine translation (NMT) is an effective deep learning approaches that produced encouraging results for high resource languages [1–3]. Sequence—to—Sequence (seq2seq) is the most widely used neural machine translation model that was applied for translating different languages, such as English and French. The encoder and decoder are the two main components of the seq2seq model, where the encoder learns to transform variable length sentences of the source language, while the decoder learns the target sentences of the destination language as output [2]. However, large parallel datasets are required for the training of neural machine translation models.

Low resource languages are languages that lacked enough parallel datasets. Notably, these languages face challenges when machine translation approaches

P. Perner (Ed.): MLDM 2018, LNAI 10935, pp. 347–356, 2018.
https://doi.org/10.1007/978-3-319-96133-0_26

are applied. As the collection of large parallel dataset is not an easy task, many researchers who worked on the translation of low resource languages considered alternative methods that do not require large parallel datasets. For example, monolingual datasets have been used in [4], while an unsupervised pretrained model was proposed in [5].

The recurrent neural network (RNN) has become the standard structure model for many learning approaches that are based on sequence modeling [6]. RNN has been used in various natural language processing applications, such as text classification [7] and language modeling [8]. Although RNN can deal with sequential data, it has a limitation when handling long sequences [6]. On the other hand, convolution neural network (CNN) is another model that can also handle sequential data. CNN model has been widely used in imaging data and some sequential tasks [9]. Some of these models used time-invariant filter function in parallel to the window for the input sequence [6,9]. CNN has several advantages such as the ability to scale long sequences of data and increase parallelism [6].

In this paper, we conducted an in-depth investigation into character neural machine translation that is based on a combination of recurrent and convolutional neural networks, in order to scale long sequences of data. The suitability of character NMT for translating low resource languages is evaluated.

2 Related Work

Recently, a number of research studies have identified challenges that face the translation of low resource languages using neural machine translation [10], and investigated alternative methods. For example, training sequence to sequence model on monolingual datasets rather than parallel data has shown an improvement in NMT [4]. Xia et al. found that a dual learning model that used recurrent neural network (RNN) language model helped in improving the performance of NMT [11]. In [5], pretrained sequence to sequence framework with language models to translate German to English was developed. The authors used monolingual corpus to train the language models [5]. Wu et al. proposed an adaptive attention-based neural network to translate Mongolian-Chinese, which is considered as a low resource language [12].

In contrast to English and other languages, limited studies have been conducted on Arabic and Vietnamese using neural machine translation. For example, translation between Arabic and Hebrew as morphologically rich languages has been proposed using character level neural machine translation with some features of morphology. This approach led to enhance the performance of machine translation [13]. Almahairi et al. used NMT to translate between Arabic and English and they compared between NMT and phrase based translation model [14]. The following section describes the proposed model.

3 Background of Neural Machine Translation

3.1 Encoder-Decoder Architecture of NMT

Neural machine translation that consists of an encoder and a decoder could be implemented using Recurrent Neural Network (RNN). RNN is capable of handling a sequence of data, such as text, which is basically a sequence of words. Figure 1 shows the architecture of the encoder and decoder. The main advantage of RNN is taking the unbounded history of previous observations into account. It keeps the important information about sentence such as the structure in memory unit called internal state and this is updated at each time step. For example, if the sentence is five words, the number of time steps will be five. There are two different types of RNN: Long Short-Term Memory (LSTM) and Gate Recurrent Unit (GRU). The main difference between them, is the structure of memory unit [11]. This model was used in numerous Natural Language Processing tasks such as, question answer system, text classification, speech recognition and text translation. However, the encoder and decoder network faces difficulty with long sentences as it needs to compress all words in the source sentence into fix length vector [1].

In order to overcome the long sentence problem, Cho and Bengio proposed a new technique, which is called attention mechanism that allows "soft" segments through the automatic search for parts of source sentence that are relevant to the predicted target word.

Neural machine translation takes the source sentence $X = (x_1, ..., x_{Tx})$ of length T_x and generates the target sentence $Y = (y_1, ..., y_{Ty})$ of length T_y, both x_t and y_t represent the index of the source and target sentences in the vocabulary, respectively.

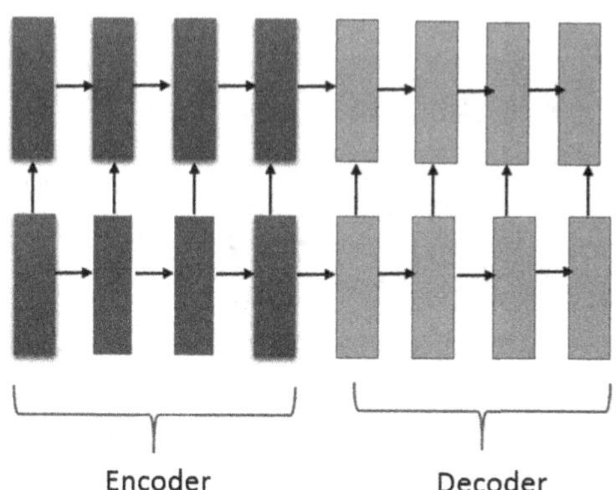

Fig. 1. The architecture of the encoder and decoder

The encoder of the attention model encodes the input source sentence into context vectors [15].

$$C = \{h_1, h_2.................., h_{T_x}\}$$

This context set is created by bidirectional recurrent neural network that contains forward RNN and reverse RNN. The aim of the forward RNN is reading the source input from the first token to the last one [11], resulting in the forward context vectors, $\{\overrightarrow{h_1}, \overrightarrow{h_{T_x}}\}$ where

$$\overrightarrow{h_t} = \overrightarrow{\Psi_{enc}}(\overrightarrow{h_{t-1}}, E_x[x_t])$$

$E_x \in R^{|V_x| \times d}$, this is the embedding matrix that contains row vectors of source symbols [15]. The reversing RNN is the opposite direction $\{\overleftarrow{h_1}, \overleftarrow{h_{Tx}}\}$ where

$$\overleftarrow{h_t} = \overleftarrow{\Psi_{enc}}(\overleftarrow{h_{t+1}}, E_x[x_t])$$

Both $\overrightarrow{\Psi_{enc}}, \overleftarrow{\Psi_{enc}}$ are referring to recurrent activation function such as gated recurrent units (GRU) or long short-term memory unit (LSTM) [15]. The forward and reverse context vectors are concatenated to form full context vectors for each position in the source sentences. For example,

$$h_t = [\overrightarrow{h_t}; \overleftarrow{h_t}]$$

The decoder, produces one symbol at a time based on the context vectors returned by the encoder [15].

3.2 Convolutional Neural Network

CNN uses convolving filters with layers which are applied to local features [16]. Convolution neural network embeds the sequence input $x = (x_1, \ldots x_m)$ in disruption space $w = (w_1, \ldots w_m)$, where $w_j \in R^f$ represents a column in the embedding matrix $D \in R^{V \times f}$.

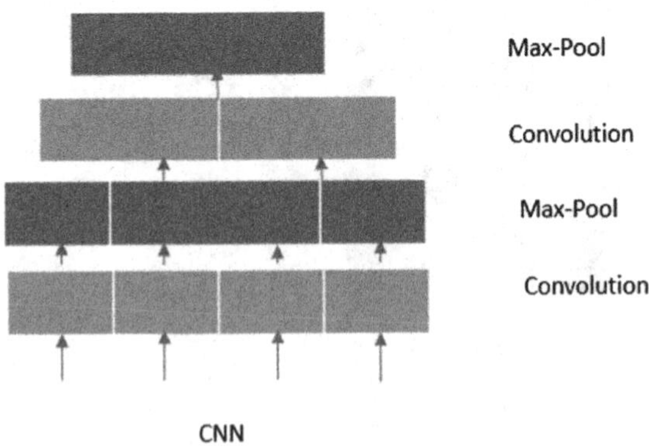

Fig. 2. Convolutional network

The encoder and decoder network share a simple block structure which computes intermediate state based on the fixed number of input elements [17]. The

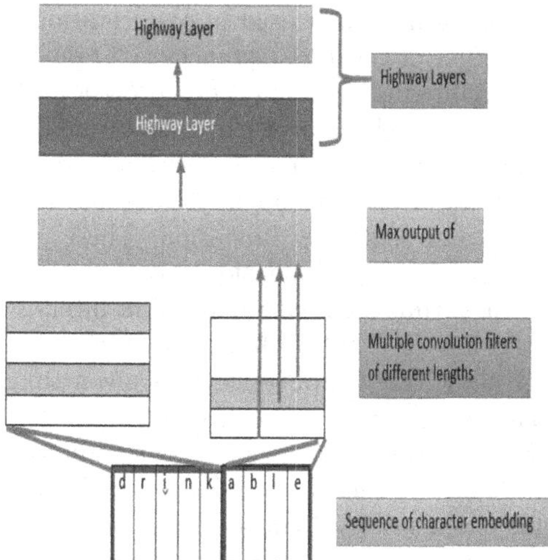

Fig. 3. CNN sequence to sequence learning at character level

output of the l-th block is denoted as $z^l = (z_1^l, \ldots, z_m^l)$ for the encoder and $h^l = (h_1^l, \ldots, h_m^l)$ for the decoder, where each block contains one dimensional convolution followed by non-linearity. The decoder with single block and kernel width k, the resulting state h_i^1 includes information over k inputs. The number of input elements increases due to stacking several blocks on top of each other. Figure 2 shows the structure of convolutional networks and Fig. 3 shows CNN sequence to sequence learning at character level.

4 Proposed Model

The proposed model aims to be an alternative translation tool for low resource languages, as the main objective behind developing this model is to have the ability to train it using small parallel datasets. The developed model is a character neural machine translation method that incorporates both RNN and CNN networks, so that it can handle long sequence of data and learn from small data. Since the subsequent time step outputs in the RNN is based on the previous hidden state, the RNN can not be parallelized and it has difficulties dealing with long sequence. On the other hand, CNN can apply time-invariant filters in parallel through the input which can be used as scale for long sequence. However, CNN has limitation of in the memory and time invariance that makes it difficult for the model to handle the order of long sequence.

Therefore, we propose a hybrid model that can deal with long sequences and their order by using RNN and CNN based on the minibatch dimensions and time-step. The loss function is used for optimizing the networks and the parameters are used as their weights. The number of iteration is the number of times the gradient is estimated and the parameters are updated. Batches size refers to number of training instances used in one iteration. Mini-batch can be used when the total number of training instances is large.

Figure 4 shows the architecture of the proposed model, which consists of two main subcomponents; the gated linear with convolution features and pooling layers. The pooling component allows parallel computing across minibatch and feature dimension, while the convolution component allows the parallel computation across minibatch and sequence dimension. These components take the input sequence $X \in R^{T \times n}$ of T n-dimensional vectors $X = (x_1, .., x_T)$. They preform convolution in the time step with bank of m filters to produce the candidate vectors from the $Z \in R^{T \times m}$of m-dimensional vectors z_t. The filters used in this model can not allow the computation of a given timestep to access the information from future timesteps. This helps to predict the next token in the sequences. The candidate vectors z passes across a tanh nonlinearity and uses convolutional filters to gain sequence of vectors for the pooling function. As we used the model for character level translation, the convolutional filters with width (k) are used to compute n-gram features at each time step.

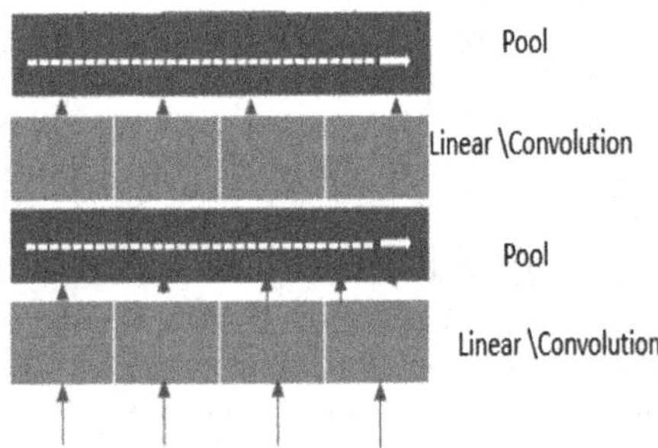

Fig. 4. Structure of the proposed model

5 Experiment and Results

This section includes details of the experiment and the produced results.

5.1 Datasets

We evaluate the model on parallel training data from transcribe TED and TEDx presentations. TIWSLT Arabic—English (Ar—En) (IWSLT2016) and English—Vietnamese (En—Vi) (IWSLT2015) spoken-domain translation [18,19] were used to train and evaluate the translation model. An addition datasets, called Tanzil, was also used to train the Ar—En translation model. This data is publicly available from the open parallel corpus (OPUS) [20]. A character level alignment was used for these languages. Table 1 shows the size of the experimental data that was used to train and evaluate the translation model.

Table 1. Experimental datasets size

Language pair	Train size	Test size
Ar-En	90.6K	934K
Tanzil	0.2 M	-
En-Vi	131K	1,080K

5.2 Experiment Setting

We trained the proposed model with sentences that consisted of a maximum of 150 words for En—Vi and 100 words for Ar—En. The encoder and decoder consist of four layers each one has 320 units. The first layer in the encoder uses a convolution filter with width k=6 and the other layers uses k = 2. We use the Adam method for optimizing the various parameters using 10 epochs and a minibatch size of 16. Adam is an optimization method used for first–order gradient-based optimization of the stochastic objective function which requires little memory. The name of this method came from adaptive moment estimation. It computes individual adaptive learning for different parameters based on the first and second moment estimation of gradients. The main advantage of this is the invariability in the magnitude of the parameter updates on rescaling the gradient during approximation of the step size using step size hyper-parameters. It works well with sparse gradients without requirement of the stationary objective [21]. In this experiment, the Adam was used based on different parameters such as stepsize $\alpha = 0.001$, $\beta_1 = 0.9$, $\beta_2 = 0.999$ and the hyper-parameters exponential decay rate for the moment estimate was set at $\epsilon = 10^{-8}$, in addition, beam search with width 8 was used to enhance the decoder.

5.3 Results

For the evaluation purpose, we apply our model to two low resource languages. The proposed model was evaluated using the BLEU method (Bilingual Evaluation Understudy) score, which was developed to evaluate the produced machine translation text [22]. While human evaluation is extensive and accurate, it is time consuming and expensive. The closer the translation to the human translation, the higher the BLEU score is. The BLEU score of 1.0 (100%) indicates a perfect match, while a score of 0.0 (0.0%) indicates a perfect mismatch. A score less than 0.15 (or 15%) indicates that the translation is not optimal and requires enhancement [22,23].

As shown in Table 2, our proposed model obtains reasonable results considering that the training was on a small amount of parallel datasets, where our translation model reached 18 BLEU score in Arabic—English (BLEU greater than 15%) and 12.95%, BLEU score in English—Vietnamese; this is not very far from the 15% threshold, and hence, the results can be considered reasonable.

It is important to mention that our proposed character level NMT model was trained on small datasets without using any other pre-trained model that

Table 2. Resulting obtained using the proposed model on Arabic-English and English-Vietnamese pairs

Language pair	Model	BLEU
IWSLT16 Arabic-English	Proposed model	18.67
	Open NMT	0.3
IWSLT15 English-Vietnamese	Proposed model	12.59
	Open NMT	0.58

Table 3. Variation in BLEU score using two different sentence lengths

Data	**Max- length**	**BLEU**
En-Vi	90	11.27
	70	10.85
Ar-En	90	17.61
	70	15.16

requires large parallel dataset. The results also indicated that the developed model was capable of producing good results on Arabic. As mentioned earlier this model can deal with longer sentences, as shown in Table 3 the model was evaluated using different sentence lengths, and we noticed that the shorter the sentence, the less accurate it was. In addition, the proposed method was evaluated against OpenNMT [24], which as this model has been used and produced promising results for high resource languages. However, when applied to our small datasets, OpenNMT did not produced acceptable BLEU scores. The main reason behind this is that the OpenNMT model was supposed to be trained on large parallel datasets which is lacked in low resource languages. Therefore, the proposed model outperformed the OpenNMT. OpenNMT uses word level as basic unit for the source and target languages. The above results indicate that character level NMT is more suitable for translating low resource languages than its word level counterpart, and that our proposed method presents a good option for translating low resource languages, particularly the ones that are morphologically rich, such as Arabic.

6 Conclusion

In this paper, we proposed a character neural machine translation model that can be applied to low resource languages. This model combines the main advantages of RNN and CNN. We have demonstrated the effectiveness of this character level NMT model in translating two low-resource languages; namely Arabic and Vietnamese.

References

1. Cho, K., Van Merriënboer, B., Gulcehre, C., Bahdanau, D., Bougares, F., Schwenk, H., Bengio, Y.: Learning phrase representations using rnn encoder-decoder for statistical machine translation. arXiv preprint arXiv:1406.1078 (2014)
2. Lamb, A., Xie, M.: Convolutional encoders for neural machine translation. WEB download (2016)
3. Sutskever, I., Vinyals, O., Le, Q.V.: Sequence to sequence learning with neural networks. In: Advances in Neural Information Processing Systems, pp. 3104–3112 (2014)
4. Gulcehre, C., Firat, O., Xu, K., Cho, K., Barrault, L., Lin, H.C., Bougares, F., Schwenk, H., Bengio, Y.: On using monolingual corpora in neural machine translation. arXiv preprint arXiv:1503.03535 (2015)
5. Ramachandran, P., Liu, P.J., Le, Q.V.: Unsupervised pretraining for sequence to sequence learning. arXiv preprint arXiv:1611.02683 (2016)
6. Bradbury, J., Merity, S., Xiong, C., Socher, R.: Quasi-recurrent neural networks. arXiv preprint arXiv:1611.01576 (2016)
7. Wang, X., Liu, Y., Sun, C., Wang, B., Wang, X.: Predicting polarities of tweets by composing word embeddings with long short-term memory. In: ACL, no. 1, pp. 1343–1353 (2015)
8. Zaremba, W., Sutskever, I., Vinyals, O.: Recurrent neural network regularization. arXiv preprint arXiv:1409.2329 (2014)
9. Zhang, X., Zhao, J., LeCun, Y.: Character-level convolutional networks for text classification. In: Advances in Neural Information Processing Systems, pp. 649–657 (2015)
10. Chen, Y., Liu, Y., Cheng, Y., Li, V.O.: A teacher-student framework for zero-resource neural machine translation. arXiv preprint arXiv:1705.00753 (2017)
11. He, D., Xia, Y., Qin, T., Wang, L., Yu, N., Liu, T., Ma, W.Y.: Dual learning for machine translation. In: Advances in Neural Information Processing Systems, pp. 820–828 (2016)
12. Wu, J., Hou, H., Shen, Z., Du, J., Li, J.: Adapting attention-based neural network to low-resource Mongolian-Chinese machine translation. In: Lin, C.-Y., Xue, N., Zhao, D., Huang, X., Feng, Y. (eds.) ICCPOL/NLPCC -2016. LNCS (LNAI), vol. 10102, pp. 470–480. Springer, Cham (2016). https://doi.org/10.1007/978-3-319-50496-4_39
13. Belinkov, Y., Glass, J.: Large-scale machine translation between arabic and hebrew-corpora and initial results. arXiv preprint arXiv:1609.07701 (2016)
14. Almahairi, A., Cho, K., Habash, N., Courville, A.: First result on arabic neural machine translation. arXiv preprint arXiv:1606.02680 (2016)
15. Firat, O., Cho, K., Bengio, Y.: Multi-way, multilingual neural machine translation with a shared attention mechanism. arXiv preprint arXiv:1601.01073 (2016)
16. Kim, Y.: Convolutional neural networks for sentence classification. arXiv preprint arXiv:1408.5882 (2014)
17. Gehring, J., Auli, M., Grangier, D., Yarats, D., Dauphin, Y.N.: Convolutional sequence to sequence learning. arXiv preprint arXiv:1705.03122 (2017)
18. Cettolo, M., Girardi, C., Federico, M.: WIT3: web inventory of transcribed and translated talks. In: Proceedings of the 16th Conference of the European Association for Machine Translation (EAMT), pp. 261–268 (2012)
19. Hong, V.T., Thuong, H.V., Le Tien, T., Pham, L.N., Van, V.N.: The English-Vietnamese Machine Translation System for IWSLT 2015 (2015)

20. Tiedemann, J.: Parallel data, tools and interfaces in OPUS. In: LREC, vol. 2012, pp. 2214–2218 (2012)
21. Kingma, D., Ba, J.: Adam: A method for stochastic optimization. arXiv preprint arXiv:1412.6980 (2014)
22. Papineni, K., Roukos, S., Ward, T., Zhu, W.J.: BLEU: a method for automatic evaluation of machine translation. In: Proceedings of the 40th Annual Meeting on Association for Computational Linguistics, Association for Computational Linguistics, pp. 311–318 (2002)
23. Neubig: Tips on building neural machine translation systems (2016). https://www.kantanmt.com/whatisbleuscore.php
24. Klein, G., Kim, Y., Deng, Y., Senellart, J., Rush, A.M.: Opennmt: open-source toolkit for neural machine translation. arXiv preprint arXiv:1701.02810 (2017)

Prediction of Re-tweeting Activities in Social Networks Based on Event Popularity and User Connectivity

Sayan Unankard(✉)

Information Technology Division, Faculty of Science, Maejo University, Chiang Mai, Thailand
sayan@mju.ac.th

Abstract. This paper proposes an approach to predict the volume of future re-tweets for a given original short message (tweet). In our research we adopt a probabilistic collaborative filtering prediction model called Matchbox in order to predict the number of re-tweets based on event popularity and user connectivity. We have evaluated our approach on a real-world dataset and we furthermore compare our results to two baselines. We use the datasets crawled by the WISE 2012 Challenge (http://www.wise2012.cs.ucy.ac.cy/challenge.html) from Sina Weibo (http://weibo.com), which is a popular Chinese microblogging site similar to Twitter. Our experiments show that the proposed approach can effectively predict the amount of future re-tweets for a given original short message.

Keywords: Re-tweets · Prediction · Micro-blog · Social networks

1 Introduction

The prediction of message propagation is one of the major challenges in understanding the behaviors of social networks. In this work, we study that challenge in the context of the Twitter social network. In particular, our goal is to predict the propagation behavior of any given short message (i.e., tweet) within a period of 30 days. This is captured by measuring and predicting the number of re-tweets.

To model the re-tweeting activities, we use the datasets crawled by the WISE 2012 Challenge from Sina Weibo, which is a popular Chinese microbloging site similar to Twitter. In Sina Weibo, retweet mechanism is different from Twitter. In Twitter, users can only re-tweet a tweet without modifying the original tweet. However, in Sina Weibo user can modify or add information from other users' in the re-tweeting path in their own re-tweet.

The dataset that to be used in this challenge contains two sets of files. Firstly, Followship network, it includes the following network of users based on user IDs. Secondly, Tweets, it includes basic information about tweets (time, user ID, messages ID), mentions (i.e., user IDs appearing in tweets), re-tweet paths, and

P. Perner (Ed.): MLDM 2018, LNAI 10935, pp. 357–368, 2018.
https://doi.org/10.1007/978-3-319-96133-0_27

Table 1. Number of original messages re-tweeted in 30 days

Number of re-tweets	Original messages		Annotated with events	
	#messages	%	#messages	%
<10	42,551,891	94.749	882,191	2.073
10–99	2,171,214	4.835	65,809	3.031
100–499	173,803	0.387	5,464	3.144
500–999	10,283	0.023	400	3.890
1,000–4,999	2,838	0.006	158	5.567
5,000–9,999	26	0.00006	2	7.692
≥10,000	11	0.00002	1	9.091
Total	44,910,066	100.00	954,025	2.124

Table 2. Number of re-tweets in 10 levels within 30 days

Level	Number of re-tweets	%
1	107,025,967	56.056
2	49,401,724	25.874
3	16,934,845	8.869
4	8,045,285	4.213
5	4,196,992	2.198
6	2,315,732	1.212
7	1,294,638	0.678
8	746,494	0.390
9	428,158	0.224
10	240,606	0.126

whether containing links. User IDs and message IDs are anonymized. Content of tweets are removed, based on Sina Weibo's Terms of Services. Some tweets are annotated with events. For each event, the terms that are used to identify the event and a link to Wikipedia[1] page containing descriptions to the event are given. For the purpose of this challenge, 369 million messages and 68 million user profiles were extracted. The sizes of the followship dataset and the microblog dataset are 12.8 GB and 64.8 GB, respectively. It should be note that the dataset is not complete but it is sufficiently large to predict the re-tweeting behavior of users on Sina Weibo.

In preparation for the challenge, we further collected some statistical information for a better understanding of the available datasets. In particular, for the followship dataset (i.e., the who is following whom relationship), we found that the majority of users have less that 10 followers (approximately 91%) as shown in Fig. 1. Additionally, for the microblog dataset (i.e., whose tweets are

[1] https://wikipedia.org.

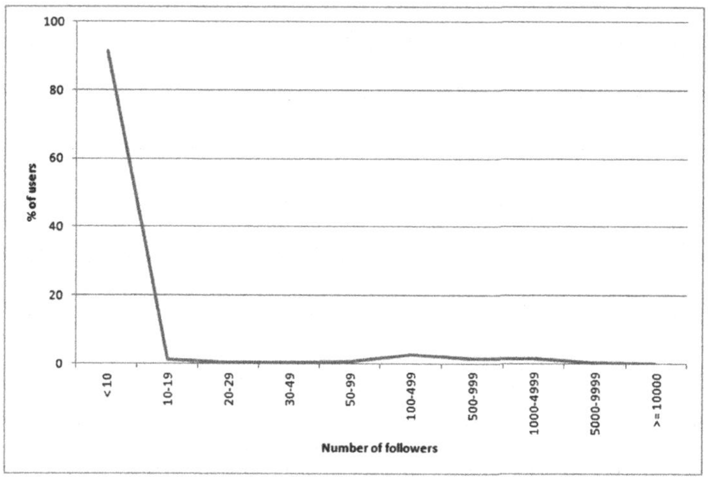

Fig. 1. User distribution based on numbers of followers

re-tweeted by whom), we ranked the distribution of the original tweets based on how many re-tweets they received within 30 days as shown in Table 1.

The table also shows the subsets of tweets that have been annotated with events. As the table shows, approximately 95% of the original tweets were re-tweeted less than 10 times, of which approximately 2% were annotated with events. In addition, most original tweets were re-tweeted in 3 levels within 30 days (approximately 91%) as shown in Table 2 and Fig. 2.

In order to understand the re-tweet activity, we also studied the re-tweet activity by day of the week and time of the day. We selected original tweets associated with events which have the number of re-tweets more than 100 for our study (6,934 messages). In Fig. 3, the graph shows the number of re-tweets per day of week. Based on a sample of tweets, Monday is the most popular day for re-tweet activity; followed by Tuesday and Friday. In Fig. 4, the chart shows the number of re-tweets per hour of the day. During the day, the most re-tweet activity happens from 10 a.m. to 12 p.m.

The contributions of this paper are summarized as follows: (1) An extensive statistical studies on the re-tweeting activities of users' behaviors in the widely used social network are provided. (2) The number of re-tweets is measured to understand the users' participation for spreading information in social network. (3) An approach to automatically predict the number of re-tweets over micro-blogs is proposed.

This paper is organised as follows, Sect. 2 is about related work. The proposed approach is presented in Sect. 3. In Sect. 4, we present the experimental setup and results, the conclusions are given in Sect. 5.

2 Related Work

Microblogging activities in social networks have been attracting growing attentions from researchers in Data Mining and Information Retrieval. One interesting

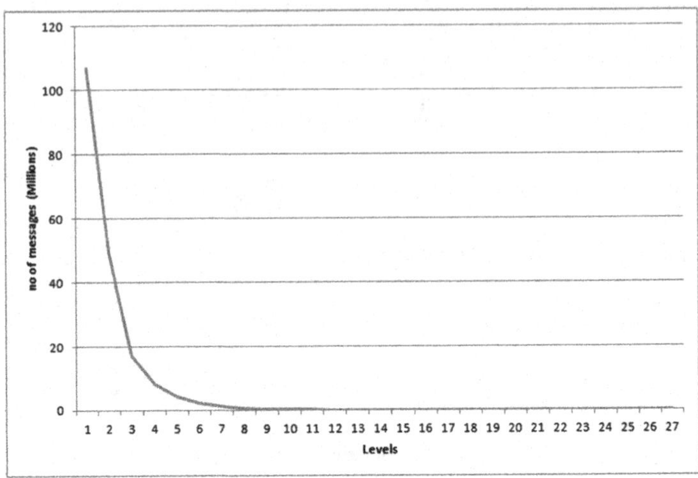

Fig. 2. Number of re-tweets in each level

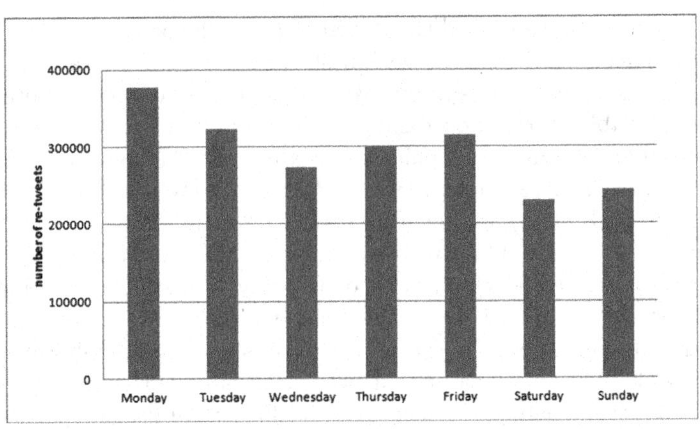

Fig. 3. Re-tweet activity by day of the week

problem is the study on the re-tweeting behaviours from an information diffusion perspective. Most works had focused on Twitter, a popular microblogging site. Insightful studies on re-tweeting behaviors can be seen from [1,2,4,9].

In [1], Boyd et al. studied the various aspects of re-tweeting. They conducted interviews with Twitter users and investigated the reasons why they re-tweet. Letierce et al. in [4] surveyed how researchers used Twitter to spread scientific messages. However, neither of them attempted to predict on whether a given message is to be re-tweeted. Galuba et al. in [2] focused on the URL propagation via re-tweets. In [9], Suh et al. gathered content and contextual features from Twitter and identified factors that impact re-tweeting. They found that URLs

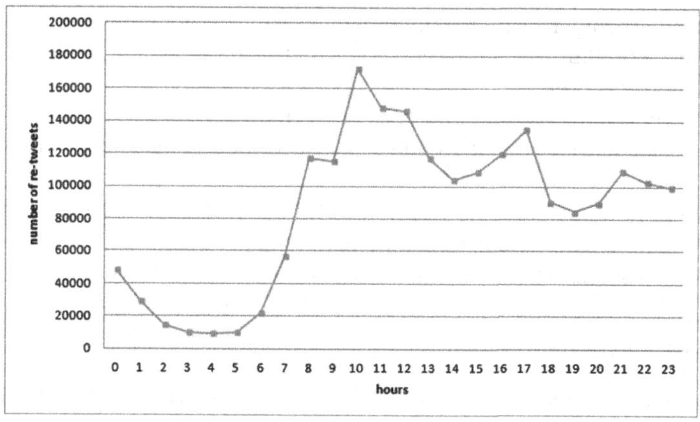

Fig. 4. Re-tweet activity by time of the day

and hashtags have strong relationships with re-tweetability and identified the number of followers and followees as important factors.

Zaman et al. in [12] adapted a probabilistic collaborative filtering model called Matchbox [8] to predict information spreading in Twitter based on features such as tweeter and re-tweeter information, and the tweet content. In [11], Yang et al. proposed a factor graph model based on users' re-tweeting history.

Recently, Petrovic et al. in [7] built a time-sensitive model based on the passive-aggressive algorithm (PA) to automatically predict re-tweets activities. Hong et al. in [3] trained a binary classifier to predict if a message will be re-tweeted or not and a multi-class classifier based on logistic regression to predict the volume of re-tweets for a given message. For the multi-class classification, they used four class labels (0: no re-tweet, 1: re-tweets less than 100, 2: re-tweets less than 10000, and 3: re-tweets more than 10000).

In [6], Peng et al. modelled the re-tweeting activities by using conditional random fields with three types of features, namely content influence, network influence and temporal decay factor. Naveed et al. in [5] argued that the tweet content is the key for re-tweeting prediction. They used logistic regression to compute re-tweet likelihood based on various interesting content features such as emotion positive/negative, exclamation/question mark, etc.

In our work, the tweet content has been removed from Sina Weibo microblog dataset pre-processed by WISE 2012 Challenge due to Sina Weibo's Term of Services.

3 Proposed Approach

3.1 Assumptions

Based on the given datasets, together with our statistical information presented in Sect. 1, we make the following assumptions:

- An event category is a group of similar events (manually grouped).
- The more popular the event category is, the more likely the tweet will be re-tweeted by a user.
- Similar events have similar re-tweet patterns.
- A user who has re-tweeted frequently in the past is likely to re-tweet in the future.
- Most users are only interested in tweets under certain event categories. Most followers are users who have similar interests.
- Users' interests and preferences are assumed to be stable.

3.2 Event Category

In WISE 2012 Challenge, the given original tweets are annotated with some social events together with their corresponding keyword lists. It is difficult to automatically group events into different categories and it is neither in our focus in this report because some events are simply labelled by personal names or by location names. Moreover, their relevant keyword lists are arbitrary and do not show clear contextual information between the keyword list and the event title. To solve this problem, we manually divide the WISE 2012 provided 46 events that have links to Wikipedia pages into 12 categories such as Natural Disaster, Celebrities, Product Release, Sports, and etc. The examples of event categories are shown in Table 5.

In order to predict the number of re-tweets, we adopt a probabilistic collaborative filtering prediction model called Matchbox which is a probabilistic model for generating personalized recommendations of items to users of a web service. Matchbox is used for the prediction of rating that users are likely to assign to items. It uses content information in the form of user and item metadata to learn correlations between them. Details of the Matchbox model can be found in [8]. This model can be applied to cope with our problem by the prediction of re-tweeting probability instead of the prediction of rating.

Matchbox is a factor graph for Bi-linear rating model. Each user and item are represented by a vector of features. Each feature is associated with a latent trait vector and the linear combination of the trait vectors for a particular user or item. An existing implementations of the Matchbox Recommender can be found at this link[2]. We adopt this model to predict whether followers of user will re-tweet the message posted by user who has posted an original tweet. For our approach, each tweet is regarded as an item while re-tweeter is considered as a user.

3.3 Tweet and Re-tweeter Features

According to datasets which have been pre-processed by WISE 2012 Challenge, we have Followship network and Tweets data without content. Although keyword lists are provided, they are arbitrary and do not show clear contextual

[2] https://docs.microsoft.com/en-us/azure/machine-learning/studio-module-reference/train-matchbox-recommender.

Algorithm 1. *PredictRetweetviaMatchbox*

```
Input: mid:message id
Output: num_r:predicted number of re-tweets
tweets = GetPrevious100Messages(mid); //Get the latest 100 messages of the
  user before the predicted message (mid) has been posted.
users = GetRetweeters(tweets);
retweets = GetRetweetHistory(tweets);
tw_vectors = CreateTweetFeatures(tweets);
usr_vectors = CreateUserFeatures(users);
model = TrainModel(tw_vectors, usr_vectors, retweets);
foreach u ∈ usr_vectors do
    predict = model.predict(u, mid);
    if predict.getProbTrue() ≥ threshold then
        num_rt = num_rt + 1;
    end
end
return num_rt;
```

information between the keywords and the event. For our approach, each tweet is regarded as an item while re-tweeter is considered as a user to train the model.

Tweet features consist of tweet id, user id who posted the original tweet, number of followers, number of followees, day of the week, time of the day and event category. Re-tweeter features include user id who re-tweeted the tweet, number of followers and number of followees. Re-tweeters are extracted from all users who have re-tweeted in the past of each tweet. The binary feedback is 1 if the re-tweeter re-tweeted the tweet within 30 days and 0 otherwise. The output of the model will be the probability of a re-tweet of the tweet by the re-tweeter.

3.4 Training Data

In order to train the model, it is required the positive binary feedback and also negative feedback. The positive feedbacks are from all re-tweet action in the past of each tweet in the same event category. For a given tweet, the negative feedbacks are from all followers in the re-tweet network who did not re-tweet a given tweet. For each test event, we train the model by random select 1,000 original tweets in the same event category as items and extract re-tweeters from re-tweet history of each tweet.

3.5 Prediction

To predict the number of re-tweets, for given original tweet and set of users if user has the high probability of a re-tweet greater than threshold, the user is likely to re-tweet the original tweet. In order to find the most suitable value for threshold, we did the prediction on different threshold values. When threshold = 0.4 it render the best performance. The algorithm is shown as Algorithm 1.

4 Experiments and Evaluations

4.1 Baselines

The two baselines were compared with our results.

Baseline 1: Regression based on Popularity and Connectivity. It is a model to predict re-tweet activities based on event popularity and user connectivity by using a naïve approach. The intuition is that a tweet is more likely to be re-tweeted if it is about a popular event and its author is highly connected with others. The prediction will be the estimation of the probabilities of these two parameters in the space (connectivity of the user and category popularity). The formula for re-tweet prediction is shown as Eq. 1.

$$NumberOfRTs = 19.950(0.024C(uid) + 0.976P(uid, category)) \quad (1)$$

where function $C(uid)$ is to find how many re-tweets a *uid* (user ID) may have based on the number of followers she has, function $P(uid, category)$ is to predict how the event category popularity influences a tweet being re-tweeted. More details can be found in [10].

Baseline 2: Classification based on User Preferences. User preferences are used to train a classifier to predict the possible number of re-tweets in 30 days for a given original tweet. Given an original tweet, the authors need to compute how possible a user will re-tweet the original tweet in the category. The candidate users are extracted from re-tweet history in a form of "who-retweet-who". The authors use $P(r, u, c)$ to denote the interestingness of candidate re-tweet user r to original user u on category c. The function is defined as Eq. 2.

$$P(r, u, c) = \sum RT(r, u, c) / \sum T(u, c) \quad (2)$$

where $RT(r, u, c)$ returns the number of re-tweets by user r from user u on category c; $T(u, c)$ returns the total number of u's tweet on category c. More details of this algorithm can be found in [10].

Table 3. Average prediction error scores

Methods	Error scores
Baseline 1 : Regression based on Popularity and Connectivity	0.700
Baseline 2 : Classification based on User Preferences	0.666
Our approach : Probabilistic collaborative filtering prediction model	**0.627**

Table 4. The 33 predicted re-tweets of our approach and baselines

Mid	Ground truth	Baseline 1	Baseline 2	Our approach
Death of steve jobs				
8872263516485596	165	228	127	428
8872961090747701	3550	135	128	312
8872983825828431	154	184	137	128
8872990233170214	121	126	140	156
Fuzhou bombings				
2700059958269443492	798	476	152	185
2700117991448817596	242	93	132	303
2700176673306864228	686	223	140	624
2701374467440601577	384	418	222	449
2701431322360449433	1271	10	148	488
Japan earthquake				
510001800832822169	576	68	157	138
510001800834928814	187	46	142	169
510001800911043840	188	46	172	42
550001800915348600	2119	43	147	463
550001805270270366	1068	5	134	40
580001800835537050	699	30	740	114
Li Na win French open tennis				
2709258383303085289	620	3	260	281
2709864654666932643	13638	33	117	52
2709870697693881414	417	25	114	246
2709871713230486085	1383	53	132	232
2709893077170155796	163	33	130	403
Xiaomi release				
8896800636296312	1230	20	119	83
8896822338137478	114	95	257	101
8896858839607761	1681	23	136	555
8896889634186199	808	4	178	185
8896952812610010	249	12	129	154
Yao Jiaxin murder case				
2243526721410152330	700	232	160	141
2243578214587694822	129	142	142	159
510001856830842390	534	170	182	159
510001856834367317	121	39	298	152
510001904903643837	1001	946	143	128
510001908564754698	3474	9	616	106
5100019107401880	1126	609	170	187
550001906873838396	4900	31	184	164

Table 5. The 12 event categories in *WISE 2012* dataset

Category	Event
Natural disaster	Earthquake of Yunnan Yingjiang
	Japan earthquake
	Yushu earthquake
	Zhouqu landslide
Product release	iPhone 4s release
	Windows Phone release
	Motorola was acquisitions by Google
	Xiaomi release
Sports	Yao Ming retirement
	Spain Series A League
	Li Na win French Open in tennis
Famous people	The death of Muammar Gaddafi
	The death of Steve Jobs
	Family violence of Li Yang
	Tang Jun educatioin qualification fake
	The death of Kim Jongil
	The death of Osama Bin Laden
Social problem	Anshun incident
	China Petro chemical Co. Ltd.
	Foxconn worker falls to death
	Guo Meimei
	Incident of self-burning at Yancheng, Jangsu
	Shanghai government's urban management officers attack migrant workers in 2011
	Yao Jiaxin murder case
	Yihuang self-immolation incident
	The death of Wang Yue
	Case of running fast car in Heibei University
Public security	Bohai bay oil spill
	Foxconn bombing in Chengdu
	Fuzhou bombings
	Shanxi
Protests	Chaozhou riot
	Mass suicide at Nanchang Bridge
	Protests of Wukan
	Qianxi riot
	Zhili disobey tax official violent
Development projects	Line 10 of Shanghai-Metro pileup
	Shenzhou-8 launch successfully
	Tiangong-1 launch successfully
Economy	House prices
	Individual income tax threshold rise up to 3500
Human right	Qian Yunhui
	Deng Yujiao incident
Accident	Gansu school bus crash
	Wenzhou train collision
Crime	Chongqing gang trials

4.2 Evaluations

For evaluation our approach, we predicted 33 test tweets and the ground truth of 33 tweets are provided by WISE 2012 Challenge[3]. For each tweet we compute the prediction error score (PE).

$$PE_i = \frac{|A_i - P_i|}{A_i} \tag{3}$$

where A_i is the actual value for tweet i and P_i is the predict value for tweet i. For each approach, the average of prediction error scores is computed.

$$Average_j = \frac{\sum_{t=1}^{N} PE_t}{N} \tag{4}$$

where N is the number of test tweets. The small number is the better prediction result. Table 3 shows the performance of our approach against baselines. Table 4 lists the predictions for the given 33 original tweets over 6 given events. In Table 3, our approach shows a better performance than others on the prediction number of re-tweets.

5 Conclusions

In this paper, we proposed an approach to automatically predict the number of re-tweets over micro-blogs. Our contributions can be summarized as: (1) We proposed a solution to estimate the volume of re-tweets for understanding the behaviors of social networks. (2) We adopt probabilistic collaborative filtering prediction model named Matchbox by the prediction of re-tweeting probability instead of the prediction of rating. (3) We provide an evaluation for the effective re-tweet prediction on a real-world dataset. Our experiments show that the proposed approach can effectively predict the number of re-tweet over the baselines. In future work, we will retrospectively study the assumptions that we have made on the given datasets and develop a hybrid approach to integrate the proposed methods.

References

1. Boyd, D., Golder, S., Lotan, G.: Tweet, tweet, retweet: conversational aspects of retweeting on twitter. In: HICSS, pp. 1–10 (2010)
2. Galuba, W., Aberer, K., Chakraborty, D., Despotovic, Z., Kellerer, W.: Outtweeting the twitterers-predicting information cascades in microblogs. In: Proceedings of the 3rd Conference on Online Social Networks, p. 3 (2010)
3. Hong, L., Dan, O., Davison, B.D.: Predicting popular messages in twitter. In: WWW (Companion Volume), pp. 57–58 (2011)

[3] http://content.wuala.com/contents/imc_ecnu/wise_challenge/A4_T2GTruth.zip?dl=1.

4. Letierce, J., Passant, A., Decker, S., Breslin, J.G.: Understanding how twitter is used to spread scientific messages. In: ACM WebSci Conference 2010, pp. 1–8 (2010)
5. Naveed, N., Gottron, T., Kunegis, J., Alhadi, A.C.: Bad news travel fast: a content-based analysis of interestingness on twitter. In: ACM WebSci Conference, pp. 1–7, June 2011
6. Peng, H.-K., Zhu, J., Piao, D., Yan, R., Zhang, Y.: Retweet modeling using conditional random fields. In: ICDM Workshops, pp. 336–343 (2011)
7. Petrovic, S., Osborne, M., Lavrenko, V.: RT to Win! predicting message propagation in twitter. In: ICWSM (2011)
8. Stern, D.H., Herbrich, R., Graepel, T.: Matchbox: large scale online bayesian recommendations. In: WWW, pp. 111–120 (2009)
9. Suh, B., Hong, L., Pirolli, P., Chi, E.H.: Want to be retweeted? Large scale analytics on factors impacting retweet in twitter network. In: SocialCom/PASSAT, pp. 177–184 (2010)
10. Unankard, S., Chen, L., Li, P., Wang, S., Huang, Z., Sharaf, M.A., Li, X.: On the prediction of re-tweeting activities in social networks – a report on WISE 2012 challenge. In: Wang, X.S., Cruz, I., Delis, A., Huang, G. (eds.) WISE 2012. LNCS, vol. 7651, pp. 744–754. Springer, Heidelberg (2012). https://doi.org/10.1007/978-3-642-35063-4_61
11. Yang, Z., Guo, J., Cai, K., Tang, J., Li, J., Zhang, L., Su, Z.: Understanding retweeting behaviors in social networks. In: CIKM, pp. 1633–1636 (2010)
12. Zaman, T.R., Herbrich, R., Stern, D.H.: Predicting information spreading in twitter. In: Computational Social Science and the Wisdom of Crowds, vol. 55, pp. 1–4 (2010)

Flow Prediction Versus Flow Simulation Using Machine Learning Algorithms

Milan Cisty(✉) and Veronika Soldanova

Faculty of Civil Engineering, Slovak University of Technology in Bratislava, Bratislava, Slovakia
milan.cisty@stuba.sk

Abstract. The paper deals with differences between two types of machine learning river flow modelling, i.e., their simulation and prediction. In this paper, "simulation" means a determination of river flows from only meteorological data. The second type of modelling, i.e., prediction, additionally includes preceding flows in the input data. Preceding flows are known at the time of making a prediction. For this reason, i.e., because less input data serve for the simulation, it is a more difficult task than the prediction, and its degree of precision is also usually lower. The authors focused on the improvement of flow simulation methodology, i.e., the determination of river flows only from climate data. Several machine learning models were tested for this purpose, and their results are compared in the paper with a conceptual hydrological model. Three options were evaluated in the paper for the improvement of the precision of the machine learning type of flows simulation: (1) the effect of the use of different types of models, (2) the impact from the expansion of input data utilizing feature engineering, and (3) improving the accuracy of the simulation by applying an ensemble paradigm. An increased degree of precision (approximately 12%) of the flow simulation was obtained after the incorporation of the above methodological enhancements to the computations (when compared to standard hydrological methods). The authors believe that the proposed methodology will be a promising alternative to the usual hydrological simulation, and it would be useful to test it in an extended study in which more streams would be evaluated.

Keywords: Flow simulation · Flow prediction · Data-driven methods

1 Introduction

Since the mid-1990s, many papers have been published in the hydrological literature which deals with the application of machine learning methods for the modelling of the rainfall-runoff process (these methods are also called "data-driven modelling"). Most of these papers only consider flow predictions [1–3] (we are mentioning only review papers due to the many published works). In the present study, the authors have followed this research and are emphasizing the existence of two types of such machine learning modelling, i.e., simulation and prediction. The difference between simulation and prediction is characterized below.

Flow *prediction* using machine learning models is usually based on input data consisting of a time series of climatic variables and on the known flow at the time the

P. Perner (Ed.): MLDM 2018, LNAI 10935, pp. 369–382, 2018.
https://doi.org/10.1007/978-3-319-96133-0_28

prediction is being made. The values of such variables are typically used in input data from several time steps before the prediction date. A flow predicted is one or more time steps (e.g., hours, days) ahead. This type of prediction has several advantages compared to a determination by other hydrology models, such as physical GIS-based models or conceptual models. The benefits of machine learning models include their simplicity, reduced amount of input data, and the higher degree of precision of the results, which in the scientific literature has been mainly demonstrated for short-term predictions [4].

In the terminology used in this paper flow *simulation* is the second type of modelling and is defined as the modelling of river flows only from data which is describing the factors directly causing it, i.e., rain, evaporation, melting of snow, and similar climate variables. In such a way understood simulation is applicable, e.g., for the generation of flows in the context of climate change impact studies. Its purpose could be to provide a long time series of a flow, which together with other data (usually simulated as well), e.g., temperatures and precipitation, can serve for statistical analyses of an expected drought, an investigation of irrigation demands, verifications of the future functioning of a water supply reservoir, etc. When comparing to the prediction a positive difference, with regard to its impact on the degree of precision, is that in a simulation (or flow generation), climate data can also be used from the same time step as is the time step for which the flow is simulated and not only from previous days (which is not possible in river flow prediction).

However, what is more important, in the context of a simulation, flows from previous days cannot be used as input data, since previous flows are not available (the entire flow time series is unknown and is going to be simulated). This difference in the amount of data that can be used as an input for prediction and simulation is substantial. This disadvantage is further underlined by the fact that flows have a strongly autocorrelative nature. From this property of river flows, it follows that the modelled flow mostly depends on its previous value. So, if the preceding flow is included in the input data, the accuracy of the calculations is much better. In contrast, the absence of data for previous flows makes a flow simulation substantially more difficult than its prediction. The result of this difference between the input data for simulation and prediction is that the precision of the simulation is more demanding to achieve.

The primary goal of this paper is an analysis of options for the improvement of flow simulation based on climate variables. In a search for the improvement of the precision of such calculations, the effects of three factors were investigated. The first is the selection of the algorithm, where a typical conceptual hydrological model and machine learning methods were compared. An analysis of the possibilities of feature engineering was the second possibility analyzed for the improvement of the calculation results. Feature engineering is constructing new input variables which are derived in various ways from primary climate data. The third option investigated was an experiment with the application of an ensemble paradigm and an analysis of its contribution to the precision of the simulated flows.

2 Materials and Methods

In the real application of the proposed method for streamflow generation, climate inputs obtained by a weather generator, specifically a daily time series of temperatures and precipitation, will be used. However, the proposed method must be verified using actual data to ensure the climatic and hydrological compatibility of all the time series. For this purpose a daily time series of the temperatures, precipitation and stream flows of the Parna Creek in the Carpathian region of Slovakia were used (Fig. 1).

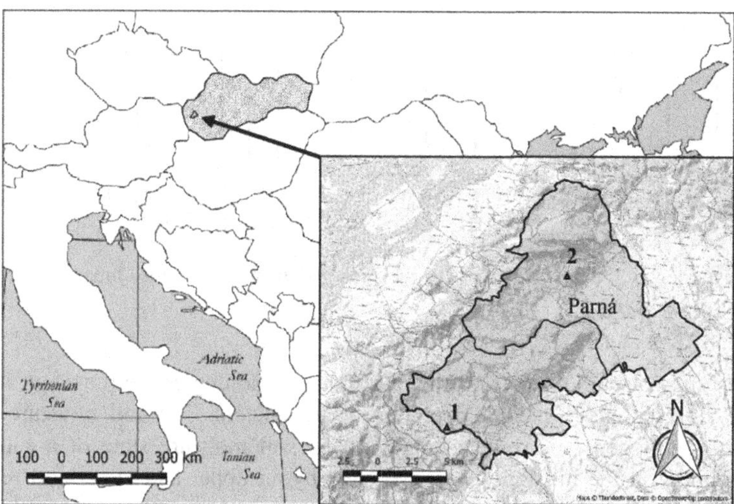

Fig. 1. Location of the test site and indication of the CarpatClim points (1, 2) from which climate data were acquired

The temperature and precipitation data were obtained from the CarpatClim [5], a publicly available geodatabase (http://www.carpatclim-eu.org). The authors therefore also verified in this article the applicability of climate data from this database to simulate water discharges in small streams, such as the Parna Creek. The CarpatClim database provides climate data in a square grid of points located in the Central Carpathian region. Figure 1 shows two of these points (labeled 1 and 2), which are located near the Parna, and from which precipitation and minimum and maximum temperatures have been obtained for the years 1961–2010. An overview of the flow regime of this creek and the regime of the climatic characteristics in its watershed is provided in Fig. 2.

In this paper one hydrological and three data driven methods for generation of flows were used. Basic principle of this methods is described in following paragraphs.

The hydrological model used in this paper was developed at the Vienna University of Technology and is freely available as a package (add-in) for R software [6]. It is a semi-distributed conceptual rainfall-runoff model, following the structure of the well-known HBV model [7]. The model runs on a daily time step and consists of a snow routine, a soil moisture routine, and a flow routing routine. The snow routine

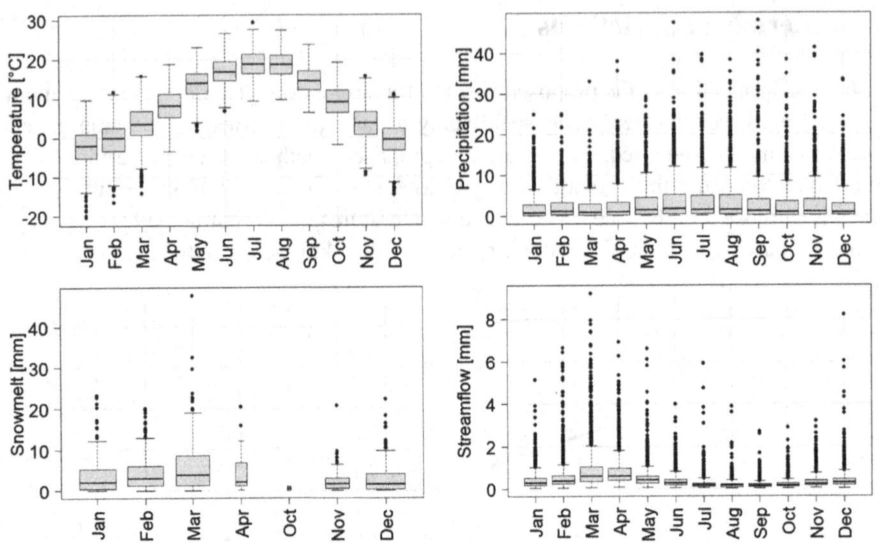

Fig. 2. Overview of the flow regime of Parna creek and the regime of the climatic characteristics in its watershed

represents snow accumulation and melting by a simple degree-day concept; it uses a degree-day factor and a melt temperature as parameters. The soil moisture routine represents runoff generation and changes in the soil moisture state of a catchment. Runoff routing on the hillslopes is represented by an upper and lower soil reservoir. Excess rainfall enters the upper reservoir and leaves by three paths, i.e., outflow from the reservoir based on a rapid storage coefficient; percolation to the lower zone with a constant percolation rate; and, if a threshold of the storage state is exceeded, by an additional outlet based on a very fast storage coefficient. Water leaves the lower zone based on a slow storage coefficient. The outflow from both reservoirs is then routed by a triangular transfer function that represents runoff routing in the streams [8]. A genetic algorithm [9] was used to calibrate this conceptual rainfall-runoff model with fifteen parameters.

In this work, three machine learning algorithms were applied, namely Random Forest, XGBoost and Deep Learning Neural Network. They are used for supervised learning problems in this study, where we use the training data (with multiple features) to predict a target variable. The Random Forest (RF) algorithm, which was initially proposed by Breiman [10], is an ensemble method that generates a set of individually trained decision trees and combines their results. The regression trees are a series of decision rules that dictate how a target variable is computed from the input (predictor) variables. A forest is a collection of trees, and an RF consists of a group or ensemble of simple tree predictors, each one of which can evaluate a target variable by using a set of predictor values. The variability in solving a given regression problem by individual members of the ensemble is realized such that RF is random in two ways. Firstly, each tree is based on a random subset of the observations (bootstrap sample), and secondly, each split in each tree is created on a random subset of all the available variables [10]. The benefit of the RF's randomness is robustness against over-fitting and good generalization abilities. Given the number of trees created, the degree of accuracy increases up to a certain point. When used as a regression method, decision trees can describe complex relationships fairly accurately among multiple variables; by aggregating the results of these regression trees into a forest, an even more accurate solution is generated. In addition to these characteristics, RF parameterization is not particularly complicated nor is RF model tuning too difficult. This study used an RF add-on package [11] with R statistical software [12]. Although it has several parameters, only two parameters specified by the user are necessary to tune to run RF: the number of trees in the forest, ntree, and the number of variables randomly sampled at each split, mtry. This study operated RF with a default value of ntree 500 and of mtry, which was found by a cross-validation procedure.

XGBoost is an abbreviation of Extreme Gradient Boosting, where the term "Gradient Boosting" was proposed in the [13]. As stated by the XGBoost algorithm author in [14], his algorithm is based on this original model. Gradient Boosting is a forwardly learning ensemble method, e.g., it builds a model in a stage-wise fashion (not in a parallel fashion, as in the case of RF). The guiding idea is that a good predictive model can be obtained through increasingly refined approximations. Gradient boosting evaluates the precision of a model in a previous stage and then develops the next model, which computes the differences between the current results computed and the known target values (i.e., not the original target values). The next models are thus mainly concentrating on the previously incorrectly computed samples. Such "boosting" continues until the desired level of accuracy is reached. In this way, gradient boosting produces a prediction model as an ensemble of weak prediction models (usually shallow decision trees). The algorithm used in this work, XGBoost, follows the principle of gradient boosting. However, there are some differences in the modelling details. XGBoost uses a more regularized model formalization to control over-fitting, which gives it a better performance in comparison with previously evolved boosting algorithms. The XGBoost developers have also made other significant performance enhancements to different features of the XGBoost implementation. These result in significant differences in speed and memory utilization due to (1) the use of sparse matrices with sparsity aware algorithms, (2) improved data structures for better utilization of the processor cache, thereby making it faster, and (3) better support for multicore processing, which reduces overall training time [14]. From the point of view of the users of boosting algorithms, when they use GBM and XGBoost for training large datasets (e.g., 5 million or more records), they can experience significantly reduced memory usage for the same dataset; it is also easier to use multiple cores to reduce training time. The cost of the advantages of using XGBoost (compared, e.g., with an RF algorithm) is that XGBoost has several parameters to tune, while RF is almost tuning-free, which is why we included both algorithms in this study. In this work, the *xgboost* R package was used. Seven parameters were tuned, so genetic algorithms were applied in the tuning process. The optimized parameters were the maximum number of iterations, learning rate, minimum loss reduction gamma, maximum tree depth, minimum child weight, subsample ratio of the training instance, and subsample ratio of the columns when constructing each tree. A description of the parameters and various recommendations for setting them can be found at XGBoost WWW [15].

A Deep Learning Neural Network (DLNN) is a tool that has significantly expanded the boundaries the real-world applicability of machine learning in recent years in computer vision, speech recognition, various recommendation systems, and predictions of some types of sequential data, such as time series. It is a new generation of artificial neural networks characterized by a "deeper" architecture compared to a multilayer perceptron, which was in use at the end of the previous century. Deep learning has been enhanced by a number of recently developed improvements. These enhancements have been published and put into practice in the last five years and cover new types of activation functions, new network architectures, improved network initialization before training, and an improvement of the training process. Deep learning is a subfield of machine learning that emphasizes learning successive layers to increasingly meaningful representations of searched patterns in the data. Even though the superiority of DLNN could probably be better shown for more complex projects than our task in this work, we wanted to test its performance in the hydrological field; to our knowledge, DLNN has rarely been tested in this area. The more complex tasks mentioned in the previous sentence means that they have a larger and more complex data structure, such as computer vision. We have used the TensorFlow software tool developed by researchers and engineers working as part of the Google "brain team". Construction of the neural network, its settings for training, and training was accomplished using the keras R package [16].

Calibration of all the models was performed on data from the above case study from Slovakia using data from the years 1961–1995. The basic inputs in all the models consisted of a time series of the average daily precipitation, evapotranspiration, air temperatures and river flows. The verification of the precision of all the models was accomplished on test data from 1996–2010. To illustrate the difference between prediction and simulation when using machine learning models, the calculation of the flow prediction for one day in advance was also performed. It includes the flow from the predication day and other previous flows among the input data. To illustrate the usefulness of using advanced machine learning models, the simulation was also accomplished with multiple linear regression.

The evaluation of the precision of the simulation included a comparison of the simulated flows with monitored data, which in our case was flows measured on the River Parna in Slovakia. The quality of the simulation was assessed using several criteria for statistical precision. For an assessment of the model's precision, the Nash-Sutcliffe coefficient was prioritized due to its frequent use in hydrological calculations, which allows for a better comparison of our results with the results of other authors.

3 Results and Discussion

In this chapter, three options for the improvement of the precision of flow simulation methods are assessed. In the first part, the effect of the selected method on the accuracy of the results of the flow simulation is evaluated. In the second subchapter, methods for the application of feature engineering and an evaluation of its impact on the precision of the calculations are quantified. In the third part, the models created in the context of a simple ensemble are described and evaluated.

The expected or required range of the precision of the simulation was identified using two calculations. The simulation using a hydrological model defined its lower limit, as the use of machine learning models has no meaning when the degree of precision is lower. By using a Deep Learning Neural Network (the most efficient among the models tested), the prediction of the flows was performed which determined the upper limit for the precision of a simulation. Such an identification of the upper limit results from the fact that the precision of prediction using machine learning models is higher than the precision of simulation, due to the integration of flows from the previous time steps into the input data. The results of these calculations, as characterized by selected statistical indicators, are given in Table 1.

3.1 Effect of the Selection of the Model

In this part, we have evaluated the simulation of flows using several models, by using their standard application. Feature engineering was not applied in these computations. This group of calculations includes calculations using the TUW hydrological model mentioned above. The calculation of the river flows was also performed using multiple linear regression (MLR) and the Random Forest (RF), Extreme Gradient Boosting (XGBoost) and Deep Learning Neural Network (DLNN) machine learning models. The optimization of the internal parameters of the machine learning models was accomplished by tenfold cross-validation. A description of the parameters necessary to tune each of the algorithms is given in the "Methods" section. The river flows, precipitation, temperatures and potential evapotranspiration data from the years 1961–1995 were used for all the models. The climate data for the linear regression and machine learning models were used in models from 20 days before the date for which the flow was simulated. The number of days affecting the calculation of the flows was acquired empirically. Usually, data from fewer days are applied in, e.g., the prediction of flows. A larger volume of the previous climate data was used in the inputs because it is the simulation that is solved in this case. Because of this the previous data on the flows are not available in this type of modelling. Additional data on the climate variables from more days should partially eliminate the lack of this information. These models are evaluated in Table 1 using standard RMSE and R2 statistics (we are not giving their description here). KGE and NSE statistics, which are frequently used in hydrology and PBIAS (percentual bias between simulated and observed values), were also used for the evaluation of the model's performances. The Nash-Sutcliffe Efficiency (NSE) is a normalized statistic that determines the relative magnitude of the residual variance compared to the measured data variance [17]. KGE is "Kling-Gupta efficiency," which was developed by Gupta [18] and later refined by Kling [19]. Both statistics can take values between minus infinity and 1; the ideal model has a value of 1. The percent bias (PBIAS) measures the average tendency of the simulated values to be larger or smaller than corresponding observed ones. The optimal value of the PBIAS is 0.0. Positive values indicate an overestimation bias, whereas negative values indicate the model's underestimation bias. From these statistical indicators, NSE is considered herein as primary, as it is the most frequently used statistic in hydrology.

Table 1. Evaluation of models that did not use feature engineering

Model	RMSE	R^2	NSE	KGE	PBIAS
TUW	0.31	0.67	0.67	0.76	−6.2
DLNN prediction	0.17	0.92	0.90	0.79	−14.8
XGBoost	0.43	0.37	0.35	0.43	21.2
RF	0.44	0.35	0.32	0.38	23.2
DLNN	0.41	0.41	0.39	0.39	5.3
MLR	0.46	0.25	0.24	0.34	11.1

RMSE – root mean square error, R^2 – coefficient of determination, NSE - Nash-Sutcliffe efficiency, KGE – Kling-Gupta efficiency, PBIAS – percent of bias

The first two lines of Table 1 contain an evaluation of two reference models that identify the theoretical minimum and maximum precision of the simulations using machine learning models. The TUW hydrological model has better values of the statistical indicators than the machine learning models. This means that the machine learning models in their basic use, are for simulation of flows (not prediction), less precise than the conceptual hydrological model. Linear regression has the worst precision indicators. Besides the hydrological model, DLNN has the best results but is also below the required minimum precision limit.

3.2 Feature Engineering and Its Influence on the Precision of a Model

Additional models evaluated in this study included feature engineering, while their input data were prepared. Constructed variables were added to the basic climate inputs for modelling the flows, which is described in this subchapter. Their construction was motivated by applying knowledge gained from the domain of hydrology. A typical example of this approach was the creation of the variable that quantifies the melting of snow, as snow (i.e., a type of precipitation which is a basic climate input) is not the direct initiator of an outflow because an outflow only comes only after the melting of snow. The potential evapotranspiration was also calculated because it is a more direct detector of water leaking from a watershed to the atmosphere than is the temperature. As the resulting flow from a basin is influenced not only by the current values of climate variables but also by their values from previous days, we also included climate data from four days before the date of the simulated flow in the model inputs. On the smaller streams with which this study deals, the outflow from the watershed is influenced by precipitation maximally from 1 or 2 preceding days. Precipitation from earlier days creates water reserves in a basin, which are a source of so-called base flow. Data from days three to four were included due to the possibility of subsequent precipitation, which makes a watershed saturated with water. Including more climate variables from previous days to the inputs did not seem to be a very useful option, which can be evaluated from the calculations in the previous subchapter (Table 1). However, as the history of the hydro-climate developments in the basin must in some way be included in the model and quantified in the input data, a variable summarizing the previous precipitation (cumRAIN) and a variable summarizing the previous evapotranspiration (cumPET) were constructed for the 60 days before the day on which the flow was simulated. The last experiment with the tuning of the input data involved a modification of the target variable, i.e., the flow. Figure 3a shows the probability distribution of the measured flow from our case study using a histogram. We can see that this variable has a right-skewed distribution, which may be unsuitable for some models.

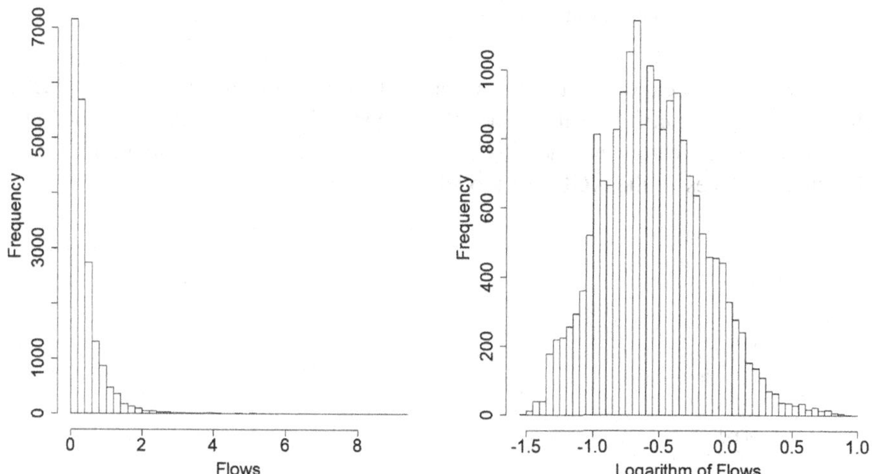

Fig. 3. Histogram of flows and histogram of logarithms of flows

For this reason, in all the models, the flow (the target variable) was replaced by its logarithm. The evaluation provided, which is in Table 2, only shows the models where such a step leads to the improved precision of a model. A histogram of the logarithms of the flows is shown in Fig. 3b, which is apparently more normally distributed.

A total of 22 explanatory variables were acquired; three of them are basic, and the remaining 19 are the result of the feature engineering.

The results of the individual calculations were evaluated using NSE and other statistical indicators and are shown in Table 2. Without using feature engineering, the precision of the models according to NSE is in a range of 0.24–0.39 (Table 1). When feature engineering was applied, the degree of precision was significantly improved and, as shown in Table 2, it is now in a range of 0.42–0.69.

An interesting point is the notable influence of the logarithm applied to the target variable (the predicted flow). This alteration significantly influenced the precision of the linear model and the precision of the XGBoost model. Although the reasons and conditions for the transformation of the target variable using a logarithm of the target were not further investigated, we consider this result, particularly regarding the XGBoost model, to be quite interesting and worth analyzing in future research.

Table 2. Evaluation of models in which feature engineering was used

Model	RMSE	R^2	NSE	KGE	PBIAS
MLR	0.41	0.48	0.42	0.46	34
MLR with log of target	0.31	0.67	0.67	0.76	−6.2
RF	0.34	0.64	0.59	0.64	27.4
XGBoost	0.33	0.66	0.61	0.62	30.6
XGBoost with log of target	0.29	0.71	0.69	0.76	14.1
DLNN	0.31	0.67	0.66	0.77	0.6

RMSE – root mean square error, R^2 – coefficient of determination, NSE - Nash-Sutcliffe efficiency, KGE – Kling-Gupta efficiency, PBIAS – percent of bias

Because we did experiment with the input data, we also evaluated the impact of individual variables on the resulting modelling. For an illustration and confirmation of the importance of feature engineering, Fig. 4 shows the relative importance of the individual variables in the XGBoost model.

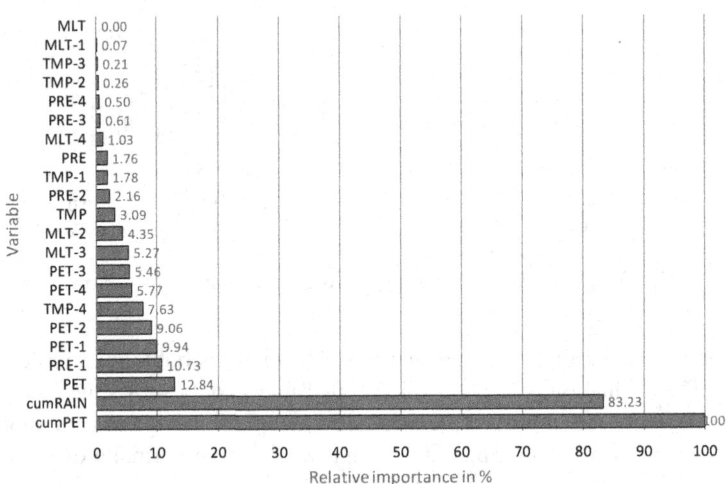

Fig. 4. Variable importance (PET – potencial evapotranspiration, PRE – precipitation, TMP – temperatures, MLT – snow melting; number indicate day before prediction)

The most significant influences on the model have both summation variables, i.e., the variables summing the previous precipitation and evapotranspiration. At this point, all the models coincided. This finding shows that entering precipitation and evapotranspiration data on a day-by-day basis for many previous days as done in the preceding subchapter is less favorable than the use of cumulative values. For further research on this phenomenon, it would be useful to optimize the number of backward days in the summation and review the effect of using a weighted sum instead of the standard sum of such data, where the older data would receive a lower weight. Such calculation experiments were not done in this study; we are mentioning them only as possible options for improving the effect of feature engineering on the precision of a model.

3.3 Ensemble Method

Ensemble regression models calculate the target variable by a combination of its multiple specifications for several models. Ensembles can be classified as those which contain many individual models (e.g., Random Forest) and ensembles composed of a smaller number of models, which are also a subject of interest of this paper. In this second type of ensemble, the precision of the participating models is the first precondition for the selection of the model into an ensemble and its success in improving an overall prediction. This degree of precision for models selected for an ensemble should be similar and as high as possible. The second precondition is that the prediction should be more precise for different domains of hydrological inputs, i.e., the correlation between the individual models in the ensemble should be as low as possible.

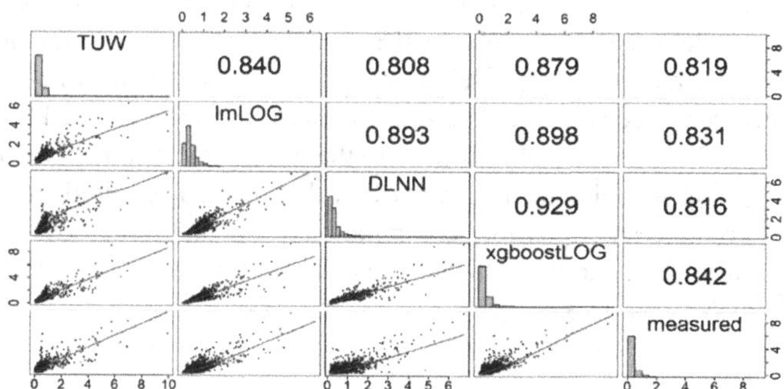

Fig. 5. Correlation between TUW model, linear regression and XGBoost with logarithm of target variable (lmLOG and xgboostLOG), DLNN, and observed flows

Based on such conditions, a conceptual hydrological model (Table 1) and three of the data-driven models tested (Table 2) were selected as members of the ensemble, i.e., the linear and the XGBoost model with a logarithm of the target variable and the DLNN model. The flows calculated through these four models have a comparable degree of accuracy but are not identical. This follows from the fact that the models are evaluated differently by various statistical indicators. The different results of the individual models are also demonstrated by Fig. 5, which summarizes the correlation of the four best models with the measured flows and the mutual correlation between these models. The correlation coefficients are in the part above the diagonal. The diagonal identifies the models and indicates the probability distribution of their results (in the form of a histogram), and the part below the diagonal expresses always the dependence of the two models (identified by horizontal and vertical projections on the diagonal) graphically.

The figure shows that the models have different histograms and therefore a different probability distribution function and different mutual relations expressed by the small charts below the diagonal. The correlation coefficients between the flows calculated by these models are in a range of 0.81–0.93.

Such a level of correlation means that each model can better simulate flows on different days, which is a precondition for the cooperation of models in an ensemble. The resulting value of the simulation can be obtained, for example, by algebraic combination of the results of the individual models or using a suitably selected meta-model with the inputs being the flows calculated by different models. Such a methodology is typical in a stacking regression [20]. The specific form and parametrization of such a metamodel must be optimized based on the results of a flow simulation from the test folds of the cross-validation by which the individual models were tuned. However, this is not possible if the hydrological model is intended to be kept in an ensemble (as it is a well-established model). This is because the hydrological model does not use cross-validation during its calibration (so we do not have test folds for it). It cannot use it because the flows are simulated in a continuous sequence of days when using a hydrological model. As a metamodel cannot be optimized, it was decided to use an average of the results of the individual models (which could be considered as a simple meta-model). The flows obtained as an average of the simulations using a hydrological model and the three selected machine learning models have an NSE of 0.75, i.e., a degree of precision which is better by almost 12% compared to the original hydrological model. If the ensemble considers the hydrological model and only two of the machine learning models, the average NSE value of such acquired simulated flows is 0.74 (there are three such possible combinations). For the three possible combinations of the hydrological model with only one selected machine learning model, the average value of NSE is 0.73.

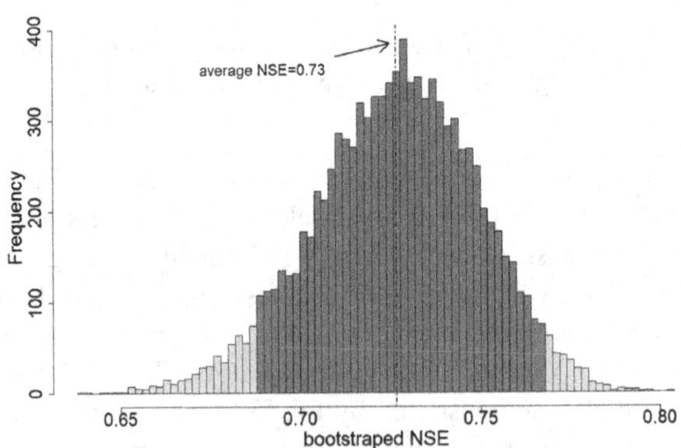

Fig. 6. NSE values of the ensemble created from bootstrap replications

Figure 6 shows a histogram of NSE values of the ensemble created in this way from 10,000 bootstrap replications selected from the four specified models. As shown in this figure, most of the NSE values have a higher degree of precision than the original hydrological model. The average NSE value from this calculation experiment is 0.73, and the 95% level of the confidence interval for this value is 0.69–0.77 (highlighted in Fig. 4 by darker color), which is quite a promising result from this approximate testing.

4 Conclusion

The authors dealt in this paper with specific features of two types of machine learning river flow modelling, i.e., with prediction and simulation. The simulation represents tasks such as the generation of flow from precipitation and temperatures in the context of climate change impact and adaptation studies or in the calculation of unknown flows in unmeasured streams using an analogy method. Data on previous flows cannot be used as inputs for such simulation tasks. In comparison with prediction, simulation usually is less precise by 10–20%; in the case study addressed in this paper, it was 17%, according to the NSE statistic. In this paper, the authors have focused on improving the precision of simulation by using constructed variables which were included in the inputs, by using the most recent machine learning models, and by the application of simple ensemble simulations. Although the precision of the individual machine learning models did not significantly exceed the precision of the hydrological model, the connection of the individual models into an ensemble showed better results by 12% than the original hydrological model. These results show the potential of the above methodology as an alternative to traditional calculation methods.

The case study which verifies the proposed procedures is focused on a specific stream located in southwest Slovakia (Central Europe). For further application of this methodology, it would be useful to accomplish some refinements, which are mentioned in the paper and verification through case studies in different locations in the future.

Acknowledgements. This work was supported by the Slovak Research and Development Agency under Contract No. APVV-15-0489 and by the Scientific Grant Agency of the Ministry of Education of the Slovak Republic and the Slovak Academy of Sciences, Grant No. 1/0665/15.

References

1. ASCE Task Committee on Application of Artificial Neural Networks in Hydrology: Artificial neural networks in hydrology. I: preliminary concepts. J. Hydrol. Eng. **5**(2), 115–123 (2000)
2. Papacharalampous, G.A., Tyralis, H., Koutsoyiannis, D.: Comparison of stochastic and machine learning methods for multi-step ahead forecasting of hydrological processes. Preprints 2017, 2017100133. https://doi.org/10.20944/preprints201710.0133.v1
3. Maier, H.R., Dandy, G.C.: Neural networks for the prediction and forecasting of water resources variables: a review of modelling issues and applications. Environ. Model Softw. **15**(1), 101–124 (2000)
4. Yaseen, Z.M., El-Shafie, A., Jaafar, O., Afan, H.A., Sayl, K.N.: Artificial intelligence based models for stream-flow forecasting: 2000–2015. J. Hydrol. **530**, 829–844 (2015)

5. Szalai, S., Spinoni, J., Galos, B., Bessenyei, M., Molar, P., Szentimrey, T.: Use of regional database for climate change and drought. In: 5th IDRC Davos 2014. Global Risk Forum GRF Davos (2014)
6. Viglione, A., Parajka, J.: TUWmodel: Lumped Hydrological Model for Education Purposes. R package version 0.1-8 (2016). https://CRAN.R-project.org/package=TUWmodel
7. Lindström, G., et al.: Development and test of the distributed HBV-96 hydrological model. J. Hydrol. **201**(1–4), 272–288 (1997)
8. Parajka, J., Merz, R., Blöschl, G.: Uncertainty and multiple objective calibration in regional water balance modelling: case study in 320 Austrian catchments. Hydrol. Process. **21**(4), 435–446 (2007)
9. Boisvert, J., El-Jabi, N., St-Hilaire, A., El Adlouni, S.E.: Parameter estimation of a distributed hydrological model using a genetic algorithm. Open J. Mod. Hydrol. **6**(3), 151–167 (2016)
10. Breiman, L.: Random forests. Mach. Learn. **45**(1), 5–32 (2001)
11. Liaw, A., Wiener, M.: Classification and regression by randomForest. R News **2**(3), 18–22 (2002)
12. R Core Team: R: a language and environment for statistical computing. R Foundation for Statistical Computing, Vienna, Austria (2017). https://www.R-project.org/
13. Friedman, J.H.: Greedy function approximation: a gradient boosting machine. Ann. Stat. **29** (5), 1189–1232 (2001)
14. Chen, T., Guestrin, C.: XGBoost: a scalable tree boosting system. In: Proceedings of the 22nd ACM SIGKDD International Conference on Knowledge Discovery and Data Mining, pp. 785–794. ACM (2016)
15. XGBoost Homepage. https://xgboost.readthedocs.io/en/latest/. Accessed 16 Mar 2018
16. Allaire, J.J., Chollet, F.: keras: R Interface to 'Keras'. R package version 2.1.4 (2018). https://CRAN.R-project.org/package=keras
17. Nash, J.E., Sutcliffe, J.V.: River flow forecasting through conceptual models part I-A discussion of principles. J. Hydrol. **10**(3), 282–290 (1970)
18. Gupta, H.V., et al.: Decomposition of the mean squared error and NSE performance criteria: implications for improving hydrological modelling. J. Hydrol. **377**(1–2), 80–91 (2009)
19. Kling, H., Fuchs, M., Paulin, M.: Runoff conditions in the upper Danube basin under an ensemble of climate change scenarios. J. Hydrol. **424**, 264–277 (2012)
20. Breiman, L.: Stacked regressions. Mach. Learn. **24**(1), 49–64 (1996)

A Method of Biomedical Knowledge Discovery by Literature Mining Based on SPO Predications: A Case Study of Induced Pluripotent Stem Cells

Zheng-Yin Hu[1], Rong-Qiang Zeng[1,2(✉)], Xiao-Chu Qin[3], Ling Wei[1], and Zhiqiang Zhang[1]

[1] Chengdu Library and Information Center of Chinese Academy of Sciences, Chengdu 610041, Sichuan, People's Republic of China
{huzy,zhangzq}@clas.ac.cn, weiling@mail.las.ac.cn

[2] School of Mathematics, Southwest Jiaotong University, Chengdu 610031, Sichuan, People's Republic of China
zrq@swjtu.edu.cn

[3] Guangzhou Institutes of Biomedicine and Health, Chinese Academy of Sciences, Guangzhou 510530, Guangdong, People's Republic of China
qin_xiaochu@gibh.ac.cn

Abstract. A large amount of valuable knowledge is hidden in the vast biomedical literatures, publications, and online contents. In order to identify the previously unknown biomedical knowledge from these resources, we propose a new method of knowledge discovery based on SPO predications, which constructs a three-level SPO-semantic relation network in the considered area. We carry out the experiments in the area of induced pluripotent stem cells, and the experimental results indicate that our proposed method can significantly discover the potential biomedical knowledge in this area, and the performance analysis of this method sheds lights on the ways to further improvements.

Keywords: Biomedical knowledge discovery · SPO
Induced pluripotent stem cells · Semantic relation network
Community detection

1 Introduction

Knowledge Discovery in Text (KDT) is the process of identifying and extracting the new, useful, potential and understandable patterns from the literatures in a credible way. With the rapid growth of biomedical literatures, Knowledge Discovery in Biomedical Literature (KDiBL) has become an important research area [8].

Information extraction plays an important part in KDT, which automatically extracts the specific terms, the corresponding characteristics and the semantic

P. Perner (Ed.): MLDM 2018, LNAI 10935, pp. 383–393, 2018.
https://doi.org/10.1007/978-3-319-96133-0_29

relations among them from the texts as the basic knowledge unit of knowledge discovery. Subject-Predication-Object (SPO) represents the semantic relationships among the knowledge units, which is widely used in the fields of knowledge organization, semantic network, knowledge discovery, and so on [3].

In this paper, we propose a new method of knowledge discovery based on SPO predications, which constructs a three-level SPO-based semantic relation network in the considered area. Then, we realize the community detection for the SPO-semantic relation network, so as to find the hidden valuable knowledge. The experimental results indicate that the proposed method can effectively discover the unknown biomedical knowledge. The performance analysis explains the behavior of our proposed method and sheds lights on the ways to further improvements.

The remaining part of this paper is organized as follows. In the next section, we briefly review the previous works related to the biomedical knowledge discovery. In Sect. 3, we investigate a new method of constructing three-level SPO-based semantic relation network to discover the potential unknown knowledge. Section 4 provides the experimental results and the performance analysis of the proposed method in the area of induced pluripotent stem cells. The conclusions are presented in the last section.

2 Literature Reviews

In this section, we present the literature reviews concentrating on the biomedical knowledge discovery.

In [1], with techniques from systems medicine, natural language processing, and graph theory, the authors created a molecular interaction network, which represents neural injury and is composed of relationships automatically extracted from the literature, in order to support the diagnosis of mild traumatic brain injury. Actually, they retrieved the citations related to neurological injury and extract the semantic predications that contain potential biomarkers. The experimental results on 99,437 relevant citations and 26,441 unique relations indicated a set of 17 potential biomarkers, which provides an opportunity to obtain more effective diagnosis than the current methods.

In [5], the authors investigated the use of deep learning methods, which have shown significant promise in identifying hidden patterns from large corpus of text in an unsupervised manner, in order to discover the hidden, interesting or previously unknown biomedical knowledge from free text resources. They used the text corpus from MRDEF file in the Unified Medical Language System (UMLS) dataset as training set to discover potential relationships. Taking a manual evaluation from a sample of the non-overlapping set, their proposed algorithm founded 32% of new relationships not originally represented in the UMLS, which provides provide a promising approach in discovering potential new biomedical knowledge from free text.

In [7], according to some semi-supervised learning methods named Positive-Unlabeled Learning (PU-Learning), the authors proposed a novel method to predict the disease candidate genes from human genome, which ia an important part of nowadays biomedical research. Since the diseases with the same phenotype have the similar biological characteristics and genes associated with these same diseases tend to share common functional properties, the proposed method detects the disease candidate genes through gene expression profiles by learning hidden Markov models. The experiments are carried out on a mixed part of 398 disease genes from three disease types and 12001 unlabeled genes, and the results indicate a significant improvement in comparison with the other methods in literature.

In [9], the authors presented a set of knowledge discovery framework to identify the unknown knowledge from the biomedical literatures based on subject-predication-object predications. Actually, they extracted the SPO predications from the biomedical literature by using UMLS corpus and SemRep. Then, they constructed the corresponding semantic network diagrams with NetMiner [10], which is applied to the field of induced pluripotent stem cells. The experimental results showed that can effectively reveal the knowledge content from the biomedical literatures.

In [11], the authors proposed a novel Sequence-based Fusion Method (SFM) is proposed to identify disease genes from human genome, which is of great importance to improve diagnosis and treatment of disease. In this method, the amino acid sequence of the proteins has been carried out to present the genes into four different feature vectors, instead of using a noisy and incomplete prior-knowledge. Then, the intersection set of four negative sets generated by distance approach is used to select more likely negative data from candidate genes, and the decision tree has been applied as a fusion method to combine the results of four independent state-of the-art predictors based on support vector machine (SVM) algorithm for the final decision. The experimental results confirm the efficiency and validity of the proposed method.

In [12], the authors proposed a method based on degree centrality that measures connectedness in a graph, in order to automatically summarize the semantic predications representing assertions in MEDLINE citations in the large graph with more than 500 citations. The experiment was carried out on the four categories of clinical concepts related to treatment of disease, the results showed that their proposed method are very competitive, in comparison with the reference standard produced manually by two physicians.

In [13], the authors presented a hybrid model for the extraction of biomedical relations that combines Recurrent Neural Networks (RNNs) and Convolutional Neural Networks (CNNs), in order to extract high-quality biomedical relations from biomedical texts. In this model, RNNs and CNNs are employed to automatically learn the features from the sentence sequence and the dependency sequences to generate the shortest dependency path for the biomedical relation extraction. The experiments are carried out on five public (protein-protein interaction) PPI corpora and a (drug-drug interaction) DDI corpus, and the experimental results indicate the proposed model can effectively boost biomedical relation extraction performance.

3 Methodology

In our work, we propose a new method of knowledge discovery based on SPO predications, which constructs a three-level SPO-based semantic relation network in a certain area. First, we present an introduction to the SPO-based semantic relation network. Then, we construct a three-level graph, which is different from the graph generated by the NetMiner. Afterwards, we investigate the method of detecting the community in the three-level graph.

3.1 Network Construction

Generally, we construct the SPO-based semantic relation network, according to four basic principles proposed by M. Fiszman et al., which are relevancy, connectivity, novelty and saliency, more information about these principles can be found in [2].

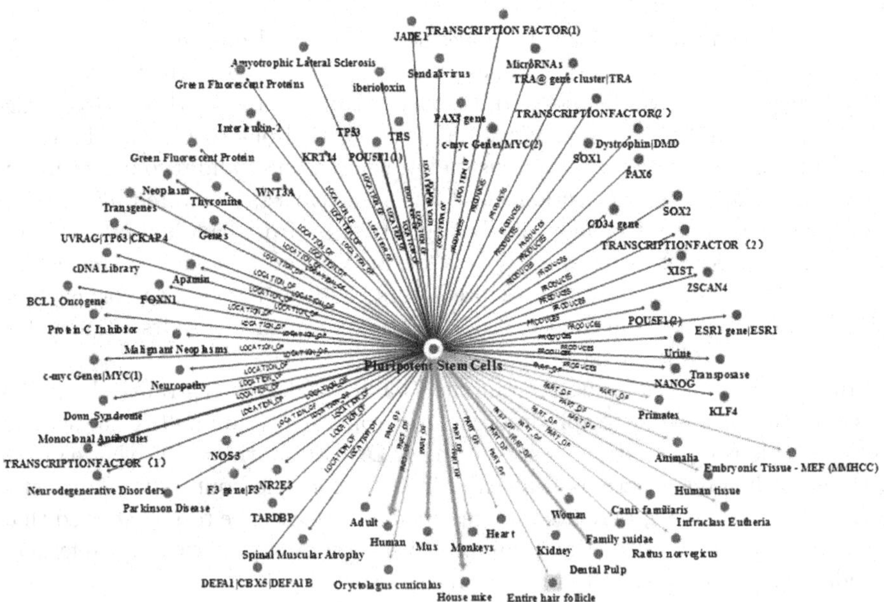

Fig. 1. An example of semantic network based on the subject of induced pluripotent stem cells [9].

Two examples of directed semantic networks of induced pluripotent stem cells are respectively illustrated in Figs. 1 and 2, which are both depicted by NetMiner. In these two figures, the node in the network represents the semantic concepts, and the corresponding color represents the type of semantic concept. In addition, the edge represents the semantic relationship with the direction from the subject to the object, the color and the width of the line represent the semantic type and the frequency of semantic description respectively.

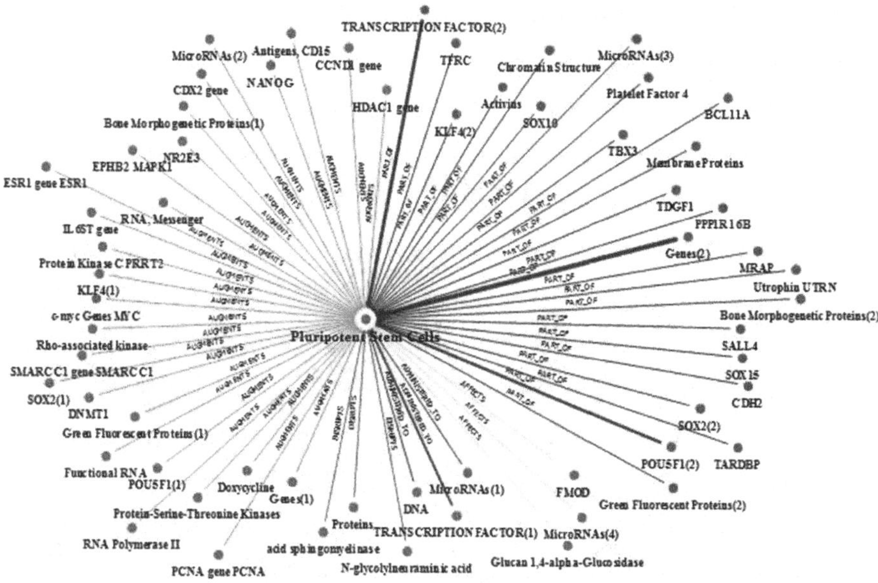

Fig. 2. An example of semantic network based on the object of induced pluripotent stem cells [9].

In Fig. 1, the Induced Pluripotent Stem Sells (IPSC) is the subject, the other nodes are the objects. Whereas, in Fig. 2, the Induced Pluripotent Stem Sells (IPSC) is the object, the other nodes are the subjects. Actually, these two figures present the semantic relation between the IPSC and the other nodes from two different angles.

However, we can only obtain the local semantic relation between one subject and the other objects (or between one object and the other subjects) in Both Figs. 1 and 2. In fact, it is very difficult to illustrate the global semantic relation between different subjects and different objects in one figure with NetMiner, which is the disadvantage of discovering the hidden biomedical knowledge. Then, it is essential that we construct the global semantic relation network for knowledge discovery.

An example of global SPO-based semantic relation network is illustrated in Fig. 3, which is composed of thousands of nodes and edges. In this figure, the grey node denotes the subject, the orange node denotes the object, and the green node denotes the semantic relation. Actually, many nodes can be both the subjects and the objects, which makes the whole network very complicated. That's to say, it is very difficult for the experts to recognize the valuable biomedical knowledge from the network.

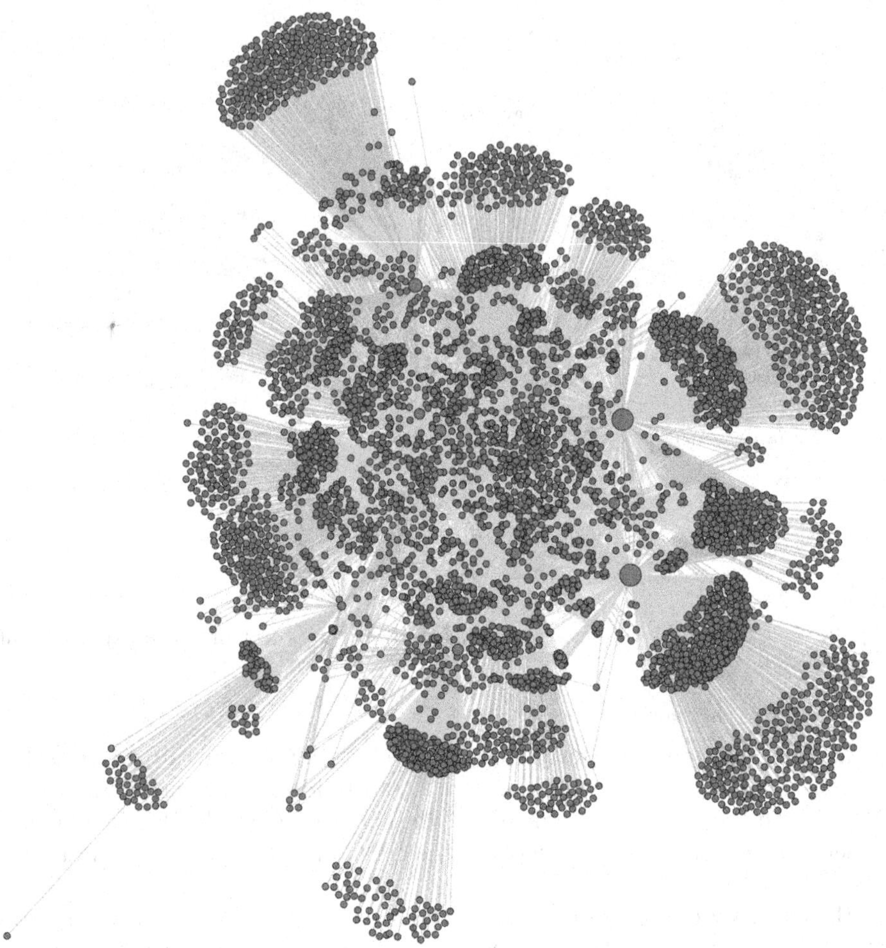

Fig. 3. An example of global SPO-based semantic relation network. (Color figure online)

3.2 Community Detection

In order to clearly recognize the valuable biomedical knowledge from global SPO-based semantic relation network, it is essential to detect the community structure, which is the intrinsic properties of networks. In our work, we take the widely accepted modularity function proposed by Newman and Girvan, which is defined as follows [6]:

$$Q = \frac{1}{2m} \sum_{vw} [A_{vw} - \frac{k_v k_w}{2m}] \delta(C_v, C_w), \tag{1}$$

Suppose the vertices are divided into the communities such that vertex v belongs to community C denoted by C_v. In Formula 1, A is the adjacency matrix of graph G. $A_{vw} = 1$ if one node v is connected to another node w, otherwise $A_{vw} = 0$. The δ function $\delta(i,j)$ is equal to 1 if $i = j$ and 0 otherwise. The degree k_v of a vertex v is defined to be $k_v = \sum_v A_{wv}$, and the number of edges in the graph is $m = \sum_{wv} A_{wv}/2$.

In addition, the modularity function can be represented in a simple way, which is formulated below [6]:

$$Q = \sum_i (e_{ii} - a_i^2), \tag{2}$$

where i runs over all communities in graph, e_{ij} and a_i^2 are respectively defined as follows [6]:

$$e_{ij} = \frac{1}{2m} \sum_{vw} A_{vw} \delta(C_v, i) \delta(C_w, j), \tag{3}$$

which is the fraction of edges that join vertices in community i to vertices in community j, and

$$a_i = \frac{1}{2m} \sum_v k_v \delta(C_v, i), \tag{4}$$

which is the fraction of the ends of edges that are attached to vertices in community i.

Then, we employ the local search procedure to effectively detect the community structure, which is presented in the Algorithm 1 [4].

Algorithm 1. Community Detection Algorithm

1: **Input**: network adjacency matrix A
2: **Output**: the best value of the modularity function
3: $P = \{x^1, \ldots, x^p\} \leftarrow$ Random_Initialization (P)
4: **repeat**
5: $x^i \leftarrow$ Local_Search (x^i)
6: **until** a stop criterion is met

In this algorithm, we randomly divide the whole network into two communities, and each smaller community is further divided into two smaller communities. Let C_i and C_j be two communities, w be a vertex from C_i or C_j, we assume that $w \in C_i$ and the corresponding change by moving vertex w from C_i to C_j can be computed as follows [4]:

$$\Delta Q(w, C_i, C_j) = \frac{k_w^j - k_w^i}{m} + \frac{k_w(a_i - a_j)}{m} - \frac{k_w^2}{2m^2}, \tag{5}$$

where k_w^i and k_w^j are respectively the number of edges connecting vertex w and the other vertices in communities C_i and C_j. While, for any vertex v in community C_i, we can also obtain the updated ΔQ value $\Delta Q'(v, C_i, C_j)$ with the formula below [4]:

$$\Delta Q'(v, C_i, C_j) = \Delta Q(v, C_i, C_j) - (\frac{k_w^2}{m^2} - \frac{2A_{wv}}{m}). \quad (6)$$

With the incremental value of the modularity function in Formulas 5 and 6, the local search procedure can chooses the best move at each step until the modularity does not improve any more. Then, we obtain the communities of the considered network.

4 Case Study

Induced pluripotent stem cells technology is one of the most important emerging frontier technologies in the biomedical field, which can nurture new stem cells with similar differentiation potential as embryonic stem cells by reprogramming the mature cells [9]. Then, it is of great significance to realize the knowledge discovery in the SPO-based semantic network. In this section, we apply our proposed method in the area of IPSC and present the experimental results with performance analysis.

4.1 Data Information

In order to carry out the experiments, we obtain the data from the PubMed Database by inputting the key words "Regenerative Medicine" from 2000 to 2014. Then, we select the literatures retrieved by the Semantic Medline Database, the type of literature is "Journal Article". The exact data information is presented in Table 1.

Table 1. The information of SPO-based semantic relation network.

	SPO
Number of subjects	2055
Number of objects	1821
Number of actions	45

In this table, we have retrieved 10,687 papers and obtained 65,042 SPO-based semantic relations, which consists of 2055 subjects, 1821 objects and 45 actions. With these information, we can construct a three-level SPO-based semantic relation network.

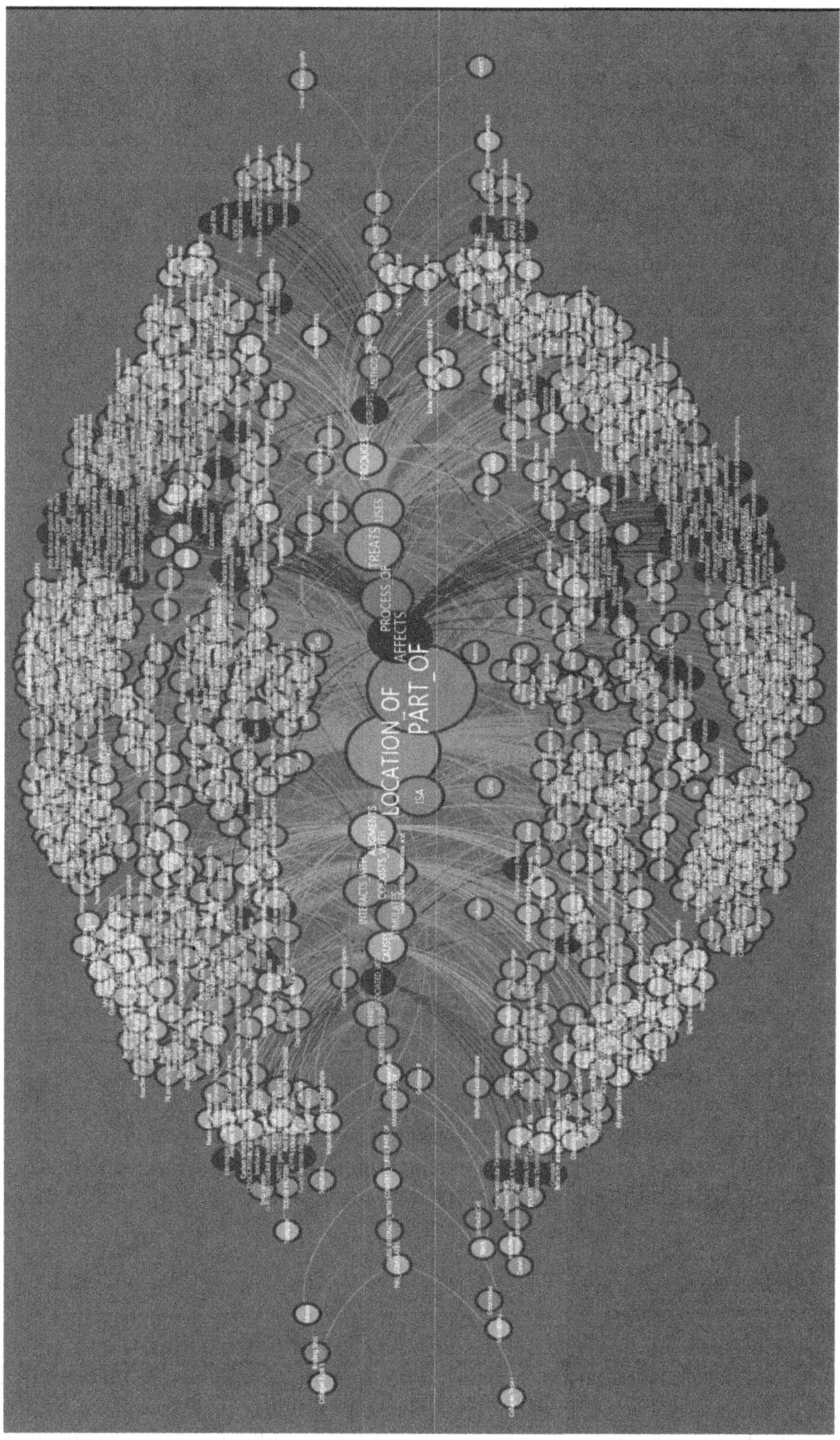

Fig. 4. The global SPO-based semantic relation network. (Color figure online)

4.2 Experimental Results

In this subsection, we present the experimental results in the area of IPSC, which are classified into different communities according to the corresponding actions. The global SPO-based semantic relation network of IPSC is illustrated in Fig. 4 below.

In this figure, the upper level and the lower level respectively represents the subjects and the objects, which are linked by the edges with different colors. The middle level represents the actions, in which the frequency is proportional to the size of the circle.

Furthermore, different communities are represented in different colors, which are composed of the subjects, the objects and the corresponding actions. For example, there is a community colored in green with the action "LOCATION_OF" in Fig. 4.

Table 2. The communities detected in global SPO-based semantic relation network.

Action	Subject	Object
LOCATION_OF	Embryo, Liver, Epidermis	Toxic effect, Purinoceptor
	Body tissue, Basement membrane	Injury wounds, Tissue Engineering
	Entire bony skeleton, Retina, ···	Cell Transformation, Neoplastic, ···
PART_OF	Mammary gland, Chorionic villi	Infraclass Eutheria, Equus caballus
	Bone Marrow Cells, Serum	Cementoblasts, Entire tendon
	Mesenchymal Stem Cells, ···	Rattus norvegicus, ···
ADMINISTERED_TO	Small Interfering, MicroRNAs	Cells, Patients, Mus
	RNA, Growth Factor, Adiponectin	Human embryonic stem cell
	High Throughput Screening, ···	Urothelial Cell, ···

The computational results are summarized in Table 2. In this table, we do not present all the found communities in the network but to provide parts of three different communities, which are colored in green (located in the center of Fig. 4), in pink (located in the center of Fig. 4) and in red (located on the left of Fig. 4).

Moreover, the subjects and the objects of the community in pink are linked by the action "PART_OF", which is the highest frequency among all the actions. From Fig. 4, we can clearly recognize the different communities, which is very helpful to realize the biomedical knowledge discovery.

5 Conclusions

In this paper, we have investigated a new method of constructing the three-level SPO-based semantic relation network for biomedical knowledge discovery. To achieve this goal, we have carried out the experiments in the area of induced pluripotent stem cells. The experimental results indicate that our proposed method can significantly discover the potential unknown biomedical knowledge in the considered area.

Acknowledgments. The work in this paper was supported by the key projects of the National Social Science Foundation of China "Theory and Applications Research of Subject-Informatics for Domain Knowledge Discovery" (Grant No: 17ATQ008), supported by the Informationization Special Project of Chinese Academy of Sciences "E-Science Application for Knowledge Discovery in Stem Cells" (Grant No: XXH13506-203), and supported by the Fundamental Research Funds for the Central Universities (Grant No. A0920502051722-53).

References

1. Cairelli, M.J., Fiszman, M., Zhang, H., Rindflesch, T.C.: Networks of neuroinjury semantic predications to identify biomarkers for mild traumatic brain injury. J. Biomed. Semant. **6**(25), 1–14 (2015)
2. Fiszman, M., Rindflesch, T.C., Kilicoglu, H.: Abstraction summarization for managing the biomedical research literature. In: Proceedings of the HLT-NAACL Workshop on Computational Lexical Semantics, pp. 76–83 (2004)
3. Keselman, A., Rosemblat, G., Kilicoglu, H.: Adapting semantic natural language processing technology to address information overload in influenza epidemic management. J. Am. Soc. Info. Sci. Technol. **61**(12), 2531–2543 (1990)
4. Lü, Z.P., Huang, W.Q.: Iterated tabu search for identifying community structure in complex networks. Phys. Rev. E **80**, 026130 (2009)
5. Rather, N.N., Chintan, O.P., Khan, S.A.: Using deep learning towards biomedical knowledge discovery. I. J. Math. Sci. Comput. **2**, 1–10 (2017)
6. Newman, M.E.J., Girvan, M.: Finding and evaluating community structure in networks. Phys. Rev. E **69**(2), 026113 (2004)
7. Nikdelfaz, O., Jalili, S.: Disease genes prediction by HMM based PU-learning using gene expression profiles. J. Biomed. Inform. **81**, 102–111 (2018)
8. Swanson, D.R.: Medical literature as a potential source of new knowledge. Bull. Med. Libr. Assoc. **78**(1), 29–37 (1990)
9. Wei, L., Hu, Z.Y., Pang, H.S., Qin, X.C., Guo, H.M., Fang, S.: Study on knowledge discovery in biomedical literature based on spo predications: a case study of induced pluripotent stem cells. Digit. Libr. Forum **9**, 28–34 (2017)
10. Workman, T.E., Fiszman, M., Hurdle, J.F., Rindflesch, T.C.: Biomedical text summarization to support genetic database curation: using semantic medline to create a secondary database of genetic information. J. Med. Libr. Assoc. **98**(4), 273–281 (2010)
11. Yousef, A., Charkari, N.M.: SFM: a novel sequence-based fusion method for disease genes identification and prioritization. J. Theor. Biol. **383**, 12–19 (2015)
12. Zhang, H., Fiszman, M., Shin, D., Miller, C.M., Rosemblat, G., Rindflesch, T.C.: Degree centrality for semantic abstraction summarization of therapeutic studies. J. Biomed. Inform. **44**, 830–838 (2011)
13. Zhang, Y., Lin, H., Yang, Z., Wang, J., Zhang, S., Sun, Y., Yang, L.: A hybrid model based on neural networks for biomedical relation extraction. J. Biomed. Inform. **81**, 83–92 (2018)

Constrained Regularized Regression Model Search in Large Sets of Regressors

Olga Krasotkina[1], Michael Markov[1], Vadim Mottl[2(✉)], Ilya Pugach[3], Dmitry Babichev[4], and Alexey Morozov[3]

[1] Markov Processes International, Summit, NJ, USA
[2] Computing Center of the Russian Academy of Sciences, Moscow, Russia
vmottl@yandex.ru
[3] Moscow Institute of Physics and Technology, Moscow, Russia
[4] Moscow State University, Leninskie Gory, Moscow, Russia

Abstract. We consider the problem of regression estimation under a complex of additional assumptions. First, the regression coefficients are assumed to be doubly constrained by individual non-negativity inequalities along with the unit-sum equality. Second, it is assumed that the number of regressors far exceeds that of samples. An additional assumption is that the regression coefficients differ from zero only within a really existing small subset of a large universe of regressors, and the search for this subset (Factor Search) is the main aim of data processing. The latter assumption ensues from the practical problem of recovering the hidden composition of an investment portfolio represented by a time series of its periodic returns (values of relative profitability). However, factor search, i.e., finding a small active subset among a huge set of correlated factors, is problematic unless some a priori information on the expected portfolio structure is available. We propose a regularized regression model based on the assumption that the portfolio under analysis is rationally composed by its administration.

Keywords: Sparse constrained regression
Returns based analysis of invest portfolios · Investment portfolio composition

1 Introduction

The method of least squares is commonly adopted in regression analysis. Let $t \in \{1, \ldots, T\}$ be indices of a finite unordered set of entities each characterized by a feature vector $\mathbf{x}_j = (x_{t,i},\ i = 1, \ldots, n) \in \mathbb{R}^n$ and a goal variable $y_t \in \mathbb{R}$. If it is assumed that there exists a linear regression $y \cong \boldsymbol{\beta}^{\mathrm{T}}\mathbf{x}$, then minimization of the residual sum of squares is an evident and in most cases excellent means to find the appropriate estimate of the regression coefficients:

$$\begin{cases} \hat{\boldsymbol{\beta}} = \arg\min \sum_{t=1}^{T} (y_t - \boldsymbol{\beta}^{\mathrm{T}}\mathbf{x}_t)^2 = \arg\min (\mathbf{y} - \mathbf{X}^{\mathrm{T}}\boldsymbol{\beta})^{\mathrm{T}}(\mathbf{y} - \mathbf{X}^{\mathrm{T}}\boldsymbol{\beta}), \\ \mathbf{y} = (y_1, \ldots, y_T) \in \mathbb{R}^T,\ \ \mathbf{X} = (\mathbf{x}_1 \cdots \mathbf{x}_T)(n \times T),\ \ \boldsymbol{\beta} = (\beta_i,\ i = 1, \ldots, n) \in \mathbb{R}^n. \end{cases}, \quad (1)$$

P. Perner (Ed.): MLDM 2018, LNAI 10935, pp. 394–408, 2018.
https://doi.org/10.1007/978-3-319-96133-0_30

This principle is commonly known as that of Ordinary Least Squares (OLS). It is clear that the straightforward solution of the respective system of linear equations $\mathbf{X}\mathbf{X}^T\boldsymbol{\beta} = \mathbf{X}\mathbf{y}$ yields the unique estimate of the regression coefficients only if the regressors are linearly independent in the data set $|\mathbf{X}\mathbf{X}^T| \neq 0$, what is possible only if $n \leq N$.

However, it is typical for many applications that the number of regressors far exceeds the number of samples $n \gg N$. In this case, it is inevitable to somehow constrain the freedom in choosing the vector of regression coefficients $\boldsymbol{\beta} \in \mathbb{R}^n$. There are two essentially different ways to implement such a contraction.

The first way is adding a regularization function $V(\mathbf{a})$ to the OLS criterion (1) to insert some a priori knowledge on the expected regression model:

$$\hat{\boldsymbol{\beta}} = \arg\min\left[V(\boldsymbol{\beta}) + (\mathbf{y} - \mathbf{X}^{\mathrm{T}}\boldsymbol{\beta})^{\mathrm{T}}(\mathbf{y} - \mathbf{X}^{\mathrm{T}}\boldsymbol{\beta})\right]. \tag{2}$$

The regularization function is a penalty meant to express the undesirability of some values of $\boldsymbol{\beta}$ – the more unlikely this vector of regression coefficient appears to the observer to be, the greater must be the penalty $V(\boldsymbol{\beta})$.

The well-known regularizations ridge $V(\boldsymbol{\beta}) = \sum_{i=1}^{n} \beta_i^2$ [1, 2], bridge $\sum_{i=1}^{n} |\beta_i|^p$ ($p \gtrless 0$) [3, 4], in particular, LASSO $\sum_{i=1}^{n} |\beta_i|$ [5], their combination Elastic Net $\sum_{i=1}^{n} (\beta_i^2 + \mu\,|\beta_i|)$ [6], and more sophisticated SCAD [7] penalize the deflection of regression coefficients from zero. Actually, the same does the matrix-based regularization $V(\boldsymbol{\beta}) = \boldsymbol{\beta}^{\mathrm{T}}\mathbf{G}\boldsymbol{\beta}$ [8]. This fact suggests considering them jointly under the name of ***central regularization***.

However, in the application problem of investment portfolio analysis that triggered this work, it is required to mathematically express a priori knowledge on the unknown portfolio composition, for which purpose we shall use a special kind of ***biased regularization*** $V(\boldsymbol{\beta}) = (\boldsymbol{\beta} - \boldsymbol{\beta}_0)^{\mathrm{T}}\mathbf{G}(\boldsymbol{\beta} - \boldsymbol{\beta}_0)$ [9], where vector $\boldsymbol{\beta}_0 \in \mathbb{R}^n$ is hyperparameter along with matrix $\mathbf{G}(n \times n)$.

But, what is especially important, another way of regularization than the penalty function (2) is most adequate to this application. The portfolio composition is to be estimated as capital sharing over a set of stock market assets or asset classes, hence, the regression coefficients must sum up to unity

$$\sum_{i=1}^{n} \beta_i = 1, \text{ or in vector form, } \mathbf{1}^{\mathrm{T}}\boldsymbol{\beta} = 1,\ \mathbf{1} = (1 \cdots 1)^{\mathrm{T}} \in \mathbb{R}^n.$$

In addition, we restrict here our consideration only to the class of mutual investment funds, which are allowed to form portfolios using only their inner capital, what mathematically results in the assumption that the regression coefficients cannot be negative $\beta_i \geq 0$, $i = 1, \ldots, n$, or $\boldsymbol{\beta} \geq \mathbf{0}$, as distinct from hedge funds that may borrow money or assets from outside sources, and coefficients may be negative $\beta_i \gtrless 0$ [10].

In terms of the regularization function (2), this restriction would be hardly treatable:

$$V(\boldsymbol{\beta}) = \{0,\ \boldsymbol{\beta} \geq \mathbf{0},\ \mathbf{1}^{\mathrm{T}}\boldsymbol{\beta} = 1; \quad \infty,\ \text{otherwise}\}.$$

It will be more convenient to implicitly add the constraints to the OLS criterion (1):

$$\begin{cases} \hat{\boldsymbol{\beta}} = \arg\min(\mathbf{y} - \mathbf{X}^{\mathrm{T}}\boldsymbol{\beta})^{\mathrm{T}}(\mathbf{y} - \mathbf{X}^{\mathrm{T}}\boldsymbol{\beta}), \\ \mathbf{1}^{\mathrm{T}}\boldsymbol{\beta} = 1,\ \ \boldsymbol{\beta} \geq \mathbf{0}, \end{cases} \quad \boldsymbol{\beta} \in \mathbb{R}^n,\ \ \mathbf{y} \in \mathbb{R}^T,\ \ \mathbf{X}(n \times T). \tag{3}$$

The constrained regression problem (3) is a particular case within the class of quadratic programming problems [11]

$$\begin{cases} \boldsymbol{\beta}^{\mathrm{T}}\mathbf{G}\boldsymbol{\beta} + \mathbf{u}^{\mathrm{T}}\boldsymbol{\beta} \to \min(\boldsymbol{\beta} \in \mathbb{R}^n), \\ \mathbf{B}\boldsymbol{\beta} = \mathbf{b} \in \mathbb{R}^k,\ \ \mathbf{D}\boldsymbol{\beta} \leq \mathbf{d} \in \mathbb{R}^m,\ \ k < n,\ \ m \leq n, \end{cases}, \tag{4}$$

where matrices $\mathbf{B}(k \times n)$ and $\mathbf{D}(m \times n)$ as well as vectors $\mathbf{b}$ and $\mathbf{d}$ are specified according to prior knowledge. Our particular case (3) is featured by especially simple constraints, one equality $\mathbf{1}^{\mathrm{T}}\boldsymbol{\beta} = 1$, $k = 1$, and n sign inequalities $\boldsymbol{\beta} \geq \mathbf{0}$, $m = n$.

There is no need to use commonly adopted general iterative methods like inner or outer point procedures to solve the constrained regression problem (3). The possibility of finitely solving this problem arises from the fact that, when the solution $\hat{\boldsymbol{\beta}} = (\beta_i \geq 0,\ i = 1,\ldots,n)$ is found and the subset of active regressors $\{i : \hat{\beta}_i > 0\}$ is defined, the optimal values of the active non-zero regression coefficients are jointly the solution of the unconstrained least squares problem within the active subset. It may seem to be unrealistic to find it among the huge number of all subsets in $\{1,\ldots,n\}$, but the total search is not required due to special properties of the optimization problem (3).

The exact penalty method considered in [12, 13] yields the solution of this problem in closed form. The essence of the method is replacing the constrained optimization problem by the unconstrained penalized function

$$\mathcal{E}_\mu(\boldsymbol{\beta}) = (\mathbf{y} - \mathbf{X}^{\mathrm{T}}\boldsymbol{\beta})^{\mathrm{T}}(\mathbf{y} - \mathbf{X}^{\mathrm{T}}\boldsymbol{\beta}) + \eta\left[\left|(\mathbf{1}^{\mathrm{T}}\boldsymbol{\beta} - 1) + \sum_{i=1}^{n}\max(0,\ -\beta_i)\right|\right].$$

It is clear that this function is a majorant of the original objective function and coincides with it at the solution point $\hat{\boldsymbol{\beta}} = (\mathbf{1}^{\mathrm{T}}\hat{\boldsymbol{\beta}} = 1,\ \hat{\beta}_i \geq 0,\ i = 1,\ldots,n)$:

$$\mathcal{E}_\eta(\boldsymbol{\beta}) \geq (\mathbf{y} - \mathbf{X}^{\mathrm{T}}\boldsymbol{\beta})^{\mathrm{T}}(\mathbf{y} - \mathbf{X}^{\mathrm{T}}\boldsymbol{\beta}),\ \mathcal{E}_\eta(\hat{\boldsymbol{\beta}}) = (\mathbf{y} - \mathbf{X}^{\mathrm{T}}\hat{\boldsymbol{\beta}})^{\mathrm{T}}(\mathbf{y} - \mathbf{X}^{\mathrm{T}}\hat{\boldsymbol{\beta}}),$$

If the penalty parameter $\eta > 0$ is large enough, any minimizer of the majorant $\mathcal{E}_\eta(\boldsymbol{\beta})$ solves the problem (3).

It is emphasized in several publications, in particular in [14], that the sign constraints, in our case $\boldsymbol{\beta} \geq \mathbf{0}$, endow the least squares criterion with the pronounced property of finding a sparse regression model, even without additional means of reducing the number of regressors.

A simple procedure, that provides non-negativity of regression coefficients and is called Active Set Algorithm, is proposed in [15] and modified in [16]. The algorithm is based on the fact that, when the solution $\hat{\boldsymbol{\beta}} = (\beta_i \geq 0,\ i = 1, \ldots, n)$ is found and the subset of active regressors $\{i : \ \hat{\beta}_i > 0\}$ is defined, the optimal values of the active non-zero regression coefficients are jointly the solution of the unconstrained least squares problem within the active subset. The iterative alternating least squares algorithm starts from an initial feasible set of regression coefficients, and, at each step, variables are identified and removed from the active set in such a way that the fit strictly decreases. The solution will be found after a finite number of iterations by simple unconstrained linear regression on the active subset. However, the computational complexity of each step is cubic relative to the size of the current active subset, and the number of iterations may be quite large.

Some algorithms specially designed for the case when active regressors are essentially correlated are considered in [14].

The paper is organized as follows. In Sect. 2 we briefly consider the practical problem of investment portfolio analysis having triggered this research. In Sect. 3 we study the specificity of the constrained regressor selection problem in large sets of correlated regressors. In Sect. 4 we experimentally study the developed mathematical and algorithmic framework. Finally, Sect. 5 concludes the paper.

2 Case Study: Returns Based Analysis of Investment Portfolios

2.1 Returns Based Portfolio Model by William Sharpe

Let the capital of an investment company under analysis be fully invested in assets of n kinds in the proportion $\boldsymbol{\beta} = (\beta_1, \ldots, \beta_n)$, $\sum_{i=0}^{n} \beta_i = 1$, $\beta_i \geq 0$. This capital share is called the portfolio of the company.

The non-negativity constraints $\beta_i \geq 0$ express the assumption that assets can be purchased only using inner capital without borrowing money or assets from outside sources, or, as it is commonly adopted to say in portfolio management, without holding short (negative) positions – most mutual funds are not allowed to do this. Investment companies which take negative (short) positions $\beta_i < 0$, that correspond to borrowed assets, are called hedge funds [10]. In this paper, we don't consider hedge funds.

Portfolio managers are, as a rule, very secretive about what they buy and sell, and the fractional structure of the portfolio $\boldsymbol{\beta} = (\beta_1, \ldots, \beta_n)$ is typically hidden from public. Such information, having been recovered, would be of great interest for those who monitor the portfolio, in particular, it would provide investors in this portfolio with an early warning.

The problem of recovering the hidden capital sharing in a portfolio from publicly available data was formulated by William Sharpe, 1990 Nobel Prize winner in

Economics [17]. His method involves analysis of the time series of periodic return on the portfolio, which the company is obliged to report, jointly with the synchronous time series of returns on the stock market assets assumed to form the portfolio.

Since an actual portfolio may contain hundreds and even thousands of instruments, Sharpe proposed to approximate the resulting portfolio return by a small number of market indexes [10] representing certain asset classes and investment styles. This approach is famous under the name of Returns Based Style Analysis, RBSA.

The periodic return is the percentage of profit from investment in a financial instrument over a certain time period. Let $z_{beg,i}$ and $z_{end,i}$ be the prices of assets at the beginning and the end of some time interval called the holding period, respectively, $z_{beg,p}$ and $z_{end,p}$ will be the cost of the portfolio as a whole. The ratio $y = \left(z_{end,p} - z_{beg,p}\right)/z_{beg,p}$ is referred to as the portfolio return for the holding period, and $x_i = \left(z_{end,i} - z_{beg,i}\right)/z_{beg,i}$ are returns of assets. If asset classes are considered, their expected returns x_i are usually estimated by special analytical companies as return indices.

For consecutive time moments $t = 1, 2, 3, \ldots$ making the succession of holding periods, for instance, stock exchange workdays, months, and quarters, the periodic returns of both portfolio and assets form a number of time series, respectively, y_t and $x_{t,i}$, $i = 1, \ldots, n$. These values can be considered as known, because any investment company must regularly publish the return of its portfolio, and returns of assets classes are regularly published in both on-line and printed financial media. In the simplest case of daily periodicity, the return of each single asset can be immediately computed from the known changes in its price.

In Sharpe's model, the periodic return of a portfolio is equal to the linear combination of periodic returns of the assets or classes of assets with coefficients having the meaning of the portfolio's shares invested in each of them at the beginning of the period under the assumption that the entire budget is fully spent on the investment:

$$\sum_{t=1}^{T}\left(y_t - \sum_{i=1}^{n}\beta_i x_{t,i}\right)^2 \to \min, \quad \sum_{i=1}^{n}\beta_i = 1, \quad \beta_i \geq 0, \; i = 1, \ldots, n. \tag{5}$$

The principle of approximation is commonly adopted in the modern investment analysis as Returns Based Style Analysis.

The constrained regression problem (5) belongs to the class of quadratic optimization problems under linear constrains (quadratic programming) [11]. Its computational complexity is polynomial with respect to the number of assets or asset classes supposedly forming the portfolio $O(n^3)$, but it is linear relative to the number of time periods on which he periodic returns were computed $O(T)$.

2.2 The Problem of Factor Search in Large Sets of Stock Market Data

A typical aim of returns data processing is factor search – finding the really existing hidden composition of the portfolio being analyzed among a very large number of potential factors – assets or asset classes (stock market indexes). Let $\mathbb{I} = \{1, \ldots, n\}$

stand for the full set of factors, then it is required to find among them a small active subset $\hat{\mathbb{I}} \subset \mathbb{I}$, $\hat{n} = |\hat{\mathbb{I}}| < n$. This means that the estimates of factor exposures, i.e., regression coefficients in the returns based portfolio model (5), must be positive inside the active subset $\hat{\beta}_i > 0$, $i \in \hat{\mathbb{I}}$, and equal zero outside it $\hat{\beta}_i = 0$, $i \notin \hat{\mathbb{I}}$. So, it is required to find the portfolio model as

$$y_t \cong \sum_{i \in \hat{\mathbb{I}}} \beta_i x_{t,i}, \;\; \sum_{i \in \hat{\mathbb{I}}} \beta_i = 1, \;\; \beta_i > 0, \; i \in \hat{\mathbb{I}}.$$

This problem should be named regressor selection in doubly constrained models. In any case, regression selection implies a kind of regularization, namely, reduction of the freedom in varying regression coefficients when minimizing the residual in the least squares criterion (5). Despite both equality and inequality constraints themselves carry quite a large bit of regularization [14], they alone are insufficient for endowing the criterion with the property of factor search when comparing the time series of portfolio returns with those of a large sets of stock market indexes.

Therefore, in the next Sect. 2.3 we propose to take into account, besides the natural constraints, also additional a priori suggestions on the rationality of the composition of the sought-for unknown portfolio from the viewpoint of its utility as financial instrument.

2.3 A Priori Assumptions on the Composition of the Portfolio Under Analysis

We proceed from the assumption that the portfolio whose returns time series $(y_t,\ t = 1,\ldots,T)$ is analyzed was composed by the team of its managers with the purpose to earn a greater return under an acceptable level of risk.

Let the returns of assets at the subsequent time periods $(\mathbf{x}_t = (x_{t,1} \cdots x_{t,n}),\ t = 1,\ldots,T$ be considered as a set of realizations of a random vector having mathematical expectation $\bar{\mathbf{x}} = E\{\mathbf{x}\} \in \mathbb{R}^n$ and covariance matrix $\boldsymbol{\Sigma} = E\Big\{(\mathbf{x} - \bar{\mathbf{x}})(\mathbf{x} - \bar{\mathbf{x}})^{\mathrm{T}}\Big\}$ $(n \times n)$. We jump here over the question whether the returns in the sequence $(\mathbf{x}_t,\ t = 1,\ldots,T)$ are independent or not, at least it is assumed that the respective time series is stationary, and its constant section is considered. Then, the portfolio return $y_t = \boldsymbol{\beta}^{\mathrm{T}}\mathbf{x}_t$ should also be considered as a stationary random process, and its section will be random variable with mathematical expectation $\bar{y} = \boldsymbol{\beta}^{\mathrm{T}}\bar{\mathrm{x}}$, namely, expected profit (return), and variance $\sigma^2 = \boldsymbol{\beta}^{\mathrm{T}}\boldsymbol{\Sigma}\boldsymbol{\beta}$, that inserts some risk of loss into the portfolio composition.

The greater the expected return the better the portfolio. On the other hand, the greater the variance of the return the greater the probability of low and even negative return and, so, the worse becomes the portfolio.

In accordance with Harry Markowitz theory of optimal portfolio composition [18], only those portfolio compositions $\left(\boldsymbol{\beta} \geq \mathbf{0},\ \mathbf{1}^{\mathrm{T}}\boldsymbol{\beta} = 1\right)$ may be considered as rational which provide the least variance for a fixed expected return $\left(\boldsymbol{\beta}^{\mathrm{T}}\boldsymbol{\Sigma}\boldsymbol{\beta} \to \min, \bar{\mathbf{x}}^{\mathrm{T}}\boldsymbol{\beta} = const\right)$, or vice versa, the maximum expected return if the variance is fixed $\left(\bar{\mathbf{x}}^{\mathrm{T}}\boldsymbol{\beta} \to \max,\ \boldsymbol{\beta}^{\mathrm{T}}\boldsymbol{\Sigma}\boldsymbol{\beta} = const\right)$. It is easy to see that both cases are covered by the condition

$$U(\boldsymbol{\beta}|\mu) = \mu\bar{\mathbf{x}}^{\mathrm{T}}\boldsymbol{\beta} - \boldsymbol{\beta}^{\mathrm{T}}\boldsymbol{\Sigma}\boldsymbol{\beta} \to \max, \quad \boldsymbol{\beta} \geq \mathbf{0},\ \mathbf{1}^{\mathrm{T}}\boldsymbol{\beta} = 1. \tag{6}$$

The quadratic function $U(\boldsymbol{\beta} \mid \mu)$ is usually called the portfolio utility [19], where parameter $0 \leq \mu < \infty$ is interpreted as risk tolerance, and, respectively, $1/\mu$ as risk aversion.

Thus, when the mathematical expectation $\bar{\mathrm{x}} = E\{\mathbf{x}\} \in \mathbb{R}^n$ and covariance matrix of the asset returns $\boldsymbol{\Sigma} = E\left\{(\mathbf{x} - \bar{\mathrm{x}})(\mathbf{x} - \bar{\mathrm{x}})^{\mathrm{T}}\right\}$ $(n \times n)$ are fixed, all the portfolio compositions, which can be reasonable from the viewpoint of impossibility to improve each of them in both expected random return and its variance, are covered by the one-parametric family (6) whose only parameter is risk tolerance $0 \leq \mu < \infty$.

By the way, it is just this fact that enables the very existence of stock market equilibrium in risk-averse community of investors, and has led finally to the famous Capital Asset Pricing Model (CAPM) [20], which was the main item in the motivation of the Nobel prize in Economics awarded in 1990 to Harry Markowitz and William Sharpe.

The portfolios that don't meet the condition (6) cannot be recognized as reasonable from the viewpoint of Markowitz theory, but it is impossible to judge which value of risk tolerance μ is better. This is inner psychological characteristic of any particular investor.

In this paper, we propose to use the class of transposed utility functions (6) as parametric family of extremely strong a priori driven regularization functions for factor search problems:

$$V(\boldsymbol{\beta}|\mu) = \boldsymbol{\beta}^{\mathrm{T}}\boldsymbol{\Sigma}\boldsymbol{\beta} - \mu\bar{\mathbf{x}}^{\mathrm{T}}\boldsymbol{\beta} \to \min. \tag{7}$$

Here $\bar{\mathrm{x}} \in \mathbb{R}^n$ and $\boldsymbol{\Sigma}\,(n \times n)$ are meant as the mathematical expectation and covariance matrix of the random vector of asset returns anticipated at the start of the investment period under the assumption that they will not essentially change during some time period in the future. The risk tolerance coefficient $0 \leq \mu < \infty$ is the hyper-parameter of this regularization function.

These statistical characteristics of the universe of stock market assets could be estimated on a sufficiently long time interval $\{t = -(T' - 1), \ldots, -1, 0\}$ that precedes the interval of portfolio observation $\{t = 1, 2\ldots, T\}$:

$$\bar{\mathrm{x}} = \frac{1}{T'} \sum_{t=-(T'-1)}^{0} \mathbf{x}_t, \quad \boldsymbol{\Sigma} = \frac{1}{T'} \sum_{t=-(T'-1)}^{0} (\mathbf{x}_t - \bar{\mathrm{x}})(\mathbf{x}_t - \bar{\mathrm{x}})^{\mathrm{T}}. \tag{8}$$

However, computing $\bar{\mathrm{x}}$ and $\boldsymbol{\Sigma}$ immediately from the portfolio observation interval

$$\bar{\mathrm{x}} = \frac{1}{T}\sum_{t=1}^{T} \mathbf{x}_t, \quad \boldsymbol{\Sigma} = \frac{1}{T}\sum_{t=1}^{T} (\mathbf{x}_t - \bar{\mathrm{x}})(\mathbf{x}_t - \bar{\mathrm{x}})^{\mathrm{T}} \tag{9}$$

will not be a fatal mistake if the universe of stock market assets can be assumed as relatively stable within a quite long time period.

3 Regularization of Factor Search in Constrained Regression Models

3.1 Central Regularization

As mentioned above in Sect. 2.2, the natural equality and inequality constraints may occur to be insufficient for endowing the regression estimation criterion (3) with the property of factor search. It will be clearly seen if we put (3) in the half scalar form:

$$\begin{cases} (\mathbf{y}-\mathbf{X}^T\boldsymbol{\beta})^{\mathrm{T}}(\mathbf{y}-\mathbf{X}^{\mathrm{T}}\boldsymbol{\beta}) = \left(\mathbf{y}-\sum_{i=1}^{n}\beta_i\boldsymbol{x}_i\right)^{\mathrm{T}}\left(\mathbf{y}-\sum_{i=1}^{n}\beta_i\boldsymbol{x}_i\right) \to \min(\beta_1,\ldots,\beta_n), \\ \sum_{i=1}^{n}\beta_i = 1,\ \ \beta_i \geq 0,\ i=1,\ldots,n. \end{cases} \tag{10}$$

The main hindrance is correlation between single regressors $\boldsymbol{x}_i = (x_{1,i}\cdots x_{T,i}) \in \mathbb{R}^T,\ i=1,\ldots,n$, which makes hardly distinguishable different combinations $[\boldsymbol{x}_i,\ i \in \hat{\mathbb{I}} \subset \mathbb{I} = \{1,\ldots,n\}]$.

This difficulty can be essentially damped if it is may be a priori assumed that regressors within the sought-for combination are weakly correlated between themselves. The less correlation between random variables $(x_i,\ i=1,\ldots,n)$, the less the variance of their linear combination $y=\sum_{i=1}^{n}\beta_i x_i$ with a fixed norm of the coefficient vector $\sum_{i=1}^{n}\beta_i^2$. Thus, if we want to find a subset of weekly correlated regressors, we have to minimize the quadratic form $\boldsymbol{\beta}^{\mathrm{T}}\boldsymbol{\Sigma}\boldsymbol{\beta}\to\min,\ \boldsymbol{\beta}^{\mathrm{T}}\boldsymbol{\beta}=const.$

This is only one kind of possible a priori assumptions on the sough-for subset of regressors, in the general case, any positive definite matrix $\mathbf{G}\ (n\times n)$ may express the user's intuition, and it must not obligatory be data-dependent. This reason leads to quadratic regularization

$$\begin{cases} \boldsymbol{\beta}^{\mathrm{T}}\mathbf{G}\boldsymbol{\beta}+c(\mathbf{y}-\mathbf{X}^{\mathrm{T}}\boldsymbol{\beta})^{\mathrm{T}}(\mathbf{y}-\mathbf{X}^{\mathrm{T}}\boldsymbol{\beta}) \to \min(\beta_1,\ldots,\beta_n), \\ \mathbf{1}^{\mathrm{T}}\boldsymbol{\beta}=1,\ \ \boldsymbol{\beta}\geq\mathbf{0}. \end{cases} \tag{11}$$

Matrix of the quadratic regularization function indicates some a priori preferable direction in the n-dimensional space of regressors. The minimum point of the quadratic regularization function is zero vector $\boldsymbol{\beta}=\mathbf{0}$, the center of the space. Therefore, we shall call such kind of regularization ***central regularization***.

3.2 Biased Regularization

3.2.1 Selectivity Control in Factor Search

However, it may happen that the a priori knowledge suggests not only the direction but also the specific value of the vector of regression coefficients. A glowing example of such a situation is when it is a priori known that the number of active regressors, for which the regression coefficients are positive $\beta_i > 0$, is smaller than the number resulting from the solution of the initial constrained problem (11), and it is required to forcibly assign zero values to some of them.

This can be done by way of choosing the quadratic regularization function as

$$(\boldsymbol{\beta} - \boldsymbol{\beta}_0)^{\mathrm{T}}\mathbf{G}(\boldsymbol{\beta} - \boldsymbol{\beta}_0) \rightarrow \min, \quad \boldsymbol{\beta}_0 < \mathbf{0},$$

where putting the bias in the negative quadrant, which is prohibited by the inequality constraints $\boldsymbol{\beta} \geq \mathbf{0}$, encourages zero values of regression coefficients. We call this quadratic form function of ***biased regularization***.

The deeper point $\boldsymbol{\beta}_0$ is pushed into the forbidden zone, the stronger will be propensity of the regularization function $(\boldsymbol{\beta} - \boldsymbol{\beta}_0)^{\mathrm{T}}\mathbf{G}(\boldsymbol{\beta} - \boldsymbol{\beta}_0) \rightarrow \min$ to reduce the number of active regressors. Expansion of the quadratic form results in regularization $\boldsymbol{\beta}^{\mathrm{T}}\mathbf{G}\boldsymbol{\beta} - \boldsymbol{\beta}_0^{\mathrm{T}}\mathbf{G}\boldsymbol{\beta} \rightarrow \min$. If we put $\boldsymbol{\beta}_0 = -\mu\mathbf{1}$, we shall have $\boldsymbol{\beta}^{\mathrm{T}}\mathbf{G}\boldsymbol{\beta} + \mu\mathbf{1}^{\mathrm{T}}\mathbf{G}\boldsymbol{\beta} \rightarrow \min$.

So, we have come to the constrained regression problem

$$\begin{cases} \boldsymbol{\beta}^{\mathrm{T}}\mathbf{G}\boldsymbol{\beta} + \mu\mathbf{1}^{\mathrm{T}}\mathbf{G}\boldsymbol{\beta} + c\left(\mathbf{y} - \mathbf{X}^{T}\boldsymbol{\beta}\right)^{\mathrm{T}}\left(\mathbf{y} - \mathbf{X}^{T}\boldsymbol{\beta}\right) \rightarrow \min(\beta_1, \ldots, \beta_n), \\ \mathbf{1}^{\mathrm{T}}\boldsymbol{\beta} = 1, \quad \boldsymbol{\beta} \geq \mathbf{0}. \end{cases} \tag{12}$$

The greater the selectivity parameter $\mu \geq 0$, the stronger is the property of the criterion (12) to shrink the subset of active regressors.

It is easy to see here a bit of analogy with the Elastic Net regression $\sum_{i=1}^{n}\left(\beta_i^2 + \mu|\beta_i|\right)$ (Sect. 1, [6]). If we had $\mathbf{G} = \mathbf{I}$, the second penalty term in (12) would be $\mu\mathbf{1}^{\mathrm{T}}\boldsymbol{\beta}$, what, on the force of the non-negativity constraints $\beta_i = |\beta_i|$, would result in

$$\mu\mathbf{1}^{\mathrm{T}}\boldsymbol{\beta} = \mu\sum_{i=1}^{n}\beta_i = \mu\sum_{i=1}^{n}|\beta_i|.$$

However, this analogy is not completely straightforward $\mathbf{1}^{\mathrm{T}}\mathbf{G}\boldsymbol{\beta} \neq \mathbf{1}^{\mathrm{T}}\boldsymbol{\beta}$, otherwise the equality constraint $\mathbf{1}^{\mathrm{T}}\boldsymbol{\beta} = 1$ would wipe out the effect of selectivity.

Selectivity parameter $\mu \geq 0$ is to be adjusted to the data set $(\mathbf{y}, \mathbf{X})$ by the principle of maximizing the generalization performance of the model, which is to be measured by an appropriate verification procedure [21].

3.2.2 Risk Tolerance Control in Portfolio Models

Another example of the necessity of biased regularization is estimation of portfolio composition under the a priori assumption of rational balance between expected return and variance of return. Indeed, the quadratic regularization function (7) can be represented as $V(\boldsymbol{\beta}|\mu) = \boldsymbol{\beta}^{\mathrm{T}}\boldsymbol{\Sigma}\boldsymbol{\beta} - \mu\bar{\mathbf{x}}^{T}\boldsymbol{\beta} = (\boldsymbol{\beta} - \boldsymbol{\beta}_0)^{\mathrm{T}}\boldsymbol{\Sigma}(\boldsymbol{\beta} - \boldsymbol{\beta}_0) + const$, where $\boldsymbol{\beta}_0 = \boldsymbol{\beta}_0 = \arg\min\left[\boldsymbol{\beta}^{\mathrm{T}}\boldsymbol{\Sigma}\boldsymbol{\beta} - \mu\bar{\mathbf{x}}^{T}\boldsymbol{\beta}\right]$. The condition $\nabla_{\boldsymbol{\beta}}\left(\boldsymbol{\beta}^{\mathrm{T}}\boldsymbol{\Sigma}\boldsymbol{\beta} - \mu\bar{\mathbf{x}}^{T}\boldsymbol{\beta}\right) = \mathbf{0} \in \mathbb{R}^n$ results in the equality

$$\boldsymbol{\beta}_0 = \frac{\mu}{2}\boldsymbol{\Sigma}^{-1}\overline{\mathbf{x}}.$$

We have obtained the biased regularization function

$$(\boldsymbol{\beta} - \boldsymbol{\beta}_0)^{\mathrm{T}}\boldsymbol{\Sigma}(\boldsymbol{\beta} - \boldsymbol{\beta}_0) \to \min, \quad \boldsymbol{\beta}_0 = \frac{\mu}{2}\boldsymbol{\Sigma}^{-1}\overline{\mathbf{x}},$$

which results in the constrained regression problem

$$\begin{cases} \boldsymbol{\beta}^{\mathrm{T}}\boldsymbol{\Sigma}\boldsymbol{\beta} - \mu\overline{\mathbf{x}}^{T}\boldsymbol{\beta} + c(\mathbf{y} - \mathbf{X}^{\mathrm{T}}\boldsymbol{\beta})^{\mathrm{T}}(\mathbf{y} - \mathbf{X}^{\mathrm{T}}\boldsymbol{\beta}) \to \min(\beta_1, \ldots, \beta_n), \\ \mathbf{1}^{\mathrm{T}}\boldsymbol{\beta} = 1, \quad \boldsymbol{\beta} \geq \mathbf{0}. \end{cases} \tag{13}$$

Risk tolerance coefficient $0 < \mu < \infty$ is a hidden characteristic of the investor having composed the portfolio under analysis, and is to be estimated along with the portfolio composition $\boldsymbol{\beta} \in \mathbb{R}^n$. This is a hyperparameter of the model, and to estimate it verification procedure is to be applied, for instance minimization of the leave-one-out criterion [21].

4 Experimental Study of the Constrained Regularized Regression Estimation Technique

4.1 Experimental Set Up

The aim of experiments is to examine the ability of the regularized factor search technique outlined in Sect. 3 to estimate the actual sparse regression model from a large set of correlated variables. As the set of regressors, we used 650 time series of monthly stock market indexes

$$\mathbf{x}_t = (x_{t,i},\ i = 1, \ldots, n) \in \mathbb{R}^n,\ n = 650, \tag{14}$$

each covering 20 years, i.e., consisting of 240 return values $t = 1, \ldots, T$, $t = 240$:

$$\mathbf{X} = (\mathbf{x}_t,\ t = 1, \ldots, T)\ (n \times T). \tag{15}$$

Having been placed in decreasing ordered, the eigenvalues of their covariance matrix $\boldsymbol{\Sigma}$ (9) quickly fall $\lambda_1 = 16369.8$, $\lambda_2 = 1435.2$, $\lambda_3 = 1076.2$, $\lambda_{20} = 134.0 < 0.01\,\lambda_1$, $\lambda_{50} = 32.2 < 0.002\,\lambda_1$, what is evidence of drastic correlation between the indexes.

The idea of experiments is attempting to recover some actual set of beta coefficients $(\beta_1^*, \ldots, \beta_m^*)$ making up a reasonable investment portfolio from the viewpoint of Markowitz risk return theory for the given risk tolerance $0 \leq \mu < \infty$ (Sect. 2.3).

In accordance with the rule (7)

$$\begin{cases} \left(\beta^*_{\mu,1},\ldots,\beta^*_{\mu,n}\right) = \arg\min\left(\boldsymbol{\beta}^T\boldsymbol{\Sigma}\boldsymbol{\beta} - \mu\bar{\mathbf{x}}^T\boldsymbol{\beta}\right), \\ \mathbf{1}^T\boldsymbol{\beta} = 1,\ \boldsymbol{\beta} \geq \mathbf{0}, \end{cases}$$

we computed 5 sets of β-coefficients, in what follows, called genuine reasonable portfolio compositions for the following values of the risk tolerance coefficient:

$$\mu_1 = 0,\ \ \mu_2 = 10,\ \ \mu_3 = 30\ \ \mu_4 = 200,\ \ \mu_5 = 1000. \tag{16}$$

As it was expected, the resulting portfolios turned out to be very sparse in the full universe of 650 stock market indexes. This means that the subset of active indexes $\mathbb{I}^*_\mu = \{i : \beta^*_{\mu,i} > 0\}$ contains a much smaller number of indexes than the entire universe $m = |\mathbb{I}^*_\mu| \ll n = 650$:

Risk tolerance	$\mu_1 = 0$	$\mu_2 = 10$	$\mu_3 = 30$	$\mu_4 = 200$	$\mu_5 = 1000$
Number of active indexes	$m_1 = 13$	$m_2 = 11$	$m_3 = 10$	$m_4 = 7$	$m_5 = 1$

For each of the genuine reasonable portfolio compositions $(\beta^*_{\mu,i},\ i \in \mathbb{I}^*_\mu)$, we computed a time series of returns of a hypothetical portfolio

$$\mathbf{y}_\mu = (y_{\mu,t},\ t = 1,\ldots,T) \in \mathbb{R}^T,\ y_{\mu,t} = \sum_{i\in\mathbb{I}^*_\mu} \beta^*_{\mu,i}x_{t,i} + \xi_t, \tag{17}$$

as the sequence of random variables with 10% noise variance $\sigma^2(\xi_t) = 0.1\left((1/T)\sum_{i\in\mathbb{I}^*_\mu} \beta^*_{\mu,i}x_{t,i}\right)$ depending on the respective sequence of regression coefficients. These five return time series along with the entire universe of indexes (15)

$$\mathbf{y}_\mu = (y_{\mu,t},\ t = 1,\ldots,T),\ \ \mu = \mu_q,\ \ q = 1,\ldots,5,\ \mathbf{X} = (\mathbf{x}_t,\ t = 1,\ldots,T), \tag{18}$$

served as the set up of our experiment.

4.2 Blind Factor Search

This blind factor search suggests that we try to recover the assumingly genuine regression model by the pure non-regularized doubly constrained least squares criterion (10)

$$(\mathbf{y} - \mathbf{X}^T\boldsymbol{\beta})^{\mathrm{T}}(\mathbf{y} - \mathbf{X}^T\boldsymbol{\beta}) \to \min(\boldsymbol{\beta}),\ \mathbf{1}^{\mathrm{T}}\boldsymbol{\beta} = 1,\ \boldsymbol{\beta} \geq \mathbf{0}, \tag{19}$$

applied to each of the five modelled time series $\mathbf{y}_\mu = (y_{\mu,t},\ t = 1,\ldots,T)$ (18). After finding the vector of estimated regression coefficients $\hat{\boldsymbol{\beta}}_\mu$ within the subset of active regressors $\hat{\mathbb{I}}_\alpha$, we computed for it the Percentage Least Squares Error indicator (PLSE)

$$PLSE = \frac{(1/T)\sum_{t=1}^{T}\left(y_t - \sum_{i\in\hat{\mathbb{I}}_\mu}\hat{\beta}_{i,\mu}x_{t,i}\right)^2}{(1/T)\sum_{t=1}^{T}\left(\sum_{i\in\hat{\mathbb{I}}_\mu}\hat{\beta}_{i,\mu}x_{t,i}\right)^2} \times 100. \tag{20}$$

The results of this analysis are presented in Fig. 1 in the form of tessellated diagrams. The diagrams comparatively show the genuine and estimated subsets of active regressors (stock market indexes) numbered in the given set of indexes (14) according to diminishing of the average return $\bar{x}_1 > \ldots > \bar{x}_i > \ldots \bar{x}_n$:

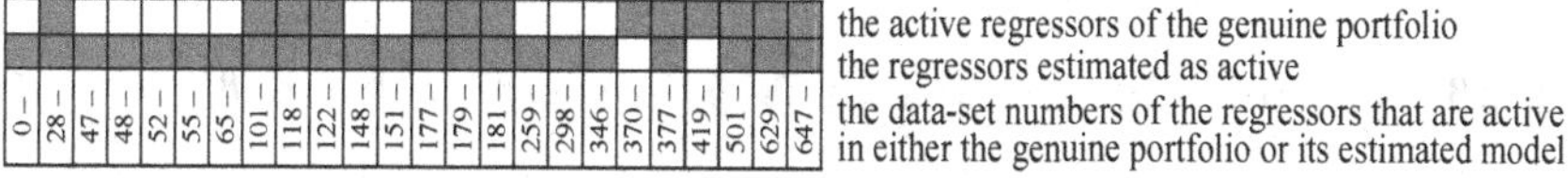

Each diagram is supplied with the Percentage Least Squares Error (20).

It is well seen from Fig. 1 that, except the trivial case of full risk tolerance $\mu \to \infty$, over 50% of indexes, being active in the actual or estimated portfolio, are wrongly identified by the blind factor search technique. The essential mismatch between the actual and the estimated portfolio compositions witnesses to the necessity of additional regularization on the basis on the firm confidence that the portfolio under analysis belongs to the class of reasonable portfolios by Markowitz.

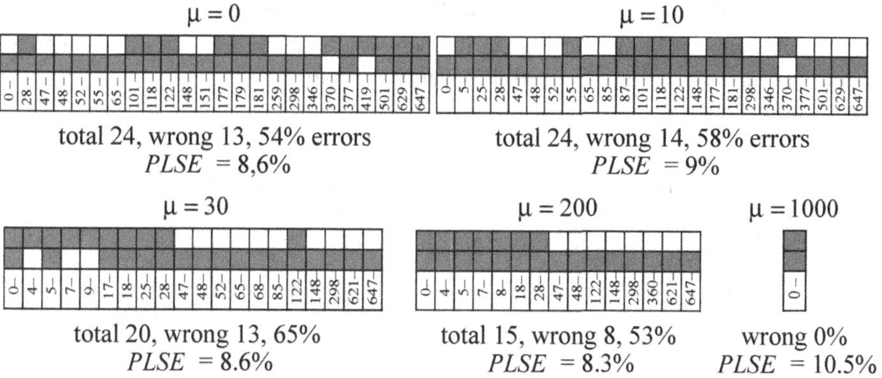

Fig. 1. Blind factor search. The genuine (upper rows) and estimated (lower rows) sets of active regressors for modelled portfolios with different risk tolerance. Gray squares ■ indicate active regressors.

4.3 A Priori Driven Factor Search

The *PLSE* indicator in Fig. 1 witnesses that each of the analyzed return time series is approximated at almost the same error level, which is a little smaller as the 10% noise variance in the modelled portfolio (17). Nevertheless, the estimated sets of active regressors essentially deflect from the modelled ones. This fact suggests to slightly push the search toward reasonable subsets, in our case, reasonable portfolio compositions. This is just that idea which is carried by the regularization function $V(\boldsymbol{\beta} \mid \mu) = \boldsymbol{\beta}^T \boldsymbol{\Sigma} \boldsymbol{\beta} - \mu \bar{\mathbf{x}}^T \boldsymbol{\beta}$ (7) to be minimized as transpose of the portfolio utility(19) to be maximized.

Instead of the pure non-regularized doubly constrained least squares criterion (19), as in Sect. 4.2, we applied to each of the five modelled time series $\mathbf{y}_\mu = (y_{\mu,t}, t = 1,\ldots,T)$ (18), $\mu = \mu_1,\ldots,\mu_5$ (16), five criteria additionally supplied with the a priori driven regularization term with different values of the risk tolerance coefficient $\mu = \mu_m$, $m = 1,\ldots,5$:

$$\boldsymbol{\beta}^T \boldsymbol{\Sigma} \boldsymbol{\beta} - \mu_m \bar{\mathbf{x}}^T \boldsymbol{\beta} + \frac{1}{T}(\mathbf{y}_\mu - \mathbf{X}^T \boldsymbol{\beta})^{\mathrm{T}} (\mathbf{y}_\mu - \mathbf{X}^T \boldsymbol{\beta}) \to \min(\boldsymbol{\beta}),\ \mathbf{1}^{\mathrm{T}} \boldsymbol{\beta} = 1,\ \boldsymbol{\beta} \geq \mathbf{0}. \quad (21)$$

Thus, we obtained five estimated vectors of regression coefficients $(\hat{\boldsymbol{\beta}}_\mu \mid \alpha_m)$ in the subset of active regressors $(\hat{\mathbb{I}}_\alpha \mid \mu_m)$. For each of them, we computed the leave-one-out indicator $(LOO_\mu \mid \mu_m)$ of the regression model within this respective active set under only the equality constraint $\sum_{i=1}^n \beta_i = 1$. Finally, we chose the risk tolerance coefficient most relevant to this specific modelled return time series $\mathbf{y}_\mu$ by the LOO criterion

$$m^* = \arg\min(LOO_\mu \mid \mu_m),\ m = 1,\ldots,5.$$

The respective regression vector $(\hat{\boldsymbol{\beta}}_\mu \mid \mu_m)$ and the active subset $(\hat{\boldsymbol{\beta}}_\mu \mid \alpha_m)$ we considered as the estimated model of the portfolio under analysis $\mathbf{y}_\mu$.

These results are shown in Fig. 2. It is obvious that the a priori driven regularization has fundamentally improved the quality of factor search.

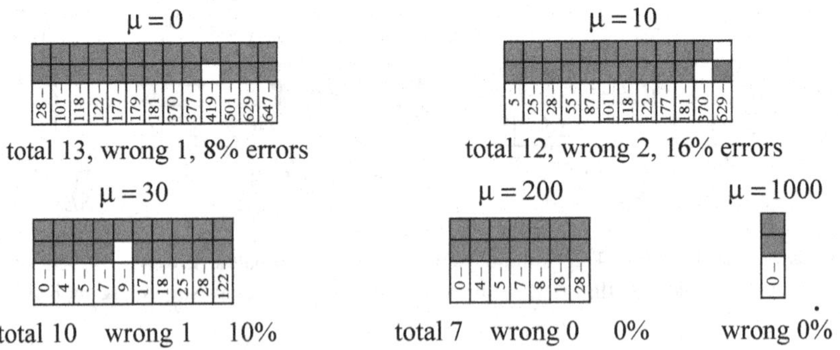

Fig. 2. A priori driven factor search. The genuine (upper rows) and estimated (lower rows) sets of active regressors for modelled portfolios with different risk tolerance μ.

5 Conclusions

We have scrutinized, if only perfunctorily, some aspects of regression analysis in large sets of regressors under two kinds of constraints – non-negativity of regression coefficients and unity sum of them. An additional assumption is that the regression coefficients differ from zero only within a really existing small subset of a large universe of strongly correlated regressors, and the factor search, namely finding this subset, is the main aim of data processing.

The large number of regressors n that essentially exceeds the number of samples T makes unacceptable the polynomial computational complexity of quadratic programming of general kind. Our contribution is the optimization procedure that exploits the linear computational complexity of separable optimization by way of iterative decomposition of the non-separable optimization problem into a succession of separable ones.

Another difficulty is that finding a small active subset among a huge set of correlated factors is very problematic unless some a priori information on the expected portfolio structure is available. We have proposed three ways of introducing additional regularization in the factor search: (1) central regularization, whose quadratic matrix should indicate some a priori preferable direction in the n-dimensional space of regressors; (2) biased regularization via penalizing the number of active regressors; (3) another kind of biased regularization that controls the average values of active regressors.

The latter two kinds of regularization are especially adequate to the needs of the typical application problem widely known under the name of Returns Based Analysis of investment portfolios. It is just this application that triggered this research.

Acknowledgement. We would like to acknowledge support from grants of the Russian Foundation for Basic Research 17-07-00436, 17-07-00993 and 18-07-01087.

References

1. Hoerl, A.E., Kennard, D.J.: Application of ridge analysis to regression problems. Chem. Eng. Prog. **58**, 54–59 (1962)
2. Vinod, H.D., Ullah, A.: Recent Advances in Regression Methods. Statistics: Textbooks and Monographs, vol. 41. Marcel Dekker Inc., New York (1981)
3. Frank, I.E., Friedman, J.H.: A statistical view of some chemometrics regression tools. Technometrics **35**, 109–148 (1993)
4. Fu, W.J.: Penalized regression: the bridge versus the LASSO. J. Comput. Graph. Stat. **7**, 397–416 (1998)
5. Tibshirani, R.: Regression shrinkage and selection via the lasso. J. Roy. Stat. Soc. Ser. B (Methodol.) **58**(1), 267–288 (1996)
6. Zou, H., Hastie, T.: Regularization and variable selection via the elastic net. J. Roy. Stat. Soc. Ser. B (Stat. Methodol.) **67**(2), 301–320 (2005)
7. Fan, J., Li, R.: Variable selection via nonconcave penalized likelihood and its oracle properties. J. Am. Stat. Assoc. **96**(456), 1348–1360 (2001). Theory and Methods

8. Kakade, S.M., Shalev-Shwartz, S., Tewari, A.: Regularization Techniques for Learning with Matrices. Cornell Univesity. Submitted on 4 October 2009. https://arxiv.org/pdf/0910.0610.pdf. Accessed 17 Oct 2010
9. Kienzle, W., Chellapilla, K.: Personalized handwriting recognition via biased regularization. In: Proceedings of the 23rd International Conference on Machine Learning (ICML 2006), Pittsburgh, Pennsylvania, USA, 25–29 June, pp. 457–464 (2006)
10. Tyson, E.: Investing For Dummies. Wiley, Hoboken (2011). 140 p.
11. Ruszczyinski, A.: Nonlinear Optimization. Princeton University Press, Princeton (2006). 448 p.
12. Zhou, H., Lange, K.: A Path Algorithm for Constrained Estimation. Cornell Univesity. Submitted on 19 March 2011
13. Janesch, S.M.H., Santos, L.T.: Exact penalty methods with constrained subproblems. Invesigacion Oper. **7**, 55–65 (1997)
14. Meinshausen, N.: Sign-constrained least squares estimation for high-dimensional regression. Electron. J. Stat. **7**, 1607–1631 (2013)
15. Gill, P.E., Murray, W., Wright, M.H.: Practical Optimization. Academic, London (1981)
16. Bro, R., de Jong, S.: A fast non-negativity-constrained least squares algorithm. J. Chemometr. **11**, 393–401 (1997)
17. Sharpe, W.F.: Asset allocation: management style and performance measurement. J. Portf. Manag. **18**, 7–19 (1992)
18. Markowitz, H.: Portfolio selection. J. Finan. **7**(1), 77–91 (1952)
19. Sharpe, W.F.: Capital asset prices with and without negative holdings. Nobel Lecture, 7 December 1990. https://www.nobelprize.org/nobel_prizes/economic-sciences/laureates/1990/sharpe-lecture.pdf
20. Sharpe, W.F.: Capital asset prices: a theory of market equilibrium under conditions of risk. J. Finan. **19**(3), 425–442 (1964)
21. Krasotkina, O., Mottl, V., Markov, M., Chernousova, E., Malakhov, D.: Methods of hyperparameter estimation in time-varying regression models with application to dynamic style analysis of investment portfolios. In: Perner, P. (ed.) MLDM 2017. LNCS (LNAI), vol. 10358, pp. 431–450. Springer, Cham (2017). https://doi.org/10.1007/978-3-319-62416-7_31

Large-Scale Targeted Marketing by Supervised PageRank with Seeds

Zhiwei (Tony) Qin[1(✉)], Chengxiang Zhuo[2], Wei Tan[2], Jun Xie[2], and Jieping Ye[2]

[1] DiDi Research America, Mountain View, CA, USA
qinzhiwei@didichuxing.com
[2] DiDi Research, Beijing, China
{zhuochengxiang,tanweiterry,xiejun,yejieping}@didichuxing.com

Abstract. In targeted marketing, the key is to spend the limited resources on as relevant a group of customers as possible to the campaign objective. We consider the business problem of selecting a group of standard/non-premier service customers and new customers to whom promotional means, e.g. issuing discount coupons are directed, with the goal of upgrading them to core premier service users. We develop a solution framework based on utilizing the anonymized interaction of user activities, within which the users are scored by their relevance to the marketing campaign objective. The links between two users are weighted, with the weights learnt in a supervised setting to ensure high relevance to the score prediction task. We modified a seeded variant of the PageRank algorithm to adept to this framework while maintaining convergence property. We demonstrate through real-world data that our framework can significantly improve the prediction relevance over conventional methods with regard to the marketing problem under consideration.

Keywords: Targeted marketing · Personalized PageRank
Graph mining · Social network

1 Introduction

Investing in growth (i.e. customer acquisition and retention) gives significant advantage in increasing market share, according to a study conducted by Forbes Insight [5]. As a relatively new industry, the ride-sharing industry especially sees this as the case. Lyft, for example, "spent \$91 million on 'sales and marketing' – which can be generally defined as the cost of new driver and rider acquisition, such as subsidies – in the first quarter, \$136 million in the second quarter and \$80 million in the fourth quarter" [2]. Unlike traditional mass marketing means, targeted marketing [4,8] puts particular emphasis on the precision of the intended audience, which promises significantly higher return-on-investment (ROI), especially with the help of an explosion of data abundance and rapid advancement of data science in recent years.

P. Perner (Ed.): MLDM 2018, LNAI 10935, pp. 409–424, 2018.
https://doi.org/10.1007/978-3-319-96133-0_31

1.1 Motivation

Service or content providers (e.g. ride-sharing, car rental, video, just to name a few) typically offer tiered services. Common program structures include a standard option, which provides economy and mainstream service (and is sometimes even free), and a premier option, which offers higher standard of service at a higher price. Some examples are Didi's ExpressCar v.s. PremierCar and Youtube's free viewing v.s. subscription-based services. While the core user groups of the two services are quite distinct, exhibiting different purchase powers and needs, there is also considerable overlapping between the two, since customers can choose to use both services. In the face of the continuing need for growth, companies often launch massive marketing campaigns to convert/upgrade standard service users to the premier service or simply to attract new customers to the premier service. Discount coupons are common promotional means. In this paper, we consider the problem of identifying a group of marketing audience who can be new or existing standard service users and are likely to convert to core premier service users with promotional means. Here, we define the core premier service users by average gross merchant value (GMV) contribution over months, which have to exceed a set level. We are also given the current core users of the premier service and a set of anonymized user interaction (social network) data. The social network covers a superset of the users of both services.

We first discuss some common approaches for targeted marketing. For campaigns involving discount coupons, one can predict the conversion probability in terms of coupon usage, i.e. $P(\text{using coupon} \mid \text{issued coupon})$. The requirement is that labels (converted or not) from previous similar campaigns are available to build a classification model. Thus, this approach may not be feasible in a cold-start situation, where no such conversion labels are available. Another issue particular to our case is that the output (i.e. conversion probability of coupon usage) does not directly align with the real objective. The select audience by this method might contain many users who simply want to enjoy the premier service at a lower price. Conversion in terms of coupon usage is insufficient to the business. The objective that provides real benefit and impact to the business bottom-line is to convert customers to regular, non-transient premier service users, which goes beyond single point of coupon usage. One can therefore predict $P(\text{convert to regular premier user} \mid \text{issued coupon})$ instead. However, labels are even harder to obtain in this case due to the long timeframe for conversion. A second simple approach for this targeted marketing problem is to select the top standard service users in terms of GMV contribution. The major concern for this method is the gap between loyalty to the standard service and the willingness to upgrade to premier service. We will discuss more in this aspect in Sect. 5.3. A third approach is to issue coupons to existing core premier users with referral bonus, which is also known as *viral marketing* [11]. The motivation is to utilize the core users' network to attract new customers. The main drawback in the context of our problem is that this approach is indirect – the effectiveness of the approach highly depends on the selected users' interest in referral bonus, and the reachability of referral coupons to the real intended audience could be low.

1.2 Overall Approach

We develop our approach with the aim to align with the business objective, to target the intended audience directly and to cover the cold-start situation mentioned above as well. User interaction network often presents useful information on closeness of relationship and similarity in preference. We treat the set of current premier service users as seed users who have certain common trait of social features and pose the business problem as one that uses the social network to infer a set of good candidates for marketing who may have similar quality to those seeds. This is built on the (reasonable) intuition that people with close social relationship with the existing premier users are more accepting to the premier service. Important to the viability of this approach is the affinity representation of a pair of users in the network. We use supervised learning to estimate the weights on the edges of the network to ensure high relevance to the marketing campaign. Subsequently, we employ a variant of PageRank to score users based on their campaign-specific affinity to the seed users. PageRank is a graph algorithm that computes a score for each vertex in the graph that measures the relative importance of that vertex based on the network topology. Finally, we recommend the top K users for promotional campaigns.

1.3 Organization

In Sect. 2, we review the PageRank algorithm and its variant which biases towards a set of seeds (Seeded PageRank). We develop a scheme in Sect. 3 to automatically learn the edge weights of the network through supervised learning. We propose a modification to Seeded PageRank to adapt to the new interpretation of the edge weights, and we show its convergence and desirable property. In Sect. 5, we demonstrate the efficacy of our proposed method through two large-scale real-world case studies.

2 PageRank Basics

We consider a graph defined by its vertex set $\mathcal{V}$ and edge set $\mathcal{E}$, $\mathcal{G} := (\mathcal{V}, \mathcal{E})$, which can represent the world-wide web or a social network, for example. Within the context of this paper, $\mathcal{V}$ is the entire set of users, and $\mathcal{E}$ represent the set of all interactions among the users. $\mathcal{I}$ is the vertex index set. There is a weight w_{ij} on each directed edge $e_{ij} \in \mathcal{E}$ connecting from vertex v_i to vertex v_j, where $v_k \in \mathcal{V}\ \forall k \in \mathcal{I}$. $W \in \mathbb{R}^{n \times m}$ denotes the matrix of the edge weights w_{ij}.

The PageRank algorithm was originally developed to rank the webpages based on their importance in the world-wide web for search engines [14]. Typically, the scores are constrained to be within a simplex so that they form a probability distribution, but they do not have to. When they do, the scores can also be interpreted as the probability that a random surfer would land on a particular webpage, with the vertices $\mathcal{V}$ being the webpages and the edges $\mathcal{E}$ being the hyperlinks between the webpages (see Fig. 1). The intuition is that important webpages have many other webpages linking to them, and a page that is referenced from important pages might also be important. The algorithm can also be more generally applied to score entities in a social network.

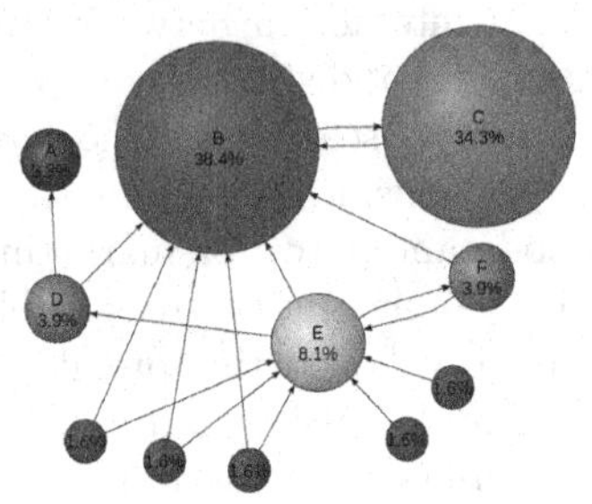

Fig. 1. Illustration of PageRank. Source: https://en.wikipedia.org/wiki/PageRank

In this paper, we consider the general case where the edges have weights $\{w_{ij}\}_{e_{ij}\in\mathcal{E}}$, also known as Weighted PageRank [18]. The score of a vertex v_i, s_i propagates to another vertex v_j according to the transition probability

$$\tilde{w}_{ij} = \frac{w_{ij}}{\sum_k w_{ik}} \tag{1}$$

i.e. w_{ij} normalized over all outbound weights from v_i. The transition probabilities collectively define the transition matrix $\tilde{W}$. $\tilde{W}$ is the row-normalized version of W.

The PageRank algorithm sends the score of each vertex to its neighbors with the weights in $\tilde{W}$. The new score of a vertex is then the weighted combination of the received scores from its neighbors. This algorithm performs outbound weights normalization since the weights outgoing from each vertex are normalized to sum to one. Figure 2 gives an illustration of this mechanism. The algorithm iteratively updates the scores $\mathbf{s}$ until they converge. The update step for iteration $t+1$ can be expressed concisely as

$$\mathbf{s}^{(t+1)} \leftarrow r\mathbf{d} + (1-r)\tilde{W}^T\mathbf{s}^{(t)}, \tag{2}$$

where $\mathbf{d}$ is a damping term (a.k.a. teleport probabilities) and r is a reset parameter. For details of PageRank, the reader is referred to [14].

2.1 Seeded PageRank

Personalized PageRank [14] is a variant of PageRank that bias towards a given vertex. Basically, the source vertex has a positive teleport probability and the

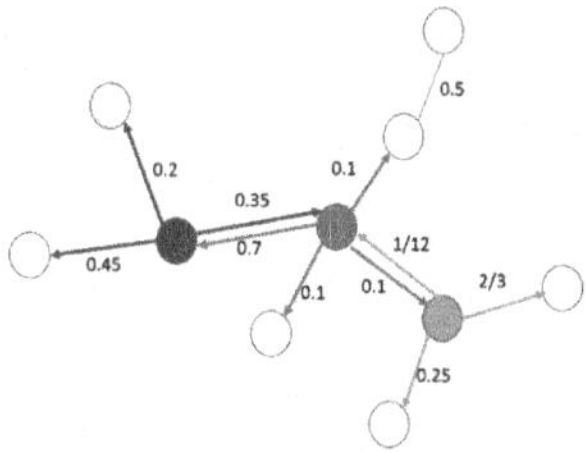

Fig. 2. Illustration of outbound weights normalization in PageRank. The colors of the edges correspond to those of the vertices. Edges of the same color are normalized together. (Color figure online)

rest of vertices in the network have zero teleport probabilities. The personalized PageRank score is then relative to a particular vertex based on the network topology. In some cases, personalized PageRank could have multiple sources [10]. We refer these sources of which energy is propagated through the network as *seeds* and call the multi-source version of personalized PageRank the *Seeded PageRank*. The scores of the non-seed vertices reflect their association strength with the seeds, even if they are not directly linked to the seeds. In the case of a social network, the score produced by Seeded PageRank indicates a person's closeness to a given group of people. See Fig. 3 for an illustration of the idea. We can view Seeded PageRank as a version of PageRank where there are vertex-specific teleport probabilities $\mathbf{d}$ and reset weights $\mathbf{r}$, so that the update Eq. (2) becomes

$$\mathbf{s}^{(t+1)} \leftarrow diag(\mathbf{r})\mathbf{d} + (I - diag(\mathbf{r}))\tilde{W}^T\mathbf{s}^{(t)}, \tag{3}$$

where $diag(\mathbf{r})$ denotes a diagonal matrix with $\mathbf{r}$ being the diagonal. It can be easily verified that the original convergence results still apply in this case, as long as $0 < \mathbf{r} \leq 1$. In our implementation, a seed has a constant score of one, and the score of a seed is kept unchanged throughout the process. Hence, $\mathbf{d}_S$ and $\mathbf{r}_S$, the entries of $\mathbf{d}$ and $\mathbf{r}$ respectively corresponding to the seeds are 1's. We set the teleport probabilities of non-seeds to 0's. The reset weight for non-seeds is set to 0.15 as typically suggested.

3 Learning Edge Weights

The quality of the PageRank algorithm and its variants considerably depends on the graph, in particular the relevance of the edge weights to the score prediction task. In many real-world cases, the edge weights are computed by some heuristics combining several known factors (or features). One obvious problem with such heuristics is how to set the combination coefficients. As the underlying network data gets updated continuously, the set of coefficients are likely to change as well, making the method very manual and unsustainable. Supervised learning models, when labels for supervision are available, present themselves as principled and scalable way of predicting the edge weights automatically based on a set of user interaction features.

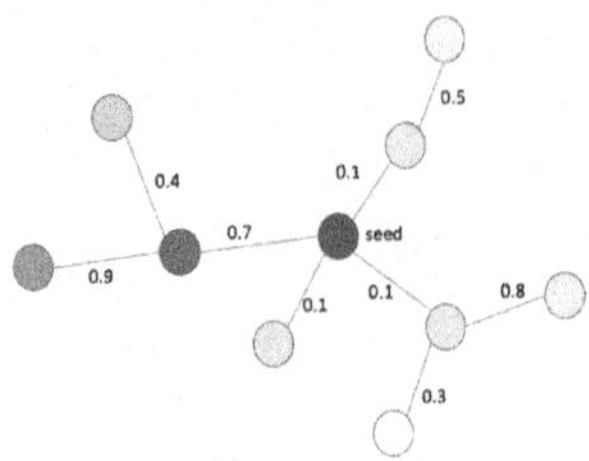

Fig. 3. Illustration of score propagation in Seeded PageRank. The center vertex is the only seed. The intensity of the color reflects the score magnitude. All edges are bi-directional. (Color figure online)

3.1 Supervised Learning Model

In this section, we lay out the key components for building a supervised learning model. Subsequently, we propose a modification to the original PageRank update inspired by the new interpretation of the edge weights.

Features. The features of the edges in a social network usually encompasses how the two connecting entities interact with each other. Some common examples include communication frequency, profile similarity, and location proximity. They can be binary or continuous. We denote the i-th feature of an edge e by $f_{e,i}$.

Edge Labels. Choosing the right edge labels is instrumental to the relevance of the learned edge weights, which directly impacts the final user scores and the recommendation. As described in Sect. 2, the edge weights determines how much of the source vertex's energy (score) can be propagated to the destination vertex. Assuming without loss of generality that edges with positive labels have higher weights than those with negative labels, then a positive label should be given to an edge connecting two vertices which are deemed similar for the purpose of the marketing campaign. Such labels can be obtained, for example, by identifying a set of non-premier service users $\mathcal{N}$ that we have high confidence that they are unlikely to be converted to premier service users. With the set of core premier services users $\mathcal{S}$ given, we can generate the edge labels $\{y_e\}$ by the following rules: $y_e = +1$, if $s(e), d(e) \in \mathcal{S}$ or $s(e), d(e) \in \mathcal{N}; y_e = -1$, otherwise, where $s(e)$ and $d(e)$ are the source and destination vertices of e respectively.

Another way of obtaining such labels is to use certain external vertex-level indicators, as long as the closeness of the two indicator values approximately reflect the proximity of the two vertices' characteristics. Then we can label an edge positive if the difference of the two vertices' indicator values is within a threshold; otherwise it is labeled negative. Since the number of edge features is typically small, not a large amount of labels are required, as demonstrated from our real-world examples.

Classification Model. We use logistic regression to train a classification model to learn the edge weights $\{w_e\}_e$, based on the edge features $\{f_{e,i}\}_{e,i}$ and labels $\{y_e\}_e$ described above. The logistic regression model is obtained by solving the following convex optimization problem, which maximizes the likelihood of the data under logistic function for positive label probabilities:

$$\min_{\mathbf{z}} \sum_{e} \log(\exp(-y_e(\mathbf{f}_e^T \mathbf{z})) + 1).$$

We have also experimented with some of the popular tree-based ensemble alternatives for classification, e.g. GBDT [6] and XGBoost [3]. However, we did not observe any advantage in prediction performance in this case.

The positive label probability (or the predicted weight) of an unlabeled edge e with features $\mathbf{f}_e$ is then given by the sigmoid function applied to the linear model $P(y_e = 1) = \frac{1}{1+\exp(-\mathbf{f}_e^T \mathbf{z})}$. The predicted weight on an edge is then the probability that both the source and destination vertices belong to the same score interval. Note that the edges features can be symmetric or asymmetric. If all features are symmetric, i.e. the features are indifferent to the order of the two vertices, there is only one predicted weight for each pair of connected vertices. In this case, we set the same weight on the two directed edges between these two vertices.

In our experiment experience, class imbalance is common in the training data generated this way. Our approach to address this problem is to under-sample the dominant class so that the numbers of examples are approximately the same for the two classes. In the case where the class imbalance is severe, we could also sample the dominant class multiple times to create multiple training sets for an ensemble of classifiers and set the class prediction probability to be the mean of the winning the class probabilities. Some more sophisticated methods based along this idea are discussed in [7].

4 Inbound-Normalized PageRank

With the edge weights defined as above, they can be approximately thought of as the probability that the score of the source vertex can be propagated to the destination vertex, since they are similar with this probability. We can thus interpret the score of a given vertex as the expectation of a score whose distribution is (empirically) defined by the scores of the neighbors. The exact expression for the update of the score of a vertex is

$$s_i^{(t+1)} \leftarrow \sum_j \hat{w_{ji}} s_j^{(t)}, \tag{4}$$

where $\hat{w}_{ji} = \frac{w_{ji}}{\max(\sum_j w_{ji}, 1)}$, i.e. the inbound weights to this vertex are normalized to make sure that they form a probability distribution. Note that the denominator of the normalization is maxed with 1. This is to avoid artificially magnifying small edge weights when the inbound weights sum is small. An extreme example of this pathological case is when a vertex v_i is weakly linked to another vertex v_j, and it is the only inbound edge for v_i. The dynamics of the score propagation is the opposite to that in PageRank in that each vertex 'pulls' the scores from its neighbors with normalized inbound weights, whereas in PageRank each vertex sends out its score to all its neighbors with normalized outbound weights. The matrix $\hat{W}$ is no longer a transition matrix as a result. Rather, the score of each vertex is an expectation of its neighbors' scores. Figure 4 illustrates this idea in contrast to Fig. 2.

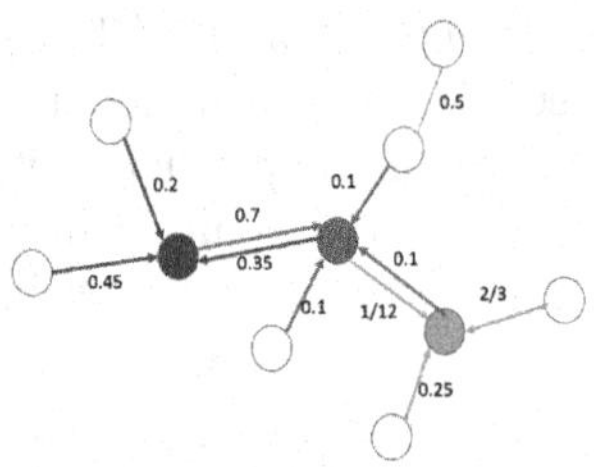

Fig. 4. Illustration of inbound weights normalization in contrast to outbound weights normalization in Fig. 2. The colors of the edges correspond to those of the vertices. Edges of the same color are normalized together. Note that the weights used to compute the new score of a vertex are different from those in Fig. 2. (Color figure online)

Similar to the Seeded PageRank update (3), we can also add a teleport probability vector $\mathbf{d}$ and a reset weight vector $\mathbf{r}$ to the inbound PageRank update, leading to the update

$$\mathbf{s}^{(t+1)} \leftarrow diag(\mathbf{r})\mathbf{d} + (I - diag(\mathbf{r}))\hat{W}^T\mathbf{s}^{(t)}, \tag{5}$$

where $\hat{W}$ is the matrix of $\hat{w}_{ji}$, i.e. the column-normalized version of W. The values of the $\mathbf{d}$ and $\mathbf{r}$ are set similarly to those of the Seeded PageRank.

4.1 Convergence

We show below that the modified algorithm retains the convergence property of the original PageRank algorithm.

Assumption 1. *Without loss of generality, we assume that the network is fully connected with currently absent edges carrying zero weights. We also assume that there are zero-score dummy vertices with just enough dummy edge weights to make the sum of inbound edge greater or equal to 1 for all vertices. To avoid a row consisting entirely of zeros, we discard all isolated vertices. We note that these assumptions are in no way restrictive and do not change the behavior of our algorithm.*

Lemma 1. *With the update scheme* (5), *the scores of all the vertices are bounded.*

Proof. Following the assumptions, the score of a given vertex is updated by a convex combination of the scores of all the other vertices in the graph (see (4)). Assuming for the moment that the seed vertices can have their scores updated as the regular vertices, then $s_i^{(t)}$ lies within the convex hull of $\{s_j^{(0)}\}_j, \forall i \in \mathcal{I}$, and $\mathbf{s}$ is clearly bounded $\forall t$. With the additional damping term and reset weights, $s_i^{(t)}$ lies within the convex hull of $\{s_j^{(0)}, d_j\}, \forall i \in \mathcal{I}$, and hence the boundedness still holds.

Another desirable property that follows is that if the seed vertices have a score of 1, then the scores of all vertices are within $[0, 1]$ and hence can be interpreted as probabilities of being potential seeds. For outbound-normalized seeded PageRank, the seeds' scores have to sum to 1 in order for the rest of the scores to be bounded within $[0, 1]$, but non-seed scores could exceed seed scores, which makes interpretation harder.

Theorem 1. *With the update scheme* (5), $\mathbf{s}^{(t)}$ *converges to a unique solution as* $t \to \infty$.

Proof. In steady state, i.e. $t \to \infty$, the update (5) becomes the fixed-point equation

$$\mathbf{s} = diag(\mathbf{r})\mathbf{d} + (I - diag(\mathbf{r}))\hat{W}^T\mathbf{s}, \tag{6}$$

from which $\mathbf{s}$ can be solved for by

$$(I - (I - R)\hat{W}^T)\mathbf{s} = R\mathbf{d}, \tag{7}$$

where $R = diag(\mathbf{r})$. Let $Q := I - (I - R)\hat{W}^T$. Under the condition $0 < r_i \leq 1$, we observe that the diagonal elements of Q satisfy

$$\begin{aligned} Q_{ii} &= 1 - (1 - r_i)\hat{w}_{ii} \\ &= 1 - \hat{w}_{ii} + r_i\hat{w}_{ii} \\ &> 1 - \hat{w}_{ii} && (8) \\ &= \sum_{j \neq i} \hat{w}_{ji}, && (9) \end{aligned}$$

where (8) follows from $r_i > 0$, and (9) follows from the fact that W^T is right-stochastic. The above inequality shows that the matrix Q is strictly diagonally dominant and is thus non-singular [9]. Therefore, $\mathbf{s}$ exists and is unique.

4.2 Full Algorithm

We now formally state the end-to-end algorithm solution for our targeted marketing problem in Algorithm 4.1.

Algorithm 4.1. Targeted Marketing via Inbound-normalized PageRank with Seeds

Given: the social network $\mathcal{G} :=< \mathcal{E}, \mathcal{V} >$ with features $\mathbf{f}_e$ associated with each edge $e \in \mathcal{E}$, the set of seed users $\mathcal{S}$, and the size of the marketing campaign K.
Generate edge labels for a subset of $\mathcal{E}$, $\mathcal{E}'$ as described in Sect. 3.1.
Train a classification model $\mathbf{z}$ based on $\mathcal{E}'$ and its labels using logistic regression.
Compute all weights of $\mathcal{E}$, W by $P(y_e = 1)$.
Initialization: $\mathbf{s}^{(0)} \leftarrow$ a random vector with $s_i^{(0)} \in (0, 1)$; $\mathbf{d} \leftarrow \mathbf{0}$; and $\mathbf{r} \leftarrow 0.15$
$\mathbf{d}_{\mathcal{S}} \leftarrow \mathbf{1}$, $\mathbf{r}_{\mathcal{S}} \leftarrow \mathbf{1}$
for $t = 1, \cdots, T_{\max}$ **do**
    $\mathbf{s}^{(t+1)} \leftarrow diag(\mathbf{r})\mathbf{d} + (I - diag(\mathbf{r}))\hat{W}^T\mathbf{s}^{(t)}$
    if $\mathbf{s}^{(t+1)}$ has converged **then**
        Break
    end if
end for
Sort $\mathcal{V}' := \mathcal{V}/\mathcal{S}$ in descending order of $\mathbf{s}^{(t+1)}$.
return $\mathcal{V}_K$, the top K entries of $\mathcal{V}'$

4.3 Related Works

There are two major lines of related literature. The first line is targeted marketing using social networks. Among the recent works, both [12,19] consider the problem of finding the key entities within the network – those who can spread the maximal influence, and they both adopt the viral marketing approach by targeting the hot nodes in the network and spreading the marketing effect through reviews and word-of-mouth. We mentioned the potential concern with this approach in our case in Sect. 1.1. Our approach, on the other hand, better utilizes the knowledge of existing set of core premier users and targets directly to the intended audience.

The second line is on the PageRank variants with edge learning. The method that we have developed is both a *node-personalized* and *edge-personalized* [15,18] PageRank, with the fixed weights on the vertices (i.e. the teleport probabilities vector $\mathbf{d}$), and we have shown how to learn the edge weights with respect to the given task in Sect. 3. [1,16] approach the edge-personalized PageRank problem with side information on the partial order of a subset of the vertices and formulate a (non-convex) optimization problem to learn the edge weights. However, the optimization algorithms are not easily parallelizable, and from their reported run-time, it appears infeasible to apply to our case where the network could contain billions of edges and hundreds of millions of vertices. We use the side information in a different way - we care only about if the scores of the two vertices of the training edges are close enough. This allows us to formulate edge learning as a classification problem, which is more distributed computing-friendly.

5 Experiments

We implemented our proposed solution framework (Algorithm 4.1) in Scala using Apache Spark[1] GraphX [17] and MLlib [13] libraries for Seeded PageRank and supervised edge learning respectively. For Seeded PageRank, we implemented the iterative approach, instead of solving the linear system (7) directly. This allows us to better leverage the distributed power of Spark GraphX. We terminate the iterations when the normalized absolute score difference between two consecutive iterations is within 1%, i.e. $\frac{\sum_{i\in\mathcal{I}}|s_i^{(t+1)}-s_i^{(t)}|}{\sum_{i\in\mathcal{I}}s_i^{(t)}} \leq 0.01$.

We applied the methods developed in the previous sections to generate the list of target users who have high potential to upgrade to the premier service and to whom promotional coupons were to be issued. The anonymized social network dataset that we have available in-house consists of 3.6 billion edges and 100 million vertices.[2] Each edge has about ten features, symmetric and asymmetric, encompassing the social interaction between the two entities (vertices). Most of the features are binary.

The algorithms were run on a Spark cluster with 500 workers and minimum configuration tuning. We observed that with our data set, the iterative algorithm was able to converge within 16 iterations to the stopping tolerance. In terms of run time, the Seeded PageRank with edge learning was usually able to finish within 20 min.

5.1 Targeted Marketing for Premier Service

We carried out off-line experiments on our main business problem described in Sect. 1 to validate the effectiveness of our proposed approach. We first identified a set of core users of the premier service $\mathcal{S}$ by historical usage frequency as defined in Sect. 1. For generating edge labels, we adopted the first method in Sect. 3.1, i.e. by identifying anther set of users $\mathcal{N}$. Note that we were using noisy labels since the identification of $\mathcal{N}$ is heuristic. For evaluation, we used the core users from last month to compute the PageRank scores, and we examined the ranking of the new core users (test seeds) from the current month within the population.

5.2 Novel Seed Entities Discovery over Social Network

To understand the extendability of our approach to other objectives, we performed a second set of experiments that are unrelated to the business problem. We resorted to a set of social activity indices $\{\hat{s_i}\}$ for a small subset of the entities $\hat{\mathcal{I}}$ from a third-party social media website. Those entities with high scores are selected as the seeds, $\mathcal{S}$. The goal is to identify entities with similar profiles as the seeds. We construct the edge labels as follows. Those edges with both

[1] https://spark.apache.org/.

[2] Unfortunately, we are unable to make the dataset public for proprietary reasons.

vertices in $\mathcal{S}$ automatically get positive labels. For an edge with the two vertices having external scores $\hat{s_i}$ and $\hat{s_j}$, the edge label is positive if $\frac{|\hat{s_i}-\hat{s_j}|}{\max(\hat{s_i},\hat{s_j})} \leq \delta_1$ or $\max(\hat{s_i},\hat{s_j}) \leq \delta_2$, where δ_1 is the closeness tolerance on the score difference, and δ_2 is the indifference threshold. This corresponds to the second edge label generation method described in Sect. 3.1. The size of the training set for edge learning is at an order of hundreds of thousand. For evaluation, we used a subset of the seeds to compute the PageRank scores of all entities, and we examined the ranking of the test seeds within the entire population. The evaluation methodology is elaborated in Sect. 5.3.

5.3 Results

We applied the Seeded PageRank with and without edge learning to the above two applications. In the case without edge learning, we used domain knowledge of the edge features to manually tune the coefficients for combining the features. For the supervised edge learning, we sub-sampled the positive class to match the size of the negative class as described in Sect. 3.1. The area under the PR curve is above 0.75 and the area under the ROC curve is above 0.70 in both applications.

To measure the general quality of the resulting PageRank scores with respect to the given seeds, we have developed evaluation metrics based on ranking the entire set of vertices $\mathcal{V}$ by their PageRank scores. We note that in our case, we do not have complete ground truth for the ranking of all the entities. All we know is that the seeds should be ranked higher than the non-seeds, i.e. the relevance of an entity is binary. Moreover, since a marketing campaign typically covers large group people, a test seed being ranked at the 1st position versus 10th position does not matter much. We are rather concerned about the overall rank distribution of the test seeds in the population. Within the set of test seeds, we focus on the upper quartile seed (ranked at 25th percentile), the median seed (ranked at 50th percentile) and the lower quartile seed (ranked at 75th percentile). We define the *upper quartile seed quantile*, q_U to be the ranking quantile of the upper quartile seed. Similarly, the *median seed quantile*, q_M is the ranking quantile of the median seed, and the *lower quartile seed quantile*, q_L is the ranking quantile of the lower quartile seed. Basically, these metrics give a three-point view of a box-plot of the seeds' ranking quantiles within the population vertices. Since the scores are typically used for selecting a subset of the entities (usually the top ones at the order of millions or tens of millions) for some application-specific tasks such as targeted marketing, two key desirable properties of a given set of scores are sought after in the evaluation: (1) the test seeds' scores should collectively be ranked high enough, and (2) the variance of the test seeds' ranks should be as small as possible. Hence, we compare the alternative methods on the median seed quantile q_M and the inter-quartile range (IQR), i.e. $q_L - q_U$. In the perfect solution, since all of the test seeds should be tied to the top rank alongside the training seeds, we have $q_M = 0$ and $IQR = 0$. It can also be easily seen that if one ranking method has better results on the above three seed quantiles than another method, it would have better *precision@n* results as well on $n = q_U, q_M, q_L$ of the itself.

We show the results for the targeted marketing application in the form of bar plots in Fig. 5. Table 1 shows the quantiles of the three quartile seeds. In addition to the PageRank variants, we also compare the results to a base-line rule-based selection method that is a common practice by business operations. The base-line considers the total amount of GMV that a customer has contributed with the standard service and selects the top ones for conversion recommendation for the premier service, based on the intuition that once a customer has been using the standard service more frequently than others, she is more likely to be converted to a premier service customer. We mentioned the concern with this assumption, and the results below confirm such concern. The scores in this case are the expenses per customer used by the rule.

Table 1. Quantiles of quartile seeds for targeted marketing

Algorithm	q_U	q_M	q_L
Seeded PageRank with edge learning and inbound normalization	2.10%	7.20%	23.40%
Seeded PageRank with edge learning	2.20%	9.07%	29.50%
Seeded PageRank without edge learning	2.10%	13.20%	56%
Base-line	5.50%	34.30%	82.80%

In Fig. 5, we present the bar plots for seed entities discovery. Table 2 shows the quantiles of the three quartile seeds.

Table 2. Quantiles of quartile seeds for seed entities discovery

Algorithm	q_U	q_M	q_L
Seeded PageRank with edge learning and inbound normalization	0.44%	1.74%	7.75%
Seeded PageRank with edge learning	0.37%	2.00%	10.00%
Seeded PageRank without edge learning	0.65%	4.51%	20.74%

From the results, it is clear that for both applications, the quantiles of the test seeds collectively rank highest and have smallest spread when we use the Seeded PageRank with inbound weights normalization and edge learning. The version with conventional outbound weights normalization comes in second. For better understanding of the results in the context of targeted marketing on a population of 100 million, we would have to issue more than 30 million coupons to reach 50% of the test seed users using the baseline method, whereas with

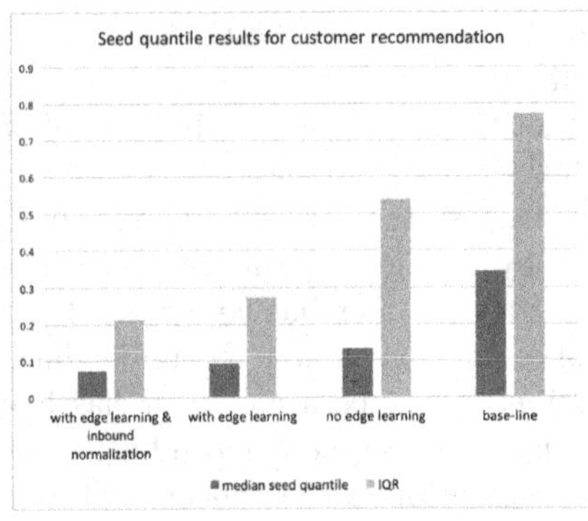

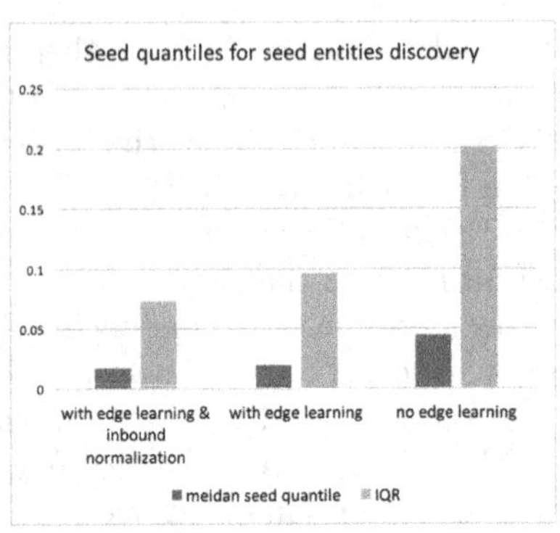

Fig. 5. Seed quantiles for targeted marketing (left) and seed entities discovery (right)

inbound-normalized seeded PageRank with edge learning, we would need to issue only about 7.2 million coupons to reach the same number of test seed users. With the conventional outbound-normalized seeded PageRank, we would have to issue close to two million more coupons compared to with our approach, and another four million more without edge learning. They demonstrate that setting the edge weights through supervised learning produces a set of scores that are better aligned with the given seeds, and using a principled graph-based approach yields better results than the simple rule-based method. In the targeted marketing case, one issue that we observed for the base-line method is that the positive correlation between high spending in the standard service and conversion to premier service is not strong enough. In fact, many users continue to be loyal to the standard service.

6 Conclusion

In this paper, we have developed an algorithmic framework based on utilizing social network for targeted marketing with a conversion horizon much longer than typical coupon usage. We have also developed a modification to Seeded PageRank to better align with the resulting interpretation of the edge weights when one uses supervised learning to predict edge weights with respect to the marketing objective. For two real-world large-scale applications, we implemented the proposed method on Spark, which allows us to scale the method to billions of edges and hundred of millions of vertices. Our experiment results demonstrate that our method produces a set of scores that are more appropriate to the given tasks than the existing base lines.

As mentioned in Sect. 4.1, an additional advantage of the inbound-normalized seeded PageRank is that the output scores can be appropriately interpreted as classification probabilities. This property makes the method easy to combine with traditional supervised learning approaches. In our use cases, for example, we could train user-level classification models, if user-level features are available.

As the output of a classifier, such as logistic regression or deep neural networks with soft-max output layer, is also a probability, combining the two outputs tend to be more straightforward than other means.

References

1. Backstrom, L., Leskovec, J.: Supervised random walks: predicting and recommending links in social networks. In: Proceedings of the Fourth ACM International Conference on Web Search and Data Mining, pp. 635–644. ACM (2011)
2. Bhuiyan, J.: Lyft is on track to turn a profit, but will need to spend more to add riders and drivers as it expands (2016). https://www.recode.net/2017/1/13/14267926/lyft-growth-revenue-losses-profitability-discounts-incentives
3. Chen, T., Guestrin, C.: XGBoost: a scalable tree boosting system. In: Proceedings of the 22nd ACM SIGKDD International Conference on Knowledge Discovery and Data Mining, pp. 785–794. ACM (2016)
4. Chen, Y., Pavlov, D., Canny, J.F.: Large-scale behavioral targeting. In: Proceedings of the 15th ACM SIGKDD International Conference on Knowledge Discovery and Data Mining, pp. 209–218. ACM (2009)
5. Forbes Corporate Communications: Investing in customer retention leads to significantly increased market share says new study (2016). https://www.forbes.com/sites/forbespr/2016/09/14/investing-in-customer-retention-leads-to-significantly-increased-market-share-says-new-study
6. Friedman, J.H.: Greedy function approximation: a gradient boosting machine. Ann. Stat. **29**, 1189–1232 (2001)
7. Galar, M., Fernandez, A., Barrenechea, E., Bustince, H., Herrera, F.: A review on ensembles for the class imbalance problem: bagging-, boosting-, and hybrid-based approaches. IEEE Trans. Syst. Man Cybern. Part C (Appl. Rev.) **42**(4), 463–484 (2012)
8. Goldfarb, A., Tucker, C.: Online display advertising: targeting and obtrusiveness. Mark. Sci. **30**(3), 389–404 (2011)
9. Golub, G.H., Van Loan, C.F.: Matrix Computations, vol. 3. Johns Hopkins University Press, Baltimore (2012)
10. Haveliwala, T.H.: Topic-sensitive PageRank. In: Proceedings of the 11th International Conference on World Wide Web, pp. 517–526. ACM (2002)
11. Leskovec, J., Adamic, L.A., Huberman, B.A.: The dynamics of viral marketing. ACM Trans. Web **1**(1), 5 (2012)
12. Li, F.H., Li, C.T., Shan, M.K.: Labeled influence maximization in social networks for target marketing. In: 2011 IEEE Third International Conference on Privacy, Security, Risk and Trust and 2011 IEEE Third International Conference on Social Computing, pp. 560–563, October 2011
13. Meng, X., Bradley, J., Yavuz, B., Sparks, E., Venkataraman, S., Liu, D., Freeman, J., Tsai, D., Amde, M., Owen, S., et al.: MLlib: machine learning in apache spark. J. Mach. Learn. Res. **17**(34), 1–7 (2016)
14. Page, L., Brin, S., Motwani, R., Winograd, T.: The PageRank citation ranking: bringing order to the web. Technical report, Stanford InfoLab (1999)
15. Weng, J., Lim, E.-P., Jiang, J., He, Q.: Twitterrank: finding topic-sensitive influential twitterers. In: Proceedings of the Third ACM International Conference on Web Search and Data Mining, pp. 261–270. ACM (2010)

16. Xie, W., Bindel, D., Demers, A., Gehrke, J.: Edge-weighted personalized PageRank: breaking a decade-old performance barrier. In: Proceedings of the 21st ACM SIGKDD International Conference on Knowledge Discovery and Data Mining, pp. 1325–1334. ACM (2015)
17. Xin, R.S., Gonzalez, J.E., Franklin, M.J., Stoica, I.: GraphX: a resilient distributed graph system on spark. In: First International Workshop on Graph Data Management Experiences and Systems, p. 2. ACM (2013)
18. Xing, W., Ghorbani, A.: Weighted PageRank algorithm. In: 2004 Proceedings of Second Annual Conference on Communication Networks and Services Research, pp. 305–314. IEEE (2004)
19. Zhang, Y., Wang, Z., Xia, C.: Identifying key users for targeted marketing by mining online social network. In: 2010 IEEE 24th International Conference on Advanced Information Networking and Applications Workshops, pp. 644–649 (2010)

Storytelling with Signal Injection: Focusing Stories with Domain Knowledge

J. T. Rigsby[1,2](✉) and Daniel Barbará[2]

[1] Naval Surface Warfare Center, 18372 Frontage Road, Dahlgren, VA 22448, USA
jrigsby1@gmu.edu
[2] CS Department, George Mason University, Mail Stop 4A5, Fairfax, VA 22030, USA
dbarbara@gmu.edu

Abstract. Given a beginning and ending document, automated storytelling attempts to fill in intermediary documents to form a coherent story. This is a common problem for analysts; they often have two snippets of information and want to find the other pieces that relate them. The goal of storytelling is to help the analysts limit the number of documents that must be sifted through and show connections between events, people, organizations, and places. But existing algorithms fail to allow for the insertion of analyst knowledge into the story generation process. Often times, analysts have an understanding of the situation or prior knowledge that could be used to focus the story in a better way. A storytelling algorithm is proposed as a multi-criteria optimization problem that allows for signal injection by the analyst while maintaining good story flow and content.

Keywords: Storytelling · Literature-based discovery
Connecting the dots · Intelligence analysis

> No one ever made a decision because of a number, they needed a story.
>
> *Daniel Kahneman,*
> *Nobel Memorial Prize in Economics, 2002*

1 Introduction

Kahneman won the Nobel Memorial Prize in Economics in 2002 for his work in prospect theory [1]. Prospect theory describes how people make choices based on risk and is grounded in probability theory, but the above quote seems to point in the opposite direction. However, the quote is not a contradiction to the mathematical nature of prospect theory; humans make decisions by rationalizing risks, creating a story of alternatives, and picking the best story based on cost-benefits analysis, but the decision does not come down to a number; the decision is based on a rational story.

P. Perner (Ed.): MLDM 2018, LNAI 10935, pp. 425–439, 2018.
https://doi.org/10.1007/978-3-319-96133-0_32

The act of finding patterns in data is often referred to as *connecting the dots* and is an important part of intelligence analysis wherein people stitch together pieces of disparate information to understand the situation. Intelligence analysts are better than computers at making leaps of intuition and *connecting the dots* when provided with a manageable set of data. Computers excel at organizing large datasets into manageable subsets that analysts can explore and comprehend. The work presented here takes advantage of both. Algorithms will be used to generate story chains; analysts will interpret the story chains to discover relationships in the data to generate a hypothesis about the actions of characters and events; the hypothesis is the rational story that leads to a decision about what to investigate next.

Previous storytelling work has focused on finding coherent stories that explain events and relationships that connect a beginning document to an ending [2–4]. Given a beginning document, d_1, and an ending document, d_n, storytelling algorithms find documents $d_2, \ldots, d_{n-1}$ to produce a story chain connecting the bookends, d_1 and d_n, to help analysts find patterns in data. The story chains are then read by an analyst to tease out narratives based on the pertinent information contained in each individual document. The full content of each document does not tell the story; rather each document in the chain contains useful tidbits of information that the analysts must pull together to form a story which is actually a hypothesis. But what happens when the analyst knows that a better story could be made if other useful pieces of information had been considered when the algorithm generated the story chain? Little research exists to deal with analyst's expectations if the returned story chain does not represent the narrative they were expecting.

A storytelling algorithm can only use the given data to make decisions about which documents should be inserted between the given bookends. But an analyst has experience and intuition that can help guide the algorithm. Allowing analysts to inject their own personal signal into the storytelling algorithm allows them to focus a story on what they consider to be pertinent. A good example is pseudonyms. People often use a fake name when they are trying to evade capture and/or detection. The analysts could read a generated story chain about particular characters and see how the generated story would have been better if the algorithm knew that person A sometimes uses an alias. Information about the person and their aliases could be signal injected in order to incentivize the algorithm to find stories involving the pseudonyms. Stories would still be generated based on information in the data but would be brought into better focus based on the analyst's knowledge. Storytelling with signal injection is novel and is the approach taken in this paper to allow analysts to create alternative stories to the ones produced from automated storytelling methods.

2 Related Work

Shahaf and Guestrin attached the phrase *connecting the dots* to storytelling and introduced characteristics of a good story chain [2]. Their follow-on efforts have primarily dealt with visualization and mapping the information into networks called *metro maps* [5–7]. Metro Maps is a visualization environment for creating and exploring summaries of information as concise structured sets of documents.

Hossain et al. developed a storytelling framework that is based on connecting documents into story chains by creating document cliques that meet a minimum distance threshold [3]. The particular document chain chosen to form a story from document cliques is based upon greedily picking the document that is closest to the ending document in each clique. Their storytelling algorithm was integrated into a visual analytics system called *Analyst's Workspace (AW)* that accomplishes document browsing with storytelling [8]. The purpose of AW is to give the analysts a data exploration environment for discovering latent connections between seemingly unrelated documents. If an analyst was displeased with a story, he or she could adjust the model parameters (*distance threshold* and *clique size*) to produce different stories for the same bookend pair of documents. This allows for alternative narrative generation that are shorter or longer in length with different documents being allowed into the story chain.

2.1 Storytelling via Optimizing Quantitative Evaluation Metrics

In [4], we developed a storytelling framework based on optimizing quantitative evaluation metrics in a greedy fashion to build story chains. The metrics, *dispersion* and *coherence*, were designed to evaluate different aspects of *story quality*. Dispersion, a measure of story flow, ascertains how well the generated story flows away from the beginning document and towards the ending document. Coherence measures how well the articles in the middle of the story provide information about the relationship of the beginning and ending document pair. Dispersion and coherence were shown to agree with human judgment as valid story quality measures, and the automatically generated story chains were shown to agree strongly with human generated stories as measured via Jaccard similarity [4].

2.2 Dispersion

Dispersion is a measure of story flow, i.e., does the generated story continuously flow away from the beginning document and towards the ending document? The story should transition from beginning to end through the middle documents in a particular order that is logical. Bi-directional (BD) dispersion is based on the assumption that backtracking towards the beginning document or a previous document in the story is detrimental to a story continuously flowing to the end. BD dispersion measures how the story disperses between all possible pairs of documents in the story and is measured in both directions. All distances in the story must increase with forward story progression and decrease in the reverse story. The BD dispersion coefficient is penalized if story flow is violated in any direction between all possible pairs of documents. This penalty is increased if documents are farther apart in the story chain, i.e., flow violations between adjacent documents are penalized less than violations between non-adjacent documents. Equation (1) shows how the BD dispersion, $\mathcal{B}$, is calculated. The documents in the story are $d_1, d_2, \ldots, d_n$ where d_1 is the beginning document, and d_n is the end. The dissimilarity matrix, D, is the ordered dissimilarity matrix from d_1 to d_n where $D(d_i, d_j)$ is the *dissimilarity* from document i to document j. The BD dispersion measure returns a value in the range $[0, 1]$, where 1 represents a story that has no backtracking towards the beginning or away from the ending document. BD dispersion is penalized in both directions, $\Leftarrow$ and $\Rightarrow$. The measure of the story progressing towards the next document is represented as $\Leftarrow$, while $\Rightarrow$ is a measure of the story moving farther away from the previous documents in the story.

$$\mathcal{B} = 1 - \frac{1}{2}\frac{1}{n-2}\sum_{i=1}^{n-2}\sum_{j=i+2}^{n}(\Rightarrow(d_i, d_j) + \Leftarrow(d_i, d_j)) \tag{1}$$

where

$$\Rightarrow(d_i, d_j) = \begin{cases} \frac{1}{n+i-j}, & \text{if } D(d_i, d_j) < D(d_i, d_{j-1}) \\ 0, & \text{otherwise} \end{cases}$$

and

$$\Leftarrow(d_i, d_j) = \begin{cases} \frac{1}{n+i-j}, & \text{if } D(d_i, d_j) < D(d_{i+1}, d_j) \\ 0, & \text{otherwise} \end{cases}$$

2.3 Coherence

Coherence is a quantitative measure that provides a numerical method to evaluate story content. The method for measuring story coherence is based on Kullback-Leibler divergence (KLD), which measures how well the vocabulary of the middle documents of the story chain $d_M = \{d_2, d_3, \ldots, d_{n-1}\}$ encode the vocabulary of the beginning and ending documents (d_1 and d_n). KLD is formalized as $KLD(p_{B \cup E}||p_M)$ from p_M to $p_{B \cup E}$, where $p_{B \cup E}$ is the normalized union of the Jelinek-Mercer smoothed probability distributions of d_B and d_E, and p_M is JM-smoothed probabilities of d_M [9].

2.4 Storytelling Algorithm

The storytelling algorithm is constructed as a multi-criteria optimization problem that maximizes dispersion and coherence simultaneously and is shown in Eq. (2). The numerator is maximized when dispersion obtains its maximum at one. Maximizing dispersion leads to stories with good flow. Maximizing coherence is equivalent to minimizing the KLD measure. Smaller KLD means the middle documents are doing a better job encoding the vocabulary of the beginning and ending documents. The parameters (α and β) are positive real numbers that prevent domination of the function by stories with medium-to-low dispersion (i.e., less than one-half) and high coherence scores (i.e., low KLD, near zero). For all experiments presented here, $\alpha = 1$ and $\beta = 1$ as in [4]. The greedy storytelling algorithm initially finds a single document that maximizes Eq. (2). The algorithm continues adding middle documents one at a time choosing the next middle document that maximizes Eq. (2). The algorithm stops iterating when the value from Eq. (2) is less than the maximum value from the previous iteration.

$$f_{\alpha,\beta}(M : d_B, d_E) = \frac{\alpha + dispersion(D([d_B \quad d_M \quad d_E]))}{\beta + KLD(p_{B \cup E} || p_M)} \tag{2}$$

3 Signal Injection

When the analyst wants to alter the generated story chain, an injection signal must be created based upon what information the analyst expected the story chain to contain. The signal is based upon the analyst's intuition about what type of information would possibly alter the story in a positive way. For example, the signal could be a list of pseudonyms of a person in the story that the analyst believes would allow the algorithm to make leaps of intuition that are not normally possible. Formally, in this paper, a signal is a probability distribution over the vocabulary of the corpus, with the words that the analyst wishes to stress receiving most of the probability mass.

In the normal storytelling optimization, Eq. (2) without signal injection, the denominator controls the effect of story content and how new documents are added to the story chain. The numerator, containing dispersion, controls the order of the story to maximize story flow from beginning to end. Having signal injection affect story flow is not a desired capability, i.e., the order of the story should not be affected by signal injection. The story should naturally flow from beginning to end as measured through dissimilarity progressing away from the beginning and towards the ending without interference from signal injection. The denominator affects the content of the story and is a natural candidate for signal injection. To maximize Eq. (2), $KLD(p_{B \cup E} || p_M)$ must be minimized. Minimizing KLD is equivalent to finding a set of middle documents, d_M with a probability distribution p_M, that minimizes the dissimilarity between $p_{B \cup E}$ and p_M. The probability distribution of $B \cup E$ is a simple weighted sum of p_B and p_E, the Jelinek-Mercer smoothed probability distributions for each document, d_B and d_E, and is shown in Eq. (3) [9].

$$p_{B \cup E} = \frac{n_b \cdot p_B + n_e \cdot p_E}{n_b + n_e} \tag{3}$$

where n_b and n_e are the total number of words in each document.

3.1 Injecting Keywords with Zipfian Distributions

When the analyst wants to alter the generated story chain, an injection signal must be created based upon what information the analyst expected the story chain to contain. Initially, the signal from the analyst to be injected takes the form of a list of keywords, K. If signal injection is to be inserted into the denominator of Eq. (2), the injection signal must take the form of a probability distribution over the corpus vocabulary, V, and have an affect upon $p_{B \cup E}$.

The natural candidate for this injection signal distribution is a Zipfian distribution because word frequencies are well-characterized by Zipf's law [10]. Zipf's law is an empirical law formulated using mathematical statistics that refers to the fact that many types of data studied in the physical and social sciences can be approximated with a Zipfian distribution, one of a family of related discrete power law probability distributions [11]. Zipf's law has been used to model population distributions for a plethora of different areas from census analysis, economics, natural language, etc.

To model lists of keywords as Zipfian distributions, we associate the individual keywords in K to the top $|K|$ terms in a Zipfian distribution, Z, and assign them a probability mass p_K which has an associated Zipf coefficient, ζ. In order to find ζ from p_K, a simple, fast divide and conquer algorithm was developed that searches for ζ in the interval $(0, 10)$ by splitting the interval on the mean and reassigning the minima or maxima of the search interval based on which side ζ should be on. This is feasible because ζ and p_K are monotonically increasing. The resulting distributions were tested with MLE and OLS methods for estimating ζ, and the algorithm accurately finds ζ out to 14 decimal places. For equation purposes, the function, $zipf$, creates Zipfian distributions for a set of ordered keywords K given p_K. For all experiments presented here $p_K = \{0.10, 0.20, \cdots, 0.80, 0.90\}$, and $zipf$ finds the associated Zipf's coefficient, ζ. The ordered set of keywords, K, to be used for signal injection are assigned the top $|K|$ probabilities based on the order of the set. The remaining probability mass is assigned randomly to the remainder of the vocabulary, V. For example when $p_K = 0.2$, the remaining 0.8 probability mass is assigned randomly to the $|V| - |K|$ terms of the vocabulary.

3.2 Storytelling with Signal Injection

Equation (2) is modified into Eq. (4) to form a storytelling optimization that allows for signal injection in the form of p_Z, a Zipfian distribution created from a keyword list, K. The distribution p_{BEZ} is a mixture of $p_{B \cup E}$, the bookend distribution, and p_Z. The equation is used in the same manner as (2); the algorithm greedily inserts one document at a time maximizing (4) until the return value is less than the value from the previous iteration.

$$f_{\alpha,\beta}(M, \eta, \zeta : d_B, d_E, K) = \frac{\alpha + dispersion(D([d_B \quad d_M \quad d_E]))}{\beta + KLD(p_{BEZ} || p_M)} \tag{4}$$

To create p_{BEZ}, there are two parameters: η, the mixing proportion in the range $[0, 1]$, and ζ, Zipf's coefficient in the range $(0, \infty)$. The role of the mixing proportion (η) is to control the amount of information used from the bookends versus the signal generated by the analyst; i.e., η controls the creation of p_{BEZ}

which is a mixture of the distributions, $p_{B\cup E}$ and $p_Z = zipf(K, \zeta)$. The decision to add a document to the middle of the story chain is based on maximizing Eq. (4). Documents that help minimize $KLD(p_{BEZ}||p_M)$ are more likely to be added, especially if they also help maximize dispersion. The proportion of probability mass assigned to $p_{B\cup E}$ is η, and the proportion assigned to the Zipfian distribution of the injected signal is $(1-\eta)$. When η is assigned to one, all of the distribution is associated to $p_{B\cup E}$ and none of the Zipfian distribution for signal injection is used. In this case, Eq. (4) reduces to storytelling with Eq. (2), story chain generation without signal injection. Because of η's ability to nullify the effect of ζ in Eq. (4), decisions about ζ were made after decisions on η. For all experiments presented here $\eta = \{0.0, 0.1, \cdots, 0.9, 1.0\}$.

$$p_{BEZ} = \eta \cdot (p_{B\cup E}) + (1-\eta) \cdot zipf(K, \zeta) \tag{5}$$

4 Data

Two datasets will be used for demonstration purposes. The first is the *Atlantic Storm (AS)* dataset. The second contains chapters from four children's classics (*4CC*). All of the datasets were merged into a single corpus for all experiments. To prepare the data for analysis, stop words, numbers, and punctuation were removed, and stemming was performed. The two datasets are very different in nature. Chapters from *4CC* are long and the importance of a single word is dampened by the large size of the documents while *AS* documents are terse and the sparing usage of terms concentrates word emphasis over a small subset of terms. The idea of using two very different datasets simultaneously throughout the experiments should test the storytelling algorithm's ability to deal with noise and data variety.

4.1 *Atlantic Storm* Data

The *Atlantic Storm* dataset has been used by others for validating storytelling concepts and algorithms [8,12]. The original data source is from [13], an unpublished manuscript. The dataset comprises 111 documents, with each document being a fictitious intelligence snippet. An example snippet can be found at the end of this subsection. The *AS* dataset was developed at the Joint Military Intelligence College as part of an evidence-based case study. The dataset has many people, organizations, and places relating to several plots. Our thorough study of the *AS* dataset has shown that the main terrorist plot is to smuggle biological agents that were purchased from Russian scientists, into the United States. Other subplots include a plan to disrupt U.S. SCADA systems with an electromagnetic pulse bomb and a collaboration of Al Qaeda with Aryans planning bombings throughout the U.S. Approximately one-fourth of the data is information about Al Qaeda members that are in the background of the subplots planning terrorist activities that do not directly belong to a particular terrorist subplot. A full categorization of *AS* terrorist subplots is as follows:

1. smuggling biological agents from Russia to the United States
   (a) smuggling from Russia to Santo Domingo
   (b) smuggling from Santo Domingo To Nassau
   (c) smuggling from Nassau to U.S.
2. disrupting U.S. SCADA systems
3. terrorist bombings within the U.S. as collaboration between Al Qaeda and God's Aryan Militia (GAM)
4. helping Al Qaeda members maintain student status within the U.S.
5. smuggling people and objects from Mexico into the U.S.

The biological agents subplot is further broken down because each leg of the smuggling is carried out by a different set of actors that are *mostly* unaware of each other. Below is an example of the first intelligence snippet in the dataset.

`CIA-01 - A Russian named Igor Kolokov was arrested in Cairo on 29 January, 2003 and charged with assault on an Egyptian police officer who had attempted to arrest him for being drunk in public. Kolokov sells medical supplies throughout the Middle East and represents a company in Moscow called Medikat. A background check on Kolokov reveals that he was formerly an administrator at the Soviet institute Vector in Strizi, near Kirov in Russia. When he was arrested, Kolokov was carrying a card with a note on it reading [in Russian]: ''Safrygin for H. Q., Peshawar.''`

4.2 Four Children's Classics

Four children's classics (*Anne of Green Gables (AGG)*, *Black Beauty (BB)*, *Peter Pan (PP)*, and *Treasure Island (TI)*)[1] were broken into chapters and assembled into a single dataset. This dataset was merged with the *Atlantic Storm* dataset to add another layer of complexity in the experiments.

5 Experiments

The first experiment is nonsensical in that it creates absurd stories using *AS* bookends with the intention of inserting *AGG* chapters into stories about terrorist plots. However, the purpose is to understand the effect of varying η and ζ and to determine their optimal settings. Follow-on experiments show human agreement for the quality of stories generated with and without signal injection.

5.1 *Atlantic Storm* Stories with *Anne of Green Gables* Signal Injection

This experiment is designed to understand the behavior of varying η and ζ from Eq. (4). Chapters of *AGG* should not naturally occur in stories where the beginning and ending documents are from the *AS* dataset, but if signal injection is working properly then one should be able to force the injection of *AGG* chapters into the middle of the story. 200 bookend sets that have at least 2 *AGG* chapters

[1] www.gutenberg.org.

and 5 *AS* intelligence snippets between them were chosen at random from the *AS* dataset. Documents were measured as being between a particular bookend set, d_B and d_E, if $\mathcal{B}(d_B, d_i, d_E) == 1$ for a document d_i; i.e., the dispersion of the story chain,$\{d_B, d_i, d_E\}$, is one. The middle sets of documents for each story were generated for the bookends by maximizing Eq. (4) while varying Zipf's coefficient and the mixing proportion. Zipf's coefficient was varied to assign a probability mass of 0.1, 0.2, ... 0.8, 0.9 to a selected set of keywords from AGG, $K = \{$"*ann*", "*marilla*","*diana*","*matthew*"$\}$; the remaining probability mass is assigned randomly to the vocabulary of the corpus. These particular keywords were chosen because they are the four most common terms in AGG chapters; Anne is the main character, and Dianna and Matthew Marilla are her best friends. Zipf's coefficient controls the probability mass of the injection signal. The mixing proportion was varied from 0.0, 0.1, ... 0.9, 1.0. The mixing proportion controls the amount of weight associated to the bookends and the signal injection; $\eta = 0.0$ signifies that all of the weight is associated to the injection signal while $\eta = 1.0$ signifies that all of the weight is associated to the bookends.

A successful injection for this experiment is a story that contains documents from *AS* and AGG in the middle but does not contain documents from *BB*, *PP*, or *TI*. If documents from *AS* and AGG are present in d_M, then the storytelling algorithm has a mixing proportion, η, that is large enough to still emphasize the importance of $p_{B \cup E}$ to add *AS* documents in the middle, but has enough weight associated to p_Z to overcome the natural tendency to not add documents from AGG. The numerator of Eq. (4) is still independent of η and ζ and forces the story to maintain high dispersion, i.e., good story flow. While the stories formed are absurd, the value of Eq. (4) still yields an overall metric of how good the stories are.

Subfigure 1a displays the average final maximum value of Eq. (4) returned for the 200 stories generated at each parameterization of η and ζ. As expected, stories with $\eta = 1$, the dark red column on the right, have the maximum values of Eq. (4) and are all the same because ζ is not allowed to affect the generated stories. This is expected because the stories with AGG chapters cannot occur in the best stories for *AS* bookends that occur at $\eta = 1$. As η decreases, values of Eq. (4) are expected to decrease. This is verified in Subfigure 1a. Subfigure 1b displays the proportion of the 200 stories that contain at least one *AS* document in the middle. Stories without *AS* documents in the middle show where values of η are too low and the generated stories start becoming absolutely absurd. Subfigure 1c displays the proportion of the 200 stories that contain at least one AGG chapter in the middle. As expected, when ζ increases and η decreases, signal injection forces AGG documents into the middle. Chapters from *BB*, *PP*, or *TI* should not be inserted into the middle set of documents in any of the stories. Subfigure 1d shows the proportion of stories that do not include chapters from *BB*, *PP* or *TI*. The subfigure reaches its maximum value when $\eta = 0.0$ and $\zeta = 0.10$, as expected, because $p_{B \cup E}$ is having no effect, and 0.90 of the probability mass of the Zipfian distribution is randomly attached to the remainder of the vocabulary which includes terminology for all the chapters of

BB, *PP* or *TI*. Subfigure 1e is the multiplication of 1a, 1b, 1c, and 1d and shows where all four qualities combine in a positive way. Subfigure 1f shows that the top values of 1e for $\eta = \{0.5, 0.6, 0.7\}$ occur at $p_K = \{0.3, 0.6, 0.9\}$ respectively. These combinations of mixing proportion (η) and Zipf's coefficient (ζ) will be used for all other experiments.

5.2 Evaluating Human Generated *Atlantic Storm* Stories

Storytelling is qualitative in nature because of the heavy dependence upon user opinion and evaluation of quality. The opinions and abilities of users vary drastically. Controlling for the dynamic nature of users and story evaluation is necessary to garner insight into judging the overall quality of computer generated stories. From previous work [4], users generated stories for nine sets of bookends using subsets of data for specific subplots of the data. Five stories were generated for Bio-agents (subplot 1), two were generated for SCADA (subplot 2), and two were generated for GAM (subplot 3); this yields a total of nine human generated stories for each user. For control purposes, 5 of the users were asked to evaluate the 45 stories for quality. Each user personally generated 9 of the 45 stories. It is expected that most of the stories will be graded as having good quality and that users would certainly evaluate their own stories as good. None of the 45 stories were evaluated as *bad* by all 5 users. Only one story was evaluated as *bad* by 4 of 5 users; this is the only story that depended completely on an object (antibiotics) as the connecting thread between intelligence snippets; all other stories were curated to include repeated usage of people to make connections between documents. Oddly, users evaluated their personal stories as *good* only at an average of 0.644 (5.8 of 9). Remarkably, users evaluated others stories as *good* on average at 0.644 (23.2 of 36) with an overall average of 0.644 (29 of 45) stories being evaluated as *good*. Because of the variance in user evaluation, a conservative simple majority rule of user votes will be used to measure if a story is *good*. This is conservative because one could argue that if a minority of users finds a story to be *good*, then it should be considered *good*.

5.3 Comparing *Atlantic Storm* Stories with and Without Signal Injection

For the next experiment, bookends were selected based on connecting two characters from the *AS* dataset. Stories were generated for each set of bookends by optimizing Eq. (4) at four different configurations of η and ζ (function of p_K): $\{\eta = 1, p_K = NA\}$, $\{\eta = 0.5, p_K = 0.3\}$, $\{\eta = 0.6, p_K = 0.6\}$, and $\{\eta = 0.7, p_K = 0.9\}$. When $\eta = 1$, p_K and ζ have no effect on the stories; this configuration is equivalent to optimizing Eq. (2) for storytelling and is considered to be storytelling without signal injection.

Eight stories were generated at the four different configurations yielding 32 stories that were evaluated by twelve users. Users were asked to evaluate the stories based on story flow and content, but the overall judgment of story quality came from being able to generate a new hypothesis about the connection of two

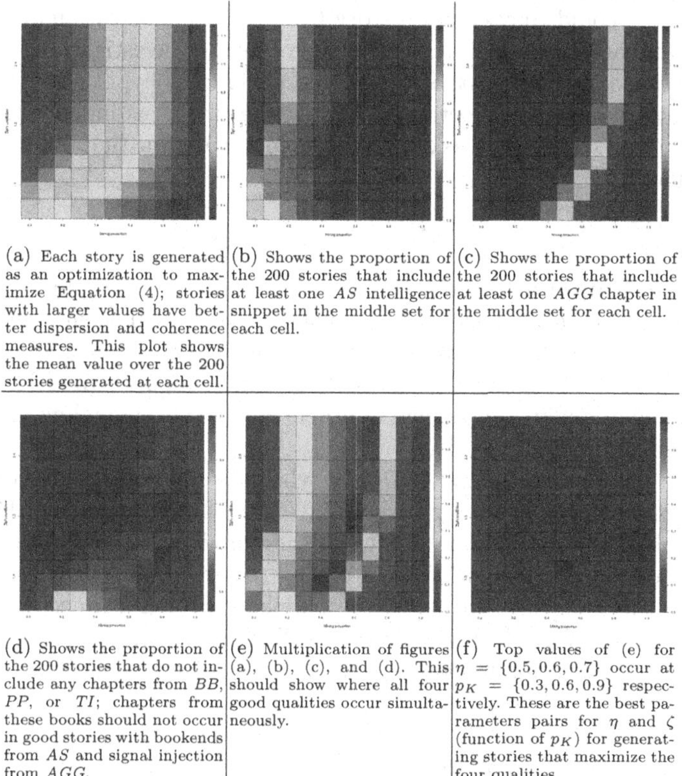

(a) Each story is generated as an optimization to maximize Equation (4); stories with larger values have better dispersion and coherence measures. This plot shows the mean value over the 200 stories generated at each cell.

(b) Shows the proportion of the 200 stories that include at least one *AS* intelligence snippet in the middle set for each cell.

(c) Shows the proportion of the 200 stories that include at least one *AGG* chapter in the middle set for each cell.

(d) Shows the proportion of the 200 stories that do not include any chapters from *BB*, *PP*, or *TI*; chapters from these books should not occur in good stories with bookends from *AS* and signal injection from *AGG*.

(e) Multiplication of figures (a), (b), (c), and (d). This should show where all four good qualities occur simultaneously.

(f) Top values of (e) for $\eta = \{0.5, 0.6, 0.7\}$ occur at $p_K = \{0.3, 0.6, 0.9\}$ respectively. These are the best parameters pairs for η and ζ (function of p_K) for generating stories that maximize the four qualities.

Fig. 1. 200 bookend sets that have at least 2 *Anne of Green Gables* chapters and 5 *Atlantic Storm* intelligence snippets between them were chosen at random from the *AS* dataset. The middle sets of documents for each story were generated for the bookends by maximizing Eq. (4) while varying Zipf's coefficient and the mixing proportion. Zipf's coefficient was varied to assign a probability mass of 0.1, 0.2, ... 0.8, 0.9 to a selected set of keywords from *AGG* ("anne", "marilla", "diana", "matthew"); the y-axis (vertical columns) vary with Zipf's; the remaining probability mass is assigned randomly to the vocabulary of the corpus. Zipf's coefficient controls the probability mass of the injection signal. The mixing proportion was varied from 0.0, 0.1, ... 0.9, 1.0; the x-axis (horizontal rows) vary with mixing proportion. The mixing proportion controls the amount of weight given to the bookends and the injection; 0.0 signifies that all of the weight is associated to the injection signal while 1.0 signifies that all of the weight is associated to the bookends. Each subfigure shows results for the 200 stories generated for each cell (parameterized by Zipf's coefficient and mixing proportion). In all cases *red* is considered to be a good quality while *blue* is considered to be bad. The middle of each story should contain documents from *AS* and *AGG* and not contain documents from *BB*, *PP*, or *TI*. Good stories should maximize Eq. (4). The best stories should contain all four good qualities. Figure (e) is the multiplication of (a), (b), (c), and (d) and shows where all four qualities combine in a positive way. Figure (f) shows that the top values of (e) for $\eta = \{0.5, 0.6, 0.7\}$ occur at $p_K = \{0.3, 0.6, 0.9\}$ respectively. These combinations of mixing proportion (η) and Zipf's coefficient (ζ) will be used for all other experiments. (Color figure online) characters, i.e., could they create a story from the chain to rationalize that the two characters are involved in a plot worthy of further investigation. Users were asked to judge a story as *good* if they could generate a hypothesis. Users were not made aware of signal injection and rated the stories as independent experiments. The users were also asked to make comments on the factors affecting their judgments.

The twelve bookend sets involved:

Connecting Ramundo Ortiz to Jose Escalante – Ramundo Ortiz and Jose Escalante are both part of the smuggling biological agents subplots 1b and 1c. Escalante is directly working with smuggling the items, while Ortiz is not involved with smuggling but provides financial support. One of Escalante's activities includes making payments to others involved in smuggling. The money for the payments is from Ortiz.

Another *AS* character is Pieter Dopple who cuts and polishes precious stones to be sold to finance illegal activities. Many of the activities in the *AS* dataset are financially supported by the selling of these polished stones. Pieter Dopple is unaware of the particulars of the terrorist plots, but is a key person when considering the financial aspects of the subplots. Stories connecting Ortiz to Escalante should involve financial overtones. For this reason, the injection signal is set to information about Dopple, and $K = \{$"*pieter*", "*dopple*", "*diamond*", "*antwerp*","*belgium*"$\}$.

Three sets of bookends were used with four different parameter configurations to generate 12 stories which are coded as DD-1, DD-2, and DD-3 in Fig. 2.

Connecting Adnan Hijazi to Omar Hanif – Adnan Hijazi is part of the student subplot 4 and helps financially support students to stay in the U.S. and conduct activities to support terrorism. Omar Hanif smuggles the biological agents from Nassau to Freeport in the Bahamas as part of subplot 1b. None of the snippets directly show that Hanif and Hijazi are aware of each other, but a snippet exists for each of them that states they visited the same jewelry shop in Cairo, Egypt on the same day. Neither snippet mentions the other's involvement. Two of the students that are being funded by Hijazi, Wali Shakur and Hamza al Midhar, *apparently* received the bio-agents from Hanif and smuggled them back into the U.S. aboard the *Viking Holiday* cruise ship; the phrase *"apparently"* is used here because none of the snippets directly corroborate Wali and Hamza receiving the package from Hanif; it is only the analyst that can make the leap of intuition that this is true. The injection signal is set to information about the two students connecting Hijazi to Hanif, and $K = \{$"*wali*", "*shakur*", "*hamza*", "*midhar*", "*viking*", "*holiday*"$\}$.

As above, three sets of bookends were used; 12 of the 32 stories are coded as HH-1, HH-2, and HH-3 in Fig. 2.

Connecting Ali Sufaat to Pyotr Safrygin – Ali Sufaat is a ship-hand and is involved in subplot 1b, smuggling bio-agents from Santo Domingo to Nassau. Pyotr Safrygin was formerly director of security at the Vector institute, a Russian biological research laboratory, and is involved in subplot 1a, smuggling bio-agents from Russia to Santo Domingo. Based on all of the snippets of information, Sufaat and Safrygin are completely unaware of each others existence or involvement. Carlos Morales is aware of both characters and is involved in orchestrating and executing smuggling subplot 1a. For this reason, the injection signal is set to information about him, and $K = \{\text{"}carlos\text{"}, \text{"}morales\text{"}\}$. Injecting this signal should allow the algorithm to find *better* stories connecting Sufaat to Safrygin.

Two sets of bookends were used; 8 of the 32 stories are coded as SS-1 and SS-2 in Fig. 2.

Results – Because the stories were judged by twelve users independently of the knowledge that signal injection had occurred, a simple metric of improved average quality can be used to test if signal injection improved a story. If a story with signal injection has a higher proportion of *good* judgments over the story from the same bookends without signal injection, it is assumed that signal injection improved the quality of the story. Figure 2 shows the proportion of stories that were rated as *good* at the four different parameter configurations by the twelve users. Based on the proportion of user judgments being greater than 0.5, stories without signal injection (parameter configuration $\{\eta = 1.0, p_K = NA\}$), 5 of 8 stories can are judged as *good*. For configuration $\{\eta = 0.7, p_K = 0.9\}$, 7 of 8 stories are assessed as *good*. For $\{\eta = 0.6, p_K = 0.6\}$ and $\{\eta = 0.5, p_K = 0.3\}$, 6 of 8 stories are *good*. Based on this simple metric, all configurations with signal injection can be considered to out perform the storytelling algorithm without injection. For configuration $\{\eta = 0.7, p_K = 0.9\}$, 6 of the 8 story sets have a larger proportion of *good* quality judgments than the story without signal injection; 1 of 8 resulted in a tie with a proportion of 0.666 being judged as *good* for story set DD-2. For $\{\eta = 0.5, p_K = 0.3\}$, 5 of 8 have a higher proportion with a single tie occurring again on story set DD-2. For $\{\eta = 0.6, p_K = 0.6\}$, 5 of 8 have a higher proportion of *good* than stories without signal injection with two ties for DD-2 and SS-2. The one story set where they all failed to have a better proportion of *good* judgments than the story without signal injection is DD-1. The comments from the users who rated DD-1 stories with signal injection as *bad* revolved around the story either being out of order, having documents that were extraneous, or both. The algorithm cannot decide to change the beginning or ending document; if a document should have been injected before the beginning or after the ending document, the algorithm is forced to inject the document in the middle if the story can maintain good dispersion. The injection signal for DD-1 incentivized the algorithm to add documents about Pieter Dopple, but the injected document had to be in the middle. The problem is that the best Pieter Dopple document that goes with the beginning and ending document is a piece of information that the users might have preferred to be before the beginning document. A new story was manually generated by sorting the document chain to begin with the Dopple document forcing the original document to be the second document in the chain. Users were asked to evaluate this story. Eleven of the twelve (0.917) users rated this story as *good* which is higher than the eight of twelve that the story without signal injection was rated as *good.* With this reevaluation, parameter configuration $\{\eta = 0.7, p_K = 0.9\}$ had a higher proportion of *good* judgments than the story without signal injection. This confirms that stories can be focused with signal injection by analysts to produce better stories that are more in line with what the user expected.

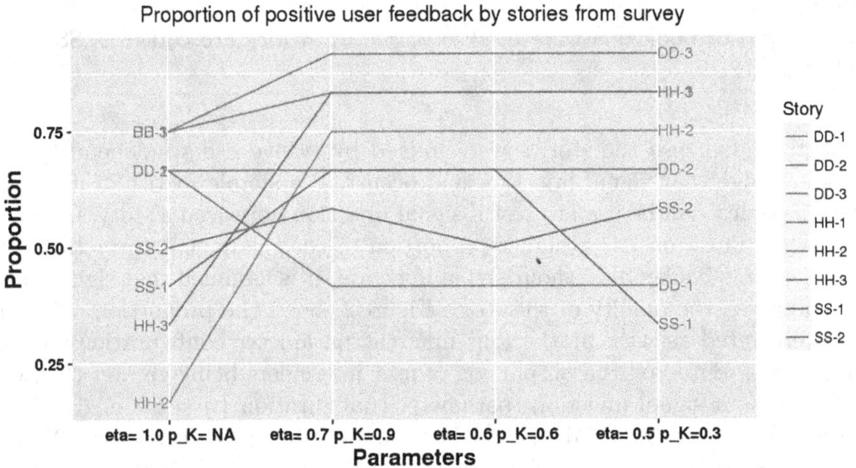

Fig. 2. Shows the proportion of users that judged a story as having *good* quality. Eight stories (DD-1, DD-2, DD-2, HH-1, HH-2, HH-3, SS-1, and SS-2) were generated without signal injection with parameter configuration $\{\eta = 1.0, p_K = NA\}$. The bookends used to generate these stories were also used to generate stories with signal injection at parameter configurations $\{\eta = 0.7, p_K = 0.9\}$, $\{\eta = 0.6, p_K = 0.6\}$, and $\{\eta = 0.5, p_K = 0.3\}$. For configuration $\{\eta = 0.7, p_K = 0.9\}$, 7 of the 8 story sets have a larger proportion of *good* quality judgments than the story without signal injection. For $\{\eta = 0.6, p_K = 0.6\}$ and $\{\eta = 0.5, p_K = 0.3\}$, 6 of 8 are better. The one story set where they all failed to have a better proportion of *good* judgments is DD-1.

6 Conclusions

We have shown that a storytelling algorithm with signal injection out performs the same algorithm without injection. This is measured based upon user judgments independent of knowledge of the injected signals. Overall, users judged stories with signal injection as having better quality than stories without injection. We further conclude that storytelling algorithms can greatly benefit from analyst's intuition and experience through signal injection. A storytelling algorithm with signal injection allows an analyst to focus a story with their personal knowledge.

Acknowledgements. This work was supported by the In-house Laboratory Independent Research (ILIR) program, the Navy Innovative Science and Engineering (NISE) program, Missile Defense Agency (MDA) contract HQ0147-17-C-7605, and the Office of Net Technical Assessment (ONTA) within the Office of the Assistant Secretary of Defense for Research and Engineering (ASD R&E).

References

1. Kahneman, D., Tversky, A.: Prospect theory: an analysis of decision under risk. Econometrica **47**(2), 263–291 (1979)
2. Shahaf, D., Guestrin, C.: Connecting the dots between news articles. In: Proceedings of the 16th ACM SIGKDD International Conference on Knowledge Discovery and Data Mining, KDD 2010, pp. 623–632. ACM, New York (2010)
3. Hossain, M.S., Gresock, J., Edmonds, Y., Helm, R., Potts, M., Ramakrishnan, N.: Connecting the dots between pubmed abstracts. PLoS ONE **7**(1), e29509 (2012)
4. Rigsby, J., Barbará, D.: Automated storytelling evaluation and story chain generation. In: 2017 IEEE 17th International Conference on Data Mining Workshops (ICDMW), November 2017
5. Shahaf, D., Guestrin, C., Horvitz, E.: Trains of thought: generating information maps. In: Proceedings of the 21st International Conference on World Wide Web, WWW 2012, pp. 899–908. ACM, New York (2012)
6. Shahaf, D., Guestrin, C., Horvitz, E.: Metro maps of science. In: Proceedings of the 18th ACM SIGKDD International Conference on Knowledge Discovery and Data Mining, KDD 2012, pp. 1122–1130. ACM, New York (2012)
7. Shahaf, D., Yang, J., Suen, C., Jacobs, J., Wang, H., Leskovec, J.: Information cartography: creating zoomable, large-scale maps of information. In: Proceedings of the 19th ACM SIGKDD International Conference on Knowledge Discovery and Data Mining, KDD 2013, pp. 1097–1105. ACM, New York (2013)
8. Hossain, M.S., Andrews, C., Ramakrishnan, N., North, C.: Helping intelligence analysts make connections. In: Scalable Integration of Analytics and Visualization, Papers from the 2011 AAAI Workshop, San Francisco, California, USA, 07 August 2011
9. Jelinek, F., Mercer, R.L.: Interpolated estimation of Markov source parameters from sparse data. In: Proceedings of the Workshop on Pattern Recognition in Practice (1980)
10. Piantadosi, S.T.: Zipf's word frequency law in natural language: a critical review and future directions. Psychon. Bull. Rev. **21**(5), 1112–1130 (2014)
11. Zipf, G.: The Psychobiology of Language: An Introduction to Dynamic Philology. M.I.T Press, Cambridge (1935)
12. Wu, H., Vreeken, J., Tatti, N., Ramakrishnan, N.: Uncovering the plot: detecting surprising coalitions of entities in multi-relational schemas. Data Min. Knowl. Disc. **28**(5–6), 1398–1428 (2014)
13. Hughes, F.J.: Discovery, proof, choice: the art and science of the process of intelligence analysis, case study 6, 'all fall down' (2005, unpublished manuscript)

Finding Active Expert Users for Question Routing in Community Question Answering Sites

Pradeep Kumar Roy[1], Jyoti Prakash Singh[1(✉)], and Amitava Nag[2]

[1] National Institute of Technology Patna, Patna, Bihar, India
pkroynitp@gmail.com, jps@nitp.ac.in

[2] Central Institute of Technology, Kokrajhar, Kokrajhar, Assam, India
amitavanag.09@gmail.com

Abstract. Community Question Answering (CQA) sites facilitate users to ask questions and get answered by fellow users interested in the topic of the question. A vast number of questions are posted on these sites every day. Some questions receive numbers of good quality answers whereas some questions fail to attract even a single answer from the community users. Also, some questions receive very late answers. The problem behind the unanswered question or late answers was that the question was not seen or not routed to the expert user or interested users. There are no identified experts of given topic on these sites. Hence, finding users who will be interested in answering a question of the specific topic and sending the question to that user is a challenging task. We have developed a system to identify the group of users who can potentially be the answerer of a given question. The group of users is identified using their past question and answers. We rank the users of the identified group considering their answering behaviour, time of posting their answers etc. The proposed methodology has several advantages such as routing questions to recently active users and at the time of their convenience. Experimental analysis shows that to get at least one answer, the question must be routed to at least eight answerers.

Keywords: Community Question Answer · Machine learning
Question routing · Expert finding

1 Introduction

Community Question Answering (CQA) sites such as Yahoo! Answers, Stack Exchange, and Stack Overflow, etc. are proved to be a good platform for the users to exchange their knowledge [3,4,10]. Every CQA sites have their own users which respond to the questions asked by other. Generally, the users search the answers to their question on Web search engines such as Google, Yahoo, Bing, etc. The search engine provides numbers of the different link for the search query, users may visit all, to search the answers to their question. But, the major issue is,

P. Perner (Ed.): MLDM 2018, LNAI 10935, pp. 440–451, 2018.
https://doi.org/10.1007/978-3-319-96133-0_33

there is no guarantee that the link contains the answers what the user is looking for. Here, CQA sites won the users faith as they can provide the good quality answers within the limited time [2,4]. CQA sites are used by a huge number of users to find answers to complex, subjective, or context-dependent questions. One of the major objectives of CQA systems is to give the most appropriate answers in the shortest possible time [2]. This is usually done by routing a new question to expert users [6,7,15]. This routing strategy varies from site to site. To provide fast and more accurate answers, the CQA system needs to route the question to proper experts who are also active [9]. To do so, the features such as the users reputation is considered as one of the major features. However, this introduces the bias in choosing an expert of a topic. Since it is not necessary that a user with more reputation point is more expert than the user having less reputation point. Because by answering the low-quality questions also a user get more number of reputation point.

In the current scenario, it might take hours or days a question in a CQA site before getting answers. This is because even a user knows well about a specific question, may not answer the question because they may not visit the system often, or the user might be confronted with many open questions. Then again, a user who answers a question may simply happen to see the question, however, is not a specialist on the question's subject. So, it might be the better option that a question is recommended to the users who are experts of the question domain as well as interested in answering them. This makes the system more reliable as the quality answers come in time. So, the main objective of this research work is to develop a method to find expert users then route questions to those experts in such a way that asker gets the best answer in the least time. Also, answerer gets only those questions to answer for which he is most capable of answering.

Rest of the paper organized as follows: Sect. 2 is the related works, the proposed method is discussed in Sect. 3. Result and discussion present in Sect. 4. Section 5 is concluding our work.

2 Related Work

Several work has been proposed by the researcher on question routing in CQA system by considering the user's authority, users reputation and other textual and non-textual features [6,10,11,15,16,18]. Dror et al. [3] addressed the problem of routing the questions as a classification problem. He finds whether a specific question will be exciting for a expert user or not. To solve this issue, they used several features from the question and answers dataset. Ji and Wang [5] design a ranking based model which consists of two different approaches, (i) Support Vector Machine (SVM) and (ii) Rank Support Vector Machine. In both the approaches, the features of questions, users, and the question-users relationship are used. SVM is only capable to classify the problem in two classes such as a user is either expert of not expert. The next approaches used to find the probability of the user expertise. The results of the Rank SVM model is used to rank the users. Yang et al. [14] considered user authority as a good predictor, as it might be possible that the authoritative user gives more authentic answers. To find the appropriate expert of a question, most of the existing work consider the best answerer or the answerer who answer the questions [4,17]. However, the list of user might not be the experts of the questions related topic. For example, as soon as a high-quality answer is provided by the experts, the other experts simply encourage the answerer by giving their opinion in term of votes. A positive vote is a good indicator that the answer is of good quality, whereas negative votes indicates the answer is not up to mark or invalid. So, most of the experts vote the posted answer and move towards unanswered questions.

Liu et al. [8] used the Yahoo! Answers dataset to finds the expert users of a topic by considering the user's reputation, their relevant subject, and the authority. Subject relevancy tells about how deep they know about the subject with respect to the target question. The reputation of a user is evaluated based on their past answers and the votes obtained by them. While the authority of the user is found from the link analysis. Li and King [7] used the query likelihood language model to propose a methodology for question routing problem, which includes user activity that predicts whether a user gives the answer in time or not and the user expertise. The expertise was evaluated from the answers already posted by them. Most of the previous works used the word matching and hence they are unable to detect the semantic. They are unable to solve the problem of the lexical gap between the user's profile and the questions [9]. To solve this issue, LDA (Latent Dirichlet allocation) or PLSA (probabilistic Latent Semantic Analysis) are used which effectively measure the similarity of the text syntactic and semantically. Xu et al. [13] said that CQA users play two different characteristics at the same time. First, when they give an answer to a topic, they are evaluated as an expert on the question's topic. Second, while asking a question to community user, sensed as lack of expertise of the question's topic.

3 Research Methodology

The primary stages of a question routing mechanism are shown in Fig. 1. When a user posts a question, the question routed to the experts by evaluating the content of the question. From the question content, first, the topic of the question was identified, then the system searches the experts for them. Then the question was routed to them. The expert users check the routed question content and if they found the question is answerable, provide the answer else leave them as it is. Even though, a large number of questions was unable to fetch a single answer from the peer users, one of the major reason behind that the question was not routed to proper experts [1].

To overcome the above issues, we propose here an algorithm for finding the potential answerer called as expert users for a given question. To achieve this, we start by creating user profiles from his data available on the CQA site. Then the question tags are grouped together, which are from the same topic. We then use the question tags and users profile to find the probable set of expert users who can be potential answerers. We then evaluate each user from the probable set and top k users are declared as experts. The methodology is shown in Fig. 2.

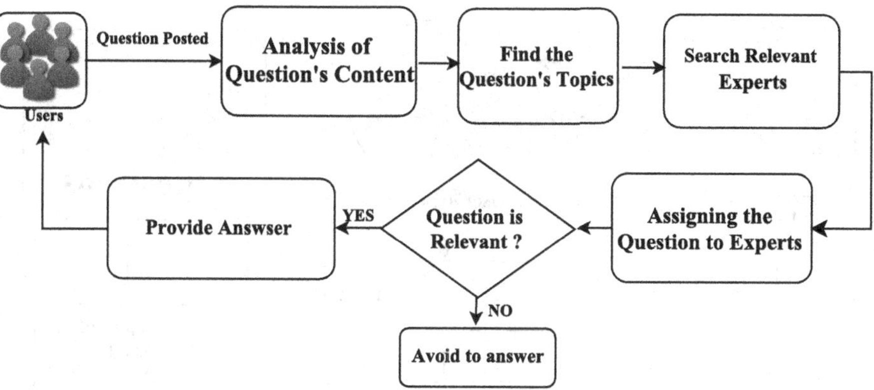

Fig. 1. Question-Answering process in a CQA site

3.1 Dataset

We have downloaded the dataset from the web link https://archive.org/details/stackexchange [12]. The dataset is in XML format contained several files. We select two xml file from them (i) Post.xml and (ii) Users.xml. From the selected xml files, the relevant attributes were extracted. The statistics of the dataset presented in Table 1.

Table 1. Statistics of programmer dataset

Number of users	210420
Number of questions	38314
Number of answers	126775
Number of tags	920

3.2 Creating User Profile

Every user in the CQA system is given a user profile which is unique. The user profile is made from the available data with the user. This data includes user's public profile which he sets up when registering on the CQA site. We also use user's quantitative data such as user's activity like number of questions asked and answers provided, number of questions and answers interacted with. The user profile of every user has following parts:

- **User ID:** This very ID is used to identify the user uniquely.
- **Number of accepted answers:** The total number of accepted answers given by that user till then.

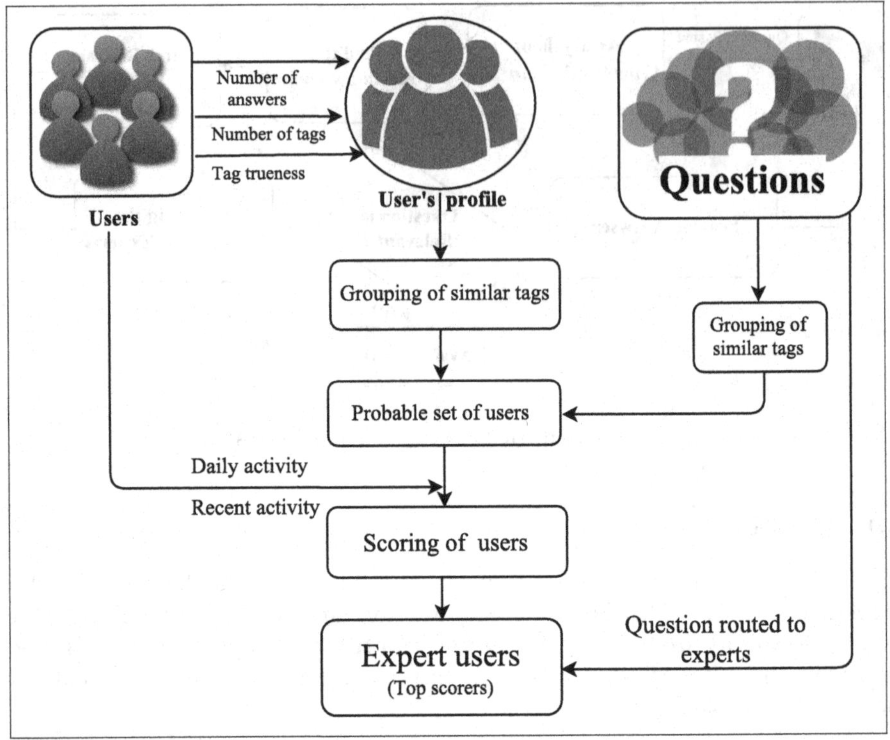

Fig. 2. Workflow of proposed question routing process

- **Number of topics of accepted answers:** The total number of unique topics the user has given any accepted answers in. This measure the width of user's expertise.
- **Number of accepted answers in each topic:** This gives a vector of numbers for each user showing the relative expertise of that user on every topic.
- **Tags Trueness (*TT*):** This attribute is the measure of user's tendency to answer questions belonging to user profile tags only. If a user answers more questions with his profile tags only then the Tags Trueness value is going to be high and vice-versa. Tag trueness is calculated by Eq. 1, a small view of calculated tag trueness value was shown in Table 2.

In Table 2, First column represents the individual users id, Second column is the Tag Trueness the value that we calculated using Eq. 1, the third column is the number of answers given by the users, and last column is the number of *Tags* of the users.

$$TT = \frac{Total\ number\ of\ ACPT}{Total\ number\ of\ AA} \tag{1}$$

Where, $ACPT$ = Accepted answers of profile tag, AA = Accepted answer.

Table 2. A sample of calculated tag trueness value

Sl. No.	User Id	Tags trueness	No. of answers	No. of tags
1	1204	0.952018279	1313	357
2	3965	0.918404352	1103	280
3	6605	0.953571429	840	327
4	9113	0.963201472	1087	314
5	51654	0.956749672	763	204
6	7422	0.925039872	627	220
7	935	0.825221239	452	148
8	4	0.949814126	538	184
9	1352	0.867647059	408	139
10	12828	0.939252336	428	165

3.3 Grouping Similar Tags

In a CQA environment, there are a lot of different tags. Some of the tags looks different but have same roots and are similar in some sense. We used this idea that if a user is expert in a topic than that user can also be an expert in topics similar to it. So, all the tags were grouped as per their similarity and because of that the effective number of tags also decreased leading to a little lesser computation and storage for the system. In our selected dataset the total number of different tags was 920 before grouping, this was decreased to 639 after grouping the similar tags together. This of similar tags was done manually at the discretion of a five member team.

3.4 Finding Probable Set of Users

We find a set of users who most probable to experts for a given question. A user is included in the set of probable users, if the user has given at least one accepted answer in at least one tag associated with that question. A user who has never before given even a single answer in any tag of that question is very unlikely to answer a new question of that tag. Because every user is not having knowledge in all the tags, so we are left with relatively very less number of users to process in further steps. The size of this probable set of users ranges from less than ten to a few hundred.

3.5 Scoring User

We have got a probable set of user. Now, we have to process these users to find the set of experts for the question. For this purpose, we assign a score to each user in the probable set of users. The higher the score, the more is the expertise of the user for that question. We use Eq. 2 for weight assignment which works as follows:

$$Score = TT * log(NPT + 10) * log(NAA + 10) * AACT * RAM \tag{2}$$

Where TT = Tag Trueness, NPT = Number of profile tags, NAA = Number of Accepted Answers, $AACT$ = Average number of Accepted Answers in Common Tags and RAM = Recent Activity Measure. The detailed description of these parameters is follow:

- Tags Trueness (TT): Tag trueness is the measure of tendency of a user to give answers that are have tag matching the profile tag. Its is calculated using Eq. 1. The higher the value of the TT value of the user, more probably that the user will answer the question.
- Number of profile tags (NPT): Number of profile tags are the measure of the width of user's knowledge. Higher the number of user's profile tags, higher is the possibility that the user will answer the question.
- Number of accepted answers (NAA): It is the count of the number of answers given by user that has been marked accepted by the questioners. The higher number of accepted answers means that the user. It is knowledgeable, so it is most probable to give an acceptable answer to the routed question too.
- Average number of accepted answers in common tags ($AACT$): Here common tags refers to the tags which are associated with the questions and are also present in the user's profile tags. Since a user may not previously have answered questions in all the tags associated with that question. Hence, average number of accepted answers takes care of it. It is a measure of the average number of questions answered by the user in the common tags, giving us a measure of the user's expertise in the question tags which are common to the user's profile tags.

  Average number of answers in common tags is calculated by Eq. 3

  $$Avgerage\ AACT = \frac{Number\ of\ AACT}{Number\ of\ CT} \tag{3}$$

  Where, CT = Common tags
- Recent activity measure (RAM): This parameter measures the history of a user has been answering questions. If a user has recently answered some questions of those tags then he has given priority over a user who has given more number of answers a long time ago. To measure this value we divide the total number of accepted answers of question tags given by the user until today in four parts by time: (i) Segment 0: Today to 3 months ago, (ii) Segment 1: 3 months ago to 6 months ago, (iii) Segment 2: 6 months ago to 12 months ago, (iv) Segment 3: 12 months ago to user's registration date.

  We assign weights to each of these four categories, giving higher weight to recent category and lower to old ones. We have assigned the following weights to these categories $[w_0, w_1, w_2, w_3]$ where $w_0 > w_1 > w_2 > w_3$ and $w_0 + w_1 + w_2 + w_3 = 1$. The higher weight given to more recent category, as the users that have been active more recently should get higher score. This measure makes a balance of trade between a number of accepted answers given by the user and how recent these answers are given. Higher the number of accepted answers, more knowledge the user has. More recent answers mean the user is more active these days for particular tags and is more probable to answer a new question on that tag. Recent activity measure (RAM) for a user is calculated using the Eq. 4:

$$RAM = \frac{a_0 w_0 + a_1 w_1 + a_2 w_2 + a_3 w_3}{Number\ of\ accepted\ answers} \tag{4}$$

Where RAM = Recent Activity Measure, a_0, a_1, a_2, and a_3 are number of accepted answers given in segment 0, 1, 2, and 3 respectively. w_0, w_1, w_2, w_3 are weight assigned to segment 0, 1, 2, and 3 respectively.

Given a question and a user, based on the expertise of the user to answer that question, and the activity of the user, *Score* is calculated. This *Score* is associated with each user. The higher the *Score* more appropriate it is to route the question to the user at the time the question is asked.

3.6 Getting Experts

Now we take into account day activity measure of ranked users. One of our purpose for this system is for users to get their answers quickly. Day activity takes care of this problem. On observing the activity of users on the CQA site from the data, we found that there is a pattern of the activity for every user based on the part of the day they are active on. Some users are more active in morning, some at night while some users are most active in evenings.

We have divided a day of 24 h into five parts based on the times of day, such as: *Morning*: 6am to 12pm, *Afternoon:* 12pm to 4pm, *Evening*: 4pm to 20pm, *Night*: 08pm 12am, and *Late night*: 12am to 6am. Also, there are trends in different days of the week. So, to capture the trend for every user we find the number of answers answered and questions asked on each weekday and in each segment of a day as shown in Fig. 3. In Fig. 3, the userId is the id of users, mor_0 is the morning of Monday, and an_0 is the afternoon of Monday, mor_1 represents the morning of Tuesday, an_1 is afternoon of Tuesday. Now, when a question is asked, its time and weekday is noted and is checked which segment of the day it is. From the ranked set of users, priority is given to those users which are more active in that day segment of the weekday. Now from these users, top ten users are taken as experts and the question is routed to these users. This way we get the experts we are required to route to.

4 Result and Discussion

The average User Topic Reputation (UTR) and Mean Reciprocal Rank (MRR) measure are used as metrics to evaluate this question routing model. Average User Topic Reputation metric gives a measure of user's reputation at answering

	userId	mor_0	an_0	eve_0	night_0	laten_0	tot_0	week_0	mor_1	an_1	...	laten_5	t
0	1204	4	29	78	78	23	212	2	6	49	...	39	
1	3965	6	56	92	51	21	226	2	9	50	...	15	
2	6605	44	20	25	18	10	117	0	46	24	...	24	
3	9113	57	42	37	24	7	167	0	80	54	...	5	
4	51654	4	43	35	28	15	125	1	4	32	...	19	
5	7422	53	26	9	4	2	94	0	58	43	...	0	
6	935	8	19	26	30	11	94	3	6	17	...	14	
7	4	11	50	25	18	14	118	1	12	47	...	7	
8	1352	16	11	8	13	13	61	0	17	22	...	12	
9	12828	20	14	15	16	8	73	0	18	26	...	14	
10	60357	5	12	10	13	3	43	3	9	19	...	1	

Fig. 3. A sample of users daily activity

a question based on their history of answering similar questions. User Topic Reputation can be calculated from Eq. 5 and the average user topic reputation can be calculated by Eq. 6.

$$UTR\ (q, u_i) = \frac{Question\ tag \cap User_i\ tag}{User_i\ tag} * User_i\ reputation \tag{5}$$

Where q is a new question with tag, and i = 1 to $|U|$ (total number of users).

$$Average\ UTR = \frac{\sum_{i=1}^{|U|} UTR\ (q, u_i)}{|U|} \tag{6}$$

We used 1924 questions for testing purpose. For every test question, we find the average User Topic Reputation of actual users who answered that question, and users which our model predicted as experts. We find that in the case of 1648 questions, the average User Topic Reputation score for predicted users was more than that of actual users who answered the questions. This sums to 85.65% of cases, in which the predicted users had better average UTR than that of actual users.

We calculate the Mean Reciprocal Rank (MRR) defined in Eq. 7. It is the average of reciprocal ranks of the predicted experts for all the questions in the test set.

$$MRR = \frac{\sum_{i=1}^{|Q|} \frac{1}{rank_i}}{|Q|} \tag{7}$$

Here $rank_i$ is the position of actual answerer in the predicted set of users, when the predicted set of users are arranged in descending order of $Score$. The calculated MRR for test set came out to be 0.1269. This means that our model needs to predict on an average ≈8 experts, so that every question is routed to the correct user.

The average Question Answering Time (QAT) measures the mean time in which the set of users answer the question when asked. The user answering time can be calculated as in Eq. 8 and average users answering time can be calculated from Eq. 9.

$$QAT\ (q, u_i) = Time_a - Time_q \tag{8}$$

Where i = 1 to $|U|$ QAT = Question Answering Time, $Time_a$ = Time of answering question q by user u and $Time_q$ = Time of posting of question q.

$$Average\ QAT = \frac{\sum_{i=1}^{|U|} QAT(q, u_i)}{|U|} \tag{9}$$

It was found that on an average 36.17% of the time the Question Answering Time (QAT) of predicted users is lower than the Question Answering Time (QAT) of actual answerers. This means that 36.17% of the time the average response time of the users routed by our algorithm, is lower than the actual answering time of the question. We also found that out of predicted set of users, at least one user is the actual answerer 29.18% of the time. This may seem low, however, in most of the cases the users with no history of answering a particular type of question, have answered the question. And these answers have been accepted, even though such users cannot be considered as experts. This is due to CQA sites provide the reliability to the users, to answers any question without any expertise of the question's topic.

5 Conclusion

We devised a strategy to route questions to users who are most suitable to answer them based on their past and recent activity on the CQA sites. Our method relied on the previous as well as current question and answer activity performed by the users. We matched a question's tags with user's tags to identify which users are most suitable to answer that question. We also took into account, user's activeness on the CQA site. Our model weighted those users more who had given more answers recently. This way we made sure that users who were more active recently, were routed more questions because they seemed to be more willing to answer them. We also tracked a day wise activity of users recording in which part of the day the user was more active so that whenever a new question is asked we routed it to the users who were active at that time only. This way we were able to successfully route questions to expert users in a CQA site. The limitation of the proposed system is we have tested our system on a single dataset. Also, the system was not tested with the dataset of another CQA system like Yahoo Answers. The question routing problem involve ranking the experts users, this may be obtained more accurately by using the deep learning model.

References

1. Asaduzzaman, M., Mashiyat, A.S., Roy, C.K., Schneider, K.A.: Answering questions about unanswered questions of stack overflow. In: 2013 10th IEEE Working Conference on Mining Software Repositories (MSR), pp. 97–100. IEEE (2013)
2. Chua, A.Y., Banerjee, S.: So fast so good: an analysis of answer quality and answer speed in community question-answering sites. J. Assoc. Inf. Sci. Technol. **64**(10), 2058–2068 (2013)
3. Dror, G., Koren, Y., Maarek, Y., Szpektor, I.: I want to answer; who has a question?: Yahoo! answers recommender system. In: Proceedings of the 17th ACM SIGKDD International Conference on Knowledge Discovery and Data Mining, pp. 1109–1117. ACM (2011)
4. Guo, J., Xu, S., Bao, S., Yu, Y.: Tapping on the potential of q&a community by recommending answer providers. In: Proceedings of the 17th ACM Conference on Information and Knowledge Management, pp. 921–930. ACM (2008)
5. Ji, Z., Wang, B.: Learning to rank for question routing in community question answering. In: Proceedings of the 22nd ACM International Conference on Information & Knowledge Management, pp. 2363–2368. ACM (2013)
6. Jurczyk, P., Agichtein, E.: Discovering authorities in question answer communities by using link analysis. In: Proceedings of the Sixteenth ACM Conference on Information and Knowledge Management, pp. 919–922. ACM (2007)
7. Li, B., King, I.: Routing questions to appropriate answerers in community question answering services. In: Proceedings of the 19th ACM International Conference on Information and Knowledge Management, pp. 1585–1588. ACM (2010)
8. Liu, D.R., Chen, Y.H., Kao, W.C., Wang, H.W.: Integrating expert profile, reputation and link analysis for expert finding in question-answering websites. Inf. Process. Manag. **49**(1), 312–329 (2013)
9. Liu, M., Liu, Y., Yang, Q.: Predicting best answerers for new questions in community question answering. In: Chen, L., Tang, C., Yang, J., Gao, Y. (eds.) WAIM 2010. LNCS, vol. 6184, pp. 127–138. Springer, Heidelberg (2010). https://doi.org/10.1007/978-3-642-14246-8_15
10. Roy, P.K., Ahmad, Z., Singh, J.P., Alryalat, M.A.A., Rana, N.P., Dwivedi, Y.K.: Finding and ranking high-quality answers in community question answering sites. Glob. J. Flex. Syst. Manag. **19**(1), 53–68 (2018)
11. Saumya, S., Singh, J.P., Baabdullah, A.M., Rana, N.P., Dwivedi, Y.K.: Ranking online consumer reviews. Electron. Commer. Res. Appl. **29**, 78–89 (2018)
12. stackexchange.com, January 2017. https://archive.org/details/stackexchange
13. Xu, F., Ji, Z., Wang, B.: Dual role model for question recommendation in community question answering. In: Proceedings of the 35th International ACM SIGIR Conference on Research and Development in Information Retrieval, pp. 771–780. ACM (2012)
14. Yang, L., Qiu, M., Gottipati, S., Zhu, F., Jiang, J., Sun, H., Chen, Z.: CQArank: jointly model topics and expertise in community question answering. In: Proceedings of the 22nd ACM International Conference on Information & Knowledge Management, pp. 99–108. ACM (2013)
15. Zhao, T., Bian, N., Li, C., Li, M.: Topic-level expert modeling in community question answering. In: Proceedings of the 2013 SIAM International Conference on Data Mining, pp. 776–784. SIAM (2013)
16. Zhao, Z., Zhang, L., He, X., Ng, W.: Expert finding for question answering via graph regularized matrix completion. IEEE Trans. Knowl. Data Eng. **27**(4), 993–1004 (2015)

17. Zhu, H., Chen, E., Cao, H.: Finding experts in tag based knowledge sharing communities. In: Xiong, H., Lee, W.B. (eds.) KSEM 2011. LNCS (LNAI), vol. 7091, pp. 183–195. Springer, Heidelberg (2011). https://doi.org/10.1007/978-3-642-25975-3_17
18. Zhu, H., Chen, E., Xiong, H., Cao, H., Tian, J.: Ranking user authority with relevant knowledge categories for expert finding. World Wide Web **17**(5), 1081–1107 (2014)

Parallel Framework for Unsupervised Classification of Seismic Facies

Jatin Bedi(✉) and Durga Toshniwal

Department of CSE, Indian Institute of Technology Roorkee, Roorkee, Uttarakhand, India
jatinbedi278@gmail.com, durgatoshniwal@gmail.com

Abstract. Seismic facies classification plays a critical role in characterizing & delineating the various features present in the reservoirs. It aims at determining the number of facies & their description for the available seismic data. During the past few decades, seismic attributes have been widely used for the task of seismic facies identification. It helps geologists to determine different lithological and stratigraphical changes in the reservoir. With the increase in the seismic data volume & attributes, it becomes difficult for the interpreters to examine each seismic line. One of the solutions given to this problem was to use some computer-assisted methods such as k-means, self-organizing map, generative topographic map and artificial neural network for analyzing the seismic data. Even though these computer-assisted methods performed well but due to the size of the 3-D seismic data the overall classification process becomes very protracted. In this paper, we introduce a parallel framework for unsupervised classification of the seismic facies. The method begins by calculating four different seismic attributes. Spark & Tensorflow based implementation of unsupervised facies classification algorithms are then used to identify the seismic facies based on the 4-D input attributes data. Further, the comparison of results (in terms of execution time & error) of Spark & Tensorflow based algorithms with already existing approach show that the proposed approach provides results much faster than previously existing MPI based approach.

Keywords: Self-Organizing Map · Facies classification
Elbow method

1 Introduction

Seismic data is defined as a 4-D time series data with three spatial dimension and one temporal dimension [9]. It consists of a set of traces collected by means of seismic surveys in which seismic sensors record a series of rays reflected back from the various surfaces beneath the earth [9]. A seismic face is defined as a unique set of seismic reflections characterized by amplitude, continuity & configuration [19]. Seismic facies classification or identification is a process of assigning a label to each sample trace or interval in the seismic data. Depending on the

P. Perner (Ed.): MLDM 2018, LNAI 10935, pp. 452–467, 2018.
https://doi.org/10.1007/978-3-319-96133-0_34

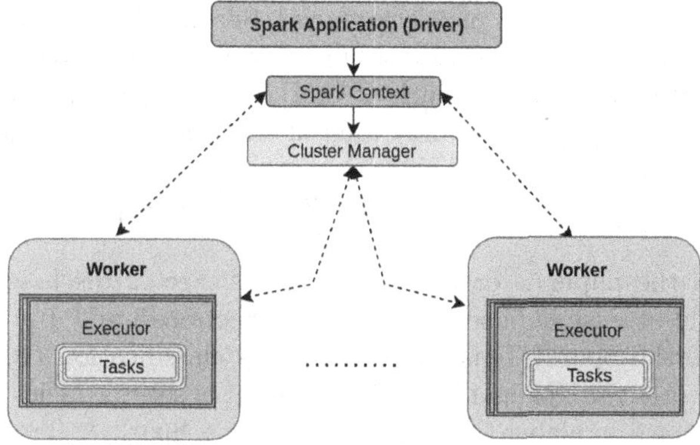

Fig. 1. Architecture of Spark

availability of training data, the process of classification can be categorized into supervised & unsupervised classification [7]. The supervised classification algorithm analyses the training data and uses it to map the new examples while unsupervised classification tries to identify the hidden patterns from the data without the need of any additional information.

The seismic facies classification plays a vital role in the characterization of reservoirs. It helps in identifying the various properties and delineating the heterogeneity present in the reservoir [2]. Traditionally, manual methods [14] were used to identify the different features such as channels, transport complexes etc. However, with the increase in the size of the seismic data or numbers of features that need to be identified, the task of manual interpretation becomes very cumbersome and time-consuming. One solution to this problem is to automate the process of facies classification by teaching the interpretation skills to the machine [14]. This process of making the machine to unfold the kind of relationship between the various patterns in the data is called as Machine Learning [11]. During the past two decades, various computer-assisted facies classification algorithm has been implemented. These algorithms used the seismic attributes to identify the pattern changes in the reservoir [1]. However, the task of determining which attribute to use for an interpretation task is very challenging and requires much expert knowledge. The automatic seismic facies identification algorithms such as K-means [3], Self-Organizing Maps (SOM) [24], Artificial Neural Network (ANN) [24] have been widely used to extract the patterns of interest from the seismic data. The output of these unsupervised classification algorithms is a map which accurately defines the spatial distribution of facies in the seismic data.

Although seismic facies classification methods do well in identifying the various properties in the reservoir, they are usually very time complex. In this era of rapid growth, both time & accuracy are essential concepts that need to be considered. Therefore, we need some methods to interpret faster without degrading the accuracy of the process. In this paper, we propose the parallelized approach for unsupervised classification of seismic facies. Spark [13] & Tensorflow [10] are open source frameworks that provide support for distributed processing by making use of the available resources such as CPU, GPU etc. K-means [3] and SOM [24] algorithms are being increasingly used in the field of geophysics for seismic facies classification. Therefore in the present work, we provide the Spark & Tensorflow based implementation of these widely used unsupervised classification algorithms.

The rest of the paper is organized as follows: Sect. 2 discusses the related work in the field of seismic facies classification. Sections 3 and 4 provides a brief introduction to the Spark, Tensorflow and facies classification techniques. The parallel frameworks for Unsupervised classification of seismic facies are discussed in Sect. 5. Section 6 explains the application of the Spark & Tensorflow based algorithms to the seismic dataset.

2 Literature Review

Sabeti and Javaherian [21] used K-means clustering algorithm to identify the facies in the seismic data. The method identifies the natural clusters present in the data by using seismic attributes. However, the main limitation of the K-means clustering algorithm is of determining the optimal value of user-defined variable K (Number of clusters). Wen [23] introduced an approach for modelling heterogeneity in the channelized reservoir by determining the spatial distribution of components of channel deposits. The seismic attributes were used for identification of the facies. Trace-based & voxel-based facies classification techniques were used to delineate the various properties in the reservoir. Further by ground truth, the author concluded that voxel-based approach provides more consistent classification results as compared to trace-based approach.

SOM is used as a popular method for extracting the similar patterns in the data with the help of seismic attributes. In 2015, Roy et al. [20] used the Self-Organizing Map (SOM) for classification of seismic facies in the Tripolitic chert reservoir. The output prototype vectors from the unsupervised SOM forms the basis for supervised classification.

Zhao et al. [25] introduced a variant of the SOM called Distance-Preserving Self-Organizing Map (DPSOM). DPSOM Preserves distance relation between the data in input space as well as in the data vectors on output SOM map. The application of this approach to the Canterbury dataset depicted that, DPSOM provides more consistent classification results. Later in 2015, Du et al. [4] combined the SOM algorithm with the Empirical Mode Decomposition (EMD) to identify the facies from the data accurately. EMD is a process used to remove noise from the data. The comparison of this combined approach with SOM showed that the combined approach better identified the variations present in the data. The correct choice of seismic attributes plays a significant effect on the overall classification process. By considering this aspect, Grana et al. [6] performed the comparison between the Bayesian classification (BC) & Expectation- Maximization (EM) methods. The comparison results of both of the methods over

North Sea dataset showed that they performed somewhat similar, but the main advantage of the EM over the BC is that, it does not require any training data. Further, this paper also introduced a framework to overcome the assumptions of the Gaussian model by providing a more accurate description of multi-model datasets.

Qi et al. [18] proposed a semi-supervised approach to classify the seismic facies. Initially, the manual interpretation method was used to find the 'N' target facies, resulting in the generation of training data. The attributes were selected for classification on the basis of their correlation & Kuwahara filtering method. The effectiveness of this approach is demonstrated by its application to the 3-D seismic volume of Louisiana, USA.

In 2015, Zhao et al. [24] reviewed six supervised & unsupervised classification algorithms including Principal Component Analysis (PCA), K-Means, SOM, GTM, Artificial Neural Network (ANN) and Support Vector Machine (SVM). These six algorithms were applied to the dataset of Canterbury Basin. On the basis of classification results, the author concluded that supervised methods provide a much better estimate of facies in the data whereas the unsupervised methods help to identify the features that may otherwise be ignored. The author also motivated the need for Big Data analysis approaches in the area of Geophysics. As the size of the seismic data or the number of features of interest we want to analyze is usually very large, so there is a need for some framework to reduce the running time of the various facies identification & analysis tasks. Therefore, in this paper, we provide the Spark and Tensorflow based algorithms for unsupervised classification of seismic facies.

3 Introduction to Spark and Tensorflow

Apache Spark [8] was introduced in 2009 by the RAD Lab at UC Berkeley. It is a successor of Hadoop Map-Reduce platform & designed with an aim to support fast and general purpose processing. The basic components for the distributed execution in the spark are shown in Fig. 1. It uses a driver/worker architecture. The driver program is responsible for job initiation & management tasks. Spark Context is used to connect to the cluster managers (either standalone or YARN). After connecting, it acquires the worker/executor nodes which are responsible for storing the data and performing different tasks over it. The spark platform allows the computations to be done in parallel. This parallel support is provided with the help of Resilient Distributed Dataset (RDD). RDD is a data structure in the spark which is defined as the immutable distributed collection of objects. The simplest way that is used in Spark to create RDD is to use parallelize() function of SparkContext.

There is basically two kind of operations that can be executed over the RDD [8]. The transformation functions create a new RDD from the existing RDD while Action operation performs computations over the RDD & then returns it. The most common transformation and action operations on RDD are Map() & Reduce(). The Map() operation takes input an RDD & a function to be applied over that RDD. The function is applied to each element of RDD and results are stored in a new RDD. Reduce() is an action operation that aggregates all the elements of RDD using some function and returns the results to driver program.

Tensorflow [10] is a framework created by Google for learning models. It was open sourced in 2015, with a vast library of functions. The core of the Tensorflow is the computational graph. The Tensorflow program's execution consists of two phases [10]. First is construction phase, in which a computational graph is created corresponding to the program and each operation in the program is denoted by a node in the graph. Second is execution phase, in which a session is created to execute the operations in the graph. The operations are distributively executed across the available computer resources such as cores, CPU, GPU cards. Tensorflow is most suitable for numerical computations and can be used in the distributed environment. Just like Spark, Tensorflow uses a master to create session and workers to execute the operations in the graph.

4 Review of Unsupervised Classification Techniques

4.1 Cross-Plotting

One of the simplest & most interactive methods used for finding and Visualizing the groups in the two-dimensional data is to cross-plot one feature against the other. In the field of geophysics, the cross plot of one seismic attribute against the another is used as an efficient method for manually identifying the natural clusters in data [24]. However, the major problem with the attributes based cross-plotting is that it can be used efficiently only when the input data is two-dimensional. When there are more than three attributes in the data, the cross-plotting becomes intractable. One of the possible solutions to this problem is to use some projection method such as PCA [15] (Principal Component Analysis), LDA [15] (Linear Discriminant Analysis), Factor Analysis [15] etc. to find the dimension of interest from the data.

4.2 K-Means Algorithm

K-means clustering [7] is the simplest clustering algorithm and is used for grouping the objects based on the information available in the dataset. It aims at assigning the data objects into different clusters depending upon the similarity between the objects. K-Means clustering algorithm initiates by randomly choosing the K initial cluster centroid where K is a user-defined parameter. Each data object is assigned to the centroid to which it is most similar and the set of objects assigned to each of the centroids forms a cluster. The similarity value between the cluster centroid and data objects is measured by using the Euclidean distance. After assignment of each data objects to some cluster, the algorithm recomputes the mean of the data objects assigned to each individual cluster and that mean value becomes the new cluster centroid. The whole process of assignment & means computation is repeated. If there are N clusters, the

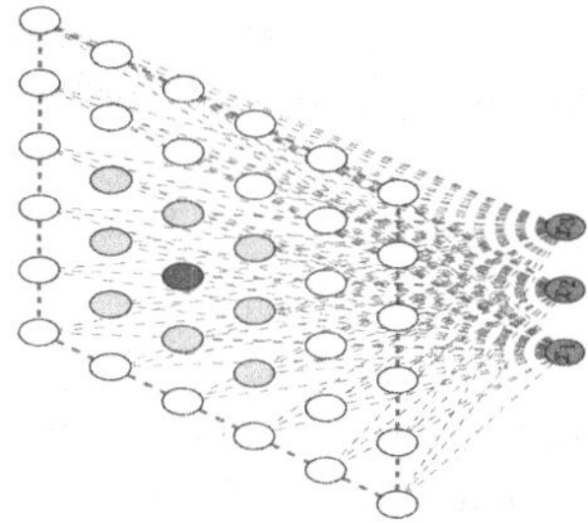

Fig. 2. Basic architecture of SOM network

algorithm will converge in N iterations. The Sum of Squared Error (SSE) is a parameter that is used to judge the quality of the clustering after each iteration. Therefore, the primary objective of the K-means algorithm is to reduce the total intra-cluster variance (SSE).

The K-means algorithm is very simple and fast algorithm. However, there are some problems associated with it. One big problem is determining the optimal number of clusters [7]. Secondly, there is a lack of proper organization which results in ambiguities in the interpretation of clustering results [24].

4.3 Self-Organizing Map

SOM [12] algorithm solves the problem associated with K-Means clustering algorithm [7] by initially over defining the number of clusters and then later joining different clusters depending on the distance criteria. SOM is a kind of artificial neural network developed by the professor Kohonen. It is a data compression method that works on the basis of competitive learning. It is named "Self-Organizing" because no prior knowledge of the characteristics of data is required to understand the SOM network fully.

Figure 2 shows the basic architecture of the SOM network. SOM provides a low dimensional representation of the input data in the form of output neurons. The output neurons are organized in the form of the 2-D grid. Each of the output neurons has location vector associated with it which determines the location of the nodes in the 2-D grid. Each individual input node is connected to all map nodes/output neurons in the SOM network. One of the biggest advantages of the SOM is that it preserves the data topology by preserving the relative distance between the input data points. The data points which are near in input data are mapped to the adjacent nodes in the output. There are three essential steps in the formation of the SOM algorithm [12] viz. Competition, Cooperation & Synaptic Weight Adaptation. These three steps are briefly described below:

Competition. Consider a m-dimensional input vector $\boldsymbol{X}$ and weights vectors $\boldsymbol{W}$ given by $\boldsymbol{X} = [x_1, x_2, \ldots, x_m]^T$ & $\boldsymbol{W}_j = [w_{j1}, w_{j2}, \ldots, w_{jm}]^T$ where j varies from 1 to L (number of output neurons in the network). During the competition phase, each of the output neurons computes a discriminant function ($\boldsymbol{W_j}^T. \boldsymbol{X}$) for a given input $\boldsymbol{X}$. The output neuron with the largest discriminant function Max($\boldsymbol{W_j}^T. \boldsymbol{X}$) will become the winner. The location vector ($i(\boldsymbol{X})$) corresponding to the winning neuron is saved to keep track of the Best Matching Unit (BMU).

Co-operation. The Cooperation phase determines the width of Topological Neighbourhood. The set of neurons falling in the neighbourhood of the winning neurons are called as "excited neurons". The Topological Neighbourhood consisting of excited neurons is determined on the basis of winning neuron ($\boldsymbol{X}$) and is given by Eq. 1 [12]

$$h_{j,i}(\boldsymbol{X}) = \exp(-\frac{d_{i,j}^2}{2\sigma_n^2}) \tag{1}$$

Where, $h_{j,i}(\boldsymbol{X})$ denotes the topological neighbourhood around i, encompassing neuron j and σ denotes the neighbourhood width.

$d_{i,j}$ determines the lateral distance [12] between the winning neuron i & excited neuron j and is given by Eq. 2.

$$\boldsymbol{d_{j,i}^2} = ||\boldsymbol{r_j} - \boldsymbol{r_i}||^2 \tag{2}$$

As we move far away from the winning neuron location the value of $d_{j,i} \rightarrow$ inf and the value of $h_{j,i} \rightarrow 0$ due to the neighborhood impact. Further, the value of sigma (σ) does not remains constant. We starts with a maximum value of the sigma and it will goes on decreasing as the iterations proceeds. This changing variable sigma is denotes by σ_n where n denotes the iteration number. The decrease in the value of width variable σ occurs in accordance with Eq. 3, where T_1 is a time constant and σ_0 denotes the initial value of sigma [12].

$$\sigma_n = \sigma_0 \exp(-\frac{n}{T_1}) \tag{3}$$

Synaptic Weight Adaptation. The next step after determining the neighborhood width of the winning neuron is to do the weight updation of the winning as well as the neighborhood neurons. In this phase of SOM, the excited neurons adjust their weights to decreases their individual values of the discriminant function in correspondence to the input pattern. The weight updates are done in accordance with Eq. 4, where η denotes the learning rate parameter [12].

$$\delta W_j = \eta . h_{j,i}(\boldsymbol{X})(\boldsymbol{X} - \boldsymbol{W_j}) \tag{4}$$

Using the discrete time formulation, the new weight of a node j is given by:

$$\boldsymbol{w_j}(n+1) = \boldsymbol{w_j}(n) + \eta_n . h_{j,i}(\boldsymbol{X})(\boldsymbol{X} - \boldsymbol{W_j}) \tag{5}$$

Just like σ, the value of learning rate parameter decays with time. The decay in the learning rate parameter w.r.t iteration no. is given by Eq. 5

$$\eta_n = \eta_0 \exp(-\frac{n}{T_2}) \tag{6}$$

η_0 denotes the initial value of the learning rate parameter. The learning rate parameter reaches its minimum value (0.37) when the value of n $\approx T_2$.

5 Parallel Architectures for Unsupervised Classification of Seismic Facies

Section 4 provides a deep insight into the working of the seismic facies classification approaches. Generally, the size of the seismic data is enormous. Therefore, it becomes difficult & time consuming to use the facies classification algorithm discussed in Sect. 4 for identification of facies in the seismic data. This is because some of the steps in the classification algorithms such as winning neuron determination, weights update, distance computation are computationally very expensive. Therefore, the present work provides a Spark Map-Reduce [8] and Tensorflow [10] based implementation of unsupervised seismic facies classification algorithms.

5.1 Spark and Tensorflow Based Implementation of K-Means Clustering Algorithm

K-means [7] is one of the most popular and accessible methods for clustering the data. Given a user-defined parameter K, the algorithm clusters the data into K clusters. With the increase in the size of the data, some of the steps in the K-means clustering algorithm such as computing the Euclidean distance, SSE becomes computationally very expensive. Therefore in this section, we use the Spark & Tensorflow framework for the seismic facies classification by reducing the computation time of K-means algorithm.

In the Spark based K-Means algorithm, the entire dataset is divided into the user-defined number of partitions. Each executor nodes receives a different partition of data. The set of initially chosen cluster centroids are broadcasted to each of the executor nodes. The different executor nodes work in parallel by computing the Euclidean distance between each element of their data partition and each cluster centroid. The output results (cluster assignments) from different partitions are combined by using the *reducebykey*() function. The new cluster centroids are chosen by computing the means of the set of point assigned to each individual clusters and the entire process repeats. In the case of Tensorflow based K-means clustering algorithm, the framework starts by generating the computational graph corresponding to the unsupervised K-means clustering algorithm. During the execution phase, the input vectors are provided to the computation graph with the help of *feed_dict*() function. The execution time of the K-means algorithm (Tensorflow) is reduced by running the various computations in the parallel with the support of the available system resources such as CPU, GPU boards etc. One of the important thing that needs to be defined while using an iterative procedure is the stopping criteria for the algorithm. In the Spark & Tensorflow based implementation of the clustering algorithm, two different stopping criteria are used. One is to specify the maximum number of iteration

Table 1. Survey parameters

Parameter name	Range
Inline	[100–750]
Crossline	[300–1250]
Z range	[0–1848]
Size (km)	[24 * 16]

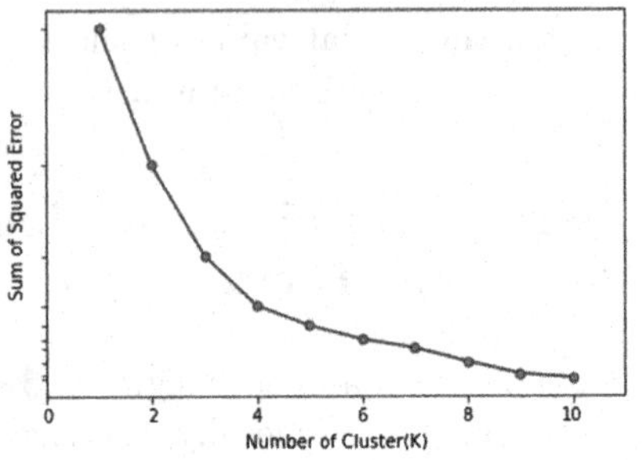

Fig. 3. Plot of K value against the SSE (Sum of Squared Error)

of the algorithm to run over data and another is to determine the convergence criteria. The K-means algorithm is said to be converged when the change in the SSE (Within clusters Sum of Squared Error) over the two successive iterations is less than some specified threshold value. The Sum of Squared error [7] is given by Eq. 7 where K denotes the number of clusters, C_i denotes i^{th} cluster centroid & *dist* is the standard Euclidean distance.

$$SSE = \sum_{i=1}^{K} \sum_{x \in C_i} dist(C_i, x)^2 \tag{7}$$

The determination of the value of user-defined parameter K (number of clusters) is one of the major problems with the K-means clustering algorithm. There are various ways viz. Elbow method [17], Gap statistics [22] methods to determine the value of K. In this paper, we used the Elbow method to decide the value of K. The Elbow method work by executing the K-means algorithm for different value of K. For each value of K, the method computes the Sum of Squared Error (SSE) value. The plot of K value against the SSE value will look like an 'Arm' and helps in determining the optimal value of K. The elbow on the plot belongs to the optimal value of 'K'. The Key idea behind the Elbow method is to choose the value of K that still has low SSE. Figure 3 shows the plot of K value against the SSE for the Dutch Sector dataset [5]. From the Fig. 3, it is pretty clear that there is an elbow at K = 4, so the optimal value of K for this dataset is 4.

5.2 Spark Based Implementation of Unsupervised Facies Classification Algorithm

Self-Organizing Map. SOM [12] has been widely used for the lower-dimensional representation of data in the form of 1-D or 2-D grid. In this section, we provide the Spark Map-Reduce based implementation of the SOM. The working of the Spark-Based SOM is explained with the help of three algorithms. Given the required input parameters the Algorithm 1 initializes the output *weight_vects* with the random values from the data. Lines 4 & 5 in the

Algorithm 1. *def train_data (data, num_iter, num_par, width, height, sigma, lr_rate)*

```
Normalize the Input 4-D Seismic Attributes Data
Initialize w(width), h(height), σ0(sigma), η0(learning rate)
Create a weight_vects of size(w,h) and Initialize it with the random values from
  the data using rand function.
Initialize the input parameters such as no. of iterations(n_iter), no. of
  partitions(n_par)
sc = SparkContext() (To initialize sparkcontext)
Rdd = sc.parallelize(data, n_par) (RDD Creation)
for i in range(n_iter) do
    w_v1=sc.broadcast(weight_vects)
    Out_weight_vects=Rdd.mappartition(train_datapar)
    Comb_Out_weight_vects=Out_weight_vects.reduce (lambda a, b : a+b)
    Normalize the Combined Weight Vector (Comb_Out_weight_vects) and rep-
  resent it    by Norm_Out_weight_vects.
    Assign weight_vects=Norm_Out_weight_vects
    Update the learning rate(η) & sigma(σ) parameters to be used in next iteraion
        according to the Eqs. 3 and 6
    Project the Output Assignments onto 2-D Color Map.
end for
```

Algorithm 2. *def train_datapar(par_data)*

```
for Each item ∈ par_data do
    weight_vects1 = weight_vects
    Call Bmu_data(item, weight_vects1, w, h) to get the location of Best Matching
  Unit    (BMU).
    Define the Gaussian Topological Neighbourhood for the Best Matching Unit
        according to Eq. 1
    Create an iterator object(itr_2) to iterate over the Entire Topological Neighbour-
  hood.
    while itr_2 not finished: do
        Update the weight of the neurons point by itr_2 object in the topological
            neighbourhood of BMU accroding to the Eq. 6
        Store the updated weights values in the weight_vects1.
        itr_2.iternext()
    end while
end for
Return the weight_vects1
```

Algorithm 1 creates a spark context and then uses it to parallelize the data by partitioning it into the user-defined number of partitions. The algorithm broadcasts the *weight_vects* to each of the executor nodes. Each executor node has a different part of data and it independently runs the Algorithm 2 for that part of data. Spark Map function is used to run the Algorithm 2 parallelly over different data partition available in executor nodes.

Algorithm 3. *def Bmu_data (x, weight_vects, width, height)*

```
Create an empty grid of width w & height h
s = np.subtract (x, weight_vects)
Create an iterator object(itr_1) to iterate over the Entire Grid(w, h)
while itr_1 not finished: do
    Calculate Euclidean distance between the input x and each weight_vects(i, j).
    Store the calcuated distance between each input x and each weight_vects(i, j)
     at corresponding location Grid(i, j), where 0 ≤ i ≤ w & 0 ≤ j ≤ h.
    itr_1.iternext().
end while
Return the location vector of the Best Matching Unit(Winning neuron).
```

Algorithm 4. *def init(width, height, n_iterations=100, learning_rate, sigma)*

```
Initialize w(width), h(height), σ0(sigma), η0(learning rate)
graph = tf.Graph()
with graph.as_default():
Initialize the grid(w,h) of weight_vects with the random values from the data using
  tf.random_normal() Function
Use tf.placeholder() function to create placeholder for the input_vect and itr_no
  Input
locts_vects = tf.constant(np.array(list(node_locations(w, h))))
for Each elem ∈ weight_vects do
    Find the Euclidean distance between input_vect & each weight_vects and use
        tf.stack to store it
    Use tf.argmin() function to return the index of best matching unit from stack.
end for
Determine the learning rate, sigma and Gaussian Topological Neighbourhood on the
  basis of iter_no using Eqs. 1, 3 and 6.
weight_delta=tf.multiply(η,tf.subtract(tf.stack([ input_vect for i in range(w*h)]),
  weight_vects))
weight_vects = tf.add(weight_vects, weight_delta)
session = tf.Session()
tf.summary.FileWriter("logs", self._graph).close()
init_op = tf.global_variables_initializer()
session.run(init_op)
end with
```

Given a data partition, the Algorithm 2 processes each element in the data. During each iteration, the algorithm picks an item from the data and computes the Best Matching Unit (BMU) corresponding to that item by calling the Algorithm 3. The Algorithm 3 uses Euclidean distance as a measure for determining the winning neuron or BMU. It returns a location vector of the winning neuron, which further forms the basis for determination of topological neighborhood.

Based on the BMU, topological neighborhood, σ and η the weight of the winning and excited neurons are updated by Algorithm 2. After processing of each element by all executor nodes, the results are returned to Algorithm 1, where

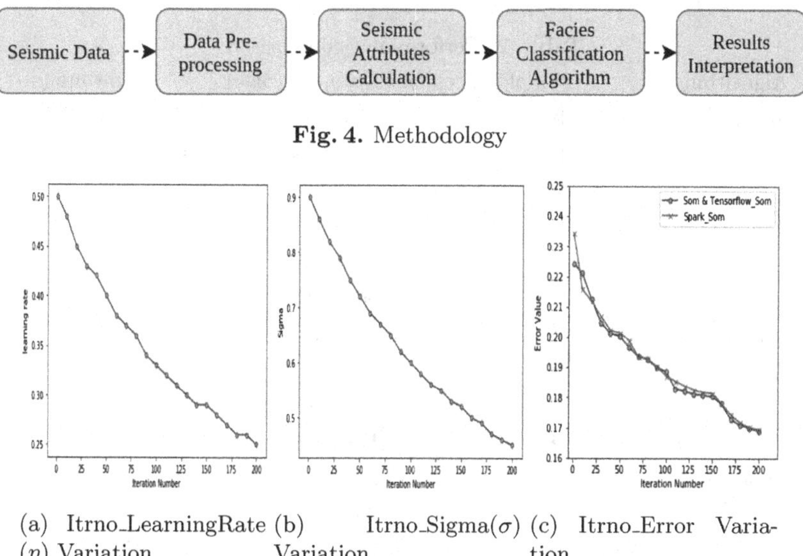

Fig. 4. Methodology

(a) Itrno_LearningRate (η) Variation (b) Itrno_Sigma(σ) Variation (c) Itrno_Error Variation

Fig. 5. Variation in SOM parameters

the Spark Reduce operation is used to combine the results from multiple executor nodes. In order to normalize the results from multiple executor nodes, we take the means of the *weight_vects*. The whole process is repeated after updating the learning rate (η) and sigma (σ) parameters on the basis of *num_iter* and other parameters.

5.3 Tensorflow Based Implementation of Unsupervised Facies Classification Algorithm

Self-organizing Map. This section provides the algorithm for Tensorflow based implementation of the SOM. As discussed in Sect. 2, the core of the Tensorflow is the computational graph in which each node manifests some operation in a Tensorflow program [10]. It provides the support for parallel processing of various numerical computations. Given the required input parameters, the Algorithm 4 (construction phase) uses the Tensorflow data structures to create the computational graph corresponding to the SOM algorithm. Line 2 in the Algorithm 4 initializes the Tensorflow graph. After initialization of computational graph, a node is created in the graph corresponding to each step (4–13) in the algorithm. Line 5 in the algorithm creates a placeholder for the variables that will be provided input during the training phase. Line 14 & 16 initializes the Tensorflow session and variables. During each iteration of the training phase, it takes all the input vectors and executes the operations in the computational graph by taking one input vector at a time. The operations in the graph are executed parallelly by making use of the available system resources such as CPU, GPU etc.

Table 2. Performance comparison

Algorithm	No. of processors	Data set (Size)	Runtime (s)
Spark_Kmeans	48	333378	13
	48	Dutch Sector dataset	22
	48	809600	25
Tensorflow_Kmeans	48	333378	9
	48	Dutch Sector dataset	13
	48	809600	15
MPI_Kmeans [24]	50	809600	65
SOM	48	36570	264
	48	333378	2280
	48	Dutch Sector dataset	5942
Spark_SOM	48	36570	120
	48	333378	1055
	48	Dutch Sector dataset	1787
	48	809600	2370
Tensorflow_SOM	48	36570	77
	48	333378	687
	48	Dutch Sector dataset	1335
	48	809600	1630
MPI_SOM [24]		809600	4125

6 Application

The Spark & Tensorflow based algorithms discussed in Sect. 5 are applied to the seismic data of the Dutch sector of North Sea [5]. Various parameters related to this survey are mentioned in Table 1. Figure 4 shows the flow chart of the methodology used for unsupervised classification of seismic facies. The very first step in every data analysis process is of collecting the data. Seismic data is gathered by means of seismic surveys. After collection of the data, the next step is to pre-process the data for removal of noise from it. The filtering [16] methods are used to make the data suitable for the analysis. In this study, a set of four seismic attributes [16] is calculated. These four attributes are selected on the basis of their importance in the facies interpretation. Given the four seismic attributes, the method constructs a 4-D feature vector to be input to the classification algorithm. Different algorithms discussed in Sect. 5 are used to do the facies classification for the input dataset based on the calculated attributes. We begin by applying the K-Means clustering algorithm. As previously discussed, determining the optimal number of clusters is a critical need for the accurate clustering. In the present work, we use the Elbow method [17] to estimate the correct number of clusters. In contrast to the K-Means, SOM algorithm projects the attributes data onto a mesh of 2-D latent space. Figure 5a, 5b shows the variations in the learning rate & sigma parameter over the first 200 iterations of the SOM algorithm. The performance of the SOM algorithm is evaluated on the basis of Quantization error which is defined as the average distance between each vector & its Best Matching Unit. Figure 5c shows the variations in the Quantization error in all implementations (SOM, Spark_Som, Tensorflow_Som) of SOM with the change in learning rate & sigma parameters. From the Fig. 5c, it is clearly visible that both spark & Tensorflow based implementation of SOM algorithm equally captures the patterns present in the data as that of the traditional implementation of the SOM while reducing the quantization error during each successive iterations.

Further, the Table 2 shows the comparison of the execution time of various implementations of unsupervised facies classification algorithms over the varying size datasets. The Table 2 also includes the results (in terms of execution time) of the already existing MPI based approach [24] for unsupervised facies classification. From the comparison of the results from the Table 2, it can be concluded that Tensorflow & Spark based implementation of SOM & K-Means algorithm provides high reduction in the execution time as compared to that of the existing MPI based approach.

7 Conclusion

Seismic facies classification is an invaluable process for identification and interpretation of large seismic volume. It helps the interpreters to delineate the heterogeneity in the reservoir. The recent developments in the field of facies classification by using seismic attributes have greatly helped the geophysicists to do the better estimation of the features in the 3-D seismic data. However, it is a very difficult and cumbersome task. This paper provides a parallel framework for automatic delineation of seismic facies. The support for parallel processing is provided by means of Apache Spark & Tensorflow framework. K-Means and SOM are the most commonly used algorithms for identification & characterization of different facies in the reservoir. Therefore in the present work, we provide Spark & Tensorflow based implementation of unsupervised seismic facies classification algorithms. The resulting error comparison of both traditional & parallel implementations shows that both approaches equally captures the features present in the data. Further, the execution time comparison of the proposed framework with the existing MPI based approach shows that the proposed framework executes faster than previously existing approaches.

References

1. Anees, M.: Seismic attribute analysis for reservoir characterization. In: 10th Biennial International Conference and Exposition (2013)
2. Chopra, S., Marfurt, K.J.: Seismic Attributes for Prospect Identification and Reservoir Characterization. Society of Exploration Geophysicists and European Association of Geoscientists and Engineers, Tulsa (2007)

3. Coléou, T., Poupon, M., Azbel, K.: Unsupervised seismic facies classification: a review and comparison of techniques and implementation. Lead. Edge **22**(10), 942–953 (2003)
4. Du, H., Cao, J., Xue, Y., Wang, X.: Seismic facies analysis based on self-organizing map and empirical mode decomposition. J. Appl. Geophys. **112**, 52–61 (2015)
5. dGB Earth Science: Open Seismic Repository (2016)
6. Grana, D., Lang, X., Wu, W.: Statistical facies classification from multiple seismic attributes: comparison between Bayesian classification and expectation-maximization method and application in petrophysical inversion. Geophys. Prospect. **65**(2), 544–562 (2017)
7. Han, J., Pei, J., Kamber, M.: Data Mining: Concepts and Techniques. Elsevier, New York (2011)
8. Karau, H., Konwinski, A., Wendell, P., Zaharia, M.: Learning Spark: Lightning-Fast Big Data Analysis. O'Reilly Media Inc., Sebastopol (2015)
9. Kearey, P., Brooks, M., Hill, I.: An Introduction to Geophysical Exploration. Wiley, Hoboken (2013)
10. Ketkar, N.: Introduction to TensorFlow. In: Ketkar, N. (ed.) Deep Learning with Python: A Hands-on Introduction, pp. 159–194. Apress, Berkeley (2017). https://doi.org/10.1007/978-1-4842-2766-4_11
11. Kodratoff, Y.: Introduction to Machine Learning. Morgan Kaufmann, Burlington (2014)
12. Kohonen, T.: The self-organizing map. Neurocomputing **21**(1), 1–6 (1998)
13. Meng, X., Bradley, J., Yavuz, B., Sparks, E., Venkataraman, S., Liu, D., Freeman, J., Tsai, D., Amde, M., Owen, S., et al.: MLlib: machine learning in Apache Spark. J. Mach. Learn. Res. **17**(1), 1235–1241 (2016)
14. Mojeddifar, S., Kamali, G., Ranjbar, H.: Porosity prediction from seismic inversion of a similarity attribute based on a pseudo-forward equation (PFE): a case study from the North Sea Basin, Netherlands. Petrol. Sci. **12**(3), 428–442 (2015)
15. Morrison, D.F.: Multivariate Analysis, Overview. Wiley Online Library (1998)
16. Opendtect: OpendTect (2016). https://dgbes.com/index.php/software. Accessed 20 May 2017
17. Pelleg, D., Moore, A.W., et al.: X-means: extending k-means with efficient estimation of the number of clusters. In: ICML, vol. 1, pp. 727–734 (2000)
18. Qi, J., Lin, T., Zhao, T., Li, F., Marfurt, K.: Semisupervised multiattribute seismic facies analysis. Interpretation **4**(1), SB91–SB106 (2016)
19. Roksandić, M.: Seismic facies analysis concepts. Geophys. Prospect. **26**(2), 383–398 (1978)
20. Roy, A., Dowdell, B.L., Marfurt, K.J.: Characterizing a Mississippian tripolitic chert reservoir using 3D unsupervised and supervised multiattribute seismic facies analysis: an example from Osage County, Oklahoma. Interpretation **1**(2), SB109–SB124 (2013)
21. Sabeti H, Javaherian A: Seismic facies analysis based on k-means clustering algorithm using 3D seismic attributes. In: Shiraz 2009–1st EAGE International Petroleum Conference and Exhibition (2009)
22. Tibshirani, R., Walther, G., Hastie, T.: Estimating the number of clusters in a data set via the gap statistic. J. R. Stat. Soc.: Ser. B (Stat. Methodol.) **63**(2), 411–423 (2001)
23. Wen, R.: 3D modeling of stratigraphic heterogeneity in channelized reservoirs: methods and applications in seismic attribute facies classification. Recorder Off. Publ. Can. Soc. Geophysicists **29**(3), 1–14 (2004)

24. Zhao, T., Jayaram, V., Roy, A., Marfurt, K.J.: A comparison of classification techniques for seismic facies recognition. Interpretation **3**(4), SAE29–SAE58 (2015)
25. Zhao, T., Zhang, J., Li, F., Marfurt, K.J.: Characterizing a turbidite system in Canterbury Basin, New Zealand, using seismic attributes and distance-preserving self-organizing maps. Interpretation **4**(1), SB79–SB89 (2016)

Recognizing Motor Imagery Tasks Using Deep Multi-Layer Perceptrons

Fernando Arce[1(✉)], Erik Zamora[2], Gerardo Hernández[1], Javier M. Antelis[3], and Humberto Sossa[1,3]

[1] Instituto Politécnico Nacional - CIC, Av. Juan de Dios Batiz S/N, Gustavo A. Madero, 07738 Ciudad de México, Mexico
fernando.arce.vega@gmail.com, gerardohernandez.hernandez@gmail.com, hsossa@cic.ipn.mx

[2] Instituto Politécnico Nacional - UPIITA, Av. Instituto Politécnico Nacional 2580, Barrio la Laguna Ticoman, Gustavo A. Madero, 07340 Ciudad de México, Mexico
ezamorag@ipn.mx

[3] Tecnológico de Monterrey Campus Guadalajara, Av. Gral Ramón Corona 2514, 45201 Zapopan, Jalisco, Mexico
mauricio.antelis@itesm.mx

Abstract. A brain-computer interface provides individuals with a way to control a computer. However, most of these interfaces remain mostly utilized in research laboratories due to the absence of certainty and accuracy in the proposed systems. In this work, we acquired our own dataset from seven able-bodied subjects and used Deep Multi-Layer Perceptrons to classify motor imagery encephalography signals into binary (Rest vs Imagined and Left vs Right) and ternary classes (Rest vs Left vs Right). These Deep Multi-Layer Perceptrons were fed with power spectral features computed with the Welch's averaged modified periodogram method. The proposed architectures outperformed the accuracy achieved by the state-of-the-art for classifying motor imagery bioelectrical brain signals obtaining 88.03%, 85.92% and 79.82%, respectively, and an enhancement of 11.68% on average over the commonly used Support Vector Machines.

Keywords: Brain computer interface · Motor imagery
Electroencephalography · Deep Multi-Layer Perceptrons
Support Vector Machine

1 Introduction

A Brain-Computer Interface (BCI) is an emergent technology that provides a non-muscular communication channel between users and computers. The system detects bioelectrical brain changes, sensed typically by Electroencephalography (EEG), and translates them into operative control signals [1,2].

Some EEG-BCI systems record and process bioelectrical brain signals while users move a specific part of their bodies classifying brain signals to control

P. Perner (Ed.): MLDM 2018, LNAI 10935, pp. 468–482, 2018.
https://doi.org/10.1007/978-3-319-96133-0_35

electronic devices [3,4]. In contrast, the most used approach for EEG-BCI is Motor Imagery (MI) wherein a user just imagines moving a specific part of his/her body, without actually any motor output (e.g., it is usual to ask users imaging moving a specific hand or foot).

MI is commonly more popular than other mental tasks due to being less tiring and more intuitive [5] (e.g., self-regulation of brain rhythms [6,7] or selective attention [8]). The system has to perceive the bioelectrical brain changes and relate them with the imagined movement to trigger an electronic device.

Since BCI provides users with a way to control an electronic device, EEG-BCI has an important impact on many real-life practical applications for disabled individuals such as "Communication and Control", "Motor Substitution", "Entertainment", and "Motor Recovery" [9] and for people with partial or complete motor impairments [3,10–13]. Notwithstanding this encouraging potential, BCI remains mainly used in research laboratories due to the absence of certainty and accuracy in the actual proposed systems [14,15]. Accordingly, the main way to accomplish this goal is by learning classifiers that relate bioelectrical brain signals with physical actions.

In literature, we can find many related works addressed to classify imagined movements. Nevertheless, most researches have achieved low classification percentages, or have reported high accuracies by selecting few individuals that obtained the best performances. Electrical signals classification from the brain remains a very challenging problem because of its large variability over subjects, high signal to noise ratio and high potential of outliers.

In this research, we study binary and ternary MI classification sceneries which are so common in BCI. Furthermore, we acquired our own dataset from seven able-bodied subjects and focused on improving the accuracy actually achieved by the state-of-the-art for BCI systems based on MI. The contributions are as follows:

1. We show experimentally for the first time that Deep Multi-Layer Perceptrons (DMLP) classify 11.68% better than a Support Vector Machine (SVM) for seven able-bodied participants in three different scenarios.
2. Through a series of experiments using our own dataset, without excluding any individuals, we demonstrate that the proposed neural network architectures outperform the actually accuracy achieved by the state-of-the-art MI bioelectrical brain signals classifiers.

The rest of the paper is structured as follows: Sect. 2 describes the methods and materials carried out to obtain and extract the features from the EEG signals. Section 3 presents the general and specific details of the used classifiers. Section 4 discusses the experimental results to assess the effectiveness of our proposal. Section 5 provides a chronological list of the most outstanding researches related to ours, together with an understandable description of each. In Sect. 6, we give our conclusions and make some recommendations for future work.

2 Methods and Materials

This section describes the experiments carried out to obtain EEG signals from human volunteers whom performed MI of the hands, the data analysis (2.1) and the feature extraction process (2.2).

2.1 EEG Dataset Description

1. Participants: Seven able-bodied subjects voluntarily participated in this study (six males and one female; age range 19–22 years; mean $\pm$ std 19.88 $\pm$ 1.13 years). None of them had a known neurological or motor disorder; moreover, they did not have prior experience with EEG recording protocols or with BCI. Experiments were carried out in accordance with the Helsinki Declaration [16]. All participants were duly informed about the experiment and they were asked to read and sign an informed consent form to participate in the study. They were also informed that they could quit the experiment whenever they wanted.
2. Data acquisition: EEG signals were acquired from 16 scalp locations using a g.USBamp biosignal amplifier and a g.GAMMAsys system with active electrodes (from g.tec medical engineering GmbH, Schiedlberg, Austria). The 16 electrodes were distributed around the sensorimotor strip according to the 10/10 international system at locations FC3, FC1, FCz, FC2, FC4, C3, C1, Cz, C2, C4, CP3, CP1, CPz, CP2, CP4 and Pz, while the ground and the reference were placed at AFz and the left ear lobe, respectively. The EEG signals were recorded at a sampling frequency of 256 Hz, bandpass filtered between 0.5 Hz and 100 Hz and power-line notch-filtered at 60 Hz.
3. Experimental procedure: Participants were seated in a comfortable armchair in front of a computer screen with both forearms resting on their lap and the hands in supination position. The experiment consisted of a cue-based MI task which was controlled by visual cues presented on the screen. The experiment was executed in trials of 9 s as illustrated in Fig. 1 (Up). Each trial started with the image of a black cross displayed in the center of the screen which indicated relaxing the body without performing or imaging any movement (relax phase). Then, an image of a left or a right arm-hand was randomly displayed in the left or right side of the screen which instructed the participant to perform the MI task with the corresponding hand (MI phase). The MI consisted of squeezing and releasing steadily a hand-sized rubber ball. Then, the image disappeared leaving a blank screen that indicated taking a break to rest and blink before starting a new trial (rest phase). The participants were asked to relax, perform the MI or rest as long as the corresponding visual cue was presented on the screen, that is, three seconds for each phase. Also, they were asked to avoid blinking or performing any movement during the relax and the MI phases. Figure 1 (Down) shows a snapshot of the experimental setup. The BCI2000 platform [17] was used to manage the presentation of the visual cues and to record the EEG signals along with synchronization events for each cue. The experiment was executed in 4 blocks of 70 trials each

(10.5 min per block and 42 min for the four blocks). Therefore, a total of 280 trials were obtained per participant, 140 for left MI and 140 for right MI as their occurrence was kept balanced. In order to avoid fatigue, participants were encouraged to rest between blocks as long as they needed.

4. Preprocessing: After the experiments, the recoded EEG signals were subjected to off-line analysis as follows. EEG signals were trimmed in trials of 9 s containing successively the relax, MI and rest phases. For each trial a time vector was constructed where the reference $t = 0$ represents the initiation of the MI phase, therefore, each trial started in $t = 3$ s and ended in $t = 6$ s. Note that the time intervals (3; 0] s, (0; 3] s and (3; 6] s correspond to the relax, MI and rest phases. EEG signals were then bandpass-filtered from 0.5 to 48 Hz using a four-order Butterworth-type filter and Common Average Reference (CAR) filtered. Finally, electrode Pz was excluded and not used in the subsequent analyzes.

The main advantage of our own dataset is that it was acquired with electrodes placed to the left and right of the participants motor cortex, and not distributed over the whole scalp. This is because MI movements trigger same brain regions than motor execution [18] and we are interested in discriminating MI movements.

2.2 Feature Extraction

The Welch's averaged modified periodogram method [19] was used to compute spectral power-based features. The spectral power was computed within the motor-related alpha and beta frequency range [8, 27] Hz at a resolution of 1 Hz using Hanning-windowed epochs of 0.5 s with an overlap of 0.5 s. This procedure resulted in a feature vector $x \in \Re^{300\times 1}$ (15 electrodes × 20 spectral power values) associated to a class label y {Rest, Left, Right} (as they were computed from the relax $(-3, 0]$ s and MI $(0, 3]$ s phases).

3 Classifiers

This section presents the general and specific details of the used classifiers. All classifiers were tested in three different scenarios:

1. Study 1: Rest *vs* Imagined $\in$ {Left, Right},
2. Study 2: Left *vs* Right, and,
3. Study 3: Rest *vs* Left *vs* Right.

3.1 Deep Multi-Layer Perceptrons

There are several types of artificial neural networks [20]. Most popular types are Multi-Layers Perceptrons (MLP), Convolutional Neural Networks (CNN), and Long Short-Term Memories (LSTM). All use the same principle to classify data. Each neuron in these models uses an affine transformation and a non-linear function to create a hyperplane to divide the input space into two regions. When

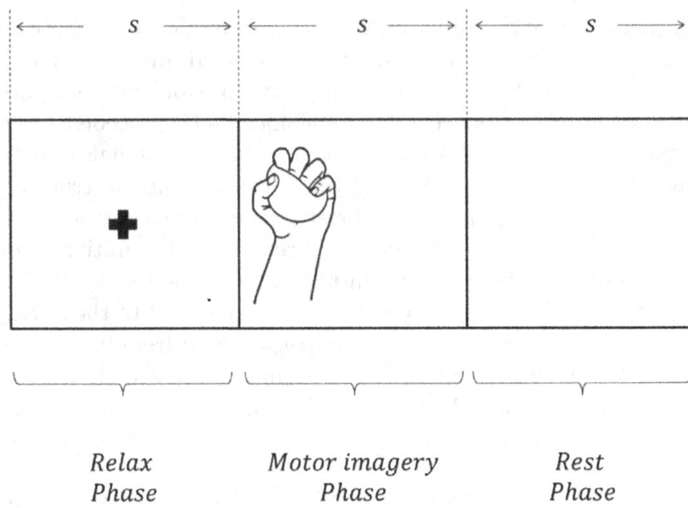

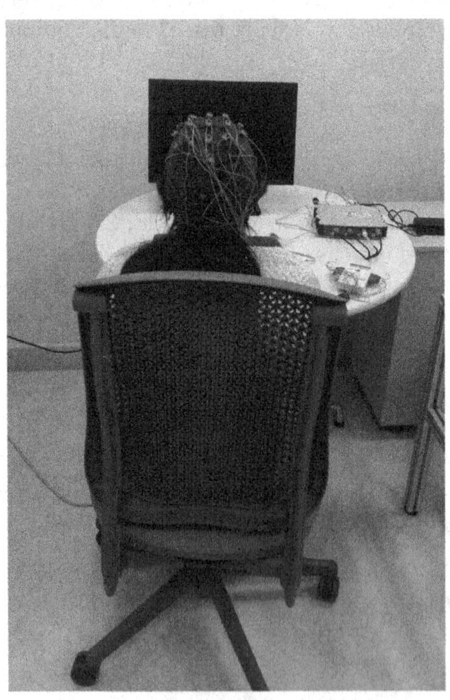

Fig. 1. Illustration of a trial temporal sequence during the experiment execution. Each trial lasted 9 s and its execution was controlled by three visual cues. Up: The first cue instructed to relax, the second instructed to perform the MI task with the left or the right hand, and the third cue indicated to rest, blink and move. Down: Snapshot of the experimental setup with a participant wearing the EEG cap and the screen where the visual cues are presented.

a set of neurons are connected to the same input and are put together in parallel, we say that they build a layer. The output of each layer is given by Eq. (1):

$$y_m = f(W_m y_{m-1} + b_m) \tag{1}$$

where y_m is the output vector of the layer m for $\forall m \in [0, 1, \ldots, M], W_m$ is the weight matrix, b_m is the bias vector, and f is the activation function. We consider that $y_0 = x$ where x is the input vector, which is not a layer. In this paper, we focus on multi-layer Perceptrons which means that layers of neurons are connected in series as Fig. 2 shows.

There is neither convolution nor recursivity in our models. Convolution is useful when the input data is an array of N dimensions, applying the same neuron in any position inside the array, to save the number of learning parameters. However, this is not relevant for our data. On the other hand, recursion allows memorizing data in a dynamic way learning relations among events through time. This memory is more important when the probability distribution of data is dynamic. We assume that our data has a static probability distribution, so we do not need recursivity.

In Fig. 2, we present the two neural architectures (based on preliminary experiments) to recognize MI tasks. Both have seven layers and two dropout layers indicated by red arrows in the figures. Dropout is a regularization method that is just applied during model training. It regularizes the learning model by turning off the neurons randomly [21]. This regularization is crucial to avoid overfitting when we have a complex learning model and few training samples. This is our case because it is really tedious to collect samples for recognizing brain signals, so we have relatively few training samples and deep neural models with seven layers.

In our models, we use four different activation functions:

$$f_{sig}(x) = \frac{1}{1 + \exp(-x)} \tag{2}$$

$$f_{tanh}(x) = \frac{\exp(x) - \exp(-x)}{\exp(x) + \exp(-x)} \tag{3}$$

$$f_{ReLU}(x) = \max(0, x) \tag{4}$$

$$f_{ELU}(x) = \begin{cases} \sigma(\exp(x) - 1) & x < 0 \\ x & x \geq 0 \end{cases} \tag{5}$$

where $x \in \Re$ and $\alpha > 0$. The logistic sigmoid (2) and hyperbolic tangent (3) are the standard functions to activate a neuron. We employed them for the model A which is for binary classification.

The ReLU (4) and ELU (5) activation functions are used in models A and B, respectively. These last functions are in the hidden layers of each model because they allow the backpropagation of gradients to reach the deepest layers without

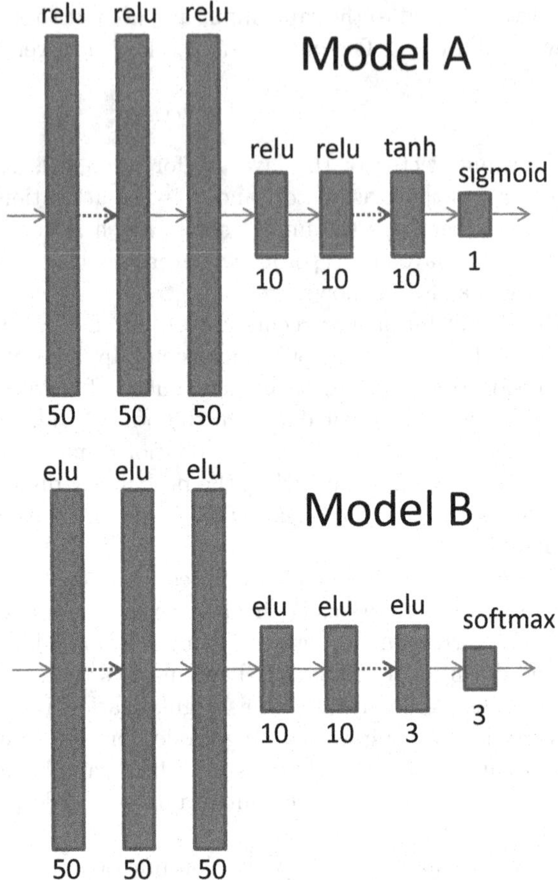

Fig. 2. Deep neural architectures for recognizing MI scenarios. Model A is used for binary classification and Model B for ternary classification. The number of neurons is at the bottom of each layer and the type of activation function is at the top.

degradation. Even when ReLU has been reported in literature for many decades before, the scientific community has recently rediscovered [22] its utility for deep neural architectures: ReLU speeds up the training and sometimes increases the performance in comparison with tanh function. On the other hand, the work [23] has shown that the function ELU leads to higher classification accuracies for image multi-classification than different types of ReLU functions.

3.2 Support Vector Machine

One of the most common classifiers used for BCI based on EEG is the well-known Support Vector Machine [24]. This technique behaves very accurately in solving nonlinear classification tasks. The architecture consists of two layers:

the first layer is for feature extraction based on kernel trick, and the second is for linear classification employing a hyperplane to separate patterns from two different classes. The purpose of the last layer is to find the hyperplane with the optimal margin among the support vectors (the patterns which are nearest to the boundary). So, it is necessary to solve the following optimization problem:

$$\min_{w,b,\zeta} \frac{1}{2}\|w\|^2 + C\sum_{i=1}^{n} \zeta_i \tag{6}$$

subject to

$$y_i \left(w^T \phi(x_i) + b\right) \geq 1 - \zeta_i \tag{7}$$

$$\zeta_i \geq 0 \tag{8}$$

where x_i is a training pattern, ϕ is a feature expansion function, w is a weights vector, b is the bias term, C is a regularization parameter, $y_i \in \{1, -1\}$ represents the class and $\zeta_i = \max\left(0, 1 - y_i \left(w^T x_i + b\right)\right)$. The performance of a SVM can be perfectly enhanced by choosing or learning a kernel function [25]. In all the experiments, the SVMs were implemented with a Radial Basic Function (RBF) kernel function, extensively applied in BCI based on EEG [5,24].

4 Results

In order to evaluate the efficacy of the proposed architectures, we computed the accuracy with the Area Under the Curve (AUC) of the Receiver Operating Characteristic (ROC) curve as a performance measure, AUC being the area between the x axis and the curve. So, the main advantage of using ROC curves is that they give us the true positive rate against the false positive rate. This is so important in BCI based on EEG systems because we do not only want to measure the average accuracy, but also the accuracy of each particular class to determine if the classifiers have some type of bias. Different AUCs were calculated using the trapezoidal approach and assessed statistically using 10-fold cross-validation technique, per subject.

Figure 3 displays the ROC curves with their respective AUCs. Analyzing the ROC curves, we can appreciate that all graphs operate above the diagonal line, near the point $(0, 1)$, which indicates the high efficacy of the besought networks. The best AUCs for Studies 1, 2 and 3 were 0.98, 0.98 and 0.96, respectively; being an AUC of 1.0 as a perfect test and an area of 0.5 as a worthless test.

Table 1 shows the experimental results achieved for our own dataset where we made a comparison between the proposed deep neural architectures and SVMs which are one of the most common classifiers used for BCI based on EEG, with their respective dropout rates used. It can be appreciated that Model A always obtained the highest accuracy for Studies 1 and 2 and Model B for Study 3. The besought models achieved an improvement of 11.68% on average over the SVM for our own dataset.

Among the three different MI scenarios, DMLP achieved the best AUC and accuracy for Study 1, showing that it is easier to classify between Rest and

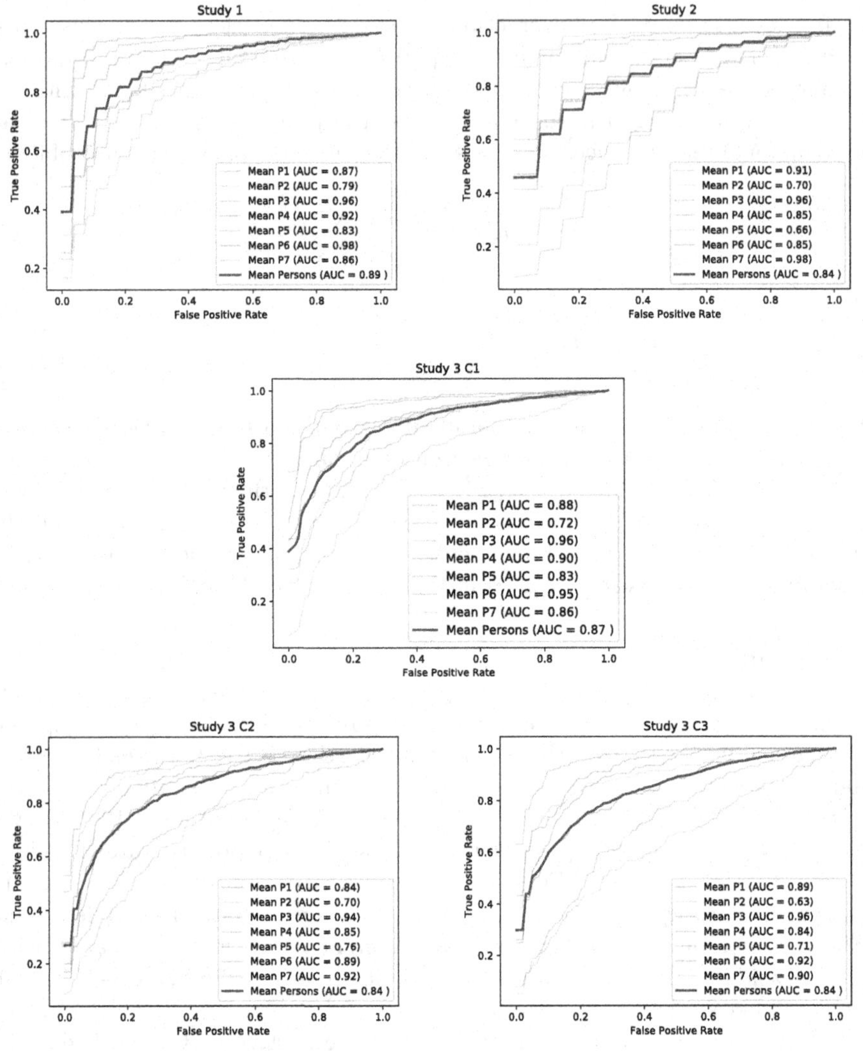

Fig. 3. ROC curves. Study 1: Rest *vs* Imagined. Study 2: Left *vs* Right. Study 3 C1: Rest *vs* Left and Right. Study 3 C2: Left *vs* Rest and Right. Study 3 C2: Right *vs* Left and Rest.

Imagined classes. On the other hand, the worst performance was obtained for Study 3 (Rest vs Left vs Right). Analyzing Study 3, we can appreciate that DMLP acquired the best performance for C1, classifying between Rest vs Left and Right classes, and the worst for C2 and C3. With all this, we can say that Left and Right classes are more correlated.

It is worth mentioning that for Model A as well as for Model B, the hyperparameters remained constant for all the participants in the dataset. In contrast,

Table 1. Results for our own datasets. The recognition accuracies are compared with SVM and DMLP classifiers demonstrated better performance in all tested cases and the three different studies. There was an improvement of 11.68% on average.

Participant	Study 1			Study 2			Study 3		
	SVM	Model A	Dropout	SVM	Model A	Dropout	SVM	Model B	Dropout
P1	75.35	**86.78**	0.60	75.00	**89.64**	0.70	66.42	**78.21**	0.75
P2	70.00	**78.21**	0.80	50.71	**75.00**	0.80	55.35	**64.82**	0.35
P3	88.21	**95.17**	0.50	88.92	**95.71**	0.70	83.21	**92.85**	0.65
P4	83.57	**91.25**	0.50	73.21	**85.00**	0.75	70.71	**80.71**	0.50
P5	70.17	**83.21**	0.70	58.57	**73.57**	0.75	58.39	**70.35**	0.75
P6	90.35	**96.25**	0.65	64.28	**86.42**	0.75	75.17	**87.14**	0.30
P7	72.67	**85.35**	0.60	90.00	**96.07**	0.70	70.89	**84.64**	0.35
Average	78.62	**88.03**		71.53	**85.92**		68.59	**79.82**	

the SVM hyperparameters were tuned for each fold to achieve the best performance.

5 Related Work

Proceeding affined researches have tested the performance of many traditional classifiers (e.g., SVM, LDA, k-nearest neighbor, and so on). In this section, we present a review of the most important works related to our research, and also provide a comprehensive explanation of each.

For instance, Blankertz et al. [26] reported an accuracy of >96% in binary classification detecting finger movements by employing SVM and variants of Fisher Discriminant. The main disadvantage of this work is that they performed physical movements, not MI. The main advantage of BCI based on MI is that these systems are available for any user.

After Blankertz et al., Schlögl et al. [27] achieved 63% accuracy using SVM for the BCI Competition 2003 dataset. In this experiment, they used five subjects to carry out four different MI tasks (e.g., left hand, right hand, foot and tongue).

Morash et al. [28] made binary and multiclass movement/imagery classification achieving 65% and 50% accuracy, respectively. They used naive Bayes algorithm to classify MI (e.g., right hand, left hand, tongue and right foot).

In 2014, Ge et al. [29] used the short-time Fourier transform to obtain multichannel information from a single-channel EEG for the IIIa from the 2005 BCI competition dataset. With this information, they trained a SVM achieving an accuracy of 75% in a four-class classification task.

Antelis et al. [30,31] identified motor states using EEG signals acquired from 18 healthy people employing a novel classification technique called Lattice Neural Networks with Dendritic Processing (LNNDP). In these experiments, they assessed the classifier performance in three different scenarios: relax vs intention, relax vs execution and intention vs execution achieving 65.26%, 69.07%

and 76.71% of accuracy, respectively. Previous to this work, in [32] they utilized the same technique (LNNDP) for hand movement classification from EEG signals.

Higher accuracies (related to our work) were reported by Vega et al. [5] where they made a comparison between different classifiers and feature selection methods. In this research, they obtained 81.45%, 77.23% and 68.71% of accuracy classifying three different MI scenarios using the available BCI Competition 2008-Graz dataset A. Best results occurred using a SVM classifier fed with Fast Correlation-Based Filter (FCBF) features.

On the part of Deep Neural Networks (DNN), these have been widely used for image recognition [22], speech recognition [33], natural language processing [34], and self-driving cars [35]. Lately, DNN have been applied for EEG signal classification based on: Convolutional Neural Networks (CNN), Stacked Autoencoders (SAE) and a combination of both.

One of the pioneering works [36] proposed a CNN with just four layers to classify steady-state visual evoked potentials. The novelty was to incorporate the Fourier transform between two layers. [37,38] employed another CNN with four layers to detect P300 waves and to extract features from different scales to classify four MI tasks, respectively. This last research achieved 100% of accuracy using just a subject.

The implementation of a CNN with three layers in a neuromorphic chip was presented in [39] for EEG recognition. We must make a special mention for the unpublished work [40] where authors applied CNNs with Long Short-Term Memory (LSTM) in order to take into account time dependencies in EEG data, yielding a significant improvement.

In An et al. [41], several Restrictive Boltzmann Machines (RBM) are trained as layers to build Deep Belief Nets (DBN) with seven to ten layers for MI tasks recognition.

Jirayucharoensak et al. [42] performed emotion recognition from non-stationary EEG signals using SAE based on RBM. They improve accuracies by around 6% respect to SVM and naive Bayes classifiers. Zheng et al. [43] created a mixed model with DBN and Hidden Markov Model (HMM) to classify emotions obtaining marginally better accuracies. In [44], stacked denoising autoencoders (SDAs) and CNNs classified EEG signals generated when individuals listening to different musical rhythms. Lastly, a similar approach was taken for MI classification [45] with eight layers of SAE and one 1D convolutional layer at the input, obtaining an improvement of 9% in terms of mean kappa value for BCI competition IV dataset 2b.

We took a different approach from the above works because we utilized deep feed-forward architectures without convolutional layers and without recursion. We proposed these architectures, based on preliminary experiments, because convolutional layers contains filters to extract learned features useful when the input data is presented in arrays of N dimensions. On the other hand, recursion unnecessary makes more difficult the training stage in comparison with feed-forward architectures, so this is main reason for avoiding it. And we showed that using spectral power-based features gives good results.

6 Conclusions

In this work, we have studied and experimentally evaluated the use of Deep Multi-Layer Perceptrons fed with power spectral features computed with the Welch's averaged modified periodogram method. All this, for recognizing MI movements assessed in three different classification sceneries, achieved superior results compared with the traditional method based on SVM. The proposed Models A and B obtained an enhancement of 11.68% on average over the commonly used SVM for our own dataset. We invite the reader to consider an improvement achieved with our proposed architectures where the hyperparameters stayed constant for all the subjects, unlike the SVM, where the hyperparameters were tuned for each fold in order to accomplish the best performance. Future work will involve the utilization of the same deep architectures to evaluate other standard datasets and the implementation of these to control external electronic devices.

Acknowledgements. E. Zamora, H. Sossa and M. Antelis would like to acknowledge the support provided by UPIITA-IPN, CIC-IPN and Tecnológico de Monterrey, respectively, in carrying out this research. This work was economically supported by SIP-IPN (grant numbers 20180180 and 20180730), and CONACYT grant numbers 65 (Frontiers of Science), 268958 and PN2015-873. F. Arce and G. Hernández acknowledge CONACYT for the scholarship granted towards pursuing their PhD studies.

References

1. Birbaumer, N.: Breaking the silence: brain-computer interfaces (BCI) for communication and motor control. Psychophysiology **43**(6), 517–532 (2006)
2. Shih, J.J., Krusienski, D.J., Wolpaw, J.R.: Brain-computer interfaces in medicine. Mayo Clin. Proc. **87**(3), 268–279 (2012)
3. Lebedev, M.A., Nicolelis, M.A.: Brain-machine interfaces: past, present and future. Trends Neurosci. **29**(9), 536–546 (2006)
4. Becedas, J.: Brain-machine interfaces: basis and advances. IEEE Trans. Syst., Man, Cybern. Part C (Appl. Rev.) **42**(6), 825–836 (2012)
5. Vega, R., Sajed, T., Mathewson, K.W., Khare, K., Pilarski, P., Greiner, R., Sanchez-Ante, G., Antelis, J.M.: Assessment of feature selection and classification methods for recognizing motor imagery tasks from electroencephalographic signals. Artif. Intell. Res. **6**(1) (2016)
6. Middendorf, M., McMillan, G., Calhoun, G., Jones, K.S.: Brain-computer interfaces based on the steady-state visual-evoked response. IEEE Trans. Rehabil. Eng. **8**(2), 211–214 (2000)
7. Wolpaw, J.R., McFarland, D.J., Neat, G.W., Forneris, C.A.: An EEG-based brain-computer interface for cursor control. Electroencephalogr. Clin. Neurophysiol. **78**(3), 252–259 (1991)

8. Farwell, L.: Talking off the top of your head: toward a mental prosthesis utilizing event-related brain potentials. Electroencephalogr. Clin. Neurophysiol. **70**(6), 510–523 (1988)
9. Millán, J.R., Rupp, R., Müeller-Putz, G., Murray-Smith, R., Giugliemma, C., Tangermann, M., Vidaurre, C., Cincotti, F., Kübler, A., Leeb, R., Neuper, C., Müeller, K., Mattia, D.: Combining brain-computer interfaces and assistive technologies: state-of-the-art and challenges. Front. Neurosci. **4**, 1–61 (2010)
10. Wolpaw, J.R., Birbaumer, N., McFarland, D.J., Pfurtscheller, G., Vaughan, T.M.: Brain-computer interfaces for communication and control. Clin. Neurophysiol. **113**(6), 767–791 (2002)
11. Schalk, G., McFarland, D.J., Hinterberger, T., Birbaumer, N., Wolpaw, J.R.: BCI 2000: a general-purpose brain-computer interface (BCI) system. IEEE Trans. Biomed. Eng. **51**(6), 1034–1043 (2004)
12. Allison, B., Dunne, S., Leeb, R., Millan, J.R., Nijholt, A.: Towards Practical Brain-Computer Interfaces: Bridging the Gap from Research to Real-World Applications. Springer, Heidelberg (2012). https://doi.org/10.1007/978-3-642-29746-5
13. Graimann, B., Allison, B.Z., Pfurtscheller, G.: Brain-Computer Interfaces: Revolutionizing Human-Computer Interaction. Springer, Heidelberg (2010). https://doi.org/10.1007/978-3-642-02091-9
14. Wolpaw, J., Wolpaw, E.W. (eds.): Brain-Computer Interfaces: Principles and Practice. Oxford University Press, New York (2012)
15. van Erp, J., Lotte, F., Tangermann, M.: Brain-computer interfaces: beyond medical applications. Computer **45**(4), 26–34 (2012)
16. World Medical Association: World medical association declaration of Helsinki: ethical principles for medical research involving human subjects. JAMA **310**(20), 2191–2194 (2013)
17. Mellinger, J., Schalk, G.: Number NIPS workshop series. In: BCI2000: A General-Purpose Software Platform for BCI Research. MIT Press (2007)
18. Lotze, M., Montoya, P., Erb, M., Hülsmann, E., Flor, H., Klose, U., Birbaumer, N., Grodd, W.: Activation of cortical and cerebellar motor areas during executed and imagined hand movements: an fMRI study. J. Cogn. Neurosci. **11**(5), 491–501 (1999)
19. Welch, P.D.: The use of fast Fourier transform for the estimation of power spectra: a method based on time averaging over short, modified periodograms. IEEE Trans. Audio Electroacoust. **15**, 70–73 (1967)
20. Hernández, G., Zamora, E., Sossa, H.: Comparing deep and dendrite neural networks: a case study. In: Carrasco-Ochoa, J.A., Martínez-Trinidad, J.F., Olvera-López, J.A. (eds.) MCPR 2017. LNCS, vol. 10267, pp. 32–41. Springer, Cham (2017). https://doi.org/10.1007/978-3-319-59226-8_4
21. Srivastava, N., Hinton, G., Krizhevsky, A., Sutskever, I., Salakhutdinov, R.: Dropout: a simple way to prevent neural networks from overfitting. J. Mach. Learn. Res. **15**, 1929–1958 (2014)
22. Krizhevsky, A., Sutskever, I., Hinton, G.E.: ImageNet classification with deep convolutional neural networks. In: Pereira, F., Burges, C.J.C., Bottou, L., Weinberger, K.Q. (eds.) Advances in Neural Information Processing Systems 25, pp. 1097–1105. Curran Associates, Inc., New York (2012)
23. Clevert, D., Unterthiner, T., Hochreiter, S.: Fast and accurate deep network learning by exponential linear units (ELUS). CoRR abs/1511.07289 (2015)
24. Lotte, F., Congedo, M., Lécuyer, A., Fabrice, L., Arnaldi, B.: A review of classification algorithms for EEG-based brain-computer interfaces **4** (2007)

25. Ali, A.B.M.S., Abraham, A.: An empirical comparison of kernel selection for support vector machines. In: Soft computing Systems: Design, Management and Applications, pp. 321–330 (2002)
26. Blankertz, B., Curio, G., Müller, K.R.: Classifying single trial EEG: towards brain computer interfacing. In: Dietterich, T.G., Becker, S., Ghahramani, Z. (eds.) Advances in Neural Information Processing Systems 14, pp. 157–164. MIT Press (2002)
27. Schlögl, A., Lee, F., Bischof, H., Pfurtscheller, G.: Characterization of four-class motor imagery EEG data for the BCI-competition 2005. J. Neural Eng. **2**(4), L14 (2005)
28. Morash, V., Bai, O., Furlani, S., Lin, P., Hallett, M.: Classifying EEG signals preceding right hand, left hand, tongue, and right foot movements and motor imageries. Clin. Neurophysiol. **119**(11), 2570–2578 (2008)
29. Ge, S., Wang, R., Yu, D.: Classification of four-class motor imagery employing single-channel electroencephalography. PLoS ONE **9**(6), 1–7 (2014)
30. Gudiño-Mendoza, B., Sossa, H., Sanchez-Ante, G., Antelis, J.M.: Classification of motor states from brain rhythms using lattice neural networks. In: Martínez-Trinidad, J.F., Carrasco-Ochoa, J.A., Ayala-Ramírez, V., Olvera-López, J.A., Jiang, X. (eds.) MCPR 2016. LNCS, vol. 9703, pp. 303–312. Springer, Cham (2016). https://doi.org/10.1007/978-3-319-39393-3_30
31. Antelis, J.M., Gudiño-Mendoza, B., Falcón, L.E., Sanchez-Ante, G., Sossa, H.: Dendrite morphological neural networks for motor task recognition from electroencephalographic signals. Biomed. Signal Process. Control **44**, 12–24 (2018)
32. Ojeda, L., et al.: Classification of hand movements from non-invasive brain signals using lattice neural networks with dendritic processing. In: Carrasco-Ochoa, J.A., Martínez-Trinidad, J.F., Sossa-Azuela, J.H., Olvera López, J.A., Famili, F. (eds.) MCPR 2015. LNCS, vol. 9116, pp. 23–32. Springer, Cham (2015). https://doi.org/10.1007/978-3-319-19264-2_3
33. Hinton, G., Deng, L., Yu, D., Dahl, G.E., Mohamed, A.R., Jaitly, N., Senior, A., Vanhoucke, V., Nguyen, P., Sainath, T.N., Kingsbury, B.: Deep neural networks for acoustic modeling in speech recognition the shared views of four research groups. IEEE Sig. Process. Mag. **29**(6), 82–97 (2012)
34. Sutskever, I., Vinyals, O., Le, Q.V.: Sequence to sequence learning with neural networks. In: Advances in Neural Information Processing Systems 27: Annual Conference on Neural Information Processing Systems 2014, Montreal, Quebec, Canada, 8–13 December 2014, pp. 3104–3112 (2014)
35. Bojarski, M., Testa, D.D., Dworakowski, D., Firner, B., Flepp, B., Goyal, P., Jackel, L.D., Monfort, M., Muller, U., Zhang, J., Zhang, X., Zhao, J., Zieba, K.: End to end learning for self-driving cars. CoRR abs/1604.07316 (2016)
36. Cecotti, H., Graeser, A.: Convolutional neural network with embedded Fourier transform for EEG classification. In: 2008 19th International Conference on Pattern Recognition, pp. 1–4, December 2008
37. Cecotti, H., Graser, A.: Convolutional neural networks for P300 detection with application to brain-computer interfaces. IEEE Trans. Pattern Anal. Mach. Intell. **33**(3), 433–445 (2011)
38. Jingwei, L., Yin, C., Weidong, Z.: Deep learning EEG response representation for brain computer interface. In: 2015 34th Chinese Control Conference (CCC), pp. 3518–3523, July 2015
39. Nurse, E., Mashford, B.S., Jimeno-Yepes, A., Kiral-Kornek, I., Harrer, S., Freestone, D.R.: Decoding EEG and LFP signals using deep learning: heading truenorth. In: Conference on Computing Frontiers, pp. 259–266. ACM (2016)

40. Bashivan, P., Rish, I., Yeasin, M., Codella, N.: Learning representations from EEG with deep recurrent-convolutional neural networks. CoRR abs/1511.06448 (2015)
41. An, X., Kuang, D., Guo, X., Zhao, Y., He, L.: A deep learning method for classification of EEG data based on motor imagery. In: Huang, D.-S., Han, K., Gromiha, M. (eds.) ICIC 2014. LNCS, vol. 8590, pp. 203–210. Springer, Cham (2014). https://doi.org/10.1007/978-3-319-09330-7_25
42. Jirayucharoensak, S., Pan-ngum, S., Israsena, P.: EEG-based emotion recognition using deep learning network with principal component based covariate shift adaptation, **2014**, Article ID 627892 (2014)
43. Zheng, W.L., Zhu, J.Y., Peng, Y., Lu, B.L.: EEG-based emotion classification using deep belief networks. In: 2014 IEEE International Conference on Multimedia and Expo (ICME), pp. 1–6, July 2014
44. Stober, S., Cameron, D.J., Grahn, J.A.: Using convolutional neural networks to recognize rhythm stimuli from electroencephalography recordings. In: Ghahramani, Z., Welling, M., Cortes, C., Lawrence, N.D., Weinberger, K.Q. (eds.) Advances in Neural Information Processing Systems 27, pp. 1449–1457. Curran Associates, Inc., New York (2014)
45. Tabar, Y.R., Halici, U.: A novel deep learning approach for classification of eeg motor imagery signals. J. Neural Eng. **14**(1), 016003 (2017)

Author Index

Printed in the United States
By Bookmasters